AF682421

THERMAL ABLATION THERAPY

THERMAL ABLATION THERAPY

THEORY AND SIMULATION

AMIRA S. ASHOUR

Department of Electronics and Electrical Communications Engineering, Faculty of Engineering, Tanta University, Tanta, Egypt

YANHUI GUO

Department of Computer Science, University of Illinois Springfield, Springfield, IL, United States

WALEED S. MOHAMED

Department of Internal Medicine, Faculty of Medicine, Tanta University, Tanta, Egypt

ELSEVIER

ACADEMIC PRESS

An imprint of Elsevier

Academic Press is an imprint of Elsevier
125 London Wall, London EC2Y 5AS, United Kingdom
525 B Street, Suite 1650, San Diego, CA 92101, United States
50 Hampshire Street, 5th Floor, Cambridge, MA 02139, United States
The Boulevard, Langford Lane, Kidlington, Oxford OX5 1GB, United Kingdom

Copyright © 2021 Elsevier Inc. All rights reserved.

MATLAB® is a trademark of The MathWorks, Inc. and is used with permission. The MathWorks does not warrant the accuracy of the text or exercises in this book. This book's use or discussion of MATLAB® software or related products does not constitute endorsement or sponsorship by The MathWorks of a particular pedagogical approach or particular use of the MATLAB® software.

No part of this publication may be reproduced or transmitted in any form or by any means, electronic or mechanical, including photocopying, recording, or any information storage and retrieval system, without permission in writing from the publisher. Details on how to seek permission, further information about the Publisher's permissions policies and our arrangements with organizations such as the Copyright Clearance Center and the Copyright Licensing Agency, can be found at our website: www.elsevier.com/permissions.

This book and the individual contributions contained in it are protected under copyright by the Publisher (other than as may be noted herein).

Notices
Knowledge and best practice in this field are constantly changing. As new research and experience broaden our understanding, changes in research methods, professional practices, or medical treatment may become necessary.

Practitioners and researchers must always rely on their own experience and knowledge in evaluating and using any information, methods, compounds, or experiments described herein. In using such information or methods they should be mindful of their own safety and the safety of others, including parties for whom they have a professional responsibility.

To the fullest extent of the law, neither the Publisher nor the authors, contributors, or editors, assume any liability for any injury and/or damage to persons or property as a matter of products liability, negligence or otherwise, or from any use or operation of any methods, products, instructions, or ideas contained in the material herein.

British Library Cataloguing-in-Publication Data
A catalogue record for this book is available from the British Library

Library of Congress Cataloging-in-Publication Data
A catalog record for this book is available from the Library of Congress

ISBN: 978-0-12-819544-4

For Information on all Academic Press publications
visit our website at https://www.elsevier.com/books-and-journals

Publisher: Mara Conner
Acquisitions Editor: Chris Katsaropoulos
Editorial Project Manager: Hilary Carr
Production Project Manager: Prem Kumar Kaliamoorthi
Cover Designer: Miles Hitchen

Typeset by MPS Limited, Chennai, India

Working together to grow libraries in developing countries
www.elsevier.com • www.bookaid.org

Dedication

To the soul of my father Prof. Salah Ahmed Ashour, who spent his life to serving people, support research and researchers, and instill morals into me and his students. He made valuable contributions to the research of Biology. Remembering his passion for research is always inspiring me to do my best, just to be like him; a dedicated researcher who always gives freely of time, advice, and expertise. For all of those who know him, he is irreplaceable.

—Amira S. Ashour

Contents

Preface

Thermal therapy in medical practice is an interesting research topic to manipulate the temperature of the tissues/body for cancer treatment. Worldwide, hot and cold therapy has specific medical applications, such as cancer and fibrosis treatment in different organs of the body. In thermal-based medicine, modern research aims to understand the effects of temperature management on the cellular and molecular structures to develop efficient and safe devices for temperature monitoring in different clinical applications. Thermal therapy includes several techniques using hyperthermia, cryopreservation, or thermal ablation. Hyperthermia can be integrated with radiotherapy clinically for cancer treatment. Furthermore, hyperthermia can be used for the treatment of strokes, brain injuries, and inflammatory diseases. In contrast, cryopreservation, which refers to cooling at temperatures below $-80^{\circ}C$, allows storage of viable cells for long periods for future use. Cryogenic freezing is used clinically for localized cancer treatment (cryoablation). In addition, lasers and other thermal methods can be used to treat obesity and metabolic disorders, such as diabetes.

Thermal ablation is a widely used evolving treatment route for cancer in numerous significant medical applications. It destroys the tumor tissues by localized heating or freezing in the different organs that cannot be treated using conventional treatment methods. Potentially noninvasive therapies based on thermal ablation, using several energy sources of electric current, include ultrasound, laser, radiofrequency, and microwaves. Over the past decade, *image-guided based thermal ablation* has been significantly extended. In ablation, high temperatures above $50^{\circ}C$ and freezing temperatures are used to destroy the localized tissues of tumors/cancers of the bone, kidney, liver, breast, brain, thyroid, and lung for treatment.

During radiofrequency ablation (RFA) an alternating electrical current of about 500 kHz generates resistive heat close to the probe. Additionally, skin surface electrodes are used as ground pads to close the electrical circuit. Generally, RFA provides superior results to control local tumors. Studies have suggested using RFA as a first-line treatment decision for renal-cell and small hepatocellular carcinomas. Nevertheless, radiofrequency (RF) heating is hampered by the local blood flow and high electrical impedance of the tissues. Microwave ablation (MWA) is used to alleviate such problems by generating volumetric/quicker heating. Furthermore, simultaneously multiple microwave antennas can be used to produce conformal or larger ablations, whereas RF electrodes entail consecutive operations that limit their effectiveness. Compared to RFA, the MWA systems provide superior safety and efficacy. In addition, cryoablation freezes the tissues to lethal levels (−20 to −40°C) for the treatment of several metastatic tumors. Thus MWA is gaining attention as an alternative to RFA due to its numerous benefits in tissue-heating physics. Microwave energy is able to propagate via desiccated and charred tissues, allowing incessant and fast volumetric heating, leading to larger ablation zones compared to RFA.

From historical times to the present day, thermal medicine has resulted in new discoveries in traditional, translational, and clinical research to break new barriers to realize enhanced therapy for patients with cancer and other diseases. This requires new strategies, simulation, and modeling of the ablation devices with knowledge of the thermal biology, tissue characteristics, and the effect of heat and cooling on normal and abnormal tissues. MR-guided therapies are speedily evolving into new clinical trial opportunities with the assistance of the new cooling, heating, and imaging platforms accompanied by heat-driven vaccine development, drug/radiation-enhancing nanoparticles, and gene expression.

Based on clinical practices, to treat several benign/malignant tumors of the liver, kidney, lung, and bone, thermal ablation technology is a rapidly evolving alternative to surgical resection. Additionally, ablation has proved its efficiency for solid tumor treatment in organs, including the uterus, pancreas, adrenal glands, breast, and prostate. Ablation can also be integrated with adjuvant therapies, including chemotherapy, nanoparticles, and radiation, to improve treatment efficacy. Consequently, thermal ablation has attracted a number of researchers to design antennas and probes with a radiation pattern that affects only the tumor tissues/cells without damaging large areas of the surrounding healthy tissues. Several ablation devices have been developed, such as the *high-intensity focused ultrasound* (HIFU), which is a noninvasive tissue-heating device with high spatial accuracy. HIFU is often combined with magnetic resonance imaging (MRI) thermometry for noninvasive temperature control and monitoring. Furthermore, integrating thermal ablation with standard cancer therapies, such as radiation therapy, chemotherapy, or immunotherapy, has attracted clinical interest. Accordingly, interest in thermal medicine (mainly thermal ablation therapy) is rapidly growing to serve different clinical applications, combined with a prompt increase in the research base, which has attracted the attention of the clinical researchers, engineers, physicians, physicists, and biotechnologists.

This book contains nine chapters that provide cutting-edge information for guiding engineers in how to use simulation software to design the applicator for developing a thermal ablation technique. It includes detailed theoretical and technical concepts of thermal ablation therapy in the different body organs. The concepts of ablation technology based on the different thermal ablation methods also are introduced in the book. Since the radiation pattern and the effect of the heat/cooling depends on the characteristics of the tissues and the size of the tumor, the book studies these factors using different energy sources. The changes in mechanical properties of tissues due to thermal denaturation are also discussed. The book emphasizes the mathematical and engineering concepts of RF and microwave (MW) energy propagation through the tissues and the heating effect, where the high heating rate created by the MW applicator can overwhelm the heat-sink effects of the neighboring vessels, generating more uniform ablation regions. The different mathematical theories which governing heat transfer and tissue damage are also explained. The design and tuning of the MW antennas to provide energy to a specific organ, such as the lung or liver, are also discussed. Computational modeling of RFA and MWA procedures for developing and implementing new efficient ablation methods in clinical systems is one of the main contents of this book. Numerical simulations using COMSOL Multiphysics software for RF and MW ablation systems with different antenna/probe designs and configurations are extensively discussed. A numerical technique for the temperature profile in the tissue finite element method (FEM) is explained. The provided simulation examples demonstrate the different antenna designs based on FEM to provide the required tools to build simulation applications related to biological materials, physics components selection (electromagnetic and heat transfer), antenna geometry, and meshing. This is efficient for predicting the propagation of the electromagnetic waves from the applicator to assess antenna performance.

The book is also discusses the deposited energy from the used electromagnetic fields, and the generated heat, in addition to the subsequent water and vapor mass, transfers in the tissues to determine the thermal amount and the extent of tissue damage. The ablation process starting from locating the tumors using ultrasound, CT, or MRI devices to determining the required heating dose and the effect on the tissues with the proper antenna selection is discussed. A bioheat model to evaluate the use of high-temperature thermal ablation therapies of diseased tissues is included, and a comparison between the different ablation techniques is conducted. The book also reports on several applications in different real-life thermal ablation therapies. This cutting-edge book highlights the simulation and challenges to design the antennas for RF and MW ablation probes.

Generally, this volume provided highlights on the medical concepts of ablation, tissues types, and characteristics, effect of heating and cooling on the different tissues and organs, the concept/advantages/disadvantages of RFA, MWA, etc., antenna design requirements, simulation scenarios of the different cases (different probes and tissues), and finally, the challenges and new perspectives in thermal ablation therapy. Following are the key features of the book:

- Includes outstanding concepts and models of thermal ablation therapy and its different devices, including the advantages and disadvantages of each. Furthermore, the mathematical models of the microwave and RF ablation theories are introduced.
- Outlines and discusses the biological responses and engineering principles by which the thermal ablation techniques can provide temperature elevation within the organs of the human body. Aspects of the different ablation techniques, including action mechanisms, required equipment, needle characteristics, and treatment techniques are included.
- Highlights the different procedures of thermal ablation, including RFA, MWA, laser ablation, and ultrasound ablation.
- Involves different simulation scenarios of liver tumors and the effect of the ablation probe design on damaged tissues and the temperature distribution within the targeted tissues.
- Includes the limitations, challenges, and future perspectives for the design of the new antenna probes for thermal ablation techniques.

Amira S. Ashour

Department of Electronics and Electrical Communications Engineering, Faculty of Engineering, Tanta University, Tanta, Egypt

Yanhui Guo

Department of Computer Science, University of Illinois Springfield, Springfield, IL, United States

Waleed S. Mohamed

Department of Internal Medicine, Faculty of Medicine, Tanta University, Tanta, Egypt

Acknowledgments

Cancer is only going to be a chapter in your life, not the whole story.

—**Joe Wasser**

Medical treatment is emergency care for symptoms that have developed over a long period of time. The symptom is the flower on a plant. Treating the symptom is picking the flower, while the plant remains untouched.

—**Gary Zukav**

We are thankful to our *parents* and *families* for their boundless support throughout our life. No words can give them the credit they deserve!

Special thanks to the Elsevier publishing team, which showed us the ropes and bestowed their trust upon us. We express our deep appreciation to the staff members at Elsevier, particularly Chris Katsaropoulos (Senior Acquisitions Editor), Hilary Carr (Content Manager), and Prem Kumar Kaliamoorthi (Production Project Manager), for their continuous support and guidance.

Last but not the least, we would like to express our gratitude to the readers, in the hope that they will find this book to be an outstanding resource in thermal ablation therapy.

1

Introduction

Chapter Outline

1.1 Cancer

Cancer is one of the deadliest diseases as classified by the World Health Organization (WHO) in recent decades due to the huge number of patients affected worldwide. The five main cancer sites include the liver, breast, lung, prostate, and colorectum, with an increase also in the occurrence of other cancer types, including in the thyroid, kidney, and pancreas (Kilfoy et al., 2009; Siegel et al., 2017; Bray et al., 2018). Accordingly, cancer treatment remains a dynamic, hot, and challenging research topic, which attracts researchers to create new therapeutic strategies and techniques. These struggling attempts against cancer over past decades have steered toward the new strategies in the fight against cancer, including thermal therapy (known as thermotherapy or hyperthermia), laser treatment, photodynamic therapy, biological therapies, gene therapy, and angiogenesis inhibitors. However, the three cornerstones of cancer treatment are chemotherapy,

Thermal Ablation Therapy. DOI: https://doi.org/10.1016/B978-0-12-819544-4.00002-2
© 2021 Elsevier Inc. All rights reserved.

radiation therapy, and surgical tumor removal to increase patient survival for specific cancer types and to give optimism to patients (Zucker et al., 2000; Tan et al., 2009; Liauw et al., 2013). However, much room for improvement remains, with efforts to develop improved optimal equipment/methods for more accurate, without complicating cancer treatment. Furthermore, an understanding of anticancer action and the biological mechanisms involved is definitely helpful in the development of therapeutic strategies and techniques.

Before we dive into the promising therapeutic strategies, including thermal therapy (hyperthermia) and ablation, some terminologies and definitions are discussed. Typically, the word "tumor" is used to describe well-defined and localized cellular growth of normal (benign "noncancerous," incapable of scattering through the body) or abnormal cells (malignant "cancer," able to spread into other tissues and form more tumors throughout the body) that has a tendency to spread in the body (Gimm et al., 2000; Vogelstein et al., 2013). Such tumors can be solid or leukemia (malignant tumor of blood), which has uncontrolled cell division mechanisms. The primary growth of a benign tumor may generate secondary spreading growths elsewhere in the body, leading to the cancer once again becoming malignant. Hence, tumor is considered the most common term to refer to any abnormal tissue/cell excessive growth that may be benign or malignant. A benign tumor can stop growing by its own, while a malignant one cannot stop independently (Washington and Leaver, 2015).

1.1.1 Cancer: definition and malignancy of cancer cells

Basically, cancer (malignancy) is a malignant process referring to the diseases where abnormal cells split uncontrollably and may invade adjacent tissues with the ability to spread to other body parts through the lymph systems and blood. The word cancer comes from the Greek "carcinos" for crab, referring to the claw-like extensions to nearby tissues.

Compared to normal cells in the body, cancer cells achieve a sort of immortality. Normal cells become cancerous after a sequence of mutations leading to continuous growth and division of the cell that is out of control. Furthermore, cancer cells spread to distant body regions and invade nearby tissues in the body, unlike normal cells. The main differences between cancer cells and normal cells are summarized as follows (Warburg, 1956; Reya et al., 2001):

- *Progress*: Normal cells develop and grow naturally during childhood, or to repair wounded tissues within a limit as needed, while cancer cells grow continuously without need and without stopping. Cancer cells are unresponsive to the signals to stop from the body or undergoing apoptosis.

- *Immortality*: Normal cells have a set lifetime as they die once reaching a certain age, where once the lengths of telomeres (structures at the end of chromosomes) are shortened, with cell division, to a specific length, the cells die. On the contrary, cancer cells resist death by restoring their telomeres without shortening with cell division, and therefore, the cancer cells become immortal.
- *Spreading (metastasizing) ability*: Normal cells are immovable due to their adhesion molecules causing them to stick to neighboring cells. On the contrary, cancer cells have broken adhesion molecules causing them to float and have free movement, spreading to other body regions.
- *Invading ability*: Normal cells react to signals from nearby structures to avoid interaction or collision based on the boundary between the cells. In contrast, cancer cells are unresponsive to such signals and spread into the regions of neighboring tissues.

Fig. 1.1 illustrates the characteristics of normal cells versus cancer cells, showing that the capability to grow, die, metastasize, and invade are significant characteristics to distinguish cancer cells from healthy normal cells.

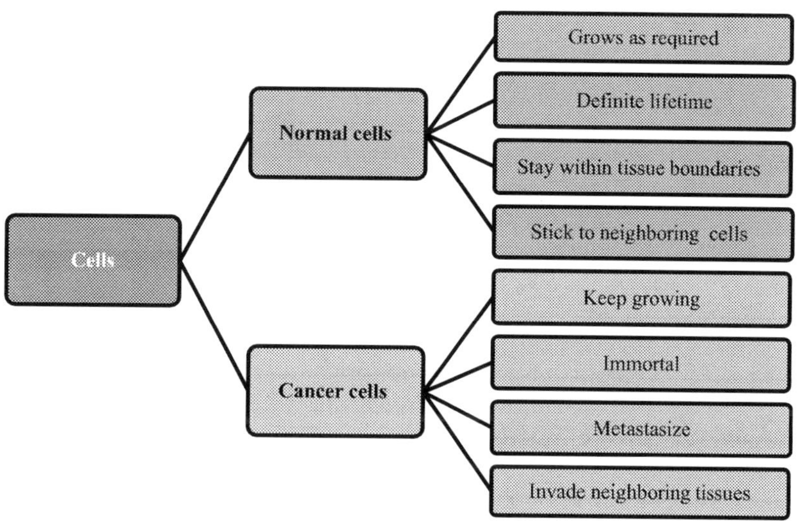

Figure 1.1 Characteristics of normal cells versus cancer cells.

1.1.2 Cancer causes and mechanics

Before exploring the main features of cancer, it is worth understanding the causes of cancer. Fundamentally, a cell transformation from normal to cancerous is triggered by agents, described as carcinogens. Some biological factors, including viruses, can increase the cancer risk in different body parts. In addition, several initiating agents, including carcinogens [such as ultraviolet light, food additives, radiation, smoking, and various chemicals] transform normal cells by damaging the deoxyribonucleic acid (DNA), hence inducing mutations. Such transformations occur gradually with exposure to the initiating agents (Reddy et al., 2003; Basu, 2018).

Accordingly, cancer cells are generated due to a series of cells' modifications, which are inherited or substances-based, causing their increased abnormality. Cancer is a multifactorial disease caused by several factors working together (Manolio, 2010). A genetic predisposition may lead to cancer, with mutations causing cells to become cancerous. The process of abnormal cancer cell transformation progressively from normal cells goes through several stages, including hyperplasia, dysplasia, and finally cancer. This cancer staging process is called differentiation. During the early stages of cancer generation, the cell appears as a normal cell of the tissue or organ, and then with disease progression, the cell increasingly becomes undifferentiated. A cancer cell can have numerous mutations (modifications), however, only a specific number of such genetic changes in cancer cells can cause the cancer to grow and divide. Accordingly, modifications that cause cancer cell growth are called driver mutations, while the other mutations are called passenger mutations. During cancer generation, proto-oncogenes (normal genes) are mutated and converted to oncogenes and drive the cancer growth, giving it its immortality. Generally, mutations of cancer cells occur in both tumor suppressor genes and oncogenes (Niculescu, 2018).

Although the immune systems in the body can recognize and detect cancer and abnormal cells, cancer cells remain active either by deactivating the immune cells or by evading detection and changing themselves in different ways to escape from the immune cells. Moreover, cancer cells are changeable after formation with continued mutations. Thus, they resist targeted therapy drugs and chemotherapy. Their continuing mutation enables them to evade the damaging effects of any treatment (Xu et al., 2015).

Generally, once a cancer becomes metastasized, its treatment will be challenging. Metastasis occurs due to cancer cells migrating to a new tissue either via neighboring tissues through the extracellular matrix or to far tissues through the blood (Egeblad and Werb, 2002). This tumor in a new position is considered to be secondary, having different genetic and molecular structures compared to the primary source tumor. The metastatic cells carry information about the type of tumor and heterogeneity. Conversely, the migrated cells in nearby tissues have information on mechanobiology and have changeable morphology. Furthermore, compared to neighboring tissues, tumors are stiffer (Gjorevski et al., 2012).

1.1.3 Cancer types

Several methods can be used to categorize cancers based on the tumor stage/grade, the DNA profile, or their origination (tissue, cell, or area).

1.1.3.1 Cancer type classification by cell origination type

Malignant cells are of several types based on the type of original cancer cell, where they originate, including carcinoma, leukemia, sarcoma, and lymphoma, and whether they arise in epithelial cells, blood cells, connective tissue, or lymphoid tissue, respectively. Carcinoma cancer arises and starts in epithelial cells, including in the cells in the layers of the covering surface, whether the external surface, such as the skin (skin cancer), or in tissues that cover internal surfaces/organs, such as the stomach (stomach cancer) (Suresh, 2007; Neville et al., 2015). Leukemia is a cancer originating in blood-generating tissues, including the bone marrow, producing huge numbers of abnormal blood cells which circulate in the blood. In contrast, sarcoma is a cancer in mesenchymal cells that begins in soft tissues and bone, including cartilage, bone, muscle, tendons, nerves, blood vessels, or fat, and any other supportive/connective tissues. Lymphoma cancer begins in immune system cells and arises from extranodal sites (e.g., testicles, spleen, or stomach) or in the lymph nodes (Shankland et al., 2012). Additionally, cancers of the central nervous system occur in the spinal cord/brain tissues. Accordingly, Fig. 1.2 demonstrates the classification of cancer cells based on the type of cell/tissue wherein the cancer originates.

Generally speaking, cancer categorization depending on the cells, including carcinomas which occur in epithelial cells, while leukemias, myelomas, and lymphomas are blood-related cancers. In addition, sarcomas occur in mesenchymal cells. Usually, the most common cancers, such as colon, lung, breast,

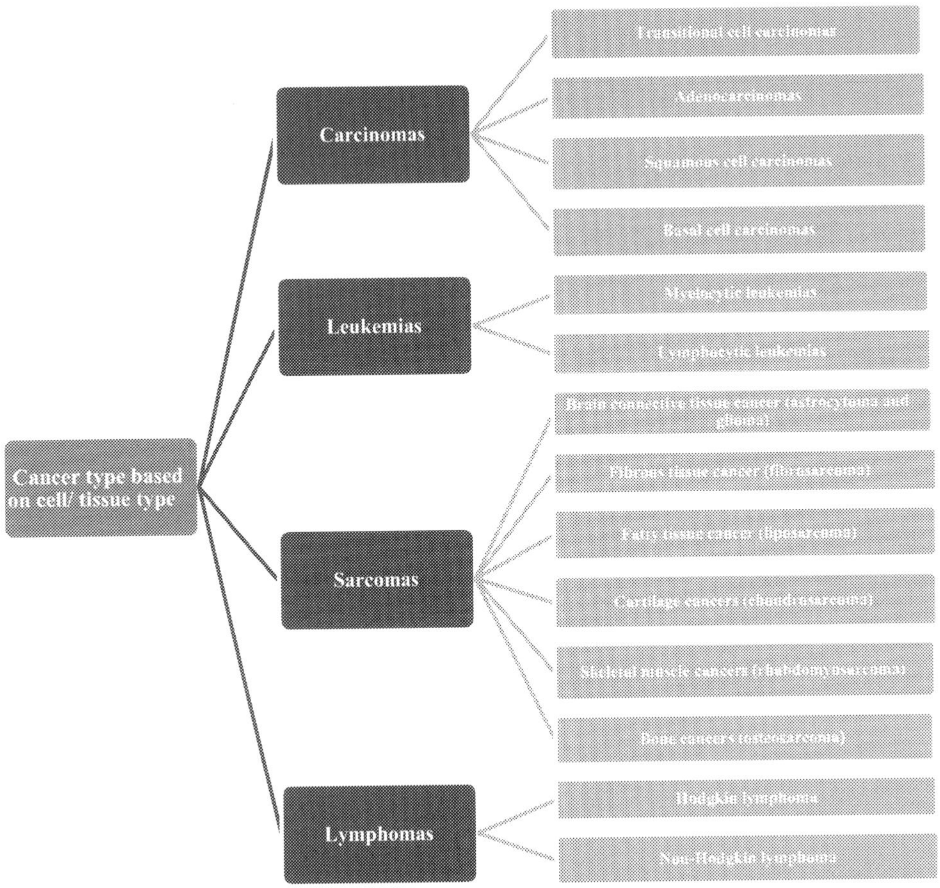

Figure 1.2 Categorization of cancer cells based on the cell/tissue types.

skin, and prostate are carcinomas (Edwards et al., 2014). However, cancers of the bones, muscles, brain, and lymph nodes are not carcinomas. Subsequently, cancer types can be categorized in terms of the body organ or system where the cancer occurs, as discussed next.

1.1.3.2 Classification of cancer types by body organ or system

Another categorization of cancer types can be conducted using the body organs or systems where the cancers arise, as reported in Fig. 1.3.

Fig. 1.3 illustrates the different cancer types in the body systems, including the respiratory system cancers, endocrine

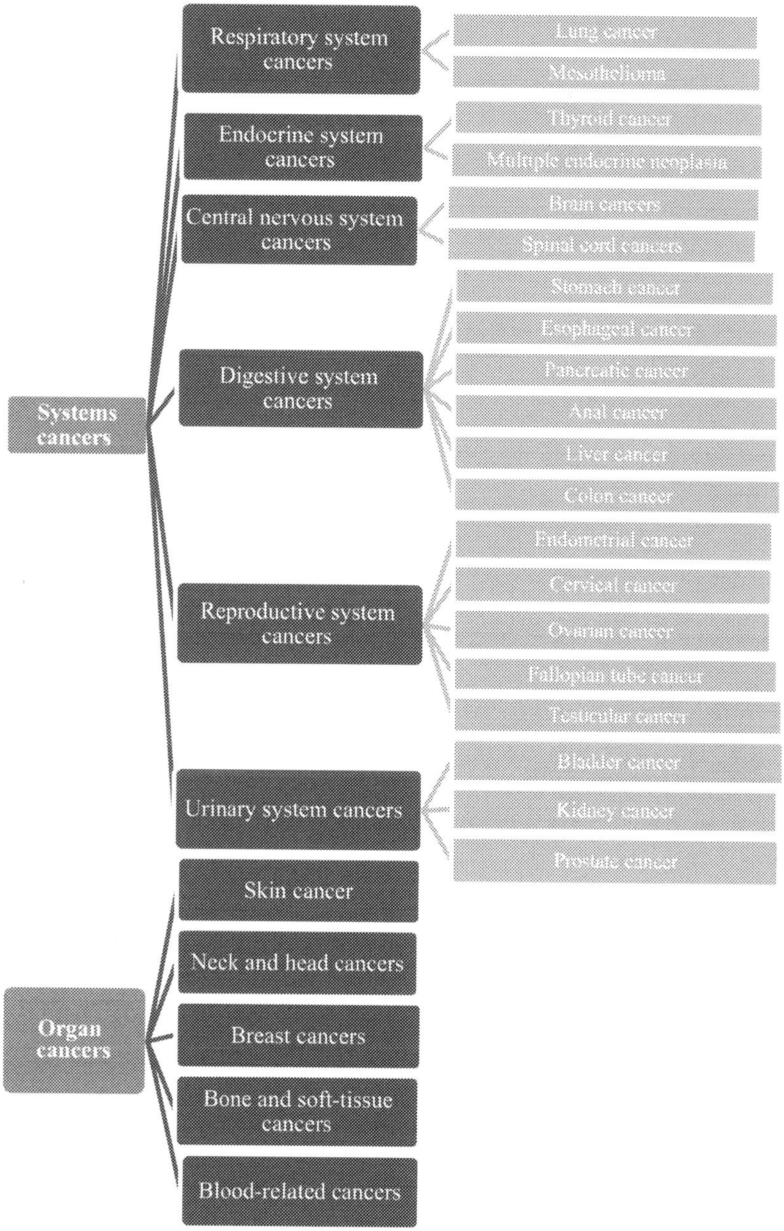

Figure 1.3 Categorization of cancer cells based on the location of the cancer in body organs/systems.

system cancers, central nervous system cancers, digestive system cancers, reproductive system cancers, and urinary system cancers. Additionally, there are several cancer types in different body parts, including skin cancers (e.g., melanoma and nonmelanoma, squamous cell carcinoma, basal cell carcinoma), neck and head cancers (cancers in the region from the vocal cords to the tongue), breast cancers, and blood-related cancers (e.g., myeloma, acute/chronic myelogenous leukemia, acute/chronic lymphocytic leukemia) (Somoano and Tsao, 2008; Chan et al., 2012).

1.2 Thermal therapy mechanisms

Theoretical physics brings an important perspective to studying biological issues

Krastan Blagoev

1.2.1 Cancer cell characteristics

Advances in biophysics and cell biology facilitate the investigation of the association between the cells/subcellular structures mechanics and cell functions, including signaling, mitosis, and locomotion/motility. The association between the microenvironment and biophysical properties of a tumor recognize the cell function and behavior. Measuring tissue rigidity has a potential role in cancer detection (Wells, 2008; Levental et al., 2009; Kumar and Weaver, 2009).

Basically, cancer is a disease of mitosis, tissue organization, proliferation, decontrolled signaling, and migration. It arises with the transformation of normal cells into cancerous ones which multiply uncontrollably. Cancer cells respond differently to the lifespan controlling signals compared to normal cells, leading to their continuing growth and increasing invasion of other body parts (Hanahan and Weinberg, 2000; Hahn and Weinberg, 2002; Lewis and Pollard, 2006). Cancers cells have several distinctive characteristics compared to normal ones, such as incapability of self-repair, having abnormal shapes and varying significantly in size, being unable to perform the functions of a normal cell, and unable to carry out normal apoptosis.

Any changes close to the cells are sensed and, accordingly, the cell transforms the physical signals to biochemical ones. This cell characteristic supports the control of cell function. Cancer arises from malfunctioning in biological cells, where the cancer cell proliferates irrepressibly and disturbs the tissue structure. The duplication and continued growth of cancer cells is because they produce identifiable signals which are transmitted using a signal transduction process (Cooper et al., 2000; Logan and Nusse, 2004).

The physical alterations in tumor tissues are associated with the biological procedure inside the tumorous cells. The increase in cell stiffness decreases the cell contractibility, which allows the interaction between the biological processes and the mechanical force on the cell (Holle et al., 2016). The variation in the mechanical characteristics in the cell indicates the presence and state of a disease/tumor. Once the characteristics of the cell change, the tumor cells start spreading to other body parts, which are observed to be a biomarker for detecting the metastatic effect of the cancer cells and increases their stiffness (Weder et al., 2014; Peinado et al., 2017). Accordingly, cell-based sensors, which are based on the biophysical cues of the cell, can be used for cancer detection (Edmondson et al., 2014).

1.2.2 Thermal characteristics of cancer cells

Temperature is considered the governing parameter for cancer cell/tissue destruction. Hence, it is necessary to highlight how heat interrelates with tissues to induce cell death. Fundamentally, heat generation in tissues induces cellular death through thermal coagulation necrosis (Taylor, 2007; Rossi-Fanelli et al., 2012).

1.2.2.1 Heating effect on cancer cells

Hyperthermia temperature ranges from 39°C to 42°C (nonlethal temperature) or >42°C (lethal temperature), where a temperature of >42°C can destroy the cells of a tumor in a temperature- and time-dependent manner. This process creates cellular changes, which are considered as a heat-induced cytotoxicity mechanism leading to cellular homeostasis loss (Hildebrandt et al., 2002; Zhang et al., 2008). Hyperthermia of cancer cells leads to aggregation and protein denaturation, which in turn cause other cellular effects on the cell cycle, including DNA synthesis/repair inhibition, arrest, transcription, and protein synthesis inactivation processes, disrupting the membrane cytoskeleton, with augmented degradation of aggregated proteins via the lysosomal/proteasomal pathways, alterations in membrane, and permeability metabolic changes. Moreover, hyperthermia decreases the plasma membrane viscosity of the cancer cell and causes lipid changes (Kampinga and Dikomey, 2001). These cellular changes due to the increased temperature lead to death of the cancer cells at temperatures ranging from 39°C to 45°C.

Furthermore, hyperthermia involves posttranslational changes, including farnelysation, acylation, glycosylation, ubiquitination, and phosphorylation. Other effects of hyperthermia include causing double-strand breaks and DNA fragmentation (Lepock, 2004). Nevertheless, the damage to the nuclear protein is considered more critically significant than the DNA impairment, where it is extensively sensitive to hyperthermia at >43°C and undergoes aggregation (Hildebrandt et al., 2002; Lepock, 2003). Accordingly, Fig. 1.4 illustrates the destructive effect of hyperthermia on cancer cells.

Furthermore, hyperthermia sensitizes cancer cells to cytotoxic sources, including radiation. On the other hand, physiological issues, such as blood flow, pH, and oxygenation influence the sensitivity of the cells to hyperthermia. The inherent sensitivity of cancer cells to heat differs pointedly amongst the different cell types as well as during the diverse phases of the cell cycle.

Typically, hyperthermia has a great impact in cancer treatment owing to the preceding reasons, as well as due to poor vascularization of tumor tissues with increased heat compared to the response of normal tissues. This leads to a disparity in heating, with a higher temperature in tumors compared to normal cells. Such a phenomenon occurs due to the dissipated heat through the circulating blood flow (Habash et al., 2006). Subsequently, hyperthermia at >42°C has the ability to damage the tumor's blood circulation, leading to cutting off of the oxygen/nutrients supplies, which causes a collapse in tumor vascularity.

Conversely, heating at temperatures ranging from 39°C to 42°C increases the blood flow in the tumor, leading to better tumor oxygenation, which in turn renders the tumors more sensitive to anticancer drugs and radiation. For example, milder

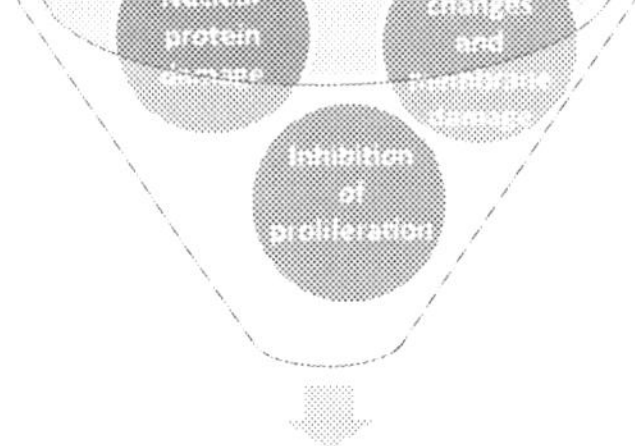

Figure 1.4 Hyperthermia-based protein damage leading to cancer cell death.

hyperthermia at a range of 40°C–41°C stimulates the immune system, therefore increasing the immune resistance leading to protection against the tumor growing (Mura et al., 2013; Oei et al., 2015). Fig. 1.5 shows the cancer cell response to a change in the hyperthermia temperature level.

Moreover, irreparable cellular destruction occurs with an increase of temperature to 46°C. Meanwhile, increasing the temperature to about 52°C reduces the required time to induce cytotoxicity (Hahn, 2012).

Moreover, proximate immediate induction occurs for protein coagulation at temperatures ranging from 60°C to 100°C, which permanently damages the cancer cells. This is considered the key to ablation therapy, which is achieved at temperatures ranging from 50°C to 100°C, through the whole target tumor tissue region. During ablation, with heat coagulation, the nearby normal fatty tissues recurrently have histological fat necrosis marks. Conversely, lethal damage appears with the existence of vital blood vessels neighboring the tumors, which are severely damaged (Schuler et al., 2014).

1.2.2.2 Freezing (cooling) effect on cancer cells

Freezing cancer cells is considered another temperature-based cancer treatment which produces hyperosmotic surroundings outside the cells. Initially, the intracellular fluid is cooled without freezing and spreads out of the cell. With continued treatment, free water is lost and the cells become dehydrated. Additionally, denaturation of the proteins occurs at about −10°C, where the cell membrane is disturbed, allowing the formation of intracellular ice that causes permanent cancer cell death (Sotsky and Ravikumar, 2002).

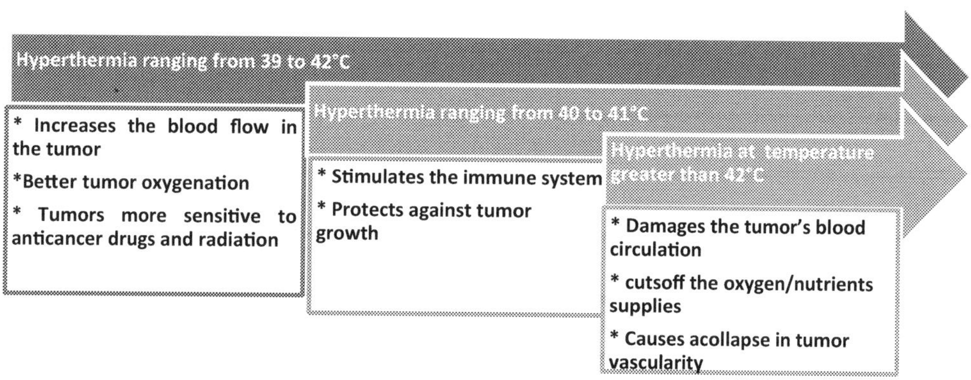

Figure 1.5 Cancer cell response versus a change in the hyperthermia temperature level.

Cryoablation enables extracellular ice melting, changing the osmotic gradient while producing hypotonic extracellular water for diffusing again to the cells and causing them to rupture. This lethal process defines the cellular damage that occurs with fast freezing. Besides the cellular-level toxic effects, cold temperature concomitantly incites local hypoperfusion and vasoconstriction causing more cell death. Typically, about 5 minutes are adequate to complete the freezing phase of the targeted tumor tissues (Hai and Tse, 2013).

Slow thawing is considered the second mechanism of cell damage over the course of cryoablation (which is performed during ablation), and is associated with the failure of delayed microcirculation. This procedure depends on vascular damage-induced ischemia of the cancer tissues during the thawing process to destroy the residual cells (Hoffmann and Bischof, 2002; Baust and Gage, 2005). With the accumulation of ice inside the microvasculature, the blood flow is diminished. Consequently, the cryoablation treatment-based cooling entails at least two consecutive freeze–thaw cycles.

Typically, cell demise occurs with cooling at temperature ranges of −5°C to −50°C, where the lethal threshold of freezing temperature is −20°C. At temperature values >40°C, the tissues of lung, prostate, breast, and liver tumors survive. Hence, the ultimate cryoablation temperature range is from −40°C to −50°C to ensure complete extracellular/intracellular freezing of water, cell destruction, cellular metabolism cessation, and finally cell death (Weber and Lee, 2005). To promote the formation of intraluminal microvascular ice, a temperature ranging from −15°C to −40°C is sufficient, where the treatment becomes more efficient with increasing time of cell exposure to this temperature (Sano et al., 2018).

Basically, the tissue cooling rate has a direct relation to the formation of intracellular ice for irreparable damage to the cells. Such a relation is considered in the design of the cooling applicators (probes) to ensure cooling rates higher than 60°C/min at the tip. The intracellular ice can be formed at cooling rates from 20 to 50°C/min, while cooling dense cellularity tissues can be achieved using slower cooling rates. In addition, the cooling rate of tissues is greatly influenced by the tip of the probe distance from the targeted tumor tissues (Mauck et al., 2009).

After the tumor tissues are exposed to the cooling process for treatment, a variety of concluded effects occur, including inflammation, local hemorrhage, cellular loss/coagulation necrosis, structural tissue damage, and initial sublethal destruction which may extended to apoptosis or repair. In addition, tissue inflammation and hyperemia may occur after completing thawing within minutes to hours. Also, the complete death occurs for cells located at the freezing margin near the cryo-applicator site. In the hypothermic zone, the peripheral tissues, which experienced temperatures from −20°C to 0°C, will still have survival potential (Mahnken et al., 2018).

1.3 Hyperthermia

A moderate increase in temperature is known as hyperthermia, which may be due to bacterial infection or pathological causes. It is known that the Egyptians over 5000 years ago used a heated stick for breast cancer treatment (Sanders and Kingsnorth, 2012; Horsman and Overgaard, 2007). This created a new opening in cancer treatment procedures using heated metals. However, such cancer treatment methods lacked direct control. Temperatures ranging from 42°C to 44°C were used using circulated water for cervical cancer treatment (Van den Tempel et al., 2016). Typically, hyperthermia is considered a heat treatment, or heat therapy, which is an alternate health management by exposing part/all of a patient's body to high temperatures (Torossian, 2008). Patients undergoing cancer therapy receive thermal therapy, including dry/moist heat, where the tumor severity dictates the treatment duration. Moreover, thermal treatment using heat can be carried out to relieve muscle and joint pain, including inflammation, arthritis, or injury, as well as to excite the sensory skin receptors to reduce the pain signals that are directed to the brain.

1.3.1 Definition of hyperthermia in clinical trials

Hyperthermia is a cancer treatment type based on exposing body tissues to high temperatures (up to 45°C), and is also called thermotherapy or thermal therapy. These high temperatures can kill/destroy cancer cells, their structures, and proteins, with slight damage to normal tissues (Hildebrandt et al., 2002; Hergt et al., 2006).

Several studies have been conducted to understand the foundations for the medical use of hyperthermia to develop technology-based clinical tools for heating tumors and measuring the heat distribution within tumors. Accordingly, hyperthermia became an active trend in clinical practice for cancer therapy due to its ability to damage/kill cancer cells by exposing the body tissues to high temperatures (42°C and above) without affecting normal tissues. In vitro studies established the noninherent differences in the sensitivity to heat of normal and tumor cells (Issels, 2008). The vascular system in solid tumors is chaotic, resulting in low pH levels and hypoxic regions in comparison to normal tissues. Accordingly, such an environment makes the hyperthermia have a more cytotoxic and destructive effect on the tumor cells within solid tumors. Furthermore, slight hyperthermia can trigger an immune system response to provide defense against tumor growth (Skitzki et al., 2009; Peer et al., 2010).

Hyperthermia can reduce the radio-resistance and drug-resistance of tumor cells, hence, it can be integrated into cancer treatment using chemotherapy and radiation therapy with a maximum temperature of 45°C (Wust et al., 2002; Peer et al., 2007). Additionally, hyperthermia's therapeutic potential at temperatures greater than 60°C is employed to destroy cancer tissues using thermal ablation. The involvement of hyperthermia in the different cancer therapeutic strategies is illustrated in Fig. 1.6.

In clinical practice, hyperthermia achieves most of its advantages in combination with radiation therapy or/and chemotherapy. It increases cancer cell sensitivity to harm or radiation as well as enhancing the effects of anticancer drugs. The most significant advantage of hyperthermia compared to chemotherapy and radiotherapy is that with hyperthermia the cytotoxicity can be directed at the tumor volume, thus decreasing the toxicity side effects. The efficacy of hyperthermia depends on the temperature and exposure duration of the treatment. Many clinical trials have studied combined hyperthermia with chemotherapy and/or radiation therapy in the management of various cancer types, such as melanoma, sarcoma, and cancers of the brain, head/neck, breast, lung, liver, bladder, and appendix (Rosenberg et al., 2001). These studies have established that a substantial tumor size reduction occurs when hyperthermia is combined with other treatment methods.

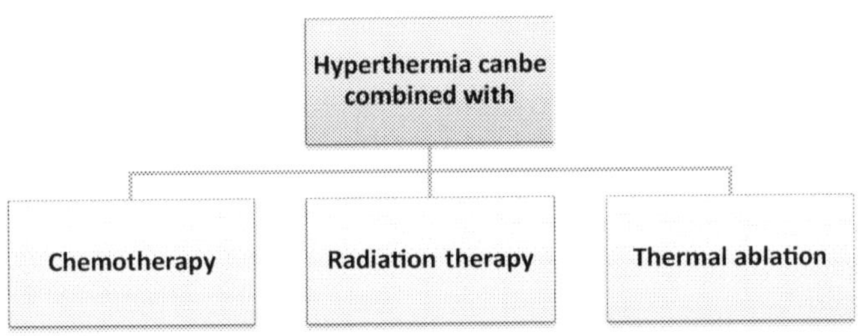

Figure 1.6 Hyperthermia combinational pairs in cancer therapy.

Subsequently, in cancer treatment, the integration between hyperthermia and radiotherapy has become efficient for the clinical outcome for oncology patients, where hyperthermia is considered the most active radiation sensitizer due to its ability to reduce tumor cell resistance to radiation. It is able to kill cancer cells under low pH and hypoxia conditions based on the heat-induced radio-sensitization mechanisms (Paulides et al., 2010).

Moreover, the combination of hyperthermia and chemotherapy has great therapeutic potential, where the confined heat delivery activates the cytotoxicity of the anticancer drugs inside the targeted tumor regions. The integration of chemotherapy with deep regional hyperthermia (RH) (hyperthermic intraperitoneal chemotherapy) is applied to treat different types of cancers, including bladder cancer (Ahmed and Zaidi, 2013; Freifeld et al., 2013) and prostate carcinoma (Tilly et al., 2005). The integration of both methods allows the targeted chemotherapy to increase the therapeutic ratio as well as radiation therapy. Several studies have proved the complementary cytotoxic effect of anticancer drugs and hyperthermia (Beuth and Schierholz, 2007). Generally, the tumor center/core in a hypoxic and low pH environment is exposed to lower drug concentrations and thus resists chemotherapy where they are present in less vascularized areas of the tumor. In such a case, hyperthermia has an essential impact in destroying cancer cells in the tumor center. In addition, in poorly vascularized areas, the temperature effect on the tumor is greater than in vascularized areas in which the circulating blood has a cooling effect. Moreover, hyperthermia at 40°C–43°C increases the vascular permeability and blood flow in the tumor (Jain, 2005) as well as the cytotoxicity.

1.3.2 Hyperthermia techniques

The impact of hyperthermia in activating the radiation therapy and chemotherapy effect on cancer tissues has encouraged scientists to develop different methods of increasing the tumor temperature. Such techniques can be classified as local hyperthermia (LH), perfusion hyperthermia, and whole-body hyperthermia (WH) (Van der Zee et al., 2008), as demonstrated in Fig. 1.7. These techniques generate hyperthermia in the cancer clinic by exposing the tumor tissues to nonionizing radiation (e.g., ultrasonic/electromagnetic fields) or conductive heat sources. Consequently, hyperthermia can be generated using noninvasive or invasive procedures through superficial application of power to increase the tumor temperature.

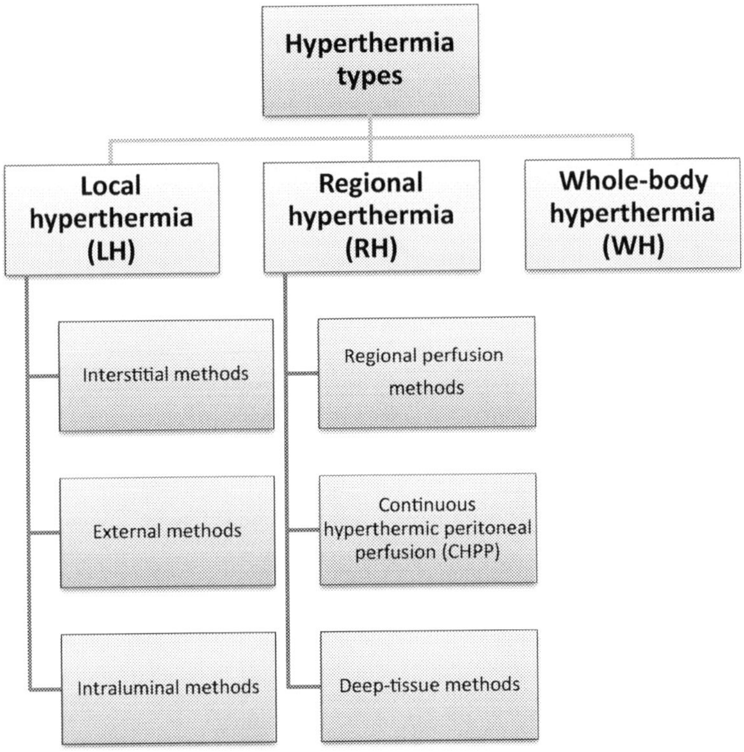

Figure 1.7 Types of hyperthermia.

1.3.2.1 Local hyperthermia

LH applies heat to small tumor regions using different methods of energy supply to heat the tumor. It generates high temperature using internal or external energy sources (heating modalities) to superficial or deep-seated internal tumors without affecting the nearby normal tissues (Short and Turner, 1980). Thus, the tumor region is externally heated by applying high-frequency waves (e.g., ultrasound or electromagnetic energy) from an external device outside the human body. Different energy types can be used to supply heat, such as ultrasound, radiofrequency, and microwave. A probe is used to accomplish the required internal heating, such as radiofrequency electrodes, microwave antenna, thin-heated wires, and hollow tubes filled with hot water. LH allows use of hyperthermia in combination with other antineoplastic therapy modalities to treat solid tumors, instead of blood diseases. However, the nonhomogeneous characteristics of the tumor regions and the variation in the blood flow influence the heat-scattering consistency in the preserved areas. This leads to limitations in obtaining a regional uniform increase in the temperature

(Nie et al., 2014; Le Renard et al., 2014). LH can be categorized based on the tumor location as follows:

- *Tumors in the deep places of the body,* such as brain tumors are treated using interstitial procedures, such as the radiofrequency ablation (RFA). These procedures heat the tumor to very high temperatures, by inserting needles or probes with a heat source into the tumor guided by imaging methods, including ultrasound, to ensure proper positioning of the probe in the tumor.
- *Tumors just beneath the skin* are treated using external methods by positioning external applicators near or around the tumor area and focusing the energy on the tumor to increase its temperature.
- *Tumors nearby or inside the body cavities,* including the rectum or esophagus, are treated using endocavitary or intraluminal approaches, where a probe carrying energy is sited inside the cavity and heat is injected into the tumor directly.

1.3.2.2 Regional hyperthermia

RH is used to heat large regions of tumor tissues, including the pelvis, stomach, liver, limbs, or body cavities (Petrovich et al., 1989) using one of three main categories:

- Perfusion hyperthermia approaches implicate regional heating using heated fluids, which spread through the tumor. Typically, lower temperature values are applied when performing perfusion with cytotoxic agents.
- Continuous hyperthermic peritoneal perfusion is used for treating cancers in the peritoneal cavity by heating anticancer drugs further from a heating device via the peritoneal cavity.
- Deep-tissue methods are used for treating cancers within the body, including bladder cancer and cervical cancer, using external applicators where radiofrequency or microwave energy is concentrated on the tumor region to increase its temperature.

1.3.2.3 Whole-body hyperthermia

To treat widespread metastatic cancers in the whole body, externally induced WH is applied to spread throughout the body. WH can be performed using numerous techniques that heat the patient to about 42°C in combination with chemotherapy and radiation (Yerushalmi, 1976). To guarantee high-quality treatment, accurate measurements of the tumor temperature are crucial, where the parameters of thermal dose depend on the temperature distribution measurements along with the tumor size and blood flow (Billard et al., 1990). Accordingly, temperature measurements in tumors are recorded using three-dimensional (3D) thermal distribution during the hyperthermia treatment. However, the main drawback of hyperthermia treatment is the risk of causing skin burns with a long hyperthermia duration (Zee et al., 2002).

1.3.3 Hyperthermia complications and side effects

If the hyperthermia temperature remains below 43°C, the normal tissues will remain unaffected. Nevertheless, owing to the regional dissimilarities in tissue features, higher temperature level may arise in different areas causing blisters and burns (Kapp et al., 2000). Perfusion methods can also cause bleeding, blood clots, and tissue swelling, and in some cases injuries to normal tissues. On the contrary, using WH causes more severe side effects, such as nausea, diarrhea, and vascular/cardiac disorders. Accordingly, there are several challenges to be overcome before using hyperthermia. Numerous clinical trials have been examined to assess the success of hyperthermia in cancer treatment and to improve the different hyperthermia methods.

1.4 Tumor ablation

Conventionally, local tumor removal requires major surgery. With the development of imaging technologies, minimally invasive tumor treatment relied upon imaging guidance to accurately place needle-like applicators to ablate the tumor tissues in a nonsurgical way with reduced morbidity (Gazelle et al., 2000). Ablation is considered a substitute treatment for surgical therapy for cancers. In clinical trials, thermal ablation is used in several cases, including brain tumors, liver/kidney tumors, breast cancers, lung tumors, heart arrhythmias, and skin lesions (Kang and Rhim, 2015). It has potential advantages compared to other surgical procedures for cancer treatment, including reduced death rates and fewer surgical complications.

Several modalities can be applied for ablative-based treatment to eradicate and destroy tumors using temperature as a peripheral excitation. The two main methods for ablation are chemical ablation using chemical agents to clot and destroy cancerous cells, and thermal ablation by overheating the cancerous tissues until they are destroyed (Yun and Kwok, 2017). With thermal ablation, an energy source is used to raise the temperature within the range of 55°C–100°C, where overheating of tissue to extraordinary temperature levels affects the biological and physiological characteristics of the tissues, leading to their destruction. Accordingly, a broad variety of ablation methods as sources of energy have been developed for overheating of the tissues, including laser ablation (LA), ultrasound-based ablation [focused ultrasound ablation (FUSA)], RFA, and microwave ablation (MWA) (Chu and Dupuy, 2014; Knavel and Brace, 2013). In cooling-based ablation, the cryoablation method uses a gas, such as argon, for cooling. Fig. 1.8 illustrates the different ablation techniques based on the temperature level.

During the ablation procedure (Fig. 1.9), the energy source penetrates the human body targeting the cancer region. Therefore an imaging-guided ablation system is used to target the tumor region (Motamedi et al., 2009; Brace et al., 2011).

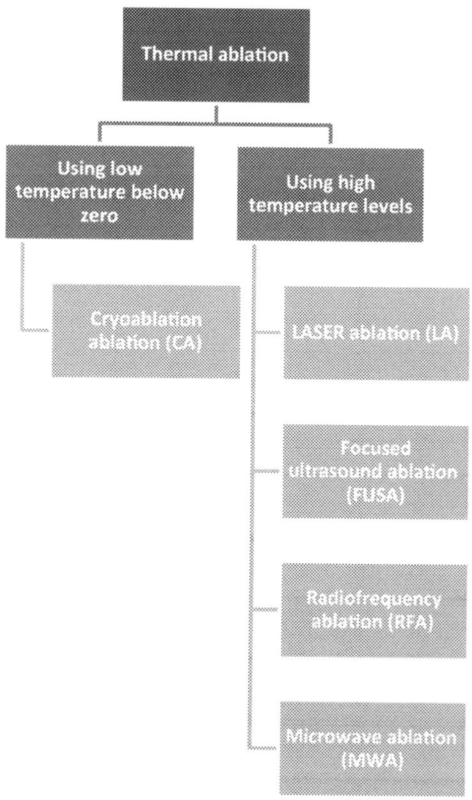

Figure 1.8 Ablation technique categories based on the temperature level.

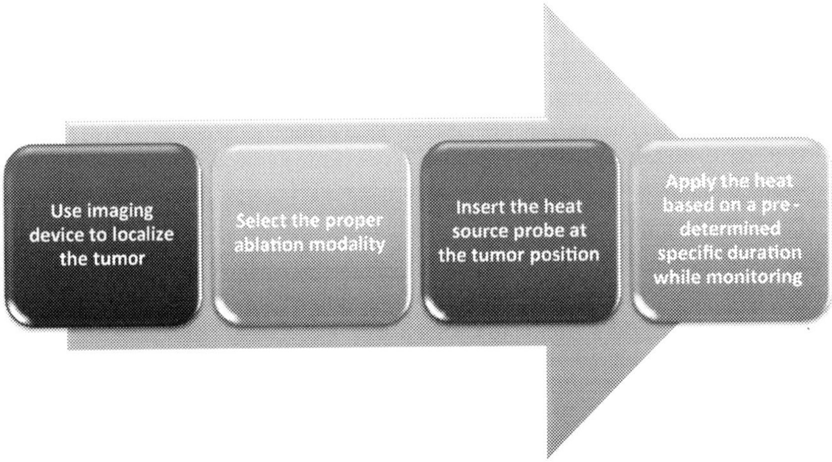

Figure 1.9 Clinical thermal ablation procedure.

Accordingly, the imaging modalities, including magnetic resonance imaging (MRI), computed tomography (CT), or ultrasound (US), are used during the placement of the probe/needle (energy source) and monitoring of the ablation procedure. This captures successive images to monitor temperature diffusion. Conversely, in cryoablation, thermosensors (thermocouples) are used to monitor the temperature of tissues.

The imaging advancements are the key factors for the success of thermal therapy. Instantaneous monitoring using US-guided imaging and visualization of the full field using CT/MRI guarantee accurate measurement of treatment efficiency, less invasive processes, and less collateral damage to nearby organs.

1.4.1 Ablation modalities

Oncology is considered the most prevalent sector in which thermal therapy can be extensively exploited. The cellular reaction and energy conversions of ablative treatments which initiate the destruction mechanisms of the tumor tissue depend on the selected ablation modality. The ultimate goal of thermal ablation procedure is cancer treatment, which creates an ablation region within an extended 1 cm at all points outside the border of the tumor as a protective margin from tumor relapse (Dewhirst et al., 2003; Stauffer and Goldberg, 2004). Typically, the ablation zone size is determined by the temperature of the ablated region and the time/duration of the ablation process.

Furthermore, in clinical applications, thermal therapy still has complications and restrictions.

Treatment based on hyperthermia can be subject to a lack of accuracy by triggering a large damaged margin around the tumor (in the surrounding healthy tissues), which is unacceptable in certain organs, for example, brain. The thermal ablative techniques, including CA, LA, FUSA, RFA, and MWA have some challenging clinical drawbacks and inconveniences, not only for physicians but also for designers/engineers. Brief descriptions of the different ablative modalities, their principles of work, limitations, and comparative studies follow.

1.4.1.1 Cryoablation ablation

In contradiction of the thermal therapies using LA, FUSA, RFA, and MWA, which are based on reaching the lethal level of cellular heat to destroy tumors, cryoablation (CA), where "cryo" means cold in Greek, uses cryogens to reach freezing temperatures to destroy targeted tissues (Baust et al., 2009). CA uses temperatures lower than zero, of about $<-42°C$ within the range of $-20°C$ to $-40°C$ (Maccini et al., 2011; Baust et al., 2014). Arnott was the first to describe cryotherapy-based oncology in 1845 for treating uterine, cervical, breast, and skin cancers using a concoction of salt and crumpled ice at $-20°C$ (Arnott, 1851). CA can be applied to localize large tumors as a primary therapeutic method. It is a slightly invasive ablation method using cryogens to produce freezing temperatures and damage the targeted tumor tissues. The mechanism of CA is intricate and arises at different temperature values lower than 0°C generated by a probe, which is injected into the tumor region. To guarantee tumor tissue cooling, the tip of the cooling probe used for CA is connected by a source of argon gas or nitrogen liquid at a temperature of $-100°C$. Nevertheless, CA undergoes complications due to tissue freezing.

The working principle of successful treatment using cryoablation is based on promptly freezing tumor tissues to reach the lethal temperature, then slowly and entirely thawing the targeted tissues. This freeze–thaw sequence is repeated until full ablation occurs with cellular demise (Baust and Gage, 2005). The freezing temperature level is reached using cryogens, namely argon gas and liquid nitrogen which have boiling points of $-185.7°C$ and $-195.8°C$, respectively, which control the feasible base temperature of the cryoablation process.

Argon gas cryogen endures heat extraction under the Joule–Thomson effect. To generate freezing temperature at the argon gas applicator tip, the physical expansion characteristics of the gas are used (Kujala and Beland, 2012). Argon gas is kept under high pressure of 3000 psi within a gas cylinder (delivery system). Accordingly, at the probe tip, the argon gas is rapidly expanding once it reaches the lower pressure. Based on the energy conservation and thermodynamics laws, the gas expansion increases the potential energy and decreases the kinetic energy, causing the local temperature to decrease. However, the probe tip temperature can be reduced to a cooling temperature of −130°C to −135°C. In contrast, using liquid nitrogen-based cryotherapy, the probe tip is supplied with liquid nitrogen using pressurized gas, where the liquid nitrogen boils and is converted to gas to extract heat from the adjacent tissues (Hewitt et al., 1997). At the applicator tip, greater freezing and lower temperatures are obtained using the liquid nitrogen system. The Leidenfrost phenomenon is applied at the metal applicator tip, where a gas vapor layer forms natural insulation due to the quick conversion of the liquid state around the probe. Thus, accumulated gas is generated around the treatment region, with the continuing gradient of slowdown in the pressure leading to blockage of the vapor (Hewitt et al., 1998). This blockage is considered the main drawback of using liquid nitrogen cryogen, and is solved using a gas bypass procedure to allow the conversion of high ratios of supplied liquid nitrogen cryogen to block the gas vapor.

In the design of the cryoablation system, the probe size depends on the targeted tumor size, where a smaller probe is recommended for percutaneous practice. However, a small cryoprobe's tip provides high base-temperature nearby region leading to small freezing temperature. Thus, multiple small cryoprobes are commonly positioned in the form of parallel arrangement with in-between separation distance for accurate temperature control (Dobak et al., 1998). Practically, the multiple cryoablation probes provided larger, more uniform, and probable iceball, which to focus on the typical shape of most of the tumors.

1.4.1.2 Laser ablation

A laser is light amplification by stimulated emission of radiation, which is a monochromatic targeted, intensive light beam. Different types of materials are used for generating laser beams, such as neodymium yttrium-aluminum-garnet, and CO_2, and also the light-emitting diodes are considered as laser sources. Moreover, a laser beam can be provided in pulsed or continuous mode (Berns et al., 1981; Kalisky, 2006; Phototherapy, 2018).

LA applies a laser energy source to target cancer tissues. The wavelength of the laser beam in LA ranges from 560 to 2100 nm, which is selected according to the desired ablation depth. Additionally, the area of the damaged cancer tissues is controlled by the characteristics of the used laser beam, including beam quality, pulse repetition rate, pulse duration, and wavelength, as well as the characteristics of the targeted tissue, including its area. LA can be used in the treatment of several cancers, especially brain tumors guided by MRI (Shi and Mao, 2017). The mechanism of LA depends on transforming the laser light to heat, through the absorption, scattering, and reflection of the laser beam, to destroy the cancerous area. Then, convection and conduction are responsible for temperature spreading (heat transfer) and overheating the targeted tumor tissues, respectively (Steiner, 2011). Typically, temperature diffusion during the LA process can be monitored using ultrasound-guided ablation or optical sensors (Tosi et al., 2014; Schena et al., 2016; Palumbo et al., 2017; Pacella and Papini, 2017). LA is a common ablation method which is safe and takes a reasonable amount of time. However, it is used only in small tumors and is expensive. The working principle of the LA procedure is to focus a laser beam on a surface to eliminate solids from the irradiated zone. Typically, the laser parameters, including pulse duration, wavelength, and energy, as well as its chemical properties, influence the tumor destruction mechanism.

1.4.1.3 Focused ultrasound ablation

Sound waves generally produce a small amount of heat while traveling through a medium and turn into heat once they are absorbed due to the sound movement of the molecules in fast moving patterns. To transform sound to heat, the ordered movement of the sound patterns is converted into disordered ones. On the other hand, ultrasound is a sound wave with frequencies higher than audible sound. It can be used in the treatment form of focused ultrasound (FUS), known as high-intensity focused ultrasound (HIFU). A set of FUS transducers is used to control the damage to the tumor tissues.

In the FUS, the sound waves use the source of the ultrasound wave beam to heat and damage the tumor tissues, without affecting the nearby healthy tissues. The use of FUS, which was originated by Lynn et al. (1942), in therapy is based on heating the tumor by sending a directed ultrasound wave beam to the tumor region. It is considered a noninvasive ablation modality for the treatment of solid tumors using intensities above 5 W/cm^2. The FUS beam is well focused on a precise targeted zone without damaging the healthy tissues (Karaboce, 2016; Lang and Wu, 2017). The FUS ablation is performed while monitoring using ultrasound or MRI devices (Crouzet et al., 2010). For thermal ablation, FUS has a variety of techniques, which can be classified based on the tumor location, namely transrectal FUS for prostate cancer treatment (Poissonnier et al., 2007), extracorporeal FUS to ablate superficial tumors (Illing et al., 2005), interstitial FUS for deep tumor ablation such as esophageal tumors (Lafon et al., 2007), and percutaneous FUS for deeper tumor ablation such as in kidney and liver tumors (Jolesz et al., 2004; Fischer et al., 2010; Al-Bataineh et al., 2012), as reported in Fig. 1.10.

In real-time monitoring of the FUS procedure, MRI can be employed to monitor the increase in temperature to reach the quantified therapeutic dose (Jolesz et al., 2005).

Generally, FUSA has the same working principle as sunlight rays while passing through a magnification glass, where it focuses the ultrasound radiation beam at a single point, triggering a substantial increase in the temperature close to the focal point using the HIFU device. Although its working principle depends on conventional US principles, FUS intensity in the focal region is much higher than with US. The intensity ranges from 100 to 10,000 W/cm^2. Both mechanical and thermal effects control the HIFU operation, where the thermal effect governs the generation of heat from the acoustic energy absorption with a rapid increase in the targeted tissue's temperature to destroying the tumor tissues using FUSA therapy (Diederich, 2005). Moreover, at high intensities, HIFU induces mechanical effects (radiation force, microstreaming, and cavitation) with the acoustic pulses which are generated due to alternating expansion/compression of tissue,

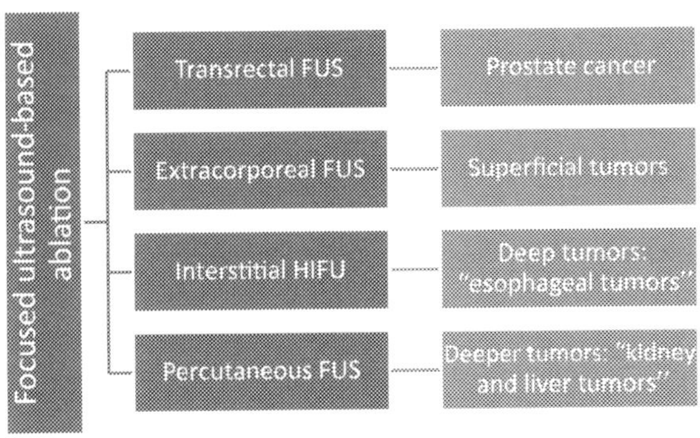

Figure 1.10 Focused ultrasound-based tumor ablation.

through which the ultrasound waves propagate (Petrou et al., 2016). The radiation force is generated with the refection and absorption of acoustic waves, while the microstreaming yields extraordinary shear forces which cause transient damage to tumor tissues (Lagneaux et al., 2002).

Lynn et al. carried out the first HIFU-based therapeutic pilot study in 1942. For treatment, it is basically based on delivering the required energy to elevate the temperature of tumor tissues while rapidly reaching a sufficient cytotoxic level to cause coagulative necrosis. Unlike solid tissues, bulky blood vessels are presumably not damaged by exposure to HIFU due to thermal energy scattering by the blood flow from the walls of vessels. This response makes HIFU safe to ablate tumors close to blood vessels, while surgical methods are risky in such instances (Gelet et al., 2000). Another advantage of using HIFU is the ability to ablate large tumors up to diameters of 3–4 cm without upper size limitations. Moreover, the lower toxicity of FUSA compared to CA and FRA as a consequence of its noninvasive nature has been established (Wu et al., 2001; Kopelman and Papa, 2007). Generally, the focusing procedure for a high-intensity US beam is realized using a multiple piston transducers arrangement, a flat transducer with the acoustic lens, or a concave self-focusing transducer.

1.4.1.4 Radio-frequency ablation

Radio frequency (RF) is the fluctuation rate of an alternating electric voltage, or current of an electric, magnetic, or electromagnetic (EM) field at a frequency ranging from 20 kHz to 300 GHz. Radio-frequency ablation (RFA) generates high temperature using high frequencies within the range of 375–500 kHz to overheat the targeted tumor regions. It is recommended clinically for cancer treatment in different organs, including the brain, breast, and liver (Mulier et al., 2012; Lee et al., 2012). RFA uses a probe which produces heat in the surrounding tissues. Thus, the temperature of the cancer tissues rises because of the resistive effect, causing cancer cell death. McGahan et al. (1990) first designed and implemented a RF monopolar electrode for liver ablation. Due to the advancements in the antenna design, different designs of RF electrodes were made, such as monopolar, bipolar, and multipolar probes with different shapes. Such RF ablation probes have needle-like shapes, including in-curved, straight, or elliptical (Ni et al., 2005). RFA is considered a nominally invasive method to eliminate different tumor types in different organs, such as the brain, liver, kidney, lung, breast, pancreas, and prostate gland (Kim et al., 2010; Singh et al., 2016). However, RFA has some restrictions including lack of accuracy, maximum temperature, and increased impedance (Ni et al., 2005).

The working principle of RFA is based on placing needle-like RF electrodes into the tumor tissues under the guidance of an imaging system. Typically, these RF probes consist of an insulated metal shaft, excluding a conductive exposed tip which provides an electrical contact with the tumor, which receives the generated RF power from the RF generator. To generate RF voltage, the RF generator consists of active and reference RF electrodes to establish an electric field path inside the body of the patient between the two electrodes. The RF frequency oscillates the electric field, causing oscillating movement of the tumor tissues with a velocity proportional to the generated field intensity (Cosman et al., 1984). To heat the targeted tissues with the required high temperature at a distance from the probe tip, the RF electrode is inserted at a depth within the tissues for sufficient time. The RFA requires a long time to increase the coagulation volume.

Different RF energy electrodes (probes) have been designed to pursue efficient RFA coagulation of larger tumor volumes in a short time. Such RF electrodes include the conventional monopolar electrode, modification of RF electrodes, pulsed RF electrodes, and clustered RF electrodes (Fig. 1.11).

Generally, the necrosis volume increases with exposure to a longer RF electrode tip. Also, using an array of electrodes increases the induced coagulation volume compared to using a single RF electrode such as the monopolar electrode (McGahan et al., 1996). Additionally, the spacing between electrodes is considered a controlling factor that affects the ablated volume and its shape.

Pulsing of energy using pulsed RF probes employed RF and other sources of energy including laser to increase the intensity of the deposited energy. In addition, such a mechanism allows the occurrence of cooling tissue neighboring the probe in the course of the slight energy deposition periods without a significant decrease in the heat level deeper in the tissues. This occurs due to the accurate balance between low- and high-energy deposition (Goldberg et al., 1999; Cohen and Foster, 2003; Foster and Glaser, 2007). To increase the energy deposition in the tumor tissues, internally cooled or multiple electrodes can be used, resulting in increasing the induced coagulation and necrosis. Recently, other therapeutic strategies integrate RF with agents, including doxorubicin chemotherapy and ethanol.

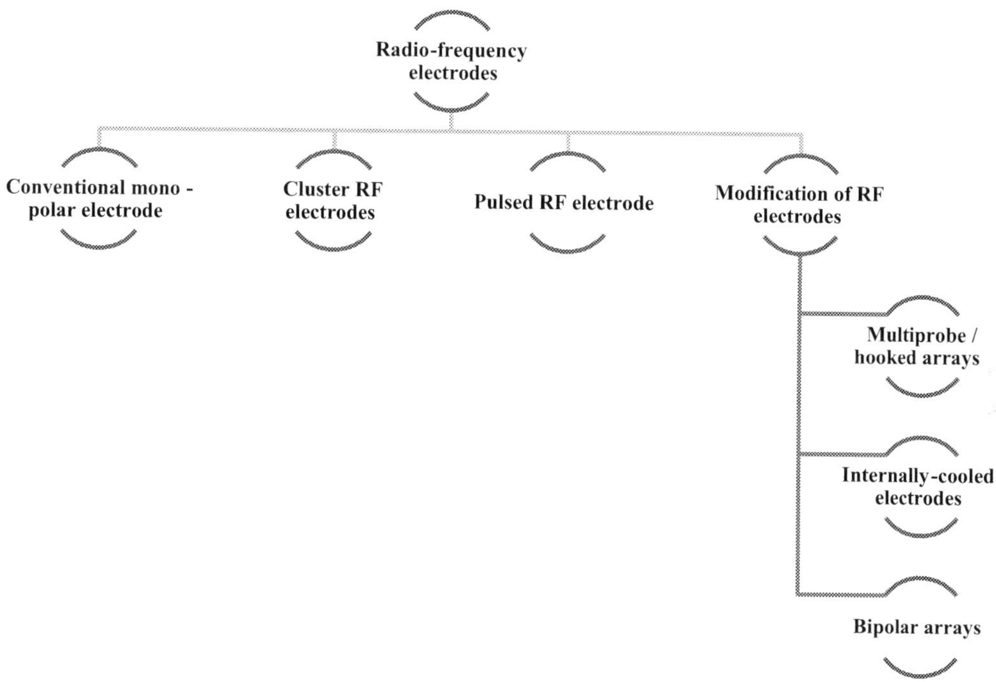

Figure 1.11 Different types of RF ablation electrodes.

1.4.1.5 Microwave ablation

To resolve the RFA drawback of increasing impedance, which decreases the RFA effectiveness, MWA has been developed. MWA uses the emitted EM signals from an antenna to heat the tumor tissues using frequencies ranging from 433 MHz to 2.45 GHz. The heating mechanism in the MWA results from the EM excitation that produces kinetic energy from the tissues' molecules, where the water molecules attempt to align with the quick EM wave oscillation. Therefore the energy absorption leads to overheating the tumor tissues even in deep regions (Chu and Dupuy, 2014).

Dissimilar to RFA, which is based on an electrical current, MWA depends on EM waves. Thus, MWA is considered an efficient ablation procedure with tissues having poor electrical conductivity, such as the lung, kidney, and bone. Moreover, MWA has the advantage of generating larger and more uniform ablation regions compared to RFA owing to the used energy nature of EM radiation. This is used efficiently for treating tumors larger than 5 cm. Likewise, to ablate tumor tissues in very deep regions and to enlarge the extent of the ablated region, multiple antennas with different designs can be used (Lubner et al., 2010). However, a well-known complication of using MWA is the fever after ablation due to the deep heating compared to RFA. In clinical practice, MWA has a promising role compared to cryoablation, LA, and RFA (Brace, 2009a,b). Momentarily, microwaves enable faster heating of larger tumors compared to the other ablation methods. MW also induces higher temperature levels in the targeted tumor and is less vulnerable to perivascular heat sinks. Accordingly, unlike RFA, MWA does not require the existence of ground pads. MWA proves its superiority to ablate different tumor types in different organs, such as the prostate, bone, kidney, lung, and liver (Bolmsjö and Vrba, 2000; Wasser and Dupuy, 2008; Liang et al., 2008; Brace, 2009a,b).

The working principle of MWA is based on inserting an antenna (applicator) into the tumor region inside the human body either laparoscopically, or percutaneously with an imaging-guided monitoring device, such as CT or US. Typically, a dielectric hysteresis process occurs during MWA, where once the MW antenna radiates microwave energy, fast rotation of proteins and water molecules in the adjacent tissues to the antenna's tip are originated. This process causes the generation of heat within the tissues. Once the temperature is raised to the ablation level (about 60°C), the tumor tissues are destroyed based on the ablation time of the maintained high temperature (Sapareto and Dewey, 1984). The common ablation duration using MWA ranges from 3 to 12 minutes (O'Rourke et al., 2007). Numerous types of antennas have been designed for MWA, including triaxial antenna (Brace et al., 2007), floating sleeve antenna (Yang et al., 2006), cap-choke antenna (Lin and Wang, 1996), and slot antenna (Ito et al., 2004).

1.4.2 Comparison between different ablation modalities

From the preceding characteristics and conditions of each ablation modality, Table 1.1 reports the main advantages and disadvantages of each method.

1.4.3 Optimal modality selection

For the best treatment and fully destruction of tumor tissues, optimal selection of the ablation modality should be considered based on several factors and by taking into account the advantages and disadvantages of each ablation method for the tumor

Table 1.1 Comparison between the different ablation techniques in terms of their advantages and disadvantages.

Modality	Cryoablation	Laser ablation	Focused ultrasound ablation	Radiofrequency ablation	Microwave ablation
Advantages	• Used with large tumors • Minimally invasive	• High accuracy • Minimally invasive • Simple design with minimum requirements	• Used with large tumors without upper limitations on the tumor size • Pointed and directed accurately to tumor • Noninvasive • Large diffusion depth • Less time • Has reduced toxicity	• Used with small tumors • Minimally invasive • Simple procedure • Less complications	• Locally targeted • Minimally invasive • Used with large tumors • Less time duration • Solves the RFA limitations
Disadvantages/ limitations	• Probe placement difficult • High rates of complications • Bleeding risk • Hemostasis limitations • Cannot be used to treat highly vascularized tumors • High recurrence rate	• Used with small tumors only and provides small ablative region • Exclusive • Several complications • Low survival rate	• Complicated requirements, such as acoustic coupling • Long ablation time • Ultrasound artifacts • Lack of reliable thermometry	• Used with small tumors only and provide small ablative region • Impedance problem • Low accuracy	• Complicated requirements, such as coaxial cable heating • Not directed • Several complications

type, size, location, and other characteristics (Hinshaw et al., 2014). Generally, for optimal thermal ablation, the probe design, material, and ability, along with the energy source of the used thermal modality, should be matched with the characteristics of the tumor region.

In terms of ablation modalities which can be either high temperature-based "hot" modality (e.g., LA, FUSA, RFA, MWA) or low temperature-based "cold" modality (e.g., CA), several factors have to be considered by comparing the performance of the different modalities in terms of the local tumor recurrence after ablation, previous morbidity rate of the patient, complete tumor ablation achievement, safety of the patient, suitability to the cancerous organ, known complications of the used modality, overall survival rate, clinical outcomes, and reported ablation modalities efficiency evidence. Table 1.2 reports a comparison of

Table 1.2 Comparison between the different characteristics of the different ablation techniques.

Modality	Cryoablation	Laser ablation	Focused ultrasound ablation	Radiofrequency ablation	Microwave ablation
Energy source	Cooling gas/liquid (argon or nitrogen) for freezing	Laser light	Ultrasound waves	Electric current at radio-frequency	Electromagnetic waves
Frequency/ wavelength	–	560–2100 nm	200 kHz–3.5 MHz	375–550 kHz	915 MHz–2.45 GHz
Heating/ cooling generation procedure	Use cooling probe to generate temperature of about −100°C	Transform the laser beam to heat	Transform the focused ultrasound beam to heat	Convert the RF alternating current to heat	Generate heat from the propagation of high-frequency EM waves
Probe/ electrode	Cryo-probe using argon or nitrogen	Fiber probe	Ultrasound array	Single/multiple probe(s)	Single/multiple antenna(s)
Generated temperature range	< −20°C to −40°C	50°C–100°C	70°C–100°C	55°C	55°C–100°C
Ablation process duration	From 7 to 30 min	About 3 min	From 1 to 10 s	About 10 min	From 1 to 5 min
Imaging-guidance device	US, CT, or MRI	US, CT, or MRI	US	US, CT, or MRI	US, CT, or MRI
Tumor type to be treated	• Liver tumor • Prostate tumor • Bone lesions • Renal tumor	• Brain tumors	• Prostate cancer • Superficial tumors (skin cancer) • Pancreatic cancer • Esophageal tumor • Liver tumors • Thyroid tumors • Uterine fibroids • Breast cancer	• Liver tumors • Brain tumors • Kidney tumors • Lung tumors • Breast tumors • Pancreas tumors • Prostate gland tumors	• Bone tumors • Lung tumors • Kidney tumors • Liver tumors

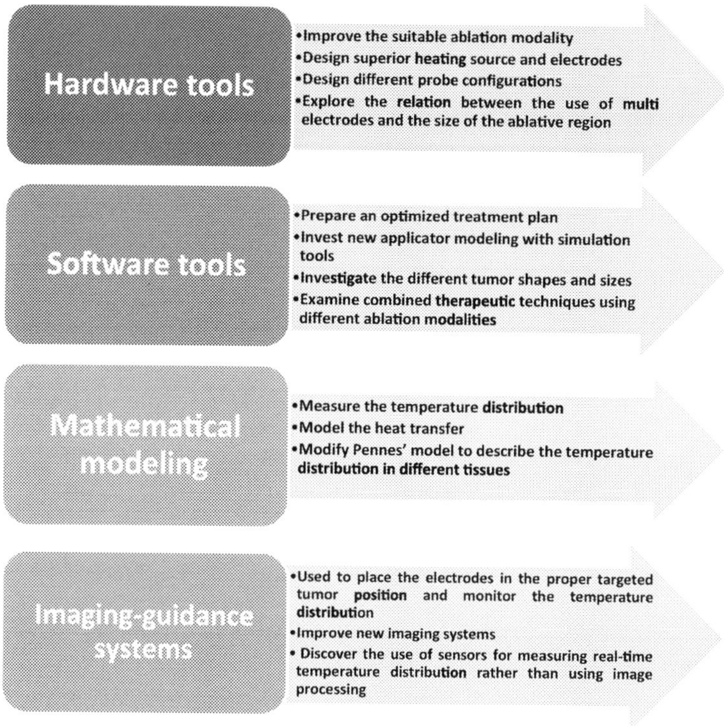

Figure 1.12 The four main factors to improve the ablation-based therapy.

the characteristics among different ablation techniques in terms of energy source, operating frequency range, cooling or heating procedure, probe type, generated temperature range, ablation process duration, used imaging-guidance device, and the reported tumor type to be treated using each ablation method.

The comparison in Table 1.2 can be used as clinical guidance as it summarizes the characteristics of the different ablation modalities. Generally, to enhance the performance of thermal therapy, several parameters must be considered, including the design of the ablation applicator, the duration selection of the ablation process, the used frequency range, the imaging-guidance system, mathematical modeling, and the software tools to predict the persuaded tissue damage along with measuring the heat distribution and heat transfer in tissues during the ablation process as reported in Fig. 1.12.

1.5 Conclusion

Ablative treatments have attracted growing attention and are considered a powerful alternative to surgical therapy for cancer treatment. Thermal ablation is employed in different clinical applications for treating several diseases, including benign prostate hyperplasia, heart arrhythmias, and liver tumors, as well as tumors in other organs, such as the brain, lung, and kidney. Prospective thermal ablation benefits include reduced morbidity and mortality rates in comparison with standard surgical resection. The integration between the biological responses and the engineering principles opened new insights to support advancements in ablation technology. Due to the specific complications concomitant with each type of ablation modality, a broad range of ablation methods have been developed, such as cryoablation, LA, FUSA, RFA, and MWA. The features of each ablation procedure, including action mechanism, required equipment, treatment method, and limitations were discussed in this chapter.

References

Ahmed, K., Zaidi, S.F., 2013. Treating cancer with heat: hyperthermia as promising strategy to enhance apoptosis. J. Pak. Med. Assoc. 63 (4), 504–508.

Al-Bataineh, O., Jenne, J., Huber, P., 2012. Clinical and future applications of high intensity focused ultrasound in cancer. Cancer Treat. Rev. 38 (5), 346–353.

Arnott, J., 1851. On the Treatment of Cancer, by the Regulated Application of an Anaesthetic Temperature. J. Churchill.

Basu, A., 2018. DNA damage, mutagenesis and cancer. Int. J. Mol. Sci. 19 (4), 970.

Baust, J.G., Gage, A.A., 2005. The molecular basis of cryosurgery. BJU Int. 95 (9), 1187–1191.

Baust, J.G., Gage, A.A., Robilottto, A.T., Baust, J.M., 2009. The pathophysiology of thermoablation: optimizing cryoablation. Curr. Opin. Urol. 19 (2), 127–132.

Baust, J.G., Gage, A.A., Johansen, T.B., Baust, J.M., 2014. Mechanisms of cryoablation: clinical consequences on malignant tumors. Cryobiology 68 (1), 1–11.

Berns, M.W., Edwards, J., Strahs, K., Girton, J., McNeill, P., Rattner, J.B., et al., 1981. Laser microsurgery in cell and developmental biology. Science 213 (4507), 505–513.

Beuth, J., Schierholz, J.M., 2007. Evidence-based complementary oncology. Innovative approaches to optimize standard therapy strategies. In Vivo 21 (2), 423–428.

Billard, B.E., Hynynen, K., Roemer, R.B., 1990. Effects of physical parameters on high temperature ultrasound hyperthermia. Ultrasound Med. Biol. 16 (4), 409–420.

Bolmsjö, M.B., Vrba, T., 2000. Microwave applicators for thermotherapy of benign prostatic hyperplasia: a primer. Tech. Urol. 6 (4), 245–250.

Brace, C.L., 2009a. Microwave ablation technology: what every user should know. Curr. Probl. Diagn. Radiol. 38 (2), 61–67.

Brace, C.L., 2009b. Radiofrequency and microwave ablation of the liver, lung, kidney, and bone: what are the differences? Curr. Probl. Diagn. Radiol. 38 (3), 135–143.

Brace, C.L., Laeseke, P.F., Sampson, L.A., Frey, T.M., van der Weide, D.W., Lee Jr, F.T., 2007. Microwave ablation with a single small-gauge triaxial antenna: in vivo porcine liver model. Radiology 242 (2), 435–440.
Brace, C.L., Hinshaw, J.L., Lubner, M.G., 2011. Thermal ablation for the treatment of abdominal tumors. JoVE (J. Visualized Exp.) 49, e2596.
Bray, F., Ferlay, J., Soerjomataram, I., Siegel, R.L., Torre, L.A., Jemal, A., 2018. Global cancer statistics 2018: GLOBOCAN estimates of incidence and mortality worldwide for 36 cancers in 185 countries. CA Cancer J. Clin. 68 (6), 394–424.
Chan, P.K., Picconi, M.A., Cheung, T.H., Giovannelli, L., Park, J.S., 2012. Laboratory and clinical aspects of human papillomavirus testing. Crit. Rev. Clin. Lab. Sci. 49 (4), 117–136.
Chu, K.F., Dupuy, D.E., 2014. Thermal ablation of tumours: biological mechanisms and advances in therapy. Nat. Rev. Cancer 14 (3), 199.
Cohen, S.P., Foster, A., 2003. Pulsed radiofrequency as a treatment for groin pain and orchialgia. Urology 61 (3), 645.
Cooper, G.M., Hausman, R.E., Hausman, R.E., 2000. The Cell: A Molecular Approach, vol. 10. ASM press, Washington, DC.
Cosman, E.R., Nashold, B.S., Ovelman-Levitt, J., 1984. Theoretical aspects of radiofrequency lesions in the dorsal root entry zone. Neurosurgery 15 (6), 945–950.
Crouzet, S., Murat, F.J., Pasticier, G., Cassier, P., Chapelon, J.Y., Gelet, A., 2010. High intensity focused ultrasound (HIFU) for prostate cancer: current clinical status, outcomes and future perspectives. Int. J. Hyperth. 26 (8), 796–803.
Dewhirst, M.W., Viglianti, B.L., Lora-Michiels, M., Hanson, M., Hoopes, P.J., 2003. Basic principles of thermal dosimetry and thermal thresholds for tissue damage from hyperthermia. Int. J. Hyperth. 19 (3), 267–294.
Diederich, C.J., 2005. Thermal ablation and high-temperature thermal therapy: overview of technology and clinical implementation. Int. J. Hyperth. 21 (8), 745–753.
Dobak III, J.D., Brown, T.L., Ghaerzadeh, K., Yu, X., CryoGen Inc, 1998. Precooling system for Joule-Thomson probe. U.S. Patent 5,758,505.
Edmondson, R., Broglie, J.J., Adcock, A.F., Yang, L., 2014. Three-dimensional cell culture systems and their applications in drug discovery and cell-based biosensors. Assay. Drug Dev. Technol. 12 (4), 207–218.
Edwards, B.K., Noone, A.M., Mariotto, A.B., Simard, E.P., Boscoe, F.P., Henley, S.J., et al., 2014. Annual Report to the Nation on the status of cancer, 1975–2010, featuring prevalence of comorbidity and impact on survival among persons with lung, colorectal, breast, or prostate cancer. Cancer 120 (9), 1290–1314.
Egeblad, M., Werb, Z., 2002. New functions for the matrix metalloproteinases in cancer progression. Nat. Rev. Cancer 2 (3), 161.
Fischer, K., Gedroyc, W., Jolesz, F.A., 2010. Focused ultrasound as a local therapy for liver cancer. Cancer J. 16 (2), 118–124.
Foster, K.R., Glaser, R., 2007. Thermal mechanisms of interaction of radiofrequency energy with biological systems with relevance to exposure guidelines. Health Phys. 92 (6), 609–620.
Freifeld, Y., Dekel, Y., Stein, A., 2013. Adjuvant methods to improve results of local bladder irrigations by chemotherapy for NMIBC. Curr. Urol. Rep. 14 (2), 71–77.
Gazelle, G.S., Goldberg, S.N., Solbiati, L., Livraghi, T., 2000. Tumor ablation with radio-frequency energy. Radiology 217 (3), 633–646.
Gelet, A., Chapelon, J.Y., Bouvier, R., Rouviere, O., Lasne, Y., Lyonnet, D., et al., 2000. Transrectal high-intensity focused ultrasound: minimally invasive therapy of localized prostate cancer. J. Endourol. 14 (6), 519–528.

Gimm, O., Perren, A., Weng, L.P., Marsh, D.J., Yeh, J.J., Ziebold, U., et al., 2000. Differential nuclear and cytoplasmic expression of PTEN in normal thyroid tissue, and benign and malignant epithelial thyroid tumors. Am. J. Pathol. 156 (5), 1693–1700.

Gjorevski, N., Boghaert, E., Nelson, C.M., 2012. Regulation of epithelial-mesenchymal transition by transmission of mechanical stress through epithelial tissues. Cancer Microenviron. 5 (1), 29–38.

Goldberg, S.N., Stein, M.C., Gazelle, G.S., Sheiman, R.G., Kruskal, J.B., Clouse, M.E., 1999. Percutaneous radiofrequency tissue ablation: optimization of pulsed-radiofrequency technique to increase coagulation necrosis. J. Vasc. Interv. Radiol. 10 (7), 907–916.

Habash, R.W., Bansal, R., Krewski, D., Alhafid, H.T., 2006. Thermal therapy, part 2: hyperthermia techniques. Crit. Rev. Biomed. Eng. 34 (6), 491–542.

Hahn, G.M., 2012. Hyperthermia and Cancer. Springer Science & Business Media.

Hahn, W.C., Weinberg, R.A., 2002. Rules for making human tumor cells. N. Engl. J. Med. 347 (20), 1593–1603.

Hai, J.J., Tse, H.F., 2013. Biophysical principles and properties of cryoablation. The Practice of Catheter Cryoablation for Cardiac Arrhythmias. John Wiley & Sons, New Jersey, pp. 1–7.

Hanahan, D., Weinberg, R.A., 2000. The hallmarks of cancer. Cell 100 (1), 57–70.

Hergt, R., Dutz, S., Müller, R., Zeisberger, M., 2006. Magnetic particle hyperthermia: nanoparticle magnetism and materials development for cancer therapy. J. Phys. Condens. Matter 18 (38), S2919.

Hewitt, P.M., Zhao, J., Akhter, J.A., Morris, D.L., 1997. A comparative laboratory study of liquid nitrogen and argon gas cryosurgery systems. Cryobiology 35 (4), 303–308.

Hewitt, P., Zhao, J., Akhter, J., Morris, D., 1998. A comparative study of liquid nitrogen and argon gas cryosurgery systems. ANZ J. Surg. 68 (7).

Hildebrandt, B., Wust, P., Ahlers, O., Dieing, A., Sreenivasa, G., Kerner, T., et al., 2002. The cellular and molecular basis of hyperthermia. Crit. Rev. Oncol. Hematol. 43 (1), 33–56.

Hinshaw, J.L., Lubner, M.G., Ziemlewicz, T.J., Lee Jr, F.T., Brace, C.L., 2014. Percutaneous tumor ablation tools: microwave, radiofrequency, or cryoablation—what should you use and why? Radiographics 34 (5), 1344–1362.

Hoffmann, N.E., Bischof, J.C., 2002. The cryobiology of cryosurgical injury. Urology 60 (2), 40–49.

Holle, A.W., Young, J.L., Spatz, J.P., 2016. In vitro cancer cell–ECM interactions inform in vivo cancer treatment. Adv. Drug Deliv. Rev. 97, 270–279.

Horsman, M.R., Overgaard, J., 2007. Hyperthermia: a potent enhancer of radiotherapy. Clin. Oncol. 19 (6), 418–426.

Illing, R.O., Kennedy, J.E., Wu, F., Ter Haar, G.R., Protheroe, A.S., Friend, P.J., et al., 2005. The safety and feasibility of extracorporeal high-intensity focused ultrasound (HIFU) for the treatment of liver and kidney tumours in a Western population. Br. J. Cancer 93 (8), 890.

Issels, R.D., 2008. Hyperthermia adds to chemotherapy. Eur. J. Cancer 44 (17), 2546–2554.

Ito, K., Saito, K., Yoshimura, H., Aoyagi, Y., Horita, H., 2004. Coaxial-slot antenna for interstitial microwave thermal therapy and its application to clinical trial, The 26th Annual International Conference of the IEEE Engineering in Medicine and Biology Society, vol. 1. IEEE, pp. 2526–2529.

Jain, R.K., 2005. Normalization of tumor vasculature: an emerging concept in antiangiogenic therapy. Science 307 (5706), 58–62.

Jolesz, F.A., Hynynen, K., McDannold, N., Freundlich, D., Kopelman, D., 2004. Noninvasive thermal ablation of hepatocellular carcinoma by using magnetic resonance imaging-guided focused ultrasound. Gastroenterology 127 (5), S242–S247.

Jolesz, F.A., Hynynen, K., McDannold, N., Tempany, C., 2005. MR imaging–controlled focused ultrasound ablation: a noninvasive image-guided surgery. Magn. Reson. Imaging Clin. 13 (3), 545–560.

Kalisky, Y.Y., 2006. The Physics and Engineering of Solid State Lasers, vol. 71. Spie Press.

Kampinga, H.H., Dikomey, E., 2001. Hyperthermic radiosensitization: mode of action and clinical relevance. Int. J. Radiat. Biol. 77 (4), 399–408.

Kang, T.W., Rhim, H., 2015. Recent advances in tumor ablation for hepatocellular carcinoma. Liver Cancer 4 (3), 176–187.

Kapp, D.S., Hahn, G.M., Carlson, R.W., 2000. Principles of hyperthermia, Holland-Frei Cancer Medicine, fifth ed. BC Decker.

Karaboce, B., 2016. Investigation of thermal effect by focused ultrasound in cancer treatment. IEEE Instrum. Meas. Mag. 19 (5), 20–64.

Kilfoy, B.A., Zheng, T., Holford, T.R., Han, X., Ward, M.H., Sjodin, A., et al., 2009. International patterns and trends in thyroid cancer incidence, 1973–2002. Cancer Causes Control. 20 (5), 525–531.

Kim, Y.S., Lee, W.J., Rhim, H., Lim, H.K., Choi, D., Lee, J.Y., 2010. The minimal ablative margin of radiofrequency ablation of hepatocellular carcinoma (>2 and <5 cm) needed to prevent local tumor progression: 3D quantitative assessment using CT image fusion. Am. J. Roentgenol. 195 (3), 758–765.

Knavel, E.M., Brace, C.L., 2013. Tumor ablation: common modalities and general practices. Tech. Vasc. Interv. Radiol. 16 (4), 192–200.

Kopelman, D., Papa, M., 2007. Magnetic resonance–guided focused ultrasound surgery for the noninvasive curative ablation of tumors and palliative treatments: a review. Ann. Surg. Oncol. 14 (5), 1540–1550.

Kujala, N., Beland, M.D., 2012. Principles of cryoablation. Interventional Oncology. Springer, New York, pp. 39–49.

Kumar, S., Weaver, V.M., 2009. Mechanics, malignancy, and metastasis: the force journey of a tumor cell. Cancer Metastasis Rev. 28 (1–2), 113–127.

Lafon, C., Melodelima, D., Salomir, R., Chapelon, J.Y., 2007. Interstitial devices for minimally invasive thermal ablation by high-intensity ultrasound. Int. J. Hyperth. 23 (2), 153–163.

Lagneaux, L., de Meulenaer, E.C., Delforge, A., Dejeneffe, M., Massy, M., Moerman, C., et al., 2002. Ultrasonic low-energy treatment: a novel approach to induce apoptosis in human leukemic cells. Exp. Hematol. 30 (11), 1293–1301.

Lang, B.H.H., Wu, A.L., 2017. High intensity focused ultrasound (HIFU) ablation of benign thyroid nodules – a systematic review. J. Ther. Ultrasound 5 (1), 11.

Le Renard, P.E., Buchegger, F., Petri-Fink, A., Hofmann, H., Doelker, E., Jordan, O., 2014. Formulations for local, magnetically mediated hyperthermia treatment of solid tumors. Nanotechnol. Res. J. 7 (1), 1.

Lee, J., Lee, J.M., Yoon, J.H., Lee, J.Y., Kim, S.H., Lee, J.E., et al., 2012. Percutaneous radiofrequency ablation with multiple electrodes for medium-sized hepatocellular carcinomas. Korean J. Radiol. 13 (1), 34–43.

Lepock, J.R., 2003. Cellular effects of hyperthermia: relevance to the minimum dose for thermal damage. Int. J. Hyperth. 19 (3), 252–266.

Lepock, J.R., 2004. Role of nuclear protein denaturation and aggregation in thermal radiosensitization. Int. J. Hyperth. 20 (2), 115–130.

Levental, K.R., Yu, H., Kass, L., Lakins, J.N., Egeblad, M., Erler, J.T., et al., 2009. Matrix crosslinking forces tumor progression by enhancing integrin signaling. Cell 139 (5), 891–906.

Lewis, C.E., Pollard, J.W., 2006. Distinct role of macrophages in different tumor microenvironments. Cancer Res. 66 (2), 605–612.

Liang, P., Wang, Y., Zhang, D., Yu, X., Gao, Y., Ni, X., 2008. Ultrasound guided percutaneous microwave ablation for small renal cancer: initial experience. J. Urol. 180 (3), 844–848.

Liauw, S.L., Connell, P.P., Weichselbaum, R.R., 2013. New paradigms and future challenges in radiation oncology: an update of biological targets and technology. Sci. Transl. Med. 5 (173), 173sr2.

Lin, J.C., Wang, Y.J., 1996. The cap-choke catheter antenna for microwave ablation treatment. IEEE Trans. Biomed. Eng. 43 (6), 657–660.

Logan, C.Y., Nusse, R., 2004. The Wnt signaling pathway in development and disease. Annu. Rev. Cell Dev. Biol. 20, 781–810.

Lubner, M.G., Brace, C.L., Hinshaw, J.L., Lee Jr, F.T., 2010. Microwave tumor ablation: mechanism of action, clinical results, and devices. J. Vasc. Interv. Radiol. 21 (8), S192–S203.

Lynn, J.G., Zwemer, R.L., Chick, A.J., Miller, A.E., 1942. A new method for the generation and use of focused ultrasound in experimental biology. J. Gen. Physiol. 26 (2), 179.

Maccini, M., Sehrt, D., Pompeo, A., Chicoli, F.A., Molina, W.R., Kim, F.J., 2011. Biophysiologic considerations in cryoablation: a practical mechanistic molecular review. Int. Braz. J. Urol. 37 (6), 693–696.

Mahnken, A.H., König, A.M., Figiel, J.H., 2018. Current technique and application of percutaneous cryotherapy. RöFo-Fortschritte auf dem Gebiet der Röntgenstrahlen und der bildgebenden Verfahren 190 (9), 836–846. © Georg Thieme Verlag KG September.

Manolio, T.A., 2010. Genomewide association studies and assessment of the risk of disease. N. Engl. J. Med. 363 (2), 166–176.

Mauck, R.L., Baker, B.M., Nerurkar, N.L., Burdick, J.A., Li, W.J., Tuan, R.S., et al., 2009. Engineering on the straight and narrow: the mechanics of nanofibrous assemblies for fiber-reinforced tissue regeneration. Tissue Eng. Part B: Rev. 15 (2), 171–193.

McGahan, J.P., Browning, P.D., Brock, J.M., Tesluk, H., 1990. Hepatic ablation using radiofrequency electrocautery. Investig. Radiol. 25 (3), 267–270.

McGahan, J.P., Gu, W.Z., Brock, J.M., Tesluk, H., Jones, C.D., 1996. Hepatic ablation using bipolar radiofrequency electrocautery. Acad. Radiol. 3 (5), 418–422.

Motamedi, D., Learch, T.J., Ishimitsu, D.N., Motamedi, K., Katz, M.D., Brien, E. W., et al., 2009. Thermal ablation of osteoid osteoma: overview and step-by-step guide. Radiographics 29 (7), 2127–2141.

Mulier, S., Jiang, Y., Wang, C., Jamart, J., Marchal, G., Michel, L., et al., 2012. Bipolar radiofrequency ablation with four electrodes: ex vivo liver experiments and finite element method analysis. Influence of inter-electrode distance on coagulation size and geometry. Int. J. Hyperth. 28 (7), 686–697.

Mura, S., Nicolas, J., Couvreur, P., 2013. Stimuli-responsive nanocarriers for drug delivery. Nat. Mater. 12 (11), 991.

Neville, B.W., Damm, D.D., Allen, C.M., Chi, A.C., 2015. Oral and Maxillofacial Pathology. Elsevier Health Sciences.

Ni, Y., Mulier, S., Miao, Y., Michel, L., Marchal, G., 2005. A review of the general aspects of radiofrequency ablation. Abdom. Imaging 30 (4), 381–400.

Niculescu, V.F., 2018. Carcinogenesis: recent insights in protist stem cell biology lead to a better understanding of atavistic mechanisms implied in cancer development. MOJ Tumor Res. 1 (1), 18–29.
Nie, L., Wang, S., Wang, X., Rong, P., Ma, Y., Liu, G., et al., 2014. In vivo volumetric photoacoustic molecular angiography and therapeutic monitoring with targeted plasmonic nanostars. Small 10 (8), 1585–1593.
O'Rourke, A.P., Haemmerich, D., Prakash, P., Converse, M.C., Mahvi, D.M., Webster, J.G., 2007. Current status of liver tumor ablation devices. Expert. Rev. Med. Devices 4 (4), 523–537.
Oei, A.L., Vriend, L.E., Crezee, J., Franken, N.A., Krawczyk, P.M., 2015. Effects of hyperthermia on DNA repair pathways: one treatment to inhibit them all. Radiat. Oncol. 10 (1), 165.
Pacella, C.M., Papini, E., 2017. Ultrasound-guided laser ablation. Advanced Thyroid and Parathyroid Ultrasound. Springer, Cham, pp. 391–401.
Palumbo, G., Tosi, D., Schena, E., Massaroni, C., Ippolito, J., Verze, P., et al., 2017. Real-time temperature monitoring during radiofrequency treatments on ex-vivo animal model by fiber Bragg grating sensors, Optical Sensors 2017, vol. 10231. International Society for Optics and Photonics, p. 102312K.
Paulides, M.M., Bakker, J.F., Linthorst, M., Van der Zee, J., Rijnen, Z., Neufeld, E., et al., 2010. The clinical feasibility of deep hyperthermia treatment in the head and neck: new challenges for positioning and temperature measurement. Phys. Med. Biol. 55 (9), 2465.
Peer, D., Karp, J.M., Hong, S., Farokhzad, O.C., Margalit, R., Langer, R., 2007. Nanocarriers as an emerging platform for cancer therapy. Nat. Nanotechnol. 2 (12), 751.
Peer, A.J., Grimm, M.J., Zynda, E.R., Repasky, E.A., 2010. Diverse immune mechanisms may contribute to the survival benefit seen in cancer patients receiving hyperthermia. Immunol. Res. 46 (1–3), 137–154.
Peinado, H., Zhang, H., Matei, I.R., Costa-Silva, B., Hoshino, A., Rodrigues, G., et al., 2017. Pre-metastatic niches: organ-specific homes for metastases. Nat. Rev. Cancer 17 (5), 302.
Petrou, A., Moris, D., Paul Tabet, P., David Wensley Richards, B., Kourounis, G., 2016. Ablation of the locally advanced pancreatic cancer: an introduction and brief summary of techniques. J. BUON 21 (3), 650–658.
Petrovich, Z., Langholz, B., Kapp, D.S., Emami, B., Oleson, J.R., Luxton, G., et al., 1989. Deep regional hyperthermia of the liver. A clinical study of 49 patients. Am. J. Clin. Oncol. 12 (5), 378–383.
Phototherapy, D., 2018. Current status of light-emitting. Lasers in Dermatology and Medicine: Dermatologic Applications. Springer, p. 285.
Poissonnier, L., Chapelon, J.Y., Rouviere, O., Curiel, L., Bouvier, R., Martin, X., et al., 2007. Control of prostate cancer by transrectal HIFU in 227 patients. Eur. Urol. 51 (2), 381–387.
Reddy, L.A.L.I.N.I., Odhav, B., Bhoola, K.D., 2003. Natural products for cancer prevention: a global perspective. Pharmacol. Ther. 99 (1), 1–13.
Reya, T., Morrison, S.J., Clarke, M.F., Weissman, I.L., 2001. Stem cells, cancer, and cancer stem cells. Nature 414 (6859), 105.
Rosenberg, S.A., Hellman, S., DeVita, V.T., 2001. Cancer: Principles and Practice of Oncology. Lippincott Williams & Wilkins.
Rossi-Fanelli, A., Cavaliere, R., Mondovì, B., Moricca, G. (Eds.), 2012. Selective Heat Sensitivity of Cancer Cells, vol. 59. Springer Science & Business Media.
Sanders, D.L., Kingsnorth, A.N., 2012. From ancient to contemporary times: a concise history of incisional hernia repair. Hernia 16 (1), 1–7.

Sano, M.B., Davalos, R.V., Robertson, J.L., Garcia, P.A., Robert, E.N.I., Virginia Tech Intellectual Properties Inc, 2018. Irreversible electroporation using tissue vasculature to treat aberrant cell masses or create tissue scaffolds. U.S. Patent 9,867,652.

Sapareto, S.A., Dewey, W.C., 1984. Thermal dose determination in cancer therapy. Int. J. Radiat. Oncol. Biol. Phys. 10 (6), 787–800.

Schena, E., Tosi, D., Saccomandi, P., Lewis, E., Kim, T., 2016. Fiber optic sensors for temperature monitoring during thermal treatments: an overview. Sensors 16 (7), 1144.

Schuler, E.L., Poliner, J.K., Nash, D.E., Neuro Code Tech Holdings LLC, 2014. Method and system for processing cancer cell electrical signals for medical therapy. U.S. Patent 8,656,930.

Shankland, K.R., Armitage, J.O., Hancock, B.W., 2012. Non-hodgkin lymphoma. Lancet 380 (9844), 848–857.

Shi, Y., Mao, Y., 2017. Magnetic resonance thermometry-guided laser interstitial thermal therapy in neurosurgery, a promising tool for dural-based lesions? World Neurosurg. 98, 836.

Short, J.G., Turner, P.F., 1980. Physical hyperthermia and cancer therapy. Proc. IEEE 68 (1), 133–142.

Siegel, R.L., Miller, K.D., Fedewa, S.A., Ahnen, D.J., Meester, R.G., Barzi, A., et al., 2017. Colorectal cancer statistics, 2017. CA Cancer J. Clin. 67 (3), 177–193.

Singh, S., Bhowmik, A., Repaka, R., 2016. Thermal analysis of induced damage to the healthy cell during RFA of breast tumor. J. Therm. Biol. 58, 80–90.

Skitzki, J.J., Repasky, E.A., Evans, S.S., 2009. Hyperthermia as an immunotherapy strategy for cancer. Curr. Opin. Investig. Drugs 10 (6), 550 (London: 2000).

Somoano, B., Tsao, H., 2008. Genodermatoses with cutaneous tumors and internal malignancies. Dermatol. Clin. 26 (1), 69–87.

Sotsky, T.K., Ravikumar, T.S., 2002. Cryotherapy in the treatment of liver metastases from colorectal cancer, Seminars in Oncol., vol. 29, No. 2. WB Saunders, pp. 183–191.

Stauffer, P.R., Goldberg, S.N., 2004. Introduction: thermal ablation therapy. Int. J. Hyperth. 20 (7), 671–677.

Steiner, R., 2011. Laser-tissue interactions. Laser and IPL Technology in Dermatology and Aesthetic Medicine. Springer, Berlin, Heidelberg, pp. 23–36.

Suresh, S., 2007. Biomechanics and biophysics of cancer cells. Acta Mater. 55 (12), 3989–4014.

Tan, M.L., Choong, P.F., Dass, C.R., 2009. Osteosarcoma–conventional treatment vs. gene therapy. Cancer Biol. Ther. 8 (2), 106–117.

Taylor, M.J., 2007. Biology of cell survival in the cold: the basis for biopreservation of tissues and organs. Advances in Biopreservation. CRC Press, Boca Raton, FL, pp. 15–62.

Tilly, W., Gellermann, J., Graf, R., Hildebrandt, B., Weißbach, L., Budach, V., et al., 2005. Regional hyperthermia in conjunction with definitive radiotherapy against recurrent or locally advanced prostate cancer T3 pN0 M0. Strahlentherapie und Onkologie 181 (1), 35–41.

Torossian, A., 2008. Thermal management during anaesthesia and thermoregulation standards for the prevention of inadvertent perioperative hypothermia. Best. Pract. Res. Clin. Anaesthesiol. 22 (4), 659–668.

Tosi, D., Macchi, E.G., Braschi, G., Cigada, A., Gallati, M., Rossi, S., et al., 2014. Fiber-optic combined FPI/FBG sensors for monitoring of radiofrequency thermal ablation of liver tumors: ex vivo experiments. Appl. Opt. 53 (10), 2136–2144.

Van den Tempel, N., Horsman, M.R., Kanaar, R., 2016. Improving efficacy of hyperthermia in oncology by exploiting biological mechanisms. Int. J. Hyperth. 32 (4), 446–454.
Van der Zee, J., Vujaskovic, Z., Kondo, M., Sugahara, T., 2008. The Kadota fund international forum 2004—clinical group consensus. Int. J. Hyperth. 24 (2), 111–122.
Vogelstein, B., Papadopoulos, N., Velculescu, V.E., Zhou, S., Diaz, L.A., Kinzler, K. W., 2013. Cancer genome landscapes. Science 339 (6127), 1546–1558.
Warburg, O., 1956. On the origin of cancer cells. Science 123 (3191), 309–314.
Washington, C.M., Leaver, D.T., 2015. Principles and Practice of Radiation Therapy-E-Book. Elsevier Health Sciences.
Wasser, E.J., Dupuy, D.E., 2008. Microwave ablation in the treatment of primary lung tumors. Semin. Respir. Crit. Care Med. 29 (4), 384–394. © Thieme Medical Publishers August.
Weber, S.M., Lee, F.T., 2005. Cryoablation: history, mechanism of action, and guidance modalities. Tumor Ablation. Springer, New York, pp. 250–265.
Weder, G., Hendriks-Balk, M.C., Smajda, R., Rimoldi, D., Liley, M., Heinzelmann, H., et al., 2014. Increased plasticity of the stiffness of melanoma cells correlates with their acquisition of metastatic properties. Nanomed. Nanotechnol. Biol. Med. 10 (1), 141–148.
Wells, R.G., 2008. The role of matrix stiffness in regulating cell behavior. Hepatology 47 (4), 1394–1400.
Wu, T., Felmlee, J.P., Greenleaf, J.F., Riederer, S.J., Ehman, R.L., 2001. Assessment of thermal tissue ablation with MR elastography. Magn. Reson. Med.: Off. J. Int. Soc. Magn. Reson. Med. 45 (1), 80–87.
Wust, P., Hildebrandt, B., Sreenivasa, G., Rau, B., Gellermann, J., Riess, H., et al., 2002. Hyperthermia in combined treatment of cancer. Lancet Oncol. 3 (8), 487–497.
Xu, X., Ho, W., Zhang, X., Bertrand, N., Farokhzad, O., 2015. Cancer nanomedicine: from targeted delivery to combination therapy. Trends Mol. Med. 21 (4), 223–232.
Yang, D., Bertram, J.M., Converse, M.C., O'Rourke, A.P., Webster, J.G., Hagness, S. C., et al., 2006. A floating sleeve antenna yields localized hepatic microwave ablation. IEEE Trans. Biomed. Eng. 53 (3), 533–537.
Yerushalmi, A., 1976. Influence on metastatic spread of whole-body or local tumor hyperthermia. Eur. J. Cancer (1965) 12 (6), 455–463.
Yun, S.H., Kwok, S.J., 2017. Light in diagnosis, therapy and surgery. Nat. Biomed. Eng. 1 (1), 0008.
Zee, J.V.D., Koper, P.C.M., Lutgens, L.C.H.W., Burger, C.W., 2002. Point-counterpoint: what is the optimal trial design to test hyperthermia for carcinoma of the cervix? Point: addition of hyperthermia or cisplatin to radiotherapy for patients with cervical cancer; two promising combinations—no definite conclusions. Int. J. Hyperth. 18 (1), 19–24.
Zhang, H.G., Mehta, K., Cohen, P., Guha, C., 2008. Hyperthermia on immune regulation: a temperature's story. Cancer Lett. 271 (2), 191–204.
Zucker, S., Cao, J., Chen, W.T., 2000. Critical appraisal of the use of matrix metalloproteinase inhibitors in cancer treatment. Oncogene 19 (56), 6642.

2

Overview of ablation techniques

Chapter outline

2.1 Introduction

Ablation is defined as the direct placement of a needle into a tumor while using sources of cold, heat, or chemical substances, to destroy the tumor. The ablation procedure is applied frequently to prevent the cancer spreading to the different organs in the

Thermal Ablation Therapy. DOI: https://doi.org/10.1016/B978-0-12-819544-4.00011-3
© 2021 Elsevier Inc. All rights reserved.

body, including the liver, kidney, or bones. Tumor ablation is usually recommended to treat tumors of the lung, bone, kidney, and liver, especially in cases where radiotherapy/chemotherapy has failed (Pereira and Salvatore, 2012). It is carried out in a nominally invasive mode with a shorter recovery time and less pain. Ablative therapy is considered a recommended first-line treatment in the case of benign liver tumors and small hepatocellular carcinomas. Ablation systems include a needle-like device and a generator to deliver the energy to the targeted tumor tissues to produce acute cellular necrosis. Laser, radiofrequency (RF), microwave (MW), and high-intensity focused ultrasound (HIFU) systems release energy to heat the tissues to at least 60°C, while tissue necrosis is effected using cryoablation systems, where the heat is reduced to less than −40°C (Pouw et al., 2010).

The produced ablation zone size and shape using the probe (antenna) in the tumor tissue depend on the probe design, thermal conduction from the active heating zone, the thermal sinks instigated by neighboring structures, such as blood vessels, and the tissue type with consideration of the changes in the properties of tissues during the ablation process. The interplay between several such factors is complex and should be studied along with keeping a safety margin around the ablated tumor region to be exposed to the temperature to produce thermal coagulation necrosis. Moreover, the rapid decay in the energy with the increased distance from the energy source compels the extent of the thermal necrosis and limits the clinical use of ablation methods for treating large tumors of diameters larger than 3–4 cm (Pilla, 2007). In the course of technology advancements, the following sections highlight the different ablation modalities.

2.2 Radiofrequency ablation

In biophysics, RF alternating electrical current (AC) refers to the integration of both electromagnetic radio transmission waves and high-frequency AC. RF generated currents have a clinical impact effect by passing through tissues, which generates thermal energy in cells due to the electromagnetic energy (EME) that is transformed first to kinetic energy (Massarweh et al., 2006). In radiofrequency ablation (RFA) this thermal energy heats the existing tissues close to the active electrodes.

In the course of the EME effects, RFA is used in clinical practice as a local ablation procedure, known as rhizotomy. RFA is considered a safe invasive procedure using open surgery, percutaneous, and laparoscopic procedures for primary/secondary tumors.

Technological advancements in RF equipment have led to varying tissue responses to RF treatment, and also joint therapies can yield developments in the clinical feasibility of large tumor clinical treatment as well as improve the ablation rate of small tumors. RFA can be used in several medical areas, including tumor destruction in different organs including bone (Cantwell et al., 2004), lung (Gandhi and Dupuy, 2005), kidney (McAchran et al., 2005), liver (Lencioni et al., 2005), breast (Agnese and Burak, 2005), and prostate (Shariat et al., 2005), and it also can be applied to eliminate cardiac arrhythmias using a catheter (Chiappini et al., 2004).

2.2.1 History of radiofrequency ablation

Historically, thermal energy was applied for ablating abnormal tissues using radiofrequency energy (RFE) technology. In the 9th century, thermal ablation was employed in cardiology, oncology, proctology, and neurology to treat lesions with limited use for musculoskeletal ablation of tissues. Different intensities of radiofrequency have been evaluated, which have a direct correlation between the percentage of the affected area (ablated) and temperature. In 1891, the basic RFA procedure was described by d'Arsonval (1891) who observed the effect of radiofrequency waves when they passed through tissues causing a rise in the tissue temperature. This raise in tissue temperature occurs without producing neuromuscular excitation (Curley et al., 2002). Beer (1910) hosted the first endoscopy management of bladder tumors by discovering the destructive effect of high-frequency current on the papillary bladder tumors by electrocoagulation. To remove malignant lesions, Clark (1911) implemented an electrosurgical device based on the oscillatory decrease in voltage and an increase in current to produce deeper heat penetration in tissues. This device was used to treat malignant lesions of the cervix, breast, and skin.

Until 1928, the use of RF in medical applications was unpopular until the invention of the Bovie knife in surgery by Cushing and Bovie (1928). The Bovie knife entails a generator of an alternating electric current (AC) at the RF range with a small sharp electrode like a knife and a ground pad. Between the Bovie knife and ground pad, this AC passes onward and inward through the patient, where the pad acts as a dispersive electrode to avoid generating heat anywhere through the patient except around the Bovie knife's tip (AC focal contact point). This AC arcs between the patient and the Bovie knife, which desiccates and chars the tissue at the tip. This device was applied to ablate tissues using a varying RF current (damped or pulsed current), while a more continuous current causes cuts in tissue. However, the area of the ablated region was limited to a few millimeters as the burned tissues adhered to the knife's tip only, which was a monopolar electrode.

Later, Organ (1976) explicated the interaction of AC physically with the tissues stating that, at low power, the AC causes ions excitement in the adjacent tissues, which leads to frictional heat that spreads by conduction into neighboring tissues. Nevertheless, at high power, the ions are rapidly destroyed through charring and dryness of the superficial tissues with minimal heat generation. Successive studies have established that efficient coagulation of tissues necessitates an increase in the local temperature to about 50°C (Goldberg et al., 1996). With exposure to RFE, observed an increase in collagen fibril diameters with shrinkage of soft tissues. Fig. 2.1 summarizes the historical advancements in the RFA procedure.

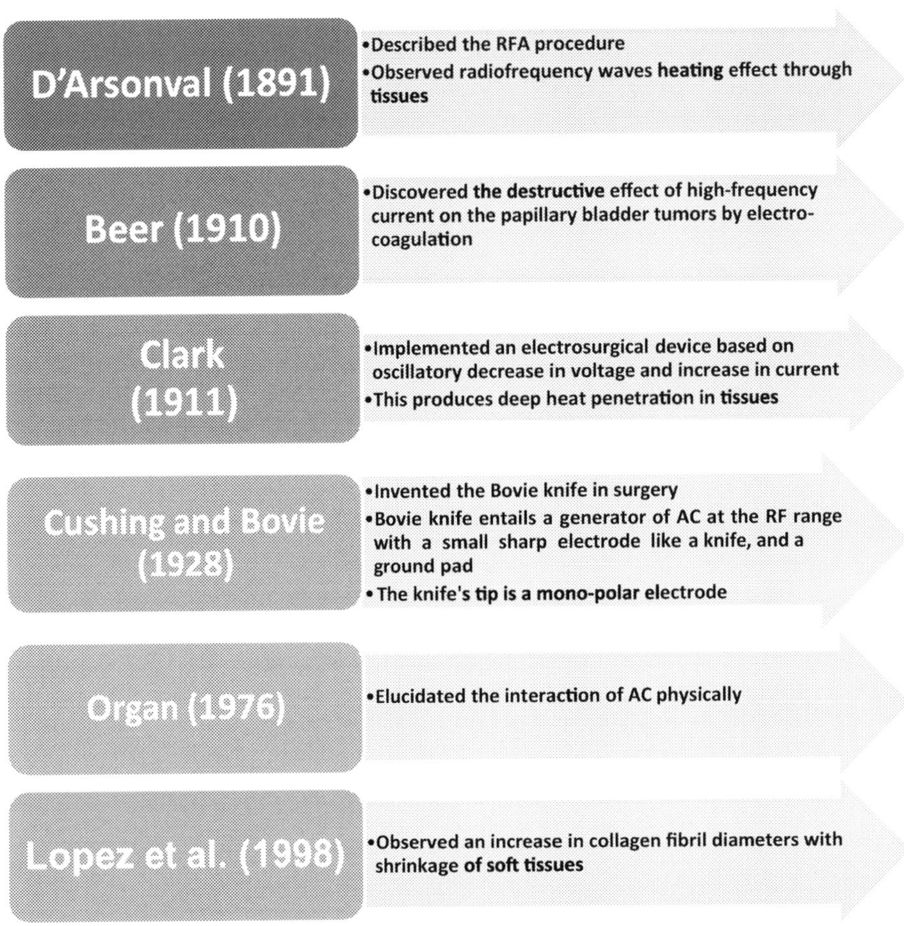

Figure 2.1 History of radiofrequency ablation.

2.2.2 Radiofrequency ablation thermodynamics and operation principle

Coagulative necrosis occurs in human cells with an increase in temperature above 50°C (LeVeen, 1997). Cell death occurs due to heating the tissue, based on time and temperature, where a few seconds are required to destroy cells at 60°C, whereas at 50°C a few minutes are needed (Dewhirst et al., 2003). During RFA, target temperatures range from 50°C to 110°C. The temperature increase in turn decreases the required time for denaturalizing the cells/tissues, leading to permanent destruction of the membrane of the cell above 60°C, followed by burning and desiccation. Accordingly, the radiofrequency working principle is based on triggering the ionic excitement of the surrounding tissues to the RF probe's tip (needle), where the probe's shaft does not generate heat. The heat is created in the tissues leading to cellular necrosis and coagulation. By applying rapid current, a small coagulation region with local tissue burning is created. The coagulation region guarantees the death of the tumor in this region, while the tissue burning (charring) prevents any further ionic agitation, which limits the volume of surrounding coagulation necrosis.

RF energy is a high-frequency AC ranging from 10 kHz to 2.59 GHz in the electromagnetic spectrum and represents long wavelengths with acceleration of the electrical charges (Pearce, 1986; Munro, 2012). RF energy has a great impact on biological tissues. The tissues of the body are relatively poor conductors due to their saturation by a saline solution. Subsequently, energy is lost with the flow of RF current through the tissues, leading to ionic agitation. The heating of tissues results from the movement of water molecules and ions and is directly proportional to the RF current density. This produced heat spreads into nearby tissues by conduction until it reaches a state of thermal equilibrium (steady state). The transformation of electrical energy to heat is governed by Joules law, which is formulated as follows:

$$\text{Energy} = \text{Resistance} * \text{Time} * (\text{Current}/\text{cross}-\text{sectional area})^2 \tag{2.1}$$

In tissues, the generated heat close to the active electrode tip is a function of resistance, time, and current density (current flow per unit area of the RF active probe). The current density decreases quickly with distance from the probe tip. Thereafter, the higher the voltage and current, the more heat is generated in the tissues. The generated RF current propagates around the electrode tip in a quite uniform way and produces a spherical region of thermal tissue damage. Additionally, the density of RF power reduces inconsistently with the square of the distance from the electrode tip, leading to a quick reduction in temperature with increasing distance from the electrode tip (Goldberg et al., 1996; Strasberg and Linehan, 2003).

For ablating large tumors, a high energy amount is required, which in turn requires high applied current density. This high-density current desiccates the tissues surrounding the electrode, instantly leading to an irregular, small ablated volume. Hence, in RFA, very high temperatures (RF energy) are impossible where the tissues start to vaporize and act as an electrical insulator. This vapor is electrically insulating and prevents any extra deposition of RF energy as it reduces the probe's active surface. Such a high temperature close to the electrode tip causes carbonized tissues to be formed, called tissue charring, which causes dielectric breakdown and limits any additional RF energy deposition (Leveillee and Hoey, 2003; McTaggart and Dupuy, 2007). Consequently, the used energy should be controlled to keep the temperature of tissues in the preferred range and limit charring and vaporization, where extreme heat deposition greater than 100°C leads to gas formation and uncontrolled charring which breaks the electrical circuit. Accordingly, for ablating large volumes, careful regulation of the energy deposition is required to avoid overheating by attaining the maximum current density.

In RFA, a high AC of about 480 kHz (typical RF range for RFA device operation is 450–500 kHz) is delivered to the tumor via the electrode (needle) to cause ionic agitation with subsequent frictional heat generation, finally causing cell/tissue death (Khan et al., 2007; Dodd et al., 2013). Nevertheless, using AC of any frequency will cause tissue heating, which is based on the frequency of the current, moreover, other events can occur, such as chemical reactions (for direct current and tissue stimulation at current frequencies lower than 10 kHz, which may arise along with the heat generation). To supply the required amount of current for heat generation, the Bovie knife concept was developed by designing a specific insulated needle to the tip of the probe to direct the current flow into the targeted tissues with the use of a grounding pad (Almeida, 2012), as illustrated in Fig. 2.2.

The needle of the RF electrode is carefully positioned deep into the targeted organ to generate a focal necrosis area. The extent of the coagulated central axis within the lesion can be enlarged by exposing these tissues to a larger area of the noninsulated needle. During ablation, several limitations of the ablated region and the echogenic response in the surrounding tissues to the RF needle are raised due to the needle design and the length

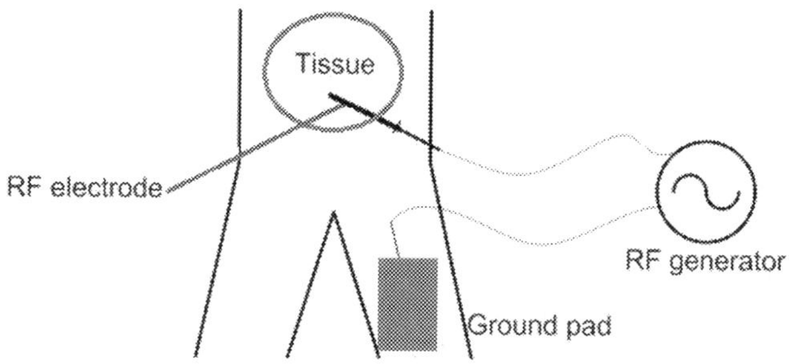

Figure 2.2 Radiofrequency procedure.

of the noninsulated needle, which affects the diameter of the ablated lesion, where the standard probe is a monopolar RF electrocautery. Inside the RF electrode, RF generator, and wires, free electrons carry charges of electric current within the tissue by ions and applying RF electric current leads to ion oscillations and movement, which produces resistive tissue heating (Gangi et al., 2005; Zhang et al., 2013). The current direction due to the ion's movement changes numerous times per second for a frequency of 500 kHz. The local thermal energy amount produced by the thermal ablation process is measured by the specific absorption rate (SAR) in W/kg. Typically, applying the RF energy necessitates the existence of a closed loop for the electrical circuit involving the RF generator, electrodes, wires, and patient. Consequently, the RF generator converts the AC (frequency of 50 or 60 Hz) from a socket into an AC of a frequency ranging between 300 and 700 kHz, which can also generate different waveforms to enable changing the effect of RF energy on the tissues (Larson et al., 2017).

The applied power for ablation in the RFA devices can be controlled using one of the following procedures, namely temperature, impedance, or power control, which are called TC, IC, and PC, respectively. In TC, a thermal sensor is integrated adjacent to the RF tip of the probe to adjust the RF power to preserve the temperature at a predetermined value (Zhou, 2011; Jeffries, 2004). In IC, the tissue impedance which is measured between the ground pad and RF probe is used to adjust the RF power based on a specific threshold. The RF power is blacked out for some time once the impedance surpasses the threshold level until the vapor settles, where tissue vaporization increases the impedance of the tissues due to the formation of tissue vapor close to the RF electrode. Then, the RF power is again applied at a lower level. In PC, the supplied electrical voltage to the RF probe is adjusted to preserve a constant RF power. Generally,

RFA uses two electrodes connected directly to the patient to close and complete the electrical circuit. These two electrodes are a small surface area active electrode located on the target region and a larger dispersive electrode located close to the electrical circuit, while in bipolar ablation, two electrodes are used (Matsudaira et al., 2003; Yu et al., 2011). RFA devices using monopolar electrodes including a large surface dispersive electrode and a small surface-active electrode. The clear divergence between active and passive electrode surfaces leads to the concentration of heat build up in the tissue near the active electrode. In contrast, both electrodes in bipolar RFA devices locate both electrodes in a way in which the electrical current passes and is concentrated between the electrodes, thus only the tissues between the two electrodes is considered to be a part of the electrical circuit. Using such bipolar RFA devices enables the precise measurements of impedance and temperature of the tissues, along with decreasing the current diversion complications (Nduka et al., 1994). For completely successful RF ablation, all the tumor tissues should be ablated with an adequate border of about 0.5–1 cm surrounding normal tissues to the tumor to reduce local tumor expansion or residual unablated tumor (Goldberg et al., 1995). Some RFA advantages include its ability to treat effectively small cancers, it does not require cutting of the skin, and it is a minimally invasive ablation procedure with lower risk to patients. Typically, the resulting necrosis size is dependent on several factors, such as ablation duration, ablation temperature, and electrode size/gauge. In addition, in RFA, there are limitations to the tumor tissue volume that can be eliminated, mainly due to restrictions with the current devices. This has become an active research area to ablate larger tumors using RFA in the future.

RFA is evaluated as the most widely used ablation procedure. Standard RFA is based on the concept of local ionic agitation which is generated in the tissues close to a noninsulated probe tip. During this process, the ions in the tissues try to track the fluctuations in the alternating current direction, which subsequently prompts focal heating. Accordingly, an electrical circuit is used in monopolar RFA systems between the RF probe and the pad attached to the patient's skin (Zhang et al., 2013). However, on the same needle tube, a neutral electrode and non-insulated electrode tip are positioned in bipolar RFA systems (Boykin, 2008). Internal cooling by flowing fluid in the RF probe and applying pulsed RF enhances the RFA procedure. For the simultaneous treatment of multiple tumor sites or large tumors, multiprobes can be used to improve the performance of the RFA system with the use of switches.

2.2.3 Radiofrequency ablation factors

The RFA method passes an electrical current of ≈500 kHz through biological tissues to heat the targeted zone with the highest power density (maximal SAR) (Haemmerich, 2010; Ulucakli and Lavery, 2013). The generated heat destroys abnormal cells using radio waves. Thereby, it is applied to reduce pain and treat malignant tumor (cancer) diseases in different body locations and other conditions. The RFA can be used for cancer treatment in different organs, such as bone, kidney, liver, pancreas, lung, breast, thyroid, and adrenal gland (Mahnken et al., 2008; D'Onofrio et al., 2016). The main principle of RFA is based on producing an electrical current using radio waves to heat a small area of tissues, thus reducing pain signals from this area. These radio waves passed through small devices called electrodes (thin needles), which carry electricity. Compared to cryoablation, the complication rate using RFA is relatively low. RFA is considered a nominally invasive process which applies electrical energy for heating to damage cancer cells. This high RFE heats the surrounding tissue, killing nearby cells. During the RFA procedure, to guide the electrode through an incision or the skin into the cancerous tissues, an imaging system is used.

By the end of the 20th century, the RFA was considered a less invasive approach for the cancer's treatment (Castro et al., 2014). For clinical use, recent RFA devices are deemed analogous to the Bovie knife and can ablate lesions of 3–4.5 cm diameter. Generally, RFA is highly efficient for small lesions with a maximum diameter of 3–5 cm.

The RF current affects tissues based on several factors, including the RF electrical exposure time, generator power, shape/size/contact area of the electrode surface, tissue characteristics (water content, impedance), and the current properties (voltage, amperage, power, output waveform, frequency) (Pearce, 2006). It has been observed that an increase in the coagulated tissue volume occurs with an increase in the contact surface area of RF electrodes, with a slow rise in the generator power, and protracted RF applied.

To overcome the problem of tissue charring/dehydration and restricted radius of clotted tissues around the RF electrode needle, various designs of RF needles, power generator, and operating parameters have been studied. In addition, the use of RFA is limited by the organ/tumor location. A large vessel neighboring the tumor to be ablated may act as a heat sink and reduce the therapeutic efficiency, where the flowing blood inside the vessel removes heat and stops the required sufficient heating of the tumor causing a "heat-sink" effect (Lu et al., 2002).

Moreover, perforation may occur during the RFA procedure in some cases, such as ablating dome lesions adjacent to the diaphragm may cause diaphragmatic perforation; also, RFA of deep lesions close to the bowel loops may cause bowel perforation. Fig. 2.3 demonstrates the different considerations and hazards of the RFA procedure.

Recent research outcomes have concluded additional factors that may modify the tissue necrosis amount (coagulation necrosis), such as (Ahmed and Goldberg, 2004):

- Total RF energy deposited into the tissue, local tissue interaction, and systemic and local heat lost. RF generators with high wattage levels raise the power deposition and increase tissue necrosis;
- Characteristics of local tissues besides tissue interaction with the RF current;
- Rapid supply of current generates charring, which decreases the ionic agitation and water contents in the tissue, thus restricting the amount of tissue necrosis;
- Continuous saline infusion procedure has an impact on RF treatment (saline-perfused RF) of larger tumor sizes to increase the coagulation diameter.

An accurate adjustment of used control parameters, including time, power, and temperature is required to perform RFA. For example, for larger tumor sizes, more time is mandatory for ablating the entire tumor. Extreme charring is adjusted by cooling the

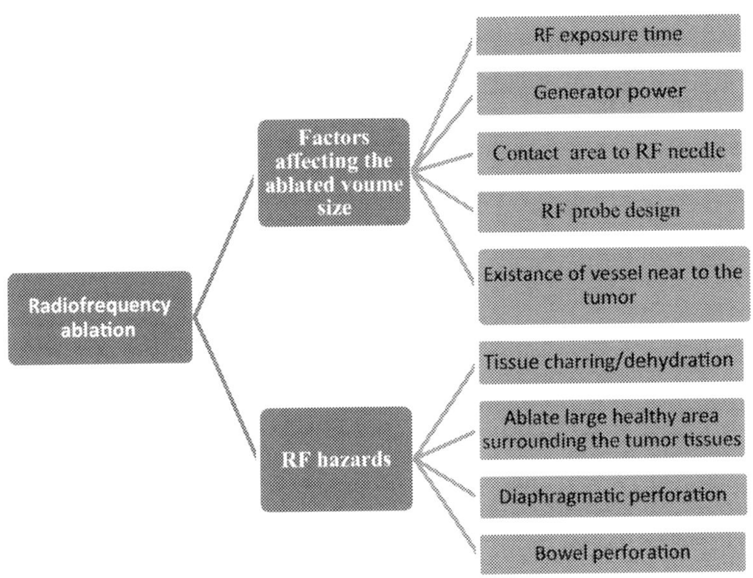

Figure 2.3 Radiofrequency considerations and hazards.

tip of the needle or inserting saline injection into the tissue (de Baere et al., 2001). Typically, most of the ablated tumors using the RFA have oval shapes. The initial RF generators generated a maximum power of 50 W, while current generators can generate power within the range of 150–200 W. The increase in the power disposition to the target area increases the destroyed tissue volume. The supplied heat amount in tissues is proportional to the current strength and the tissue resistance, while it is inversely proportional to the distance from the needle tip.

2.2.4 Limitations and challenges of radiofrequency ablation

RFA is most popular thermal procedure of all available thermal modalities due to its safety, efficiency, and feasibility. Some inherent shortcomings due to the RF heating remain, even with the continued advancements in RFA systems.

The electrical current flow is key to the RFA heating procedure. RFA occurs by passing an electric current through ionic fluid which permeates the tissues between the ground pad and electrode. Subsequently, the temperature of the tissues is raised due to the high current density neighboring the electrode, causing tissue death. Once the temperature reaches 100°C, the water in tissues starts to vaporize, leading to a rapid decrease in the tissue's electrical conductivity, and then an impedance spike appears before the active heating process is ended. Since multiple-probe RFA devices require electric current to switch between the different probes, each probe reaches 100°C less than 100% of the time. Due to the termination of heating once the temperature reaches 100°C, this off-time allows the tissues to rehydrate and reestablish an electrical current route, but over time.

Typically, RFA applies a frequency from 450 to 500 kHz to damage the tissues in the FRA electrode's proximity by creating friction, resulting in heating. To avoid tissue charring and nonconductivity, the produced heat by RFA is preserved at from 50°C to 100°C (Poulou et al., 2015; Kim, 2018), where the burnt tissues act as an insulator preventing the RF energy from being transferred to nearby tissue, so restricting the ablation volume. To exploit the RFA ablation volume, optimal protocols can be applied by using systematic and slow energy deposition. Moreover, to apply RFA for the treatment of large tumors, multiple overlapping ablations are required to generate a sufficiently large safety margin (Sartori et al., 2008).

Another limitation that RFA faces occurs in perivascular regions due to the heat-sink effect, where the RFA is blocked by high local recurrence rates owing to the inability of the RF current to produce heat faster than the cooling effect of the flowing blood. Tumors extremely close to large vessels are only partly ablated because of the heat-sink effect, where the formed thermal energy by ablation is incompletely shunted away from the lesion by the cooler blood. Thus, ground pads are required with bipolar or multipolar RFA devices.

2.3 Microwave ablation

Dissimilar to RFA, microwave ablation (MWA) allows several applicators to be used concurrently as it does not depend on any electrical circuit (Dupuy, 2009). MWA can treat larger tumor volumes and in a faster manner compared with other heat-based modalities, such as laser ablation and RFA. Microwave technology in MWA provides the RFA benefits along with other advantages which increase its efficiency in tumor treatment. These benefits include a faster ablation time, damaging larger tumors, providing higher intratumoral temperatures, achieving optimal heating of tumors, and producing less routine pain. Furthermore, MWA improves the convection profile, facilitates the use of multiple applicators, and does not involve the placement of grounding pads (Skinner et al., 1998; Garrean et al., 2009; Wieland, 2010; Facciorusso et al., 2016).

In the electromagnetic spectrum, MW radiation allocates regions between radio waves and infrared radiation with frequencies in range of 900–2450 MHz. Typically, electromagnetic (EM) radiations have electric charges like polar water molecules in the human body. The electric charges on the EM waves flip between negative and positive. Since the water molecules are asymmetric; the charge alignment on the atoms is such that the oxygen side has a negative charge, while the hydrogen side has a positive charge. At high frequencies, the microwave oscillations lead to a change in the signs of electric charges on the molecules backward and forward about a billion times in a second, causing an interaction with a water molecule (Crocetti et al., 2010). The water molecules are hit by MW radiation, which flips the electrical charges, and is based on the frequency of MW energy. To maximize this interaction, the MW radiation is adjusted to the natural frequency of water molecules. The speed of the flipping process of the molecules causes heat that depends on the dynamic movement speed of molecules, which increases the temperature of water molecules. Consequently, electromagnetic MW heats the tissues by stirring water molecules in the nearby tissues, generating heat and friction, hence resulting in cellular death through coagulation necrosis. This concept is the key point of using MW in ablation, where MWA indicates using electromagnetic approaches to cause tumor damage by using devices with frequencies of no less than 900 MHz (Shibata et al., 2002).

Clinically, MWA is considered a promising procedure, which is potentially able to treat a wider range of tumor cases in different organs. A theoretical advantage of MWA is that MW energy can provide adequate contact of independent lesions with the MW source. Nevertheless, tentative studies have established an exponential decay of the penetration of EM fields into tissues with a steep decay using MW frequencies. This advantage can hypothetically develop the flexibility of MWA, particularly in organs where muscular valleys and rims cannot be reached using traditional RFA due to the capability of designing MW antennas with elastic linear applicators.

2.3.1 History of microwave ablation

In medicine, microwaves are used efficiently for tumor ablation to destroy and burn the tumor tissues. In the early 1980s, microwave coagulation was used to create hemostasis (Tabuse et al., 1985). Microwaves are used to produce thermal coagulation of tissues by inducing an alternating electric field of 2450 MHz (ultra-high speed), which leads to rotating of the water molecules. In 1986, the first percutaneous MWA-aided liver biopsy was carried out (Alexander et al., 2015). Seki et al. (1994) presented the first clinical MWA. With the significant advances in technology, this first MWA device has been used in different clinical applications. These technological advances include developments in MW applicators, internal cooled applicators (Knavel et al., 2012; Morrison et al., 2013), and spatial and synchronous power distribution to several antennas (Harari et al., 2016). Fig. 2.4 illustrates the historical progress in the MWA technology.

Since then, the term "microwave ablation" has been used instead of the terms "microwave coagulation therapy" and "percutaneous microwave coagulation therapy" (Ahmed, 2014). In addition, MWA is used for tissue/tumor ablation in several instances, including colorectal metastatic cancer, breast cancer, hepatocellular carcinoma, liver tumors, adrenal malignant carcinoma, renal hamartoma, renal cell carcinoma, abdominal tumors, and lung cancer (Panel, 2012). Different systemic/local treatment methods have been conducted for each malignancy to

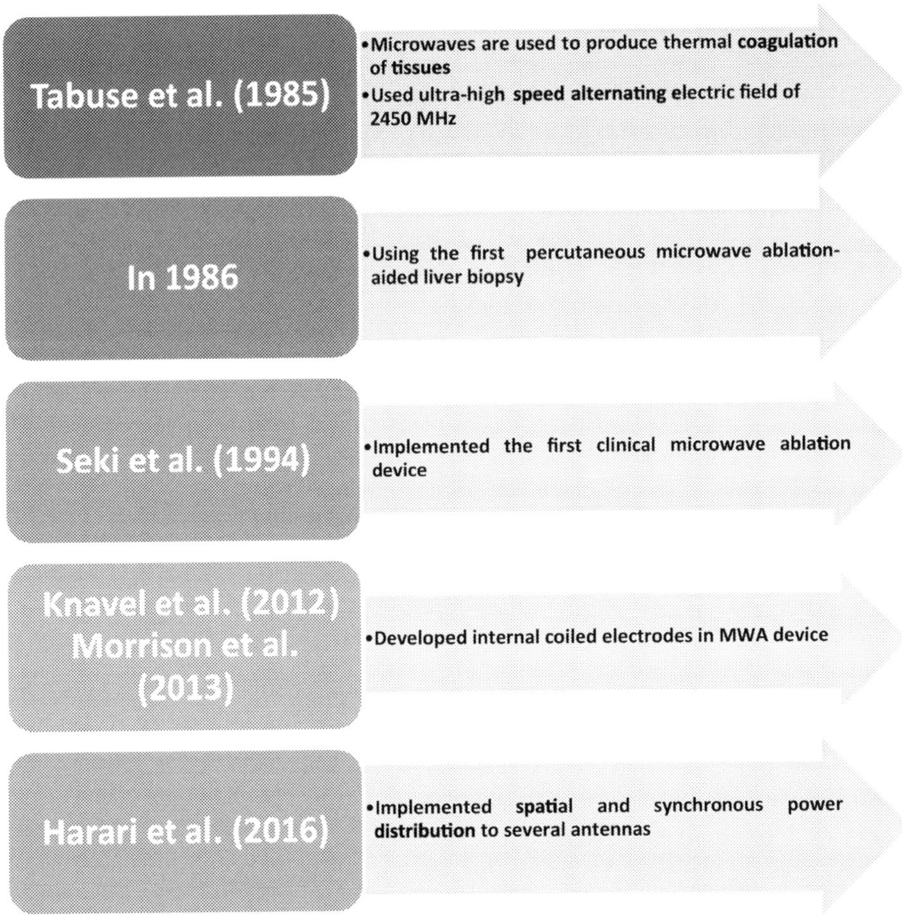

Figure 2.4 History of microwave ablation.

provide enriched local control for a minimally invasive process while conserving the organ's normal tissues, reducing the hospitalization period, and cutting morbidity.

To increase the ablated area size with larger ablation boundaries and to decrease the progression of the local tumor, many MWA generators and devices with higher energies have been industrialized. Numerous models of small antennas have been designed for MWA without damaging healthy tissues by tuning the used power. Early-generation MWAs have the same safety and efficiency as RFA, with superiority over RFA in terms of the shorter required ablation time and larger ablation areas of greater than 3 cm (Simon et al., 2005; Boutros et al., 2010; Qian et al., 2012; Lubner et al., 2013; Di et al., 2014).

2.3.2 Microwave ablation thermodynamics and operation principle

Microwaves are EM radiation with wavelengths ranging from 1 m to 1 mm, where frequencies range from 300 MHz to 300 GHz, corresponding to wavelengths of 100 to 0.1 cm, respectively. Unlike RFA, which heats the tissues using ion flow, MWA uses microwave exposure to oscillate the polar water molecules to produce frictional heating. Such an MWA procedure ultimately generates tissue necrosis in solid tumors, where the MW propagates through different tissue types, even if the tissue has high impedance, low thermal conductivity, or low electrical conductivity (Lubner et al., 2010). Moreover, MWs can penetrate through the desiccated or charred tissues that accumulate around the ablation electrode.

Accordingly, both RFA and MWA destroy tissues through thermal induction of coagulative necrosis, where the MW frequencies in MWA are from the RF spectrum. However, these two ablative modalities vary in their energy deposition mechanisms, where RFA has imperfect efficiency in ablating low electrical conductivity tissues, such as adipose tissues, as it needs an electrically conductive path to transmit the resistive heat. In contrast, the antenna of the MWA device emits an electromagnetic field of 915 or 2450 MHz to nearby tissues, which generates heat greater than 150°C through dielectric polarization (Kim, 2018). In addition, MWA is more efficient in tissues with a high water content. To generate heat at lethal temperatures in MWA a dielectric hysteresis is used to destroy tissues by applying an EM field ranging from 900 to 2500 MHz. The EM field forces the polar water molecules in tissues to unceasingly realign with the fluctuating electric field, which increases their kinetic energy, and therefore the tissue temperature.

The dielectric hysteresis process is responsible for the microwave heating phenomenon, where the applied EM field alternates polarity and flips the molecules with inherent dipole moments (e.g., water molecules) with the field, which realign the molecules continuously with the applied EM field billions of times/second. Since the water molecule dipoles cannot flip and alternate their polarities as fast as the field speed, some of the MW energy is absorbed by the material and transformed to heat. The rate of generated heat Q_h is proportionate to the square of the magnitude of applied electric field E, which is expressed as follows:

$$Q_h = \sigma |E|^2 \tag{2.2}$$

where σ is effective conductivity (S/m) that measures the MW absorption. Since the biological tissues have high water contents, they have high conductivity and ability to absorb MW energy (Duck, 1990). Thus, MW energy can be introduced into the body by interstitial devices (Lin and Wang, 1996; Yang et al., 2006). For focal tumor ablation, interstitial antennas or an antenna array is preferred to avoid the restrictions of the spatial resolution and field penetration. These interstitial antennas receive power from the MW generator via a distribution system.

At sufficient MWA power, the generated heat is about 100 times faster than in RFA. However, the same heating procedure heats the tip of the antenna and the cables used for power distribution. This generated undesirable heat has less severity than the generated heat in the tissues. Distinct to the RF electrodes which couple the electrical current to the tissues during the RFA; antenna electrodes are used in MWA to couple the MW energy to tissues. In MWA, the antenna radiates EM fields which generate a microwave-based heating effect around the antenna, finally leading to tissue ablation (Kanazawa et al., 2020). Unlike RFA, where the impedance spike cuts off the RF heating effect, in MWA, tissue desiccation does not prevent the propagation of the EM field. Accordingly, ablation of bone and lung tissues can be performed effectively using microwaves.

Currently, numerous commercial microwave systems are present worldwide (Hoffmann et al., 2013). Typically, the components of an MWA system consist of a generator, transmission cable (power distribution system), and applicator (electrode/ antenna), as illustrated in Fig. 2.5.

A detailed description of the MWA system is as follows:

- *Microwave generator*: The generator is a vital controller that permits the required power time for treatment and controls the temperature by supplying the suitable MW energy. The allowed frequencies to be used by the generator, based on the Federal Communications Commission (FCC), are 915 MHz or 2.45 GHz, which are higher than the frequencies used in RFA. These frequencies allow the emission of MW radiation through antennas in the human body without using ground pads. Specifically, the 915 MHz frequency produces deeper penetration of the MW field compared to using the 2.45 GHz frequency for a large volume. The MWA noninvasive device can access inaccessible regions in the human body. This generator is placed adjacent to the patient with the required safety conditions.

  The MW generator generates power using solid-state amplifier sources or a magnetron (Brace et al., 2007):

  - A magnetron produces EM energy by quickening electron movement via a magnetic field inside a resonant cavity,

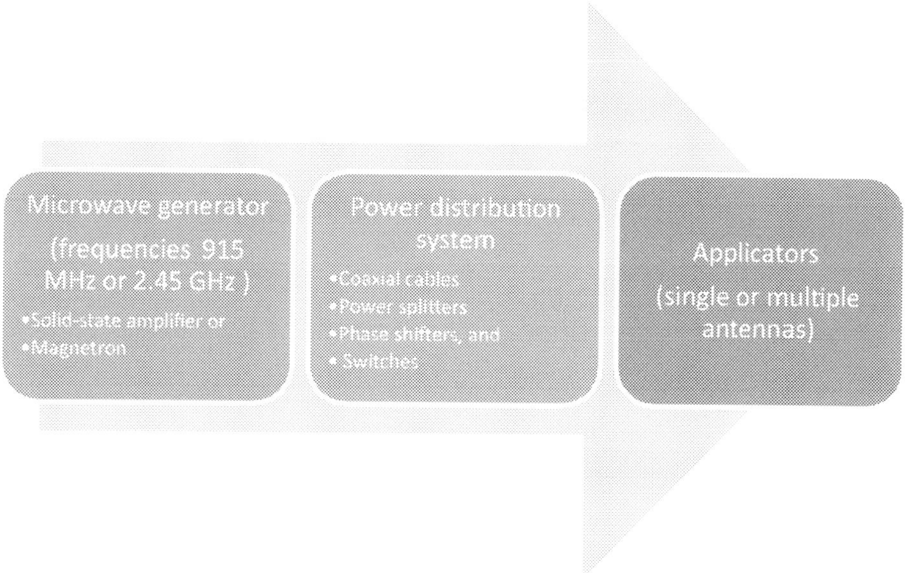

Figure 2.5 Standard components of a microwave ablation system.

which controls the output frequency based on the geometry of the cavity (Brace, 2009). The magnetron has low cost and high reliability, as well as providing high output power greater than 10 kW, and efficiency greater than 70%. Nevertheless, it has a large size owing to its need to include a heavy, large transformer to produce a high-voltage power supply, which can be engaged to power several antennas using a single source.

- A solid-state generator produces staged power, where each stage amplifies the power of the preceding stage using a transistor-based amplifier. It is characterized by good robustness, high stability, and reasonable output power of less than 150 W. Compared to the magnetron generator, the solid-state generator has higher cost and low efficiency of less than 30%, while it is more controllable and has a smaller size.

- *The power distribution system* entails different components, namely cables, switches, phase shifters, and power splitters. The transmission cables are movable and robust to transfer, carry, and distribute the MW power from the generator to the antenna with low energy loss. Coaxial cables are the most commonly used owing to their perfect propagation characteristics, which involves dielectric material and inner/outer conductors. Different characteristics of such cables are considered including diameter and cooling ability to transfer the power safely without any failure or overheating. However, by decreasing the diameter of the cable, the power handling capability decreases rapidly.

Unlike RFA, MW generators can power a number of antennas using the same source. The other components of the power distribution system in the MWA system (i.e., switches, phase shifters, and power splitters) are employed to control the duty cycle, amplitude, and phase to power multiple antennas. The switches can be used to select the activated antennas. The phase shifters offer a phased-array procedure of multiple antennas for hyperthermia and tumor ablation (Lin et al., 2000) to connect the EM waves to decrease or increase the heating amount. In addition, the power splitter divides equally any input power into multiple output channels. Typically, in MWA, only four output channels can be provided by the power splitter.

- *Applicators* contain single or multiple antennas, which are called disposable elements, electrodes, or probes; each antenna is used to create a large ablation zone. These MW antennas may have different designs and are used to transfer the MW energy to the tumor tissues, which are traditional applicators with active tips of lengths from about 0.6 to 4 cm with several diameters. Some properties are considered for the design of such antennas, including the return loss (reflection coefficient) and radiation pattern (Poggi et al., 2015). To exploit the energy transfer, the return loss from an antenna to tissue should be minimized. Consequently, the design of the MW applicator is a balance of antenna diameter, heating pattern of tissues, and power efficiency, with the applicator design tradeoff to obtain the required ablation. For example, the reflected energy from an antenna decreases the heating of tissues, whereas an increase in the undesired heating of the antenna burns the skin and ablates large surrounding healthy zones to the tumor itself. To decrease the reflection coefficient and at the same time focus the radiated energy, different antenna designs can be used.

In some cases, energy may scatter near the border of a single antenna centralized in the targeted tumor. In such a case, an ablation-based overlapping approach is essential to treat the entire tumor with adequate margins using several antennas. Multiple proximity applicators can be activated concurrently without switching. For instance, applicators can be phased and positioned to overlap the EM field during the ablation procedure by including centripetal energy convergence from the margin to the center of the tumor. The ablation strategy using simultaneous multiple antennas is based on activating the different antennas at the same time without using switching, where the EM fields from the individually separated antennas are independent. Thus, to increase the generated heat as well as the thermal synergy between different EM fields, multiapplicators can be used during the ablation process to ablate numerous tumors instantaneously. During electrode activation, the multiapplicators can be powered constantly without swapping from one antenna to the others. Therefore the increase in heating is proportional to the number of antennas, where the constructive phase allows effective heating and generates higher temperatures compared to using a single antenna (Planché et al., 2013).

In addition, a cooling system is used to solve the produced shaft heating which is triggered by the reflected power. An active, robust shaft cooling appliance can diminish several risks and is used while delivering high power (Sun et al., 2012). The MW antenna transfers the energy into tissue, where the antenna's geometry and design control the power coupling efficiency and active heating zone. The design of the MW antenna is a tradeoff between the pattern of tissue heating, antenna diameter, and power effectiveness to reach a precisely desired ablation. Different antenna designs can be used in MWA devices, such as triaxial, slotted, choked, monopole, dipole, and needle-like antennas (Lin et al., 1996; Nevels et al., 1998; Pisa et al., 2001; Ito et al., 2004; Meredith et al., 2005; Bertram et al., 2006; O'Rourke et al., 2007).

Microwaves have the ability to propagate via different types of tissues, such as lung and bone, which have high impedance, low thermal conductivity, and low electrical conductivity. Furthermore, MW penetrates the desiccated/charred tissues that tend to accumulate around all hyperthermic ablation electrodes, leading to imperfect power transfer for non-MW energy systems. The MWA procedure is similar to RFA, which involves direct localization of an electrode (antenna) in the targeted tumor under imaging/modality guidance, which creates a hyperechoic region surrounding the electrode's needle (Vogl et al., 2017). Typically, MWA can be used to support open surgery, percutaneous, and laparoscopic treatment procedures. A thin MW antenna (14.5 gauge) is directly placed into the tumor. Electromagnetic MW is produced from an MW generator through an electrode needle (antenna) to the targeted tumor area. Each generator can produce a power of 60 W at 915 MHz. Unlike RFA, in MWA the ground pads are not required due to the EM waves' inherent properties. A thermocouple is placed separately to measure the intratumor temperatures. An interstitial antenna is used for coupling the energy from the generator power with the tissues to generate the path for the MW energy to radiate into the tissues (Yang et al., 2007). Moreover, MWA can ablate larger volumes compared to RFA, even if both strategies use similar applicators (Yu et al., 2011), where with MWA, higher temperature levels up to 180°C can be reached with the ability for rapid temperature increase (Brace et al., 2009). Comparing MWA with RFA, it is less disposed to thermal diffusion and the effect of heat-sink, with higher energy deposition. Heat-sink refers to heat dissipation through blood vessel perfusion (Goldberg et al., 1998; Yu et al., 2008; Kim, 2018). Conversely, thermal diffusion also depends on the characteristics of tissues which are changeable with heating.

2.3.3 Microwave ablation factors

MW frequencies are used in MWA within the range of 900–2500 MHz to oscillate the polar water molecules in tissues. This oscillation increases the kinetic energy of the molecules, which in turn increases the tissue temperature, leading to coagulative necrosis of tumors. Similar to RFA, the heating mechanism varies significantly, meanwhile, MW energy is emitted into the tissues through an antenna that controls the tumor heating. In MWA, an MW generator emits an EM wave through the exposed, noninsulated part of the antenna. Electromagnetic MW agitates the water molecules in the nearby tissues, generating heat and friction, thereby inducing cellular death through coagulation necrosis.

For heating, unlike RF which requires an electrically conductive path, MW propagates through tissues even with high impedance, low thermal conductivity, or low electrical conductivity, such as charred tissues (Schramm et al., 2006; Brace et al., 2009). Since during fast ablation, MW can produce very high temperatures in the tumor region, MWA improves the treatment ability, convection profile, ablates larger areas, and produces less heat-sink effect compared to RFA (Ahmed et al., 2014). In contrast to RFA, MWA has several advantages in ablating large tumors, tumors located around large blood vessels, and highly perfused regions, while the energy of RF is limited. Compared to RF energy, MW energy yields to higher temperature values, with faster heating leading to larger necrosis areas of a better demarcated ablation region. Nevertheless, MWA has several technical limitations, including large-diameter probes, shaft heating, low power, nonspherical ablation volumes, and requirement for multiple insertions to ablate small regions, as reported in Fig. 2.6.

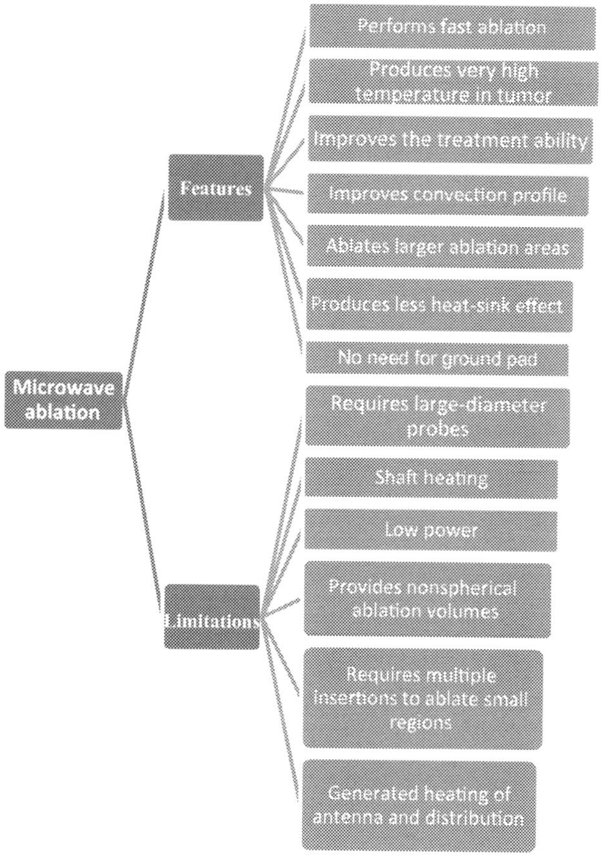

Figure 2.6 Advantages and limitations of microwave ablation.

To increase the amount of delivered power, a cooling jacket can be used around the antenna to decrease cable heating (Wang et al., 2008). To diminish the heating-back effects, a choke coil can be used in the antenna's distal portion. In addition, effective conductivity and relative permittivity are considered the most significant properties of MWA (Brace, 2009). The effects of these properties on the MWA process are explained as follows:

- Effective conductivity indicates the absorption ability of tissues to MW energy. The increase in water content and in conductivity, nonetheless absorbs MW, and vice versa, the conductivity decreases in the tissues with low water content, although the microwave propagation increases (Rosen et al., 2002; Furse et al., 2009). In organs with high water content levels, such as liver and kidney, the electrical conductivity is high, which permits quicker absorption of the MW energy, however, it decreases the field penetration. Conversely, heating gradually increases the microwave propagation by creating tissue desiccation and then conductivity reduction.

- Relative permittivity (absolute permittivity) refers to the tissue's ability to receive an electric field (Panagopoulos et al., 2002). At a specified frequency, the relative permittivity defines the applied field's wavelength, which affects the propagation of MW energy through the tissues and determines the design of the used antenna. High permittivity levels cause short wavelengths. Therefore it is known that better microwave diffusion is attained in tumors compared to normal tissues, due to the higher permittivity in the tumor tissues. Hence, dissimilarities between tumor tissues and surrounding tissues, in terms of permittivity, might permit better MWA-based treatment.

Theoretically, several factors affect the mechanical, thermal, and electrical tissue properties. For MWA, the tissue water content is considered the most significant factor. Tissue heating affects dramatically the electrical properties of the tissues, especially with temperatures above 100°C. In MWA, tissue desiccation occurs with the boiling of water and its conversion to gas due to the increase in temperature, whereas the current cannot pass through tissues in RF with increased temperature above 100°C. This blocking of the current in RFA occurs due to boiling of the water, which is required for ion flow, and which increases the impedance (Vogl et al., 2011). However, this phenomenon does not affect the MW penetration as MWs can pass through gas or tissues and cause heating at any water content and temperature level (Sidoff et al., 2016). Consequently, organs with high water contents increase the rate of heat generation, but decreasing the EM field penetration. In contrast, during ablation, since the conductivity and permittivity decrease in the dehydrated tissues, the MW propagation changes (Brace et al., 2008).

Another factor is perfusion, which affects the volume of ablation by reducing the heat away from the periphery of the ablation region in highly perfused organs (Cornelis et al., 2011). Accordingly, the heat-sink effect describes the complexity of heating the tumor, which is adjacent to blood vessels greater than 3 mm. Since MW provides fast heating, this decreases the negative effect of perfusion. Thermal steadiness occurs at the MWA boundaries, where the heat loss rate due to perfusion is greater than the generated heat rate at the periphery of the ablation zone. This steady state depends on the quantity and type of delivered energy, the distance of the tumor from the ablation electrode, the treatment duration, and the type and vascularization of tissues (Cornelis et al., 2011; de Baere and Deschamps, 2014).

2.3.4 Limitations and challenges of microwave ablation

MWA has the benefits of RFA and provides additional benefits due to the MW heating physics. Such advantages of MWA over RFA include its concurrent, multiple electrode ability with each active antenna, no ground pads needed, less sensitive to heat-sinks, propagation through poor conductive tissues, deeper energy penetration, greater intratumoral temperatures, larger ablation volumes, and quicker heat generation. Furthermore, MWA is able to generate very high temperatures above 100°C, and is effective in ablating tissues with high impedance, such as charred and desiccated tissues (Wright et al., 2003; Haemmerich et al., 2005). Due to the fast heat generation abilities of MWA leading to its reduced susceptibility to thermal sinks, MWA is a good candidate to ablate tumors positioned close to main blood vessels.

MWA is frequently used for the treatment of tumors in the lungs, kidney, and liver, which exceeds 4 cm in diameter. More uniform and larger ablation regions using MWA might more efficiently damage neoplastic cells compared to RFA. This management can be achieved intraoperatively or percutaneously using computed tomography (CT) or ultrasound (US) for guidance. For surgical treatment, the tips of MW applicators have been developed, however this causes a larger ablation zone. The innovative generations of MWA systems allow instantaneous temperature monitoring at the applicator tip to reduce the overheating hazard and provide adjustable output energy consistent with the ablation zone size (Saccomandi et al., 2015).

However, MWA still suffers from challenging limitations caused by deficits in the present MWA systems, including using multiple antennas which necessitates multiple generators, undesirable heating along with the antenna which leads to ablation of tear-drop-shaped zones, and the required large diameters of the antenna to about 14 gauge. Analogous to RFA, the MWA procedure includes direct placement of the antenna tip in the tumor, however, the MWA antenna tip is thicker. In addition, decreasing the cable surface area may increase the cable heating, and lead to greater power loss (Bertram et al., 2006). Another MWA limitation is the complexity in distributing the MW energy because it requires more bulky wires compared to the small ones that are used to feed energy to RF probes. MW power is intrinsically more problematic for generating and delivering energy efficiently and safely to tissues compared to RF. In addition, the distribution cables are heat prone while carrying large power amounts during

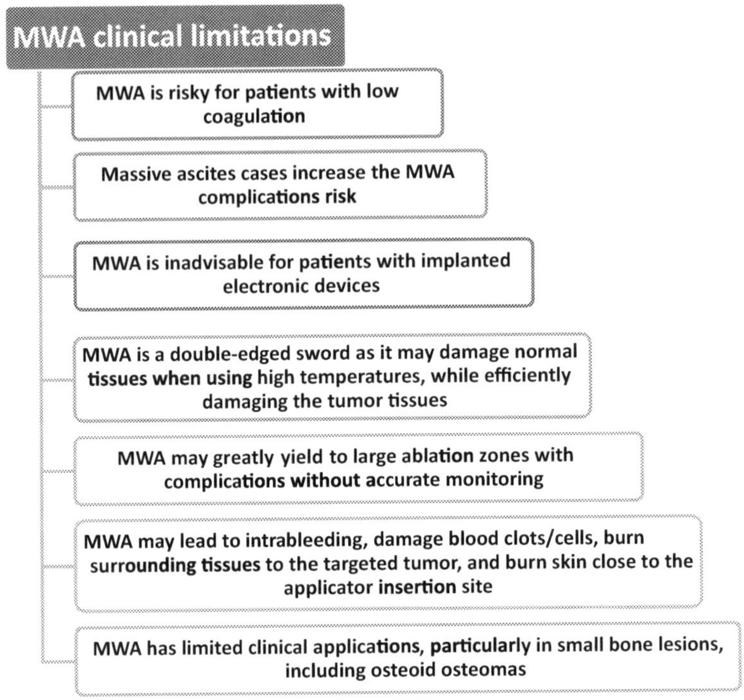

Figure 2.7 MWA clinical limitations.

the MWA of large tumor volumes. Also, MWA is more exclusive and expensive compared to LTA and RFA. Also, in MWA, there remains some impulsiveness in the shape and size of the ablated zone, which is associated with technical factors (Sawicki et al., 2017). Finally, from a clinical point of view, MWA has several clinical limitations as reported in Fig. 2.7.

The final limitation in Fig. 2.7 is due to the required large-diameter probes, shaft heating, and underpowered systems.

2.4 Laser ablation

Lasers are termed after the electronic substance, gas, liquid, or solid, which can be used to generate light. Typically, LASER light refers to Light Amplification by Stimulated Emission of Radiation, which is unlike regular light. Unlike regular light that spreads out in all directions and has several different wavelengths, laser light has a single wavelength which is concentrated in a very narrow beam. Accordingly, a laser beam can be focused on an specific region of solid or tissue to eliminate a layer/layers of a specific material. This process is called laser ablation, which can be used in therapy with the progress made in the technology of fiber lasers, where a high level of precision, effectiveness, and accuracy are achieved (Sliney and Trokel, 1993; Alam et al., 2003; Sliney and Mellerio, 2013; Absten and Joffe, 2013).

For cancer treatment, lasers have two roles, namely for shrinking or destroying a targeted tumor with heat, or for activating a photosensitizing agent (chemical agent) to destroy cancer cells in photodynamic therapy (PDT) (Castano et al., 2005; Agostinis et al., 2011). Generally, a laser focused as an intense light beam can be used to heat and damage small tumor zones, where interstitial laser ablation (ILA) is considered a hyperthermic ablation procedure. Among the traditional ablation techniques in thermal therapy, LA has been proved to have the impacted opportunity of being directed through an elastic and small fiber in deep-lying organs. Generally, three laser types can be applied for treatment, namely neodymium:yttrium-aluminum-garnet (Nd YAG) lasers, carbon dioxide (CO_2) lasers, and argon lasers.

Laser therapy is conducted using an elastic endoscope, which is a lighted/thin tube used to explore the internal tissues of the body for further ablation. LA therapy is considered a nominally invasive treatment process using laser technology, when laser radiation of high fluence, short wavelength interacts with the tumor to be ablated. For accurate LA treatment, LA is performed under ultrasound-guided visualization to target the tumor tissues with the positioned laser probe to be inserted into the tumor region (Alamoud et al., 2019). Once the LA beam is emitted through the inserted probe in the cancerous cells, thermal energy is generated to damage the tumor tissues.

2.4.1 History of laser ablation

Solon et al. (1961) carried out the first use of a laser in the clinical domain. They explored the emitter capabilities of high intensity laser light for use as an energy source in different biomedical applications. Afterward, Dougherty et al. (1977) presented the use of phototherapy in malignancy to increase the depth of penetration using higher porphyrin doses, light intensity, and a fiber optics approach. Gardner et al. (1982) treated solid tumors in mice using a dose of Nd YAG laser light. Later, Bown (1983) showed that a succeeding laser band of energy density after the Nd YAG laser band has a significant impact on tumor therapy in the presence of photosensitizing agents. The high energy levels of the laser light have heating effect on the biological cells instead of only being locally absorbed by the cells. At this laser light energy level, the alterations in laser energy vary from slow warming of neighboring regions to evaporation of the exposed surfaces of

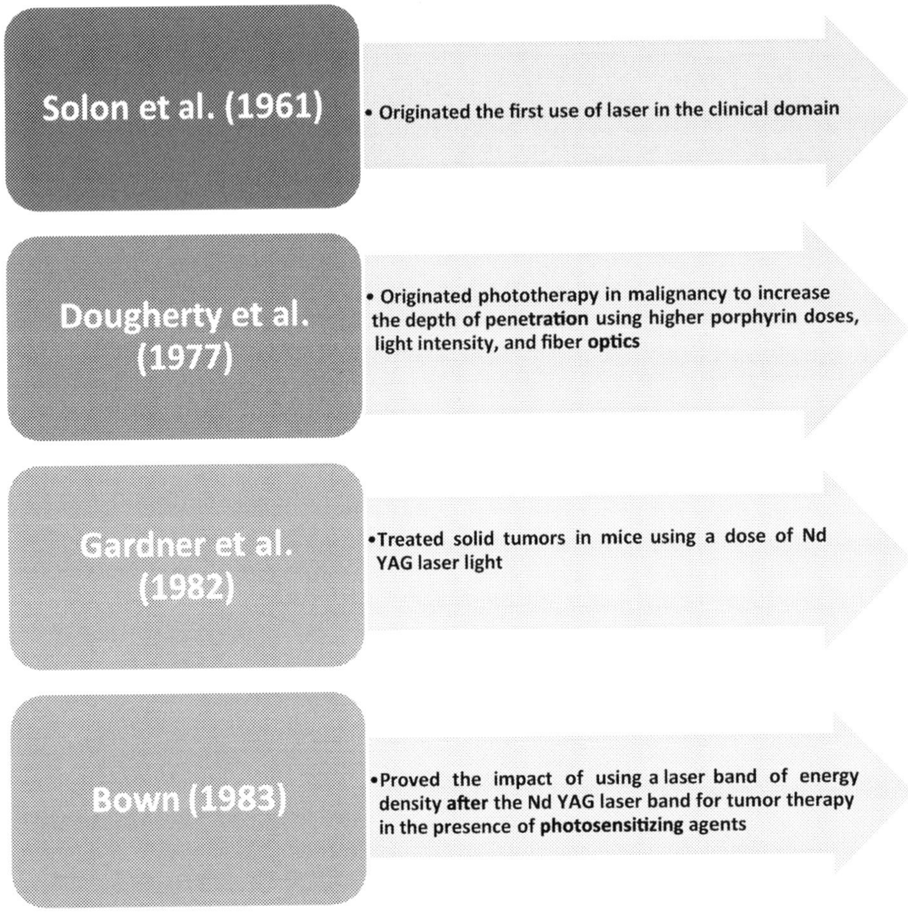

Figure 2.8 History of laser light ablation in therapy.

several organs. Since then, the thermal effect of laser light has been extensively used in tumor therapy, whereby at the end of the 1980s, ablation use of the laser light for pancreas, liver, prostate cancer, and brain cancers originated (Van Hillegersberg et al., 1994; Muschter and Hofstetter, 1995; Bown et al., 1998). A summary of the preceding historical studies of laser light ablation origination in therapy is reported in Fig. 2.8.

2.4.2 Laser ablation thermodynamics and operation principle

Laser light has two main properties that characterize the use of laser beams in thermal therapy applications. These properties of the laser beam include the light collimated character and a high degree of laser beam mono-chromaticity. The property of collimation provides the opportunity to obtain narrow beams with large energy densities. Consistent with the Fraunhofer diffraction relationship, the laser beam optimal divergence angle is restricted by the diameter of the laser source and the wavelength of the emitted light (Brauch et al., 2000).

In laser ablation therapy, the level of damage depends on the dissipated energy (Bown et al., 1980). In all directions, light from these lasers scatters in the tissues, leading to analogous changes adjacent to the region beneath the beam. Subsequently, soft tissues are heated by their absorption of a small amount of energy. Thermal contraction occurs in the targeted zone by increasing the amount of energy due to water vaporization within the tissues deprived of extensive cell necrosis. At the same time, higher energy amounts destroy the cells in situ, eventually vaporizing the cells and leaving a hole.

Three main laser types can provide appropriate output beams, namely argon ion laser, Nd YAG laser, and the carbon dioxide (CO_2) laser. In addition, laser diodes can be used instead of the Nd YAG laser owing to their superior characteristics, including being less expensive, portable, and more compact (Jackson and Lauto, 2002; Kingslake, 2012). Each of these laser types has a different wavelength, as demonstrated in Fig. 2.9.

The generated beams by these laser types have thermal effects based on the differences in the wavelength, leading to differences in the tissue volume that can be heated before vaporizing the superficial cells. This determines the suitable application for each laser type. The CO_2 laser is commonly used for the treatment of early-stage cancers. Similar to CO_2, the argon lasers, which have a larger wavelength compared to the CO_2 lasers, are thus transmitted to a short distance into the targeted tissues. Thus, argon lasers are used

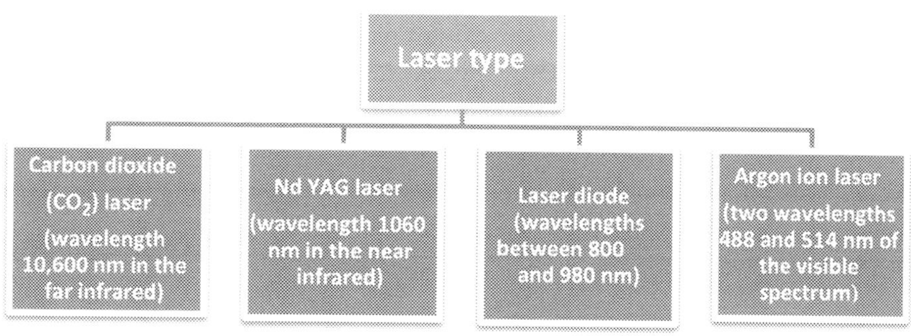

Figure 2.9 Laser types and their wavelengths.

for the treatment of eye tumors, skin cancers, and colonoscopies for polyp removal (McCaughan et al., 1983; Shields and Shields, 2009). In addition, they can be used with light-sensitive drugs in PDT to destroy tumor tissues. In contrast, light from Nd YAG lasers can drive deeper into the targeted tissues compared to the CO_2 and argon lasers. Thus, they can treat quickly any blood clots.

Significantly, in soft tissues, the absorption properties differ based on the applied wavelength of the laser beam. Generally, in pigmented cells, the argon ion and Nd YAG laser beams are absorbed more, while in cells full of water, the CO_2 laser beam is more strongly absorbed (Blayden and Mott, 2013). Accordingly, if adequate energy is absorbed by the tissues in each of these cases, the outer cells will be damaged, and any additional application of laser beam will create a hole in the tissues. An additional difference between the effects of the different laser types is the size of the damaged surrounding areas. To calculate the effect of applying a laser beam with the exposure time, it is found that slow heating of larger volumes happens with long exposure times and a low power level. However, if a low energy level is used, the degree of ablation decreases with the Nd YAG and argon lasers.

With laser ablation, the soft tissues absorb the energy from the applied beam, where a small amount of energy will only warm up the tissues. Then, by increasing the energy, thermal shrinkage of the targeted region occurs due to water vaporization within the tissues without cell necrosis. The continuing of high energy levels will destroy the tissues in situ, and eventually vaporize the cellular structures leading to a hollow (Apitz and Vogel, 2005).

For tumor therapy, using CO_2 laser heats small volumes, while using a large beam spot size contracts superficial cells to stop capillary oozing (Goldman, 1981). Accordingly, the CO_2 laser can be used as a laser knife owing to its extremely localized effect, where the cells under the beam can be vaporized with only slight destruction of neighboring areas. This CO_2 laser beam can be applied to evaporate the superficial layers of tissues at any required depth or to cut tissues. In contrast, both the Nd YAG and argon ion lasers have two advantages compared to the CO_2 laser, namely the wavelengths of these lasers have lower absorption coefficients leading to a larger heating volume of tissues compared to using a CO_2 laser, and the beams from Nd YAG and argon ion lasers can be combined with lithe fiber-optic endoscopes, where these laser beams can be transmitted through single elastic fibers (Fried et al., 2001; Staninec et al., 2009; Majumdar and Manna, 2012).

In LA, the generated laser light beam at a specific wavelength is applied to the targeted tissues using a fiber-optic applicator. To increase the ablation volume, the laser applicators with a diffusing or scattering tip can be used. The tissues absorb or scatter the laser light energy, which is then converted into heat (Pacella et al., 2011). Like RFA, with decreasing scattering of laser energy in the desiccated tissues, LA is hampered by its self-limiting nature. Hence, several LA devices have applicator arrays for increasing the ablation volume at the expense of increasing the intrusiveness, complexity, and cost. Consequently, to perform LA for tumor tissues, laser light, and a carriage medium to transmit the laser beam to the tissues are required. Hence, the LA system includes a power source, an elastic optical fiber as a medium, and reflecting mirrors. The mirrors are used to provide a monochromatic light at a definite wavelength based on the laser characteristics and its interaction with tissues. The main two parts of the LA system can be described as follows:

- *Power supply*: During LA of the tumor tissues, the laser settings are selected, where the laser mode can be a pulsed or continuous mode (Nikfarjam and Christophi, 2003; Saccomandi et al., 2012). The difference between the two laser modes is the power of the generated laser beam. In pulsed mode, a high laser power greater than 100 W is used, where the release of laser energy is in a series of pulses, in contrast, in the continuous mode, the laser power is low with values of 2 or 3 W up to 30 W, where the laser energy is released continuously rather than in a pulsed/irregular mode.
- *Optical fiber*: The LA applicator is an optical waveguide with an emitting end, which is called bare fiber which is used in the course of the first use of LA. The design of the applicator's emitting surface controls the geometry of the damaged tissues, the temperature on the surface of the tissues, and the emitted power density. Several applicators are designed, where the optical applicator's emission properties affect the ablated tissue geometry. These different applicators include zebra applicators, cylindrical fiber tip, and sapphire-tipped fibers (Heisterkamp et al., 1997; Sturesson, 1998; Mensel et al., 2006). To controlled a large ablated volume, multiple bare fibers or applicators with cooled tips can be designed (Saccomandi et al., 2014). Bare-tip fiber is inserted via needles into the targeted tumor; in addition, one or more fibers can be used based on the tumor diameter.

Commonly, LA is used for the treatment of superficial cancers, such as skin cancer, as well as to treat early stages of cancers, including small-cell lung, vulvar, vaginal, and penile cancers. LA therapy is applied via a flexible endoscope, which is fitted with an elastic optical fiber tube. At the fiber tip, the transmitted laser light increases the temperature of the targeted tissues and damages them.

Accordingly, laser trabecular ablation (LTA) exploits laser devices that transform from electrical into light energy to heat the tumor tissues until they are destroyed by coagulative necrosis. In the near-infrared spectrum, the diode laser and Nd YAG laser are the most commonly used types owing to their optimal light penetration ability. Light is supplied through an elastic bare-tip fiber of 300–600 μm diameter, which provides spherical thermal tumor ablation of 12–15 mm diameter. In addition, a multisource device or beam-splitting device tolerates using concurrently up to four fibers to deliver laser light to each fiber (Francica et al., 2007; Di et al., 2014).

2.4.3 Laser ablation factors

Laser light interaction with living tissues has several factors that affect LA. The LA influence at each point in the treated volume depends on the tissue's response to the absorbed energy, the light intensity, and the tissue's absorption characteristics at the used laser wavelength. Nevertheless, it is impossible to preserve a constant light intensity in a finite tissue volume, where the energy absorption in the superficial levels reduces the intensity as the light path is deeper in the tissues. Moreover, it is challenging to preserve a constant light intensity across the surface of the treated organ (Anderson and Ross, 1994; Loschenov et al., 2000; Ntziachristos, 2010). Consequently, deliberation of the laser light effects is divided into two levels, namely interaction in small volumes under the laser beam directly at which the light intensity is known (e.g., tumors in the skin), and interactions over larger volumes that include different changes and exist in deep regions of the body at which the light intensity of the laser beam becomes extremely low. Thus, the light effect of a specific intensity depends on the used wavelength. Laser at extremely low intensity has an insignificant ablation effect in tumor therapy. Such very low intensity may occur while treating large tumors, where the intensity can descend through penetration of the laser beam with a set distance from the main beam (Jarm et al., 2010; Karimi et al., 2017).

For tumor therapy, lasers are used to deliver light of high intensity to a small targeted volume under accurate control. The tissue response depends on the light intensity and wavelength as well as the absorption characteristics of the targeted organ. The used laser beam then has thermal/heating effects on the tissues, including vaporization, necrosis, and local fibrosis. The laser–tissue interaction can be defined in terms of three phenomena, namely absorption, reflection, and scattering, where the absorbed light by the tissues is transformed into heat. Irreversible damage to the targeted tissues occurs by short exposure to the laser at a temperature greater than 60°C or by lengthy exposure to the laser beam at temperatures between 45°C and 55°C (Izzo, 2003). Subsequently, LA is affected by several factors during the generation of heat in the tissues. These factors include the physical characteristics of the tissues, the wavelength of the laser light, the emission properties of the optical applicators, and the laser settings (i.e., laser energy, laser power, and exposure time) (Jacques et al., 2013). These factors, along with the factors that affect the success of the clinical LA procedure, are reported in Fig. 2.10.

During the treatment, the preceding factors should be considered to ablate the tumorous tissues without destroying the healthy nearby ones. The laser light effect on tissues depends on the laser light wavelength and the laser operation mode. Therefore the different laser types can be deployed for tumor ablation to obtain different effects based on the tumor volume and location (depth). These laser types mainly differ due to their inconsistent wavelengths, and hence the absorption length (Schneider et al., 2016). In particular the absorption length is considered the key factor to determining the required wavelength and type of laser beam. It is defined as the required tissue depth to absorb 63% of incident light, where it is the function of the optical properties of the targeted tumor and its nearby healthy tissues (Jacques et al., 2013). Additionally, the blood vessel proximity as well as the tissue's

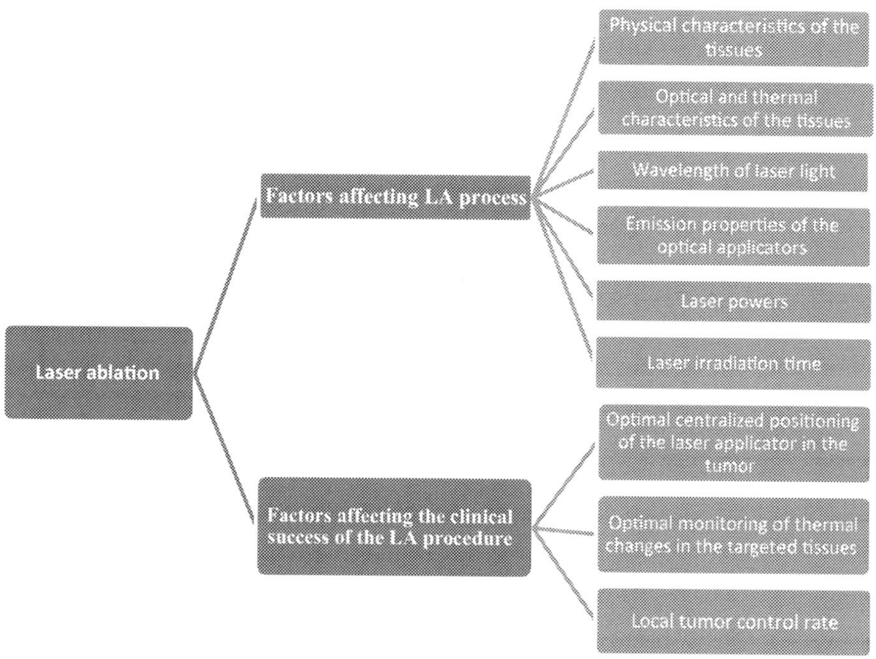

Figure 2.10 Factors affecting the LA process and its clinical success.

thermal/optical characteristics control the ablation volume and thermal diffusion of the light energy.

For superficial treatments, LA using laser beams which are intensely absorbed wavelengths by the targeted tissues is used. In contrast, for treatment of deep tumors, the used lasers should have a high optical depth of penetration abilities. For required deep penetration, lasers emitting infrared light can be used. Nd YAG with 1064 nm wavelength and laser diodes with 800–980 nm wavelengths, each with about 10 cm absorption length (Cobb, 2006), are employed in clinical practice. Conversely, the CO_2 laser at 10,600 nm wavelength and has low penetration depths ranging from 10 μm to about 1 mm, and thus it is deployed in superficial treatment only. Also, low-level laser therapy (LLLT) can be applied to treat lymphedema and severe mouth sores (Schwarz et al., 2008).

2.4.4 Limitations and challenges of laser ablation

Compared to the other ablation methods, LA is considered less robust, but has the same safety and efficiency as RFA with a shorter treatment time (Francica et al., 2012). The feasibility and low-cost requirement (cost-effectiveness) are considered the key advantages to LA as it uses a very fine needle to insert the fibers in the tumor, making it a safe procedure. Additionally, the prospect of using multiple fibers (from one to four fibers) achieves ablation for different tumor sizes and enables the treatment of tumors that differ in size or with multiple small tumors in the same region. LA is a minimal heat transfer (low energy waste) procedure, where much less of the heat's energy is conveyed to the nearby regions which guarantees less damage to the surrounding healthy tissues (i.e., less destruction occurs to the normal tissues) (Vuylsteke and Mordon, 2012). During the LA procedure, patients usually have less scarring, swelling, bleeding, and pain compared to the other ablation methods.

However, LA has several clinical limitations, as follows:

- The difficulty of placing the fibers correctly in the tumor, especially in the case of using multiple fibers due to the thin tip of the used needle;
- LA efficiency is limited by the heat-sink effect;
- The ablation size is small and limited because the laser light is simply absorbed/scattered and has restricted tissue penetration leading to a maximum ablation area of about 1–2 cm^2 (Davalos et al., 2005);
- Some laser treatments have short therapeutic effects as the treatment may not last long.

2.5 High-intensity focused ultrasound ablation

Ultrasound has diagnostic and therapeutic applications in addition to those for different cancer diseases. For the treatment of metastatic disease and primary solid tumors, HIFU is considered an extracorporeal and noninvasive procedure. HIFU is also called FUS (focused ultrasound surgery) and MRgFUS (MRI-guided focused ultrasound) (Orsi et al., 2010; Damianou, 2004). It is a nonionization and noninvasive treatment procedure using US transducer. Unlike the ultrasound imaging modality, HIFU uses very high energy of increased intensity of 720 mW/cm^2 compared to 100–10,000 W/cm^2, where the US transducer concentrates the sound waves (Zhou, 2011). Within the body, this focused US beam generates heat of about 65.6°C at the specific region of interest, which leads to the absorption of an acoustic wave triggering fast ablative heating of the targeted tissues. Such a process destroys the targeted tissues due to heating the tissues quickly to cytotoxic levels (Jolesz and Mcdannold, 2008). For monitoring the heating degree and planning the treatment, magnetic resonance images are captured in real time.

Typically, HIFU ablation is based on the thermal mechanism for cell damage as well as inducing mechanical effects in tissues. The mechanical effect of the HIFU process occurs due to the formation of cavities caused by the acoustic pulses, which expand and contract. The rapid growth of the cavities and their collapsing releases shock waves in the nearby tissues producing mechanical cell damage (Tezel and Mitragotri, 2003; Nikfarjam et al., 2005).

2.5.1 History of high-intensity focused ultrasound ablation

HIFU was discovered in the 1940s (Ter Haar, 1995), however, in thermal ablative technologies for clinical practice, HIFU is considered an innovative ablative procedure. In the 1970s, ultrasound was used to induce hyperthermia by increasing the temperature of tissues to about 43°C in about 1 hour (Ter Haar et al., 1980). The developments in the imaging technologies have played a significant role in the resurgence of HIFU in thermal therapy applications, especially to treat lesions in structurally precarious locations as well as solid malignant lesions in the uterine fibroids, prostate, breast, liver, and pancreas (Orsi et al., 2010; Khokhlova and Hwang, 2011).

For tumor treatment, a renaissance of HIFU followed in the 1990s with the improvement of modern technologies in real-time imaging, energy delivery modes, and transducer design. The first use of HIFU beams was by Lynn et al. (1942), who designed an effective focused ultrasound generator to create focal heating to damage the liver tissues of an animal model with slight effects on the surrounding healthy tissues. Then, Fry et al. (1955) carried out the first attempt at using HIFU with humans for neurologic disorders. Jolesz and Jakab (1991) verified the use of an MRI scanner for guiding the HIFU transducer. Later, HIFU became an efficient candidate for fast tumor treatment due to its improved ability to cause immediate cell death of internal tissue volumes (Gelet et al., 2000). Recently, for a variety of tumors, various studies have explored HIFU as a potentially nonionizing, active therapy procedure using evidence-based practice strategies for treatment of eye, liver, brain, uterus, prostate, breast, pancreas, and thyroid cancers (Beerlage et al., 1999; Thuroff et al., 2003; Uchida et al., 2006; Stewart et al., 2003; Schlachetzki et al., 2002; Ter Haar et al., 1989; Lizzi et al., 1978; Xie et al., 2008; Wu et al., 2005; Esnault et al., 2011). To increase the efficiency of HIFU, MRI is used to monitor the changes in the tissues during the ablation process; also, elastography with color Doppler can be used. Accordingly, the fusion of HIFU with imaging can be employed to accomplish spatially precise treatment under monitoring, which is an essential step in the therapeutic protocol (Diederich, 2005).

2.5.2 High-intensity focused ultrasound ablation operation principle

Sound is a physical phenomenon which generates sound waves/vibrations and is characterized by the frequency (number of vibrations per second). It propagates over matter and cannot travel through a vacuum. Sound waves transmit energy from one location to another. Audible sound has frequencies from 20 Hz to 2 kHz, while any sound of higher frequencies is called ultrasound and is inaudible to humans (Edwards, 2010). Ultrasound is produced by generating mechanical vibrations by converting electrical pulses, which are converted back to electrical energy. To perform such a process, an ultrasound transducer is used to convert electrical pulses (energy) to mechanical (acoustic) energy, and the echo (reverted sound back) to an electrical signal using a polarized material piece with attached electrodes to two of its opposite surfaces. Accordingly, an electrostriction phenomenon occurs, where the molecules of the polarized material are aligned with any applied electric field causing induced dipoles inside the molecular construction of the material and changes in the material dimensions. Furthermore, a piezoelectric effect occurs, where the polarized material, such as barium titanate ($BaTiO_3$) or quartz (SiO_2), produces an electric field with a change in the material dimensions due to the imposed mechanical force. Typically, for most acoustic transducers, the active element is a piezoelectric ceramic, which can produce different wave modes based on the used cutting procedure.

This transducer is used in traditional diagnostic devices at frequencies from 2 to 20 MHz. HIFU depends on the same concept as conventional US. It has a spherical geometry for precise ablation of the targeted tumor creating necrosis at every depth from zero to the focal length of the used transducer without damaging nearby healthy tissues. HIFU applies to the bioeffects which occur with the insonation of human tissues with HIFU waves leading to changes in the sonographic energy into heat. The heat energy is instigated by friction which results from the interaction of the vibrating tissue over time. Cavitation occurs by exposing the tissues to HIFU energy for a long time (Damianou, 2004). When the HIFU energy hits the tumor, the cavitation is settled from gas bubbles produced within the tissues due to the pressure of the rarefaction or expansion. The formation probability of the cavitation increases with increasing of hitting the bubble by the ultrasound beam. During the rarefaction phase, when the bubble collapses after rapid progress, inertial cavitation occurs. Significant cellular damage occurs due to the transient cavitation which bursts the bubbles within the tissues. Additionally, by targeting a tumor with HIFU, a returning echo arises which can be used to detect cavitation.

The HIFU transducer includes a piezoelectric element to convert electrical signals into mechanical vibrations (Evans et al., 2007). The oscillating piezoelectric materials generate ultrasound waves in the receiving tissues (medium). They can handle comparatively high power levels and focus the resultant US beam at a small volume. The mechanical vibrations are focused at the transducer's focal point, where heat energy is established by the friction of the vibrating tissues. Focusing can be realized geometrically by using a curved lens and a plane transducer, or a curved spherical-section transducer. In the focal area, the tissue temperature increases, causing a finite necrotic region. The progression of HIFU transducers has enhanced the treatment times, beam control, and the capability to control the beam distortion. For tumor treatment, a single transducer or a phased-array transducer can be used. A single transducer uses frequencies of 0.5–10 MHz to focus the beam at the sphere's center, while a phased-array transducer is electronically directed.

The ultimate concepts of the sonographic procedure include spatial peak temporal average (SPTA) intensity and transmission frequency (Powers and Kremkau, 2011). An increase in output and/or frequency has a direct effect on the local tissue temperature, where the temperature of the tissues increases with an increase in frequency and intensity. However, due to attenuation, tissues located deeper from the surface will have a lower level of heat, where the tissues can absorb energy. This absorbed energy tends to diffuse to a lower strength as it moves away from the source. In addition, the beam intensity is governed by the tissue density, for example, soft tissues absorb less energy than bones, where the bone thermal index (TIB) and the soft tissue thermal index (TIS) can be calculated. Moreover, during the sonographic process, it is essential to identify the related pressure effects along with the thermal effects in the tissues. These pressure effects occur due to a sound wave which produces pressure changes during its propagation through the tissues (medium). The amplitude of the sound wave is proportional to the pressure against the tissue based on the operating frequency and intensity of the transducer. As the sound wave propagates, cavitation may occur where the tissues consist of fluids. Transient cavitation may occur with an increase in the pressure/amplitude of the sound wave, causing significant obliteration of the tissue. Accordingly, a mechanic index (MI) for measuring the pressure threshold to be destructive is calculated.

The standard of HIFU ablation is the ultrasound energy absorption during its diffusion in the tumor tissues leading to coagulative thermal necrosis, which causes induced cavitation destruction of the tumor cells. Unlike US, the time-averaged intensities in HIFU within the focal region are about some orders higher than 100–10,000 W/cm^2, with peak rarefaction pressures and peak compression pressures equal to 20 and 70 MPa, respectively (Crouzet et al., 2015).

During the ablation process, beams of HIFU (concentrated pulses) are directed using a piezoelectric transducer to focus on the targeted tumor tissues for depositing locally high levels of sound energy. Heat is generated inside the tumor due to the mechanical vibration of the focused sound waves. Thus the transducer delivers high sound pressures to a focal point deprived of damaging unwanted surrounding healthy tissues. The produced high pressure causes cavitation and heating to about 65°C–85°C. During the acoustic wave propagation over the tissues, the absorbed portion is converted to heat. The focused beams can penetrate deep into tissues leading to heating even a tiny tumor

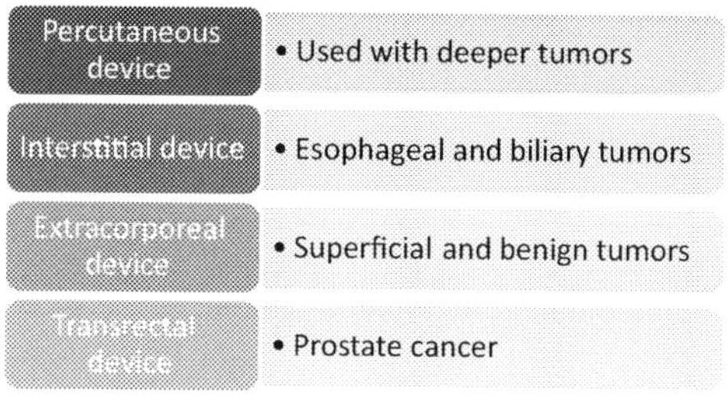

Figure 2.11 The main four HIFU devices.

region. The produced tissue destruction is a function of the tissue temperature and the thermal dose/duration, where an elliptically shaped necrosis region is formed with a lengthy axis parallel to the US beam. The previous procedure is followed to develop the main HIFU devices: percutaneous, interstitial, extracorporeal, and transrectal (Deardorff and Diederich, 2000; Zhou, 2011; Kim et al., 2008; Saldanha et al., 2010), as reported in Fig. 2.11.

Typically, tissue situated at a depth away from the surface has a lower level of heat due to the attenuation in the intensity, where the ability of the tissue to absorb energy and the diffusion ability of the energy decreases with the lower strength at distant positions from the energy source. In addition, the beam intensity is affected by tissue density. For clinical applications, the HIFU transducer should be optimally designed using the progress of the piezoactive materials with particular acoustic properties. For example, composites of piezoactive elements and lead zirconate titanate-type ceramics can be driven at high power for clinical applications. Other parameters affecting the design of the HIFU devices are the transducer diameter, the distance at which the focal volume is positioned, and the radius of curvature of the curved transducers. In the case of using transducers arrays, the spatial distribution of the transducers and the number/size of each transducer along with their relationship to the operating frequency are considered and evaluated to realize the proper acoustic power while avoiding the grating lobes triggered by spreading out the sound energy from the transducer in unwanted directions.

2.5.3 High-intensity focused ultrasound ablation factors

HIFU is a noninvasive ablation modality, which has the potential to handle solid organ tumors using focused extracorporeal US beams (Wu, 2006). Nevertheless, there are various limitations and factors related to its use. HIFU beams can create burns in the tissues lying between the target tumor and the transducer. Limitations are raised due to the absorption, refraction, or reflection of the acoustic energy along the beam path or neighboring the targeted ablation region. Thermal damage can arise at undesirable deep penetration over the tumor or at shallower regions of the tumor (Kennedy et al., 2003). In the course of HIFU therapy, anatomical or external obstacles can affect the direction of the US beams or shift the ablation region from the focal area to a prefocal site, increasing the estimated ablation zone. Such factors have to be allowed for prior to HIFU ablation as they cause complications and limit the ablation success.

Several external factors affecting the preparation of HIFU treatment include the existence of pancreatic, biliary, or duodenal stents, where the stents may be metallic or plastic and affect the US energy. These factors can obstruct the acoustic energy propagation and shift the ablation region to the prefocal region, leading to thermal injury with increased heating of nearby untargeted tissues. Clinically, several anatomical concerns which interact with US beams include the soft tissues, and the skin has variable densities and attenuation.

The transducer's aperture is considered a significant factor that has a role while delivering the HIFU ablation. The damage to the soft tissue adjacent to the tumor is reduced by using a transducer with a wide aperture which also allows highly concentrated acoustic energy at the ablation focus. However, a wider aperture encounters anatomical barriers and interference, while a smaller/narrower aperture reduces the probable obstacles to the energy path. Using a narrow aperture an intense US beam delivers very high thermal ablation energy to a small focal point which may induce intraprocedural pain. At present, using HIFU is restricted to small ablative regions less than 1 cm due to the difficulty in monitoring, difficult targeting, and multiple recurrent treatments.

2.5.4 Disadvantages and challenges of high-intensity focused ultrasound ablation

Since the US waves are quickly attenuated in bones and reflected between air–gas and soft tissue media, the incidence of ribs, gaseous bowel, and lung in the interface of the FUS tumor-targeted region is challenging. Sonication via the cranium is also problematic due to the variable thickness, high attenuation, and density of the skull (Malietzis et al., 2013). Furthermore, the US beam propagates inconsistently in nonuniform soft tissues. Consequently, a fitting acoustic window is required for the US beam to propagate via the body to the targeted tumor. Using the HIFU ablation technology provides various benefits, such as:

- Noninvasive treatment procedure;
- Nonionizing radiation;
- Faster recovery time;
- Less risk; and
- Greater precision compared to other ablation modalities.

In addition, HIFU is considered a low-cost procedure, and US has the ability to end bleeding; thus the HIFU ablation process can be repeated an infinite number of times. HIFU can be applied for the treatment of uterine fibroids, liver, prostate, breast, and renal tumors. It is relatively harmless for ablating tumors close to large blood vessels, where the blood flow distributes the thermal energy from the wall of the vessels. However, use of HIFU suffers from several disadvantages, including:

- The limitation in US penetrance through various tissue types as it works efficiently only with superficial tumors;
- Its limited use to ablate regions subject to respiration motion due to its sonic shadowing;
- Vapor formation, desiccation, and coagulation are detrimental to US energy propagation;
- High-intensity US waves undergo reflection and scattering, leading to damage to tissues adjacent to the targeted zone.

According to the advantages and disadvantages of the HIFU, it can be applied in several clinical practice applications, including ophthalmology, urology, gynecology, oncology, and neurosurgery. Compared to other ablation systems, such as laparoscopic radiofrequency and cryotherapy, HIFU ablation reports reduced toxicity. Clinically, US beam reflection, diffraction, and scattering should be cautiously accounted for or prevented. Also, acoustic coupling of the HIFU transducer to the skin during treatment is essential to avoid skin burns.

The design of the HIFU system limits its easily use in the clinical practice compared to standard treatment methods (Christian et al., 2014; Dubinsky et al., 2008). Clinical uses of HIFU are ongoing to improve and expand its applications, accompanied by progressively clinically applicable fast treatment time leading to its adoption as a routine part of multimodal therapy for several tumors.

2.6 Other ablation modalities

2.6.1 Cryosurgical ablation

Cryoablation (cryotherapy) damages tumors by reaching cytotoxic temperatures to cool the tissues based on the Joule-Thomson effect (Mala, 2006). During cryosurgery, repeated freeze–thaw–freeze cycles are performed to avoid unwanted tissue damage with 0°C to −20°C temperatures. The drop in temperature level damages tissues due to extracellular crystallization and failure of the metabolism, producing cell dehydration and forming intracellular ice crystals. This cooling effect achieves intra- and extracellular ice crystal creation with tissue osmosis. In the first freezing stage, extracellular ice crystals are formed, which raises the osmolarity in the extracellular space and shifts the fluid from the intracellular to the extracellular part. During the thawing stage, the osmotic gradients lead to water spreading into the intracellular part with intracellular ice crystal formation leading to complete cell damage. Moreover, during thawing, post-freezing damage occurs.

During the cryoablation procedure, the main parameters affecting tissue destruction are the lowest achieved temperature and the cooling rate of freezing near the cryoprobe surface. Damage to the tissues occurs at about −40°C. Numerous cryoablation approaches are settled to increase the ablation size to a diameter of 10 cm for treating large tumors (Fuchshuber, 2012). Cryoablation has several advantages including conception of the ablation region, which is correlated with ice balls, the preservation of the collagenous construction of the ablated tissues, and the ability to treat large liver tumors >3 cm. It is efficient for the treatment of cardiac arrhythmias with reduced side effects and shortened recovery duration. Cryotherapy is usually applied to treat metastatic osseous lesions and renal/prostate/lung cancers. However, it has imperfect achievement in the curing of ventricular arrhythmias owing to the limited ablation ability under high heat loads. It is also associated with small ablated regions, severe cryoshock, and high bleeding risk due to the deficiency of coagulation during the cryotherapy process.

2.6.2 Nanoparticle-based ablation

Since the well-known thermal ablation procedures do not differentiate normal and tumor tissues, there are several strategies incorporating nanoparticles to provide precise selective heat delivery to tumors. Nanotechnology-based cancer treatment is considered one of the interstitial thermotherapy forms to select the amount of heat deposition to the tumor (Beik et al., 2016). The tumor tissues are either destroyed directly using thermal ablation or sensitized in association with radiotherapy and chemotherapy based on the treatment duration and realized intratumoral temperature. Nanomaterials, including nanocomposites-assisted and nanoparticles-assisted thermal therapy, offer various benefits over traditional techniques. Due to the optical and magnetic properties of the nanomaterials they can activate the heat elevation by absorbing electromagnetic, RF waves, or near-infrared light (NIR) in the tumor zones for heat destruction selectivity of the tumor while preventing heat generation in the surrounding healthy regions.

For multimodal therapy integration and repeated treatments, nanoparticles stay in place during the treatment region (Cabral and Baptista, 2014). Advancements in nanotechnology lead to the manufacturing of a plethora of different nanoparticle surface modifications, core physicochemical properties, shapes, and sizes. The nanoparticle shape and size determine the performance of nanoparticle-based ablation by controlling the penetration ability of the heat in the blood vessels, affecting the macrophage uptake rate, and reaching the targeted region. Different nanoparticles include carbon nanotubes, silica–gold nanoshells, magnetic/gold nanoparticles, and nanorods, which produce heat along with the generated heat by the other ablation modalities, such as RF and MW. Nanoparticle-mediated thermal therapy is applied with the alternating magnetic field, RF field, or near-infrared light as an energy source, as illustrated in Fig. 2.12 which shows a couple of alternatives for the used nanoparticles and the conventional ablation method.

Discrepancies in the nanoparticle composition and structure affect the ability of tumor targeting and diffusion accompanied by photothermal heating efficacy, which varies with nanomaterial toxicity. Among the challenges to applying nanomaterials in

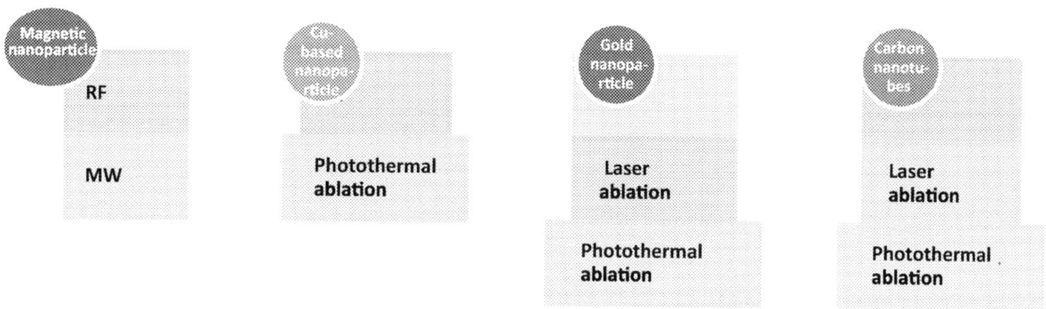

Figure 2.12 Pairs of ablation techniques and nanoparticle types.

thermal therapy are the lack of noninvasive methods for biodistribution assessment, deficiency of treatment planning tools for predicting the temperature distributions, and nanoparticle inhomogeneous distribution. Consequently, investigations and additional studies are an active research area.

2.7 Conclusion

Recently, thermal ablation therapy has become more central in tumor treatment to avoid the complications of surgery and to bridge the transplantation of organs in some cases. Thermal ablation destroys tumor tissue in a nominally invasive fashion by producing a cytotoxic temperature for a short duration, avoiding destruction of neighboring vital tissues/structures. The common ablation methods, such as laser, ultrasound, microwave, and radiofrequency, damage the tissues by elevating the temperature of the tissues locally above 50°C. On the other hand, cryoablation uses a sub-zero temperature level to destroy tumor tissues by freezing. These ablative methods offer a minimal (laparoscopically/percutaneously) noninvasive method for cancer therapy.

While RFA is considered the more practiced thermal ablation method, both microwave and laser ablation methods have proved their efficiency and are more suitable in several cases based on the location and size of the tumor. In interventional radiology and oncology, the selection of ablation method depends on the physician's experience and predilection along with tumor parameters, where each ablation method has its working principles, practical characteristics, advantages, and restrictions. Equally, contraindications in the individual patient's case to the available ablation method should be considered, such as the existence of a pacemaker which restricts the use of RFA, for example. Accordingly, the choice of the used ablation modality must consider the ablation procedure limitations, the patient's characteristics, and the features of the tumor.

Numerical and simulation schemes are recommended to study the temperature profiles and heating patterns to measure the thermal damage during typical treatment. For each energy source, different applicator structures can be tested and evaluated to show the relationship between the geometry of the probe and the energy reduction in defining the thermal and heating damage profiles. In addition, the theoretic model can study the effects of perfusion collapse due to heating in different tissue types. Thus, several key points can be summarized as follows:

- Different ablation modalities have been developed with various specifications;

- The use of nanomaterials is recommended to guarantee the selection of the exact zone for ablation;
- Mono-/multiapplicator-based methods should be distinguished from multibit-polar-/multiapplicator-based methods;
- The selection of the most appropriate modality depends less on the used energy type than its dependency on the spatial distribution of the energy delivery strategy;
- The tissue damage profile during the ablation process should be considered;
- One of the main factors that should be considered is the shape and size of the ablation zones with the various tissue types;
- Antenna cooling abilities to optimize the use of the power generator are significant;
- Wavelength/field control, thermal control, and predicting the ablation zone during the ablation procedure should be considered during the setting process;
- Insufficient settings may cause random ablation volume limits to the clinical efficiency with increasing risk of local complications;
- Simulation and numerical studies are recommended to develop new ablation systems, evaluate the existing ones, and find the most appropriate ablation method for a specific tumor characteristic.

In addition to the preceding points, heat transfer in the tissues can also be modeled using the bioheat transfer equation in view of the thermal iso-effect dose. Novel quantitative measures, including the ablation resistance score, were stated to improve treatments and obtain superior analysis of the thermal ablation regions. Different tissue properties affect the ablation process, such as tissue perfusion/type, adjacent heat sinks, the presence of blood vessels, the absorption rate of energy through the thermal ablation, and the dielectric effects.

Numerous studies have been conducted to evaluate the feasibility and safety of thermal ablation methods by focusing on the treatment of different tumor types. The outcomes of such trials have suggested the safety of ablation in controlling locally primary/metastatic tumors. Nevertheless, some active research topics should be considered to solve the existing restrictions of thermal ablative efficiency, including insistent residual tumor growth at the ablation margin, the incapability of efficiently treating large tumors, and the inability to achieve 100% tumor destruction. In clinical applications, thermal ablation is applied mostly to treat different organ sites, including thyroid, liver, pancreas, breast, prostate, kidney, lung, brain, and bone. Reduced mortality and morbidity are considered the potential benefits of using thermal ablation therapy compared to typical surgical procedures.

References

Absten, G.T., Joffe, S.N., 2013. Lasers in Medicine: An Introductory Guide. Springer.

Agnese, D.M., Burak Jr, W.E., 2005. Ablative approaches to the minimally invasive treatment of breast cancer. Cancer J. 11, 77–82.

Agostinis, P., Berg, K., Cengel, K.A., Foster, T.H., Girotti, A.W., Gollnick, S.O., et al., 2011. Photodynamic therapy of cancer: an update. CA: A Cancer J. Clinicians 61 (4), 250–281.

Ahmed, M., Solbiati, L., Brace, C.L., et al., 2014. Image-guided tumor ablation: standardization of terminology and reporting criteria–a 10-year update. Radiology. 273 (1), 241–260.

Ahmed, M., 2014. Image-Guided Tumor Ablation: Standardization of Terminology and Reporting Criteria – A 10-Year Update. RSNA 2014. J. Vasc. Interv. Radiol. 25, 1706–1708.

Ahmed, M., Goldberg, S.N., 2004. Radiofrequency tissue ablation: principles and techniques. Radiofrequency Ablation for Cancer. Springer, New York, NY, pp. 3–28.

Alam, M., Dover, J.S., Arndt, K.A., 2003. Energy delivery devices for cutaneous remodeling: lasers, lights, and radio waves. Arch. dermatology 139 (10), 1351–1360.

Alamoud, A.H., Baillot, E., Belabbas, C., Ardestani, H.S.S., Bahai, H., Baldit, A. and et al., 2019. Continuous and Pulsed Experiments with Numerical Simulation to Dissect Pituitary Gland Tumour by Using Liquid Jet. *arXiv preprint arXiv:1903.04328.*

Alexander, E.S., Wolf, F.J., Machan, J.T., et al., 2015. Microwave ablation of focal hepatic malignancies regardless of size: A 9-year retrospective study of 64 patients. Eur. J. Radiol. 84, 1083–1090.

Almeida, T.R.D.O.D., 2012. *Development of a Stent-Based Electrode for Radio Freqency Thermal Ablation Procedure* (Doctoral dissertation).

Anderson, R.R., Ross, E.V., 1994. Laser-tissue interactions. Cutaneous Laser Surgery: The Art and Science of Selective Photo-thermolysis. Mosby-Year Book, St Louis, pp. 1–18.

Apitz, I., Vogel, A., 2005. Material ejection in nanosecond Er: YAG laser ablation of water. liver, skin. Appl. Phys. A 81 (2), 329–338.

Beer, E., 1910. Removal of neoplasms of the urinary bladder. JAMA 54, 1768.

Beerlage, H.P., Van Leenders, G.J., Oosterhof, G.O., Witjes, J.A., Ruijter, E.T., Van De Kaa, C.A., et al., 1999. High-intensity focused ultrasound (HIFU) followed after one to two weeks by radical retropubic prostatectomy: Results of a prospective study. Prostate 39, 41–46.

Beik, J., Abed, Z., Ghoreishi, F.S., Hosseini-Nami, S., Mehrzadi, S., Shakeri-Zadeh, A., et al., 2016. Nanotechnology in hyperthermia cancer therapy: From fundamental principles to advanced applications. J. Controlled Rel. 235, 205–221.

Bertram, J.M., Yang, D., Converse, M.C., Webster, J.G., Mahvi, D.M., 2006. Antenna design for microwave hepatic ablation using an axisymmetric electromagnetic model. Biomed. Eng. Online 5, 15.

Bertram, J.M., Yang, D., Converse, M.C., Webster, J.G., Mahvi, D.M., 2006. A review of coaxial-based interstitial antennas for hepatic microwave ablation. Crit. Reviews™ Biomed. Eng. 34 (3).

Blayden, J., Mott, A., 2013. Soft-tissue lasers in dental hygiene. John Wiley & Sons.

Boutros, C., Somasundar, P., Garrean, S., Saied, A., Espat, N.J., 2010. Microwave coagulation therapy for hepatic tumors: review of the literature and critical analysis. Surg. Oncol. 19, e22–32.

Bown, S.G., 1983. Phototherapy of tumors. World J. Surg. 7 (6), 700–709.

Bown, S.G., 1998. New techniques in laser therapy. Bmj 316 (7133), 754–757.

Bown, S.G., Salmon, P.R., Storey, D.W., Calder, B.M., Kelly, D.F., Adams, N., et al., 1980. Nd YAG laser photocoagulation in the dog stomach. Gut 21, 818.

Boykin, R.D., 2008. *Characterization of radiofrequency ablation zones for the purpose of treatment planning* (Doctoral dissertation, The University of Texas Health Science Center at San Antonio).

Brace, C.L., 2008. Temperature-dependent dielectric properties of liver tissue measured during thermal ablation: toward an improved numerical model. Conf. Proc. IEEE Eng. Med. Biol. Soc. 2008, 230–233.

Brace, C.L., 2009. Radiofrequency and microwave ablation of the liver, lung, kidney, and bone: what are the differences? Curr. Probl. Diagn. Radiol. 38, 135–143.

Brace, C.L., 2009. Microwave ablation technology: what every user should know. Curr. Probl. diagnostic radiology 38, 61–67.

Brace, C.L., Hinshaw, J.L., Laeseke, P.F., Sampson, L.A., Lee, F.T., 2009. Pulmonary thermal ablation: comparison of radiofrequency and microwave devices by using gross pathologic and CT findings in a swine model. Radiology 251, 705–711.

Brace, C.L., Laeseke, P.F., Sampson, L.A., Frey, T.M., van der Weide, D.W., Lee Jr., F.T., 2007. Microwave Ablation with Multiple Simultaneously Powered Small-gauge Triaxial Antennas: Results from an in Vivo Swine Liver Model. Radiology. 244, 151–156.

Brauch, U., Loosen, P., Opower, H., 2000. High-power diode lasers for direct applications. High-Power Diode Lasers. Springer, Berlin, Heidelberg, pp. 303–368.

Cabral, R.M., Baptista, P.V., 2014. Anti-cancer precision theranostics: a focus on multifunctional gold nanoparticles. Expert. Rev. Mol. diagnostics 14, 1041–1052.

Cantwell, C.P., Obyrne, J., Eustace, S., 2004. Current trends in treatment of osteoid osteoma with an emphasis on radiofrequency ablation. Eur. Radiol. 14, 607–617.

Castano, A.P., Demidova, T.N., Hamblin, M.R., 2005. Mechanisms in photodynamic therapy: part two—cellular signaling, cell metabolism and modes of cell death. Photodiagnosis photodynamic Ther. 2 (1), 1–23.

Castro Jr, A., Jenkins, L., Ekwenna, O., Shields, J., Salas, N., Leveillee, R., 2014. Instrumentation and technique: hyperthermal ablation: radiofrequency and microwave ablation. Percutaneous Ren. Surg. 237.

Chiappini, B., Di Bartolomeo, R., Marinelli, G., 2004. Radiofrequency ablation for atrial fibrillation: different approaches. Asian Cardiovasc. Thorac. Ann. 12, 272–277.

Christian, E., Yu, C., Apuzzo, M.L., 2014. Focused ultrasound: relevant history and prospects for the addition of mechanical energy to the neurosurgical armamentarium. World Neurosurg. 82, 354–365.

Clark, W.L., 1911. Oscillatory dessication in the treatment of accessible malignant growths and minor surgical conditions. J. Adv. Ther. 29, 169–183.

Cobb, C.M., 2006. Lasers in periodontics: a review of the literature. J. periodontology 77 (4), 545–564.

Cornelis, F., Grenier, N., Moonen, C.T., Quesson, B., 2011. In vivo characterization of tissue thermal properties of the kidney during local hyperthermia induced by MR-guided high-intensity focused ultrasound. NMR Biomed. 24, 799–806.

Crocetti, L., Bozzi, E., Faviana, P., Cioni, D., Della Pina, C., Sbrana, A., et al., 2010. Thermal ablation of lung tissue: in vivo experimental comparison of microwave and radiofrequency. Cardiovascular interventional radiology 33 (4), 818–827.

Crouzet, S., Rouviere, O., Lafond, C., Chapelon, J.-Y., Gelet, A., 2015. Focal high-intensity focused ultrasound (hifu). Technical aspects of focal therapy in localized prostate cancer. Springer.

Curley, S.A., Cusack Jr, J.C., Tanabe, K.K., Stoelzing, O., Ellis, L.M., 2002. Advances in the treatment of liver tumors. Curr. Probl. Surg. 39, 449–571.

Cushing, H., Bovie, W.T., 1928. Electro-surgery as an aid to the removal of intracranial tumors. Surg. Gynecol. Obstet. 47, 751–784.

d'Arsonval, M.A., 1891. Action physiologique des courants alternatifs. CR Soc. Biol. 43, 283–286.

D'Onofrio, M., Ciaravino, V., De Robertis, R., Barbi, E., Salvia, R., Girelli, R., et al., 2016. Percutaneous ablation of pancreatic cancer. World J. Gastroenterology 22 (44), 9661.

Damianou, C., 2004. MRI monitoring of cavitation produced by high intensity focused ultrasound. Cyprus J. Sci. Technol. .

Davalos, R.V., Mir, L.M., Rubinsky, B., 2005. Tissue ablation with irreversible electroporation. Ann. Biomed. Eng. 33, 223–231.

de Baere, T., Deschamps, F., 2014. New tumor ablation techniques for cancer treatment (microwave, electroporation). Diagn. Interv. Imaging 95, 677–682.

de Baere, T., Denys, A., Wood, B.J., Lassau, N., Kardache, M., Vilgrain, V., et al., 2001. Radiofrequency liver ablation: experimental comparative study of water-cooled versus expandable systems. Am. J. Roentgenology 176 (1), 187–192.

Deardorff, D.L., Diederich, C.J., 2000. Axial control of thermal coagulation using a multi-element interstitial ultrasound applicator with internal cooling. IEEE Trans. ultrasonics, ferroelectrics, frequency Control. 47, 170–178.

Dewhirst, M.W., Viglianti, B.L., Lora-Michiels, M., Hanson, M., Hoopes, P.J., 2003. Basic principles of thermal dosimetry and thermal thresholds for tissue damage from hyperthermia. Int. J. Hyperth. 19 (3), 267–294.

Di Costanzo, G.G., Francica, G., Pacella, C.M., 2014. Laser ablation for small hepatocellular carcinoma: state of the art and future perspectives. World J. Hepatol. 6, 704–715.

Di Vece, F., Tombesi, P., Ermili, F., Maraldi, C., Sartori, S., 2014. Coagulation areas produced by cool-tip radiofrequency ablation and microwave ablation using a device to decrease back-heating effects: a prospective pilot study. Cardiovasc. Intervent Radiol. 37, 723–729.

Diederich, C.J., 2005. Thermal ablation and high-temperature thermal therapy: overview of technology and clinical implementation. Int. J. Hyperth. 21, 745–753.

Dodd III, G.D., Dodd, N.A., Lanctot, A.C., Glueck, D.A., 2013. Effect of variation of portal venous blood flow on radiofrequency and microwave ablations in a blood-perfused bovine liver model. Radiology 267 (1), 129–136.

Dougherty, T., Boyle, D., Weishaupt, K., Gomer, C., Borcicky, D., Kaufman, J., et al., 1977. Phototherapy of human tumors. In *Research in photobiology*. Springer, Boston, MA, pp. 435–446.

Dubinsky, T.J., Cuevas, C., Dighe, M.K., Kolokythas, O., Hwang, J.H., 2008. High-intensity focused ultrasound: current potential and oncologic applications. Am. J. roentgenology 190, 191–199.
Duck, F.A., 1990. Physical Properties of Tissue: A Comprehensive Reference Book. Academic Press, London.
Dupuy, D.E., 2009. Microwave Ablation Compared with Radiofrequency Ablation in Lung Tissue – Is Microwave Not Just for Popcorn Anymore? Radiology 251 (3).
Edwards, H., 2010. Ultrasound physics and technology: How, why and when. SAGE Publications Sage UK, London, England.
Esnault, O., Franc, B., Menegaux, F., Rouxel, A., De Kerviler, E., Bourrier, P., et al., 2011. High-intensity focused ultrasound ablation of thyroid nodules: first human feasibility study. Thyroid 21, 965–973.
Evans, K.D., Weiss, B., Knopp, M., 2007. High-intensity focused ultrasound (HIFU) for specific therapeutic treatments: a literature review. J. Diagnostic Med. Sonography 23, 319–327.
Facciorusso, A., Di Maso, M., Muscatiello, N., 2016. Microwave ablation versus radiofrequency ablation for the treatment of hepatocellular carcinoma: A systematic review and meta-analysis. Int. J. Hyperth. 32 (3), 339–344.
Francica, G., Iodice, G., Delle Cave, M., Sarrantonio, R., Lapiccirella, G., Molese, V., et al., 2007. Factors predicting complete necrosis rate after ultrasound-guided percutaneous laser thermoablation of small hepatocellular carcinoma tumors in cirrhotic patients: a multivariate analysis. Acta Radiol. 48, 514–519.
Francica, G., Petrolati, A., Di Stasio, E., Pacella, S., Stasi, R., Pacella, C.M., 2012. Effectiveness, safety, and local progression after percutaneous laser ablation for hepatocellular carcinoma nodules up to 4 cm are not affected by tumor location. AJR Am. J. Roentgenol. 199, 1393–1401.
Fried, D., Ragadio, J.N., Akrivou, M., Featherstone, J.D., Murray, M.W., Dickenson, K.M., 2001. Dental hard tissue modification and removal using sealed transverse excited atmospheric-pressure lasers operating at lambda = 9.6 and 10.6 um. J. Biomed. Opt. 6 (2), 231–239.
Fry, W.J., Fry, F., Barnard, J., Krumins, R., Brennan, J., 1955. Ultrasonic lesions in mammalian central nervous system. Science 122, 1091.
Fuchshuber, P.R., 2012. Principles and Safety of Radiofrequency and Cryo Ablation. The SAGES Manual on the Fundamental Use of Surgical Energy (FUSE). Springer.
Furse, C., Christensen, D.A., Durney, C.H., 2009. Basic introduction to bioelectromagnetics. CRC press.
Gandhi, N.S., Dupuy, D.E., 2005. Image-guided radiofrequency ablation as a new treatment option for patients with lung cancer. Semin. Roentgenol. 40, 171–181.
Gangi, A., Basile, A., Buy, X., Alizadeh, H., Sauer, B., Bierry, G., 2005. Radiofrequency and laser ablation of spinal lesions, Seminars in Ultrasound, CT and MRI, Vol. 26. WB Saunders, pp. 89–97, No. 2.
Gardner, W.N., Hugh-Jones, P., Carroll, M.A., Hewitt, E.R., Hewitt, H.B., Whimster, W., 1982. Quantitative analysis of effect of neodymium-YAG laser on transplanted mouse carcinomas. Thorax 3 (7), 594.
Garrean, S., Hering, J., Saied, A., Hoopes, P.J., Helton, W.S., Ryan, T.P., et al., 2009. Ultrasound monitoring of a novel microwave ablation (MWA) device in porcine liver: lessons learned and phenomena observed on ablative effects near major intrahepatic vessels. J. Gastrointest. Surg. 13 (2), 334–340.
Gelet, A., Chapelon, J., Bouvier, R., Rouviere, O., Lasne, Y., Lyonnet, D., et al., 2000. Transrectal high-intensity focused ultrasound: minimally invasive therapy of localized prostate cancer. J. Endourology 14, 519–528.

Goldberg, S.N., Gazelle, G.S., Halpern, E.F., Rittman, W.J., Mueller, P.R., Rosenthal, D.I., 1996. Radiofrequency tissue ablation: importance of local temperature along the electrode tip exposure in determining lesion shape and size. Acad. Radiol. 3, 212–218.

Goldberg, S.N., Hahn, P.F., Tanabe, K.K., et al., 1998. Percutaneous radiofrequency tissue ablation: does perfusion-mediated tissue cooling limit coagulation necrosis? J. Vasc. Interv. Radiol. 9 (1 Pt 1), 101–111.

Goldberg, S.N., Gazelle, G.S., Dawson, S.L., Rittman, W.J., Mueller, P.R., Rosenthal, D.I., 1995. Tissue ablation with radiofrequency: effect of probe size, gauge, duration, and temperature on lesion volume. Academic radiology 2 (5), 399–404.

Goldman, L. (Ed.), 1981. The Biomedical Laser. Springer-Verlag, Berlin-Heidelberg-New York.

Haemmerich, D., Lee Jr, F.T., 2005. Multiple applicator approaches for radiofrequency and microwave ablation. Int. J. Hyperth. 21 (2), 93–106.

Haemmerich, D., 2010. Biophysics of radiofrequency ablation. Crit. Reviews™ Biomed. Eng. 38 (1).

Harari, C.M., Magagna, M., Bedoya, M., et al., 2016. Microwave ablation: comparison of simultaneous and sequential activation of multiple antennas in liver model systems. Radiology. 278 (1), 95–103.

Heisterkamp, J., van Hillegersberg, R., Sinofsky, E., Ijzermans, J.N.M., 1997. Heat-resistant cylindrical diffuser for interstitial laser coagulation: Comparison with the bare-tip fiber in a porcine liver model. Lasers Surg. Med. 20, 304–309. Available from: https://doi.org/10.1002/(SICI)1096-9101(1997)20:3.

Hoffmann, R., Rempp, H., Erhard, L., Blumenstock, G., Pereira, P.L., Claussen, C.D., et al., 2013. Comparison of four microwave ablation devices: an experimental study in ex vivo bovine liver. Radiology 268, 89–97.

Ito, K., Saito, K., Yoshimura, H., Aoyagi, Y., Horita, H., 2004. Coaxial-slot antenna for interstitial microwave thermal therapy and its application to clinical trial. Conf. Proc. IEEE Eng. Med. Biol. Soc. 4, 2526–2529.

Izzo, F., et al., 2003. Thermal Ablation Techniques: Microwave and Interstitial Laser Ablation of Liver Tumors. Ann. Surg. Oncol. 10, 491–497.

Jackson, S.D., Lauto, A., 2002. Diode-pumped fiber lasers: A new clinical tool? Lasers Surg. Med.: Off. J. Am. Soc. Laser Medicine Surg. 30 (3), 184–190.

Jacques, S.L., 2013. Optical properties of biological tissues: A review. Phys. Med. Biol. 58, R37–R61.

Jarm, T., Cemazar, M., Miklavcic, D., Sersa, G., 2010. Antivascular effects of electrochemotherapy: implications in treatment of bleeding metastases. Expert. Rev. Anticancer. Ther. 10 (5), 729–746.

Jeffries, T.E., 2004. Laser ablation inductively coupled plasma mass spectrometry. Wilson Wilson's Compr. Anal. Chem. 42, 313–358.

Jolesz, F.A., Jakab, P.D., 1991. Acoustic pressure wave generation within an MR imaging system: potential medical applications. J. Magnetic Reson. Imaging 1, 609–613.

Jolesz, F.A., Mcdannold, N., 2008. Current status and future potential of MRI-guided focused ultrasound surgery. J. Magnetic Reson. Imaging: An. Off. J. Int. Soc. Magnetic Reson. Med. 27, 391–399.

Kanazawa, K., Noritake, K., Takaishi, Y., Kidera, S., 2020. Microwave Imaging Algorithm Based on Waveform Reconstruction for Microwave Ablation Treatment. IEEE Trans. Antennas Propag. .

Karimi, M., Sahandi Zangabad, P., Baghaee-Ravari, S., Ghazadeh, M., Mirshekari, H., Hamblin, M.R., 2017. Smart nanostructures for cargo delivery: uncaging and activating by light. J. Am. Chem. Soc. 139 (13), 4584–4610.

Kennedy, J.E., Ter Haar, G., Cranston, D., 2003. High intensity focused ultrasound: surgery of the future? Br. J. radiology 76, 590–599.

Khan, M.R., Poon, R.T., Ng, K.K., Chan, A.C., Yuen, J., Tung, H., et al., 2007. Comparison of percutaneous and surgical approaches for radiofrequency ablation of small and medium hepatocellular carcinoma. Arch. Surg. 142 (12), 1136–1143.

Khokhlova, T.D., Hwang, J.H., 2011. HIFU for palliative treatment of pancreatic cancer. J. Gastrointest. Oncol. 2, 175.

Kim, C., 2018. Understanding the nuances of microwave ablation for more accurate post-treatment assessment. Future Oncol. 14 (17), 1755–1764.

Kim, Y.-S., Rhim, H., Choi, M.J., Lim, H.K., Choi, D., 2008. High-intensity focused ultrasound therapy: an overview for radiologists. Korean J. radiology 9, 291–302.

Kingslake, R. (Ed.), 2012. Applied optics and optical engineering, Vol. 6. Elsevier.

Knavel, E.M., Hinshaw, J.L., Lubner, M.G., et al., 2012. High-powered gas-cooled microwave ablation: shaft cooling creates an effective stick function without altering the ablation zone. AJR Am. J. Roentgenol. 198 (3), W260–W265. Available from: https://doi.org/10.2214/AJR.11.7971.

Larson, E.J., Ford, C.G. and Waskiewicz, A.M., Covidien L.P., 2017. *Systems and methods for measuring tissue impedance through an electrosurgical cable.* U.S. Patent 9,655,670.

Lencioni, R., Della, P.C., Bartolozzi, C., 2005. Percutaneous image-guided radiofrequency ablation in the therapeutic management of hepatocellular carcinoma. Abdom. Imaging 30, 401–408.

LeVeen, R., 1997. Laser hyperthermia and radiofrequency ablation of hepatic lesions. Semin. Interv. Radiol. 205, 313–324.

Leveillee, R.J., Hoey, M.F., 2003. Radiofrequency interstitial tissue ablation: wet electrode. J. endourology 17 (8), 563–577.

Lin, J.C., Hirai, S., Chiang, C., Hsu, W., Su, J., Wang, Y., 2000. Computer simulation and experimental studies of SAR distributions of interstitial arrays of sleeved-slot microwave antennas for hyperthermia treatment of brain tumors. IEEE Trans. Microw. Theory Tech. 48, 2191–2198.

Lin, J.C., Wang, Y.J., 1996. The cap-choke catheter antenna for microwave ablation treatment. IEEE Trans. Biomed. Eng. 43 (6), 657–660.

Lizzi, F.L., Coleman, D.J., Driller, J., Franzen, L.A., Jakobiec, F.A., 1978. Experimental, ultrasonically induced lesions in the retina, choroid, and sclera. Investig. Ophthalmol. & Vis. Sci. 17, 350–360.

Loschenov, V.B., Konov, V.I., Prokhorov, A.M., 2000. Photodynamic therapy and fluorescence diagnostics. LASER PHYSICS-LAWRENCE 10 (6), 1188–1207.

Lu, D.S., Raman, S.S., Vodopich, D.J., Wang, M., Sayre, J., Lassman, C., 2002. Effect of vessel size on creation of hepatic radiofrequency lesions in pigs: assessment of the "heat sink" effect. Am. J. Roentgenology 178 (1), 47–51.

Lubner, M.G., Brace, C.L., Hinshaw, J.L., Lee, F.T., 2010. Microwave tumor ablation: mechanism of action, clinical results, and devices. J. Vasc. Interv. Radiol. 21, S192–S203.

Lubner, M.G., Brace, C.L., Ziemlewicz, T.J., Hinshaw, J.L., Lee Jr., F.T., 2013. Microwave ablation of hepatic malignancy. Semin. Intervent Radiol. 30, 56–66.

Lynn, J.G., Zwemer, R.L., Chick, A.J., Miller, A.E., 1942. A new method for the generation and use of focused ultrasound in experimental biology. J. Gen. Physiol. 26, 179.

Mahnken, A.H., Bruners, P., Günther, R.W., 2008. Techniques of interventional tumor therapy. Dtsch. Arzteblatt Int. 105 (38), 646.

Majumdar, J.D., Manna, I. (Eds.), 2012. Laser-assisted fabrication of materials, Vol. 161. Springer Science & Business Media.

Mala, T., 2006. Cryoablation of liver tumours–a review of mechanisms, techniques and clinical outcome. Minim. Invasive Ther. & Allied Technol. 15, 9–17.

Malietzis, G., Monzon, L., Hand, J., Wasan, H., Leen, E., Abel, M., et al., 2013. High-intensity focused ultrasound: advances in technology and experimental trials support enhanced utility of focused ultrasound surgery in oncology. Br. J. Radiology 86, 20130044.

Massarweh, N.N., Cosgriff, N., Douglas, P.S., 2006. Electrosurgery: history, principles, and current and future uses. J. Am. Coll. Surg. 202 (3), 520–530.

Matsudaira, K., Nakagawa, H., Wittkampf, F.H., Yamanashi, W.S., Imai, S., Pitha, J.V., et al., 2003. High incidence of thrombus formation without impedance rise during radiofrequency ablation using electrode temperature control. Pacing Clin. electrophysiology 26 (5), 1227–1237.

McAchran, S.E., Lesani, O.A., Resnick, M.I., 2005. Radiofrequency ablation of renal tumors: past, present, and future. Urology 66, 15–22.

McCaughan Jr, J.S., Guy, J.T., Hawley, P., Hicks, W., Inglis, W., Laufman, L., et al., 1983. Hematoporphyrin-derivative and photoradiation therapy of malignant tumors. Lasers Surg. Med. 3 (3), 199–209.

McTaggart, R.A., Dupuy, D.E., 2007. Thermal ablation of lung tumors. Tech. Vasc. interventional radiology 10 (2), 102–113.

Mensel, B., Weigel, C., Hosten, N., 2006. Minimally Invasive Tumor Therapies. Laser-Induced Thermotherapy. Springer, Berlin/Heidelberg, Germany, pp. 69–75.

Meredith, K., Lee, F., Henry, M.B., Warner, T., Mahvi, D., 2005. Microwave ablation of hepatic tumors using dual-loop probes: results of a phase I clinical trial. J. Gastrointest. Surg. 9 (9), 1354–1360.

Morrison, P., 2013. Devices and equipment in interventional oncology and their operation. In: Dupuy, D.F.Y., McMullen, W. (Eds.), Image-Guided Cancer Therapy. Springer, New York, NY, pp. 179–199.

Munro, M.G., 2012. Fundamentals of electrosurgery Part I: Principles of radiofrequency energy for surgery. In: Feldman, L.S., Fuchshuber, P., Jones, D. B. (Eds.), The fundamental use of surgical energy (FUSE) manual. Springer, New York, pp. 15–60.

Muschter, R., Hofstetter, A., 1995. Interstitial laser therapy outcomes in benign prostatic hyperplasia. J. Endourol. 9 (2), 129–135.

Nduka, C.C., Super, P.A., Monson, J.R., Darzi, A.W., 1994. Cause and prevention of electrosurgical injuries in laparoscopy. J. Am. CollSurg 179, 161–170.

Nevels, R.D., Arndt, G.D., Raffoul, G.W., Carl, J.R., Pacifico, A., 1998. Microwave catheter design. IEEE Trans. Biomed. Eng. 45 (7), 885–890.

Nikfarjam, M., Christophi, C., 2003. Interstitial laser thermotherapy for liver tumours. Br. J. Surg. 90 (9), 1033–1047.

Nikfarjam, M., Muralidharan, V., Christophi, C., 2005. Mechanisms of focal heat destruction of liver tumors. J. Surgical Res. 127, 208–223.

Ntziachristos, V., 2010. Going deeper than microscopy: the optical imaging frontier in biology. Nat. methods 7 (8), 603.

Organ, L.W., 1976–1977. Electrophysiologic principles of radiofrequency lesion making. Appl. Neurophysiol. 39, 69–76.

O'Rourke, A.P., Haemmerich, D., Prakash, P., Converse, M.C., Mahvi, D.M., Webster, J.G., 2007. Current status of liver tumor ablation devices. Expert. Rev. Med. Devices 4 (4), 523–537.

Orsi, F., Arnone, P., Chen, W., Zhang, L., 2010. High intensity focused ultrasound ablation: a new therapeutic option for solid tumors. J. cancer Res. therapeutics 6, 414.

Pacella, C.M., Francica, G., Di Costanzo, G.G., 2011. Laser ablation for small hepatocellular carcinoma. Radiol. Res. Pract. 2011, 595627.
Panagopoulos, D.J., Karabarbounis, A., Margaritis, L.H., 2002. Mechanism for action of electromagnetic fields on cells. Biochemical biophysical Res. Commun. 298 (1), 95–102.
Panel, T.E., 2012. Guidelines on the diagnosis and treatment of primary liver cancer (2011 edition) Chin. Clin. Oncol. 1 (1).
Pearce, J.A., 2006. Electrosurgical unit. In: Webster, J.G. (Ed.), Encyclopedia of medical devices and instrumentation. Wiley, New York, pp. 156–177.
Pearce, J.A., 1986. Review of electrical field relations. Electrosurgery. Wiley Medical Publications, New York, pp. 224–234.
Pereira, P.L., Salvatore, M., 2012. Standards of practice: guidelines for thermal ablation of primary and secondary lung tumors. Cardiovascular interventional radiology 35, 247–254.
Pilla, A.A., 2007. Mechanisms and therapeutic applications of time-varying and static magnetic fields. Biol. Med. Asp. electromagnetic fields 3.
Pisa, S., Cavagnaro, M., Bernardi, P., Lin, J.C., 2001. A 915-MHz antenna for microwave thermal ablation treatment: physical design, computer modeling and experimental measurement. IEEE Trans. Biomed. Eng. 48 (5), 599–601.
Planché, O., Teriitehau, C., Boudabous, S., Robinson, J.M., Rao, P., Deschamps, F., et al., 2013. In vivo evaluation of lung microwave ablation in a porcine tumor mimic model. Cardiovascular interventional radiology 36 (1), 221–228.
Poggi, G., Tosoratti, N., Montagna, B., Picchi, C., 2015. Microwave ablation of hepatocellular carcinoma. World J. Hepatology 7 (25), 2578.
Poulou, L.S., Botsa, E., Thanou, I., Ziakas, P.D., Thanos, L., 2015. Percutaneous microwave ablation vs radiofrequency ablation in the treatment of hepatocellular carcinoma. World J. Hepatol. 7 (8), 1054–1063.
Pouw, R.E., Wirths, K., Eisendrath, P., Sondermeijer, C.M., Ten Kate, F.J., Fockens, P., et al., 2010. Efficacy of radiofrequency ablation combined with endoscopic resection for Barrett's esophagus with early neoplasia. Clin. gastroenterology hepatology 8, 23–29.
Powers, J., Kremkau, F., 2011. Medical ultrasound systems. Interface Focus. 1, 477–489.
Qian, G.J., Wang, N., Shen, Q., Sheng, Y.H., Zhao, J.Q., Kuang, M., et al., 2012. Efficacy of microwave versus radiofrequency ablation for treatment of small hepatocellular carcinoma: experimental and clinical studies. Eur. Radiol. 22, 1983–1990.
Rosen, A., Stuchly, M.A., Vander Vorst, A., 2002. Applications of RF/microwaves in medicine. IEEE Trans. Microw. Theory Tech. 50 (3), 963–974.
Saccomandi, P., Schena, E., Caponero, M.A., Di Matteo, F.M., Martino, M., Pandolfi, M., et al., 2012. Theoretical analysis and experimental evaluation of laser-induced interstitial thermotherapy in ex vivo porcine pancreas. IEEE Trans. Biomed. Eng. 59 (10), 2958–2964.
Saccomandi, P., Schena, E., Giurazza, F., Del Vescovo, R., Caponero, M.A., Mortato, L., et al., 2014. Temperature monitoring and lesion volume estimation during double-applicator laser-induced thermotherapy in ex vivo swine pancreas: A preliminary study. Lasers Med. Sci. 29, 607–614.
Saccomandi, P., Schena, E., Massaroni, C., Fong, Y., Grasso, R.F., Giurazza, F., et al., 2015. Temperature monitoring during microwave ablation in ex vivo porcine livers. Eur. J. Surgical Oncol. (EJSO) 41 (12), 1699–1705.
Saldanha, D.F., Khiatani, V.L., Carrillo, T.C., Yap, F.Y., Bui, J.T., Knuttinen, M.G., et al., 2010. Current tumor ablation technologies: basic science and device review. Seminars in interventional radiology. © Thieme Medical Publishers, pp. 247–254.

Sartori, S., Tombesi, P., Macario, F., Nielsen, I., Tassinari, D., Catellani, M., et al., 2008. Subcapsular liver tumors treated with percutaneous radiofrequency ablation: a prospective comparison with nonsubcapsular liver tumors for safety and effectiveness. Radiology 248, 670–679.

Sawicki, J.F., Shea, J.D., Behdad, N., Hagness, S.C., 2017. The impact of frequency on the performance of microwave ablation. Int. J. Hyperth. 33 (1), 61–68.

Schlachetzki, F., Hölscher, T., Koch, H.J., Draganski, B., May, A., Schuierer, G., et al., 2002. Observation on the Integrity of the Blood-Brain Barrier After Microbubble Destruction by Diagnostic Transcranial Color-Coded Sonography. J. Ultrasound Med. 21, 419–429.

Schneider, C., Johnson, S.P., Walker-Samuel, S., Gurusamy, K., Clarkson, M.J., Thompson, S., et al., 2016. Utilizing confocal laser endomicroscopy for evaluating the adequacy of laparoscopic liver ablation. Lasers Surg. Med. 48 (3), 299–310.

Schramm, W., Yang, D., Haemmerich, D., 2006. Contribution of direct heating, thermal conduction and perfusion during radiofrequency and microwave ablation. Conf. Proc. IEEE Eng. Med. Biol. Soc. 1, 5013–5016.

Schwarz, J.P., Spackman, J.R., Fahey, D.W., Gao, R.S., Lohmann, U., Stier, P., et al., 2008. Coatings and their enhancement of black carbon light absorption in the tropical atmosphere. J. Geophys. Research: Atmospheres 113 (D3).

Seki, T., Wakabayashi, M., Nakagawa, T., et al., 1994. Ultrasonically guided percutaneous microwave coagulation therapy for small hepatocellular carcinoma. Cancer. 74 (3), 817–825. doi:10.1002/1097-0142(19940801)74:3 < 817::AID-CNCR2820740306 > 3.0.CO;2-8.

Shariat, S.F., Raptidis, G., Masatoschi, M., Bergamaschi, F., Slawin, K.M., 2005. Pilot study of radiofrequency interstitial tumor ablation (RITA) for the treatment of radio-recurrent prostate cancer. Prostate 65, 260–267.

Shibata, T., Iimuro, Y., Yamamoto, Y., et al., 2002. Small hepatocellular carcinoma: comparison of radiofrequency ablation and percutaneous microwave coagulation therapy. Radiology 223, 331–337.

Shields, C.L., Shields, J.A., 2009. Ocular melanoma: relatively rare but requiring respect. Clin. dermatology 27 (1), 122–133.

Sidoff, L., Dupuy, D.E., 2016. Clinical experiences with microwave thermal ablation of lung malignancies. Int. J. Hyperth. 1–9.

Simon, C.J., Dupuy, D.E., Mayo-Smith, W.W., 2005. Microwave ablation: principles and applications. Radiographics 25 (Suppl 1), S69–S83.

Skinner, M.G., Iizuka, M.N., Kolios, M.C., Sherar, M.D., 1998. A theoretical comparison of energy sources—microwave, ultrasound and laser—for interstitial thermal therapy. Phys. Med. Biol. 43, 3535–3547.

Sliney, D.H., Mellerio, J., 2013. Safety with lasers and other optical sources: a comprehensive handbook. Springer Science & Business Media.

Sliney, D.H., Trokel, S.L., 1993. Introduction to Medical Laser Systems. Medical Lasers and Their Safe Use. Springer, New York, NY, pp. 1–22.

Solon, L.R., Aronson, R., Gould, G., 1961. Physiological implications of laser beams. Science 134 (3489), 1506–1508.

Staninec, M., Darling, C.L., Goodis, H.E., Pierre, D., Cox, D.P., Fan, K., et al., 2009. Pulpal effects of enamel ablation with a microsecond pulsed $\lambda =$ 9.3-μm CO2 laser. Lasers Surg. Med.: Off. J. Am. Soc. Laser Medicine Surg. 41 (4), 256–263.

Stewart, E.A., Gedroyc, W.M., Tempany, C.M., Quade, B.J., Inbar, Y., Ehrenstein, T., et al., 2003. Focused ultrasound treatment of uterine fibroid tumors: safety and feasibility of a noninvasive thermoablative technique. Am. J. Obstet. Gynecol. 189, 48–54.

Strasberg, S.M., Linehan, D., 2003. Radiofrequency ablation of liver tumors. Curr. Probl. Surg. 40, 451–498.
Sturesson, C., 1998. Interstitial laser-induced thermotherapy: Influence of carbonization on lesion size. Lasers Surg. Med. 22, 51–57.
Sun, Y., Cheng, Z., Dong, L., Zhang, G., Wang, Y., Liang, P., 2012. Comparison of temperature curve and ablation zone between 915-and 2450-MHz cooled-shaft microwave antenna: results in ex vivo porcine livers. Eur. J. radiology 81, 553–557.
Tabuse, K., Katsumi, M., Kobayashi, Y., et al., 1985. Microwave surgery: hepatectomy using a microwave tissue coagulator. World J. Surg. 9, 136–143.
Ter Haar, G., 1995. Ultrasound focal beam surgery. Ultrasound Med. & Biol. 21, 1089–1100.
Tezel, A., Mitragotri, S., 2003. Interactions of inertial cavitation bubbles with stratum corneum lipid bilayers during low-frequency sonophoresis. Biophysical J. 85, 3502–3512.
Ter Haar, G., Sinnett, D., Rivens, I., 1989. High intensity focused ultrasound-a surgical technique for the treatment of discrete liver tumours. Phys. Med. & Biol. 34, 1743.
Ter Haar, G., Stratford, I., Hill, C., 1980. Ultrasonic irradiation of mammalian cells in vitro at hyperthermic temperatures. Br. J. radiology 53, 784–789.
Thuroff, S., Chaussy, C., Vallancien, G., Wieland, W., Kiel, H.J., Le Duc, A., et al., 2003. High-intensity focused ultrasound and localized prostate cancer: efficacy results from the European multicentric study. J. Endourology 17, 673–677.
Uchida, T., Ohkusa, H., Nagata, Y., Hyodo, T., Satoh, T., Irie, A., 2006. Treatment of localized prostate cancer using high-intensity focused ultrasound. BJU Int. 97, 56–61.
Ulucakli, M.E. and Lavery, C.S., 2013, November. The effect of perfusion rate on the radiofrequency ablation of liver tumors. In *ASME 2013 International Mechanical Engineering Congress and Exposition*. American Society of Mechanical Engineers Digital Collection.
Van Hillegersberg, R., Van Staveren, H.J., Kort, W.J., Zondervan, P.E., Terpstra, O. T., 1994. Interstitial Nd: YAG laser coagulation with a cylindrical diffusing fiber tip in experimental liver metastases. Lasers Surg. Med. 14 (2), 124–138.
Vogl, T.J., Naguib, N.N., Lehnert, T., Nour-Eldin, N.E., 2011. Radiofrequency, microwave and laser ablation of pulmonary neoplasms: clinical studies and technical considerations – review article. Eur. J. Radiol. 77, 346–357.
Vogl, T.J., Nour-Eldin, N.-E.A., Hammerstingl, R.M., Panahi, B., Naguib, N.N., 2017. Microwave ablation (MWA): basics, technique and results in primary and metastatic liver neoplasms–review article. RöFo-Fortschritte auf dem Geb. der Röntgenstrahlen und der bildgebenden Verfahr. 1055–1066. © Georg Thieme Verlag KG.
Vuylsteke, M.E., Mordon, S.R., 2012. Endovenous laser ablation: a review of mechanisms of action. Ann. Vasc. Surg. 26 (3), 424–433.
Wang, Y., Sun, Y., Feng, L., Gao, Y., Ni, X., Liang, P., 2008. Internally cooled antenna for microwave ablation: results in ex vivo and in vivo porcine livers. Eur. J. Radiol. 67, 357–361.
Wieland, I. (2010). *Development of a 1.8 mm Percutaneous Applicator with Closed Cycle Cooking for Microwave Tumour Ablation* (Doctoral dissertation, University of Bath).
Wright, A.S., Lee Jr, F.T., Mahvi, D.M., 2003. Hepatic microwave ablation with multiple antennae results in synergistically larger zones of coagulation necrosis. Ann. Surg. Oncol. 10 (3), 275–283.

Wu, F., 2006. Extracorporeal high intensity focused ultrasound in the treatment of patients with solid malignancy. Minim. Invasive Ther. & Allied Technol. 15, 26–35.

Wu, F., Wang, Z.-B., Zhu, H., Chen, W.-Z., Zou, J.-Z., Bai, J., et al., 2005. Extracorporeal high intensity focused ultrasound treatment for patients with breast cancer. Breast cancer Res. Treat. 92, 51–60.

Xie, B., Ling, J., Zhang, W., Huang, X., Zhen, J., Huang, Y., 2008. The efficacy of high-intensity focused ultrasound (HIFU) in advanced pancreatic cancer. Chin. J. Clin. Oncol. 5, 183–186.

Yang, D., Bertram, J.M., Converse, M.C., et al., 2006. A floating sleeve antenna yields localized hepatic microwave ablation. IEEE Trans. Biomed. Eng. 53, 533–537.

Yang, D., Converse, M.C., Mahvi, D.M., Webster, J.G., 2007. Measurement and analysis of tissue temperature during microwave liver ablation. IEEE Trans. Biomed. Eng. 54 (1), 150–155.

Yu, J., Liang, P., Yu, X., Liu, F., Chen, L., Wang, Y., 2011. A comparison of microwave ablation and bipolar radiofrequency ablation both with an internally cooled probe: results in ex vivo and in vivo porcine livers. Eur. J. Radiol. 79, 124–130.

Yu, N.C., Raman, S.S., Kim, Y.J., Lassman, C., Chang, X., Lu, D.S., 2008. Microwave liver ablation: influence of hepatic vein size on heat-sink effect in a porcine model. J. Vasc. Interv. Radiol. 19 (7), 1087–1092.

Yu, J., Liang, P., Yu, X., Liu, F., Chen, L., Wang, Y., 2011. A comparison of microwave ablation and bipolar radiofrequency ablation both with an internally cooled probe: results in ex vivo and in vivo porcine livers. Eur. J. radiology 79 (1), 124–130.

Zhang, B., Moser, M., Zhang, E., Zhang, W.J., 2013. Radiofrequency ablation technique in the treatment of liver tumours: review and future issues. J. Med. Eng. & Technol. 37 (2), 150–159.

Zhang, L., Wang, N., Shen, Q., Cheng, W., Qian, G.J., 2013. Therapeutic efficacy of percutaneous radiofrequency ablation versus microwave ablation for hepatocellular carcinoma. PLoS one 8 (10), e76119.

Zhou, Y.-F., 2011. High intensity focused ultrasound in clinical tumor ablation. World J. Clin. Oncol. 2, 8.

3

Mathematics and finite element method of thermal ablation therapy

Chapter outline

3.1 Introduction

Thermal ablation is a powerful treatment method for benign and cancerous diseases in different organs, such as the uterus, bone, brain, prostate, lung, kidney, liver, and other tissues. Generally, thermal therapy with high temperature is widely used to treat the reshaping, tissue modification, and cardiac arrhythmias (Chakravarty et al., 2008). Its goal depends on raising or decreasing the temperature within the target tumor region to a

Thermal Ablation Therapy. DOI: https://doi.org/10.1016/B978-0-12-819544-4.00006-X
© 2021 Elsevier Inc. All rights reserved.

critical value to induce the coagulative necrosis without damaging the surrounding healthy tissues (Machi et al., 2001). Additionally, these ablation processes are executed with image-guidance modalities to adjust the location of the applicator in the correct tumor region and to monitor the treatment procedure. The most common clinical treatment modalities include radiofrequency ablation (RFA), microwave ablation (MWA), ultrasound ablation, laser ablation, and cryoablation (Ahmed and Goldberg, 2002).

Consequently, several *governing equations of the generated wave* are used according to the used ablation procedure, as follows.

3.1.1 Laplace equations with radiofrequency ablation

RFA, which has been considered as an interventional treatment technique in recent years, is used in several medical applications, such as tumor destruction in different organs locations and cardiac arrhythmia elimination (Haemmerich et al., 2003). To develop and investigate modern techniques, and improve the currently available techniques, theoretical models using computer simulations are considered a strong powerful tool to rapidly provide important information related to the thermal and electrical behavior of the ablation with minimum cost (Berjano, 2006). Generally, RFA uses a probe with a pair of electrodes, including an active probe with small surface area, which is located on the tumor, and a greater dispersive probe to lock the electrical circuit (Hocquelet et al., 2017). Moreover, bipolar ablation is achieved using a bipolar probe with two active electrodes (Laquiere et al., 2016). Furthermore, RFA is of use in some other surgical fields for treating other pathologies, such as articular cartilage therapeutic heating, minimizing gastric reflux by applying external RF power into the gastroesophageal junction, thermal remodeling for oral cavity tissue to cure obstructive apnea during the sleep, and refractive errors correction in ophthalmology (Berjano, 2006).

Commonly, the theoretical models which are used for ablation simulation mainly consist of six steps, including the following (Ewertowska et al., 2018):

- The physical situation representing the simulation model must be observed and simplified.
- The appropriate mathematical equations, which express the electrical and thermal phenomena, are arranged.

- The boundary conditions concerned with the electrical and thermal parameters are determined.
- The physical characteristics related to the biological tissues with the materials contained within the model are obtained.
- An appropriate numerical method is chosen to realize a solution analytically or computationally.
- The postprocessing related to the generated calculated results is conducted.

Accordingly, RFA is operated within a frequency range of 450–550 kHz, where the externally applied RF voltage, which causes the ionic agitation, friction, and tissue heating, is governed by the Laplace equation, which is a second-order linear partial differential equation describing numerous physical processes as a special case of the Poisson equation. This equation formulates several potential field properties, such as the potential gradient, which equals zero in the absence of any energy source. The Laplace equation is expressed as follows (Jiang et al., 2017):

$$\nabla.(\sigma \nabla V) = 0 \tag{3.1}$$

where V is the used RF voltage (volt), and σ is the tissue electrical conductivity (S/m), which is temperature dependent for more accurate results as described in the following expression (Singh and Repaka, 2017b):

$$\sigma(T) = \sigma_b(1 + 0.02(T - T_b) \tag{3.2}$$

where σ_b is the tissue's constant electrical conductivity at the standard temperature (T_b). Also, the computed electric field intensity E (V/m) is obtained using the following equation (Singh and Repaka, 2017a):

$$E = -\nabla V \tag{3.3}$$

Hence, the generated current density J (A/m^2) inside the tissue is calculated using the following expression (Alba-Martínez et al., 2011):

$$J = \sigma E \tag{3.4}$$

Consequently, using Eqs. (3.3) and (3.4), the localized generated power density Q_{RF} (w/m^3) is computed as follows (Singh et al., 2015b):

$$Q_{RF} = J.E = \sigma.E^2 \tag{3.5}$$

In addition to the electrical and thermal equations are mentioned, it is essential to specify the boundary conditions, and RFA is usually accomplished using a fixed potential. Hereafter, the boundary conditions related to the electrical equation are expressed using two types, including (Audigier et al., 2017):

- Neumann boundary condition, where a null current is set at the tissue–air interface, planes, symmetry axis, and points far away from the heating region.
- Dirichlet boundary condition, where there is a constant potential at the electrodes, including a nonzero voltage on the active probe, and a zero voltage on the dispersive probe. In contrast, in the case of ablation with constant current, a nonzero current value is set on the active probe, and the same value with a negative sign is set on the dispersive probe.

Usually, RFA is integrated with a closed-loop control to keep the impedance lower than a preset value. Furthermore, the boundary conditions related to the thermal equation are expressed using three types, including (Audigier et al., 2015):

- Neumann boundary condition, in which a null thermal flux is set at the plane, and symmetry axis;
- Dirichlet boundary condition, in which a fixed temperature is set at the same active probe for modeling an interior cooled probe, and the points far away from the heated region;
- Forced or free, in which thermal convection is modeled using the thermal transfer coefficient, for example, at the blood–endothelium interface within the cardiac chambers, and the tissue–air interface.

Additionally, an initial temperature value is set for the transient thermal analysis; this selected value must equal the selected one in the experimentations for later comparison with the computer simulation. Clinical RFA is almost modeled with normothermic temperature values of 35°C, 36°C, and 37°C (Berjano et al., 2005; Berjano and Hornero, 2004; Tungjitkusolmun et al., 2002). Occasionally, hypothermic temperature values of 32°C are sometimes set for modeling the surgical cardiac RFA (Berjano and Hornero, 2005). Moreover, in the other models with computer results compared with ex vivo experiments, the initial temperature value is considered like the ambient temperature value of 20°C–25°C (Ekstrand et al., 2005).

3.1.2 Electromagnetic equations with microwave ablation

Ablation by microwave (MW) power is considered an efficient technique for handling tumors in different organs, including the esophagus, colon-rectum, liver, stomach, prostate, and lung in males, and the cervix, colon-rectum, stomach, lung, and breast in females (Guerrero López et al., 2017). MWA heats tissues with higher water content, such as breast carcinoma, more than those with lower water content, like the breast glandular or adipose tissues (Brace, 2008). Additionally, it provides several benefits, such as lower ablation times, the availability of using various microwave probes to treat several tumors simultaneously, greater constant intratumoral temperatures, and improved convection profile. Also, the grounding pads are not required, hence this protects patients from skin burn risk (Zhou and Arellano, 2018). Furthermore, MWA can treat larger zones than other treatment methods using almost the same size applicator. Also, the obtained ablated shape/size is nearly consistent, and largely unaffected by the heat-sink outcome, which is generated from vascular structures that exist near the target tumor (Zanus et al., 2011).

Hence, absorption and propagation through MWA are controlled by Maxwell's equations as expressed in the following equations (Prakash, 2010):

$$\nabla.D = \rho \tag{3.6}$$

$$\nabla.B = 0 \tag{3.7}$$

$$\nabla xE = \frac{\partial B}{\partial t} \tag{3.8}$$

$$\nabla xH = J_c + J_d = J_c + \frac{\partial D}{\partial t} \tag{3.9}$$

where D is the electric field density (C/m^2), ρ is the charge density (C/m^2), B is the magnetic field density (T), H is the magnetic field intensity (A/m), J_c is the conduction current density (A/m^2), and J_d is the displacement current density (A/m^2). The electromagnetic fields are radiated from a microwave antenna through biological tissue, derived from the Maxwell's equations, and formulated as follows (Wu et al., 2016):

$$-K^2\mu_r\left(\varepsilon_r - \frac{j\sigma}{\varepsilon_o w}\right)\overline{E} + \nabla x(\nabla x\overline{E}) = 0 \tag{3.10}$$

where K is a free space propagation constant (1/m), μ_r is the medium differential permeability at the selected operating frequency (H/m), w is the angular frequency at the selected operating frequency (rad/s), and ε_r is the tissue relative permittivity at the selected operating frequency ($\varepsilon_o = 8.854{*}10^{-12}$F/m) for the vacuum. Moreover, the dielectric tissue properties, including the electrical permittivity (ε), and the electrical conductivity

(σ) are mainly time dependent, and expressed with linear functions as described in the following equations (Tehrani et al., 2020):

$$\sigma(T) = s_1\left(1 - \frac{1}{1 + e^{(s_2 - s_3 T)}}\right) \tag{3.11}$$

$$\varepsilon(T) = r_1\left(1 - \frac{1}{1 + e^{(r_2 - r_3 T)}}\right) \tag{3.12}$$

where s_1, s_2, s_3, r_1, r_2, and r_3, are the coefficients related to the dielectric properties with temperature dependence, which varies based on the tumor type and tissue region. The tissue dielectric properties decrease with an increase in the temperature due to water evaporation occurred during MWA. In addition, the electric and magnetic fields are derived as follows (Soltani et al., 2020):

$$\overline{E(r)} = \frac{\propto}{r} e^{j(wt - kz)} \hat{r} \tag{3.13}$$

$$\overline{H(r)} = \frac{\propto}{r z_w} e^{j(wt - kz)} \hat{\emptyset} \tag{3.14}$$

where r represents the electric field direction, $\emptyset$ represents the magnetic field direction, Z represents the propagation direction, z_w is the conductor wave impedance (Ω), $\propto$ is the integration constant, and k is the wave number. Typically, $\propto$ is computed using the following equation (Keangin and Rattanadecho, 2018):

$$\propto = \sqrt{\frac{p_{in} z_w}{\pi \ln(r_o / r_{in})}} \tag{3.15}$$

where p_{in} is the input MW power (watt), and r_{in}, and r_o are the inner/outer radius of the antenna, respectively. The propagation of constant k can be formulated as follows (Keangin et al., 2011):

$$k = \frac{2\pi}{\lambda} = \frac{w}{c_l} \tag{3.16}$$

where c_l is the light speed in free space (m/s), and λ is the wavelength (m), which is given by: $\lambda = \frac{c_l}{f\sqrt{\mu_r \varepsilon_r}}$ (Vogl et al., 2018). In addition, the wave impedance of the cable dielectric Z_w is calculated using the following expression (Keangin et al., 2011):

$$Z_w = \sqrt{\frac{\mu_o}{\varepsilon_o \varepsilon_r}} = Z_c \frac{1 + \Gamma}{1 - \Gamma} \tag{3.17}$$

where μ_o is the free space permeability, Z_c is the characteristics impedance of the antenna, that is adjusted to avoid the impedance mismatch between the antenna and the cable, and Γ is the reflection coefficient, which is expressed as follows (Rubio et al., 2015):

$$\Gamma(dB) = 10\log_{10}\left(\frac{p_r}{p_{in}}\right) \tag{3.18}$$

where p_r is the reflected power in watt. Accordingly, the specific absorption rate (SAR) in (watt/kg), which indicates the average power amount absorbed through a unit tissue mass, is computed as follows (Chiang et al., 2013):

$$\text{SAR} = \frac{\sigma}{2\rho}|\overline{E}|^2 \tag{3.19}$$

The slots spacing existing in the microwave antenna depends on the effective wavelength in the biological tissue (λ_{eff}) at the selected operating frequency (Kaur and Maini, 2014):

$$\lambda_{eff} = \frac{c}{f\sqrt{\varepsilon_r}} \tag{3.20}$$

where f is the operating frequency obtained from the microwave generator (Hz). The boundary conditions related to the heat flux are expressed using the Neumann boundary condition with the energy balance conjunction as described using the following equation (Cho et al., 2006):

$$K\frac{\partial T}{\partial n} = h(T - T_s) \tag{3.21}$$

where h is the coefficient of heat transfer at regions close to the model's boundary (watt/m^3K), and T_s is the surrounding environment temperature. Additionally, tissue shrinkage is due to thermal ablation, where the MWA affects 50% of the volumetric tissue contraction at a temperature above 102.1°C (Liu and Brace, 2017). Therefore the volume deformation can be expressed using a linear elastic model as follows (Tehrani et al., 2020):

$$E_{el} = E_{tot} - E_{el} \tag{3.22}$$

Likewise, the thermal strain, that is obtained from the thermal ablation, is explained using the following equation (Liu and Brace, 2017):

$$E_{th} = \aleph_t(T - T_i) \tag{3.23}$$

where T_i is the initial temperature (K), and $\aleph_t$ is the expansion coefficient, which is given by (Tehrani et al., 2020):

$$\aleph_t = \frac{0.03\,TTI^{0.4684}}{T - 285} \tag{3.24}$$

where TTI is a temperature–time integration function, which is given by (Liu and Brace, 2017):

$$TTI(T,t) = \begin{array}{ll} 0 & at\ T \leq 12^{\circ}\mathrm{C} \\ 0.16 \int_0^t (T - T_o)dt & at\ 12^{\circ}\mathrm{C} < T \leq 44.1^{\circ}\mathrm{C} \\ 0.3 \int_0^t (T - T_0)dt & at\ 44.1^{\circ}\mathrm{C} < T \leq 102.1^{\circ}\mathrm{C} \\ 0.542\ (T - T_0)dt & at\ T > 102.1^{\circ}\mathrm{C} \end{array} \tag{3.25}$$

3.1.3 Formulas for laser ablation

Laser beams are used in several applications, such as the heating procedure followed by cooling for modifying surface structures, melting the material layers for alloying, and modifying their properties in some cases, such as corrosion, wear, cutting, drilling, and the ablation of the material due to its several known advantages, including the least heat-affected zones (HAZ), high directionality, and high power density (Yamashita et al., 2006; Phipps, 2007). The HAZ are based mainly on the laser's optical properties, such as the beam spot size, beam power, and the optical properties of the material, such as the absorptivity (Zhang et al., 2020). Hence, the mathematical model, which is based on the laser equations, is considered a strong tool to simulate the ablation process without needing the trial and error procedure, that is time consuming and expensive (Vasantgadkar et al., 2010). Generally, the laser beams are focused using a lens, which converges a minimal beam waist generated around its focal point. Thus, the laser generates heat inside the material workpiece with absorption efficiency, which depends on:

- The beam characteristics, such as the spatial intensity profile;
- The incoming beam divergence and wavelength;
- The operating conditions; and
- The properties of the material.

The laser beam is specified based on its shape, defocusing distance, divergence, beam radius, and intensity distribution, that is assumed for a Gaussian beam shape in the TEM_{00} mode expressed using the following equation (Majumdar and Xia, 2007):

$$I = I_o \frac{R_c(0)^2}{R_c(z)^2} \exp^{\frac{-x^2 - y^2}{R_c(0)^2}} \tag{3.26}$$

where I is the power density of the beam laser (watt/m^2), I_o is its power density at the laser beam's center, R_c is the beam radius (m), which expresses the distance computed from the laser beam center until a drop occurs in the beam intensity by $1/e^2$ from the value computed at the laser beam center, and $R_c(0)$ is the effective beam radius at the lens focal plane.

However, the laser beam extends and propagates through the target medium, and its intensity distribution almost stays stationary with the beam width (Wani and Kant, 2016). Therefore the laser Gaussian beam divergence that expresses the beam expansion through the medium is described as follows (Gilbert et al., 2014):

$$R_c(z) = R_c(0)\sqrt{\left(\frac{z+W_L}{R_c(0)^2\pi/\lambda}\right)^2 + 1} \quad (3.27)$$

where W_L is the distance between the material surface (m) and the lens focal plane, which has two ranges of values, including lower than zero, in which the lens's focal plane is under the material surface, and a range with values higher than zero. The diameter of the laser beam varies with changes in the defocusing distance, and both are considered the most important parameters for the laser beam, which mainly affect its density intensity (Borisov et al., 2000). Additionally, the ablated depth resulting from the incident laser beam fluence Δh(m) is calculated using the following equation (Gilbert et al., 2014):

$$\Delta h = a' + b'\log(\theta) \quad (3.28)$$

where θ is the actual fluence, which concerns the target surface gradient, and the beam out of focus, and a', b' are constants depending on the laser type and workpiece material. Subsequently, the reference laser fluence (θ_o), depending on the laser system, has a maximum value when the spot diameter of the laser has a minimum size, and the pulse energy has a possible maximum value.

Commonly, the fluence at a determined element inside the material workpiece, which expresses the total incident energy on a specific area (A), is computed using the following expression (Cheng et al., 2016):

$$\theta = \frac{P_{av}}{\delta x\, \delta y\, f} \quad (3.29)$$

where P_{av} is the average power of the laser (watt), the beam fluence can be stated as a function of x and y locations inside the workpiece ($\theta(x, y)$), when δx and δy tend to zero. Hence, the maximum laser fluence that is obtained from the beam distribution of energy through a laser spot with diameter

τ (m) is described using the following equation (Zhao and Shin, 2013):

$$\theta_{\max} = \frac{8P_{av}}{\pi\,\tau^2 f} \tag{3.30}$$

3.1.4 Formulas for ultrasound ablation

Intense ultrasound energy is used to damage tumorous tissues through absorption inside the tumor and raising the temperature to specific higher values. This therapy ablation treatment procedure is known as high-intensity focused ultrasound (HIFU) (Soneson and Myers, 2010). HIFU therapy is applied to ablate numerous tumors of different organs, including uterine fibroids, breast, prostate, liver, soft-tissue sarcomas, and pancreas. It outperforms the conventional treatment technique due to its fewer complications, nonionization, and noninvasion (Garrean et al., 2009). Moreover, it is relatively safe for ablating tumors adjacent to the great blood vessels (Solovchuk et al., 2012). In addition, a short ablation duration is sufficient to raise the temperature to values which are adequate to destroy tumor cells (Qin et al., 2012). Moreover, there are some main advantages of the HIFU technique over other treatment modalities, such as its ability to control the deposition power pattern, which achieves accurate and efficient thermal targeting, and enhanced energy penetration (Hijnen et al., 2012). Its applicators consist of linear tubular arrays with an independent power sector, and curvilinear or planar transducers to control the heating over the applicator length and along the angular expanse.

The theoretical thermal models of HIFU include solving the physical equations which govern the external energy deposition generated by the ultrasound generator, and the heat transfer within the tissue to specify the profile of the transient temperature and estimate the tissue damage under heating (Hou et al., 2011).

Consequently, the ultrasound pressure P, which is used to model the wave propagation with finite amplitude in the soft tissue, is expressed by the linear Westervelt wave equation as follows (Solovchuk et al., 2012):

$$\nabla^2 P + \frac{\beta}{{c_0}^4 \rho_0}\frac{\partial^2 p^2}{\partial t^2} + \frac{€}{{c_0}^4}\frac{\partial^3 p}{\partial t^3} - \frac{1}{{c_0}^2}\frac{\partial^2 p}{\partial t^2} = 0 \tag{3.31}$$

where $\nabla^2 P$ and $-\frac{1}{{c_0}^2}\frac{\partial^2 p}{\partial t^2}$ define the lossless wave propagating obtained at the sound small signal's speed, $\frac{€}{{c_0}^4}\frac{\partial^3 p}{\partial t^3}$ is essential to contain the ultrasound pressure equation owing to the fluid viscosity and thermal conductivity, and € is the acoustic diffusivity. In the case of soft tissues, the acoustic diffusivity to compute the viscousness and thermal losses inside the fluid is described using the following equation (Sedaghatkish et al., 2020):

$$€ = \frac{2\Psi {c_0}^3}{w^2} \tag{3.32}$$

where Ψ is the acoustic absorption coefficient. Furthermore, some HIFU numerical simulation studies neglect the nonlinear effect for simplification of the analysis (Solovchuk et al., 2013). The transducer surface of a small element dS is supposed to continuously vibrate with the same velocity over the direction perpendicular to the transducer surface as follows (Solovchuk et al., 2012):

$$V = V_0\, e^{iwt} \tag{3.33}$$

Hence, the velocity potential V_p, which is deduced by transforming the linear wave equation to the integral diffraction equation, is defined using the following equation (Liu et al., 2007):

$$V_{\overline{p}} = \iint \frac{V}{2\pi d} e^{-\Psi d - ikd} dS \tag{3.34}$$

where d is the distance between the field point $\overline{p}$, and the source point on the determining element of the transducer surface. The pressure at point $\overline{p}$ is calculated as follows (Yin et al., 2006; Solovchuk et al., 2012):

$$P_{\overline{p}} = ik\rho c_u V_{\overline{p}} \tag{3.35}$$

where ρ is the density of tissue (kg/m^3) and c_u is the ultrasound speed in the tissues (m/s). The power deposition of the ultrasound wave per unit volume of the target tissue Q_u is expressed using the following equation (Solovchuk et al., 2013):

$$Q_u = 2\Psi I_u \tag{3.36}$$

where I_u is the localized acoustic intensity, which is calculated using the following expression (Solovchuk et al., 2012):

$$I_u = \frac{P_{\overline{p}}^2}{2\rho c_u} \tag{3.37}$$

Eqs. (3.27) and (3.28) concern the attenuation and diffraction effects rather than the nonlinear effects. Accordingly, the attenuation and absorption coefficients in the tissues increase with increasing frequency of the ultrasound wave. On the other hand, the ultrasound penetration depth is limited in its frequency (Stephens et al., 2012).

The boundary condition of HIFU has a fixed/initial temperature value of the tissue 37°C, and on the modeled tissue edges. Moreover, the boundary condition of the convective heat transfer, which is applied to the internal catheter wall, is expressed as follows (Prakash and Diederich, 2012):

$$\vec{n}.K\nabla T = h(T_{\infty} - T) \tag{3.38}$$

where $h = 500^{\text{watt}}/_{\text{m}^3\ \text{K}}$, $T_{\infty} = 25°\text{C}$.

3.1.5 Formulas for cryosurgical ablation

Cryoablation is an efficient treatment technique used to damage tumors through fast thawing/freezing at the cryoprobe tip. It uses the Joule–Thomson effects with the high pressure of helium gas and argon gas under the guidance modality (Georgii et al., 2019). To analyze the characteristics of the freezing procedure, such as the limited radius, a freezing function $F(t)$ can be defined to indicate the freezing radius using the following expression (Nakatsuka et al., 2010):

$$F(t) = Ae^{Bt} + C \tag{3.39}$$

where A and B are lower than zero, and C is a free parameter, which is greater than zero. The freezing function behaves nonlinearly, and in the case of infinity time, its first term is neglected and the function extends to the term C, which is equivalent to the limited ice-ball radius. The boundary conditions of the freezing procedure using a cryoprobe with an outer radius R_{cp} are described using the following equations (Nakatsuka et al., 2010):

$$T = \begin{matrix} T_{cp} & for\ 0 \le r \le R_{cp} \\ T_o & for\ R_m \le r \end{matrix} \tag{3.40}$$

where T_o is the body temperature (°C), T_{cp} is the cryoprobe active tip temperature (°C), and R_m is the radius determined from a point with temperature T_o (m). Therefore the freezing procedure is controlled by the energy balance relation at the freezing internal point ($r = R_f$) as described in the following equation (Nakayama et al., 2008):

$$\rho_f R_f h_{sf} \frac{dR_f}{dt} = \left(R_f K_f \frac{\partial T}{\partial r}\right)_{r=R_f} - \left(R_f K_u \frac{\partial T}{\partial r}\right)_{r=R_f} - \rho_u c_u R_f (T_o - T_f) \frac{dR_f}{dt} \tag{3.41}$$

where the subscripts u and f are related to unfrozen and frozen, respectively, and h_{sf} is the dormant heat of the solidification.

Thus, the temperatures within the frozen and unfrozen domains are computed using the following expressions (Dodi et al., 2014):

$$\frac{T - T_{cp}}{T_f - T_{cp}} = \frac{\ln(r/R_{cp})}{\ln(R_f/R_{cp})} \quad \text{for } R_{cp} \le r \le R_f \tag{3.42}$$

$$\frac{T - T_o}{T_f - T_o} = \left(1 - \frac{r - R_f}{R_m - R_f}\right)^2 \quad \text{for } R_f \le r \le R_m \tag{3.43}$$

Furthermore, the term $(R_f K_f \frac{\partial T}{\partial r})_{r=R_f}$, which expresses the heat flux within the frozen region, can be approximated providing the freezing speed is not very fast, as explained in the following equation:

$$\left(R_f K_f \frac{\partial T}{\partial r}\right)_{r=R_f} = K_f \frac{T_f - T_{cp}}{\ln(R_f/R_{cp})} \tag{3.44}$$

Using the same method, the term $-(R_f K_u \frac{\partial T}{\partial r})_{r=R_f}$, which expresses the heat flux within the unfrozen region, can be obtained as described in the following equation (Nakatsuka et al., 2010):

$$\left(R_f K_u \frac{\partial T}{\partial r}\right)_{r=R_f} = 2K_u(T_o - T_f)\left(R_f \sqrt{\frac{1}{6}\left(3\frac{q_m}{K_u(T_o - T_f)} + \frac{w_b}{\alpha_u}\right)} + \frac{1}{8}\left(\frac{\frac{w_b}{\alpha_u} + \frac{6q_m}{K_u(T_o - T_f)}}{\frac{w_b}{\alpha_u} + \frac{3q_m}{K_u(T_o - T_f)}}\right)\right) \tag{3.45}$$

where α_u is known as the thermal diffusivity, which is calculated as follows (Chen et al., 2002):

$$\alpha_u = \frac{K}{c\rho} \tag{3.46}$$

To obtain more accurate results, the tissue properties used in the bioheat equation, including definite heat, thermal conductivity, and density, are temperature dependent, as described in the following equations (Chan and Ooi, 2016):

$$c = \begin{cases} c_f & at\ T < T_f \\ c_f(1 - \vartheta) + \vartheta c_u + \dfrac{h_{sf}}{\rho(T_u - T_f)} & at\ T_f \le T \le T_u \\ c_u & at\ T > T_u \end{cases} \tag{3.47}$$

$$K = \begin{matrix} K_f & at\ T < T_f \\ K_f(1-\vartheta) + \vartheta K_u & at\ T_f \leq T \leq T_u \\ K_u & at\ T > T_u \end{matrix} \tag{3.48}$$

$$\rho = \begin{matrix} \rho_f & at\ T < T_f \\ \rho_f(1-\vartheta) + \vartheta \rho_u & at\ T_f \leq T \leq T_u \\ \rho_u & at\ T > T_u \end{matrix} \tag{3.49}$$

where ϑ is the phase-transition function, which determines the liquid fraction within the mushy region as described in the following equation (Georgii et al., 2019):

$$\vartheta = \frac{T - T_f}{T_u - T_f} \tag{3.50}$$

3.1.6 Formulas for electroporation ablation

Electroporation ablation is a minimally invasive and nonsurgical technique used for tumor ablation. It depends on applying a wave of several 100 volts per unit centimeter in a 100 μs with short pulses into the organ cells (Pech et al., 2011). The electroporation method was studied in in vitro and in vivo trials. Interestingly, it is known that the electroporation treatment technique affects the organ cell membrane (Sweeney et al., 2016). It provides the ability to introduce nonpermeant molecules, such as DNA plasmid, bleomycin, and cytotoxic drugs, and ions, through the living cells, which is called reversible electroporation (IRE). Reversible electroporation directly kills the cells in the target tumor using a nonthermal mechanism. Other ablation techniques, such as RFA, MWA, radiotherapy, and cryotherapy, cause dramatic damage to the neighboring structures to the target tumor, including blood vessels, bile ducts, and nerves (Pillai et al., 2015). In contrast, IRE, electrochemotherapy (ECT), and electroporation-based therapies (EBTs) have many advantages, including the following (Gallinato et al., 2020):

- The organ scaffold is maintained due to the thermal mechanisms with a short pulse duration; hence, these therapies are very suitable for tumors which are close to vital organs, enhancing healing.
- Electroporation ablation preserves the main vessel integrity, disrupts neovascularization, and achieves tumor hypoxia.
- Electroporation efficiently reduces bleeding due to the vascular lock effect.
- Antigens create a protected response in contrast to other cancer cells, and metastasis, which is far from the target tumor because the proteins are not denatured (Thomson et al., 2015).

In addition, such electroporation techniques have some limitations and technical difficulties, as explained in the following points (Rombouts et al., 2017):

- Unlike the standard ablation treatment techniques that deal with only one needle, EBTs require two needles as a minimum but usually three or four needles for ablation of tumors with more complex shapes;
- The prior determination concerning the treated region is more complex than the standard ablation techniques;
- Also, EBTs suffer from unsuitable post-treatment evaluation procedures for the treatment performance.

Generally, the electroporation model contains an organ tissue defined as a conductive medium that has conductivity based on the electric field amplitude ($-\nabla V$). It is mathematically explained using the following equations (Gallinato et al., 2020):

$$-\nabla.(\sigma||\nabla V||\nabla V) = 0 \quad \text{in domain } \emptyset \tag{3.51}$$

$$\partial_n V|_{b_{out}=0} \tag{3.52}$$

$$V|_{\mathcal{A}^{\pm}} = g^{\pm} \tag{3.53}$$

where b_{out} is the boundary related to the domain $\emptyset$, which is derived from the activated needles, $\mathcal{A}^{\pm}$ expresses the activated cathode, and anode needles, which are used for the delivery of the pulse, and $g^{\pm}$ denotes the Dirichlet data at the $\mathcal{A}^{\pm}$ needle. Furthermore, the conductivity classical explanation of the target tissue contains a sigmoid function with four parameters, as described in the following equation:

$$\sigma(\lambda) = \sigma_{non} + \frac{\sigma_f - \sigma_{non}}{2}(\text{erf}(S_{non}(\lambda - E_{th})) + 1) \tag{3.54}$$

where σ_{non} is the nonelectroporated tissue conductivity, σ_f is the tissue conductivity, S_{non} is the nonlinearity slope, and E_{th} is the specified threshold amplitude of the electroporation. Hence, the electrical intensity, which flows within one needle, that is, $\mathcal{A}^{+}(I_{\mathcal{A}^{+}})$ can be accurately and efficiently computed through calibration of the conductivity of the organ tissues from the initial test pulses, and their registered intensities as explained in the following expression (Gallinato et al., 2019):

$$I_{\mathcal{A}^{+}} = \int_{\mathcal{A}^{+}} \sigma(||\nabla V||)\partial_n V \, dx \tag{3.55}$$

The previous integral direct computation causes numerical errors because the intensity is over the needle. To prevent the generated error, $I_{\mathcal{A}^{+}}$ must be computed over the potential closed isosurface (S_{u_p}), where u_p is the chosen potential value. Therefore the intensity is expressed using Green's formula as follows (Gallinato et al., 2019):

$$I_{\mathcal{A}^+} = \int_{S_{u_p}} \sigma(||\nabla V||)\partial_n V \, dx \tag{3.56}$$

The surface S_{u_p} is approximated with the mesh cell faces. Hence, the flux over these faces is computed. Moreover, the conductivity harmonic mean and the electric field are calculated at the centers of the faces using the identified conductivity, and the values of the potential are calculated at the adjacent mesh center (Neal et al., 2012).

3.2 Heat transfer formulas

3.2.1 Heat transfer process in body tissues

The heat balance occurs when the gained heat rate equals the expenditure heat rate, otherwise, the extra heat changes the local internal temperature and heat transfer through the conductive tissues during the occurrence of a temperature difference. Hence there is a significant need for a simplified assumption, which describes the thermal in vivo status of the organism (Blazejczyk et al., 2012). The thermal ablation treatment technique is considered an application of heat transfer in biological tissues and fluid flow in blood vessels, which requires precise mathematical models. This advanced heat transfer model is a function of the properties of the system and depends on several factors (Orr and Eberhart, 1995), such as:

- The organism thermal inertia;
- The organism geometry;
- The physiological and thermophysical properties of several tissues, and their reliance on the location, time, and temperature;
- The heat production resulting from the radiative absorption and/or metabolic processes;
- The exchange of generated heat with the neighboring environment;
- The heat conduction due to the thermal gradients;
- The blood flow role concerned with the countercurrent exchange of the generated heat between the great blood vessels, and the thermal energy transport circulating through the blood;
- The thermoregulatory mechanisms, and their functions in the organism, including increased metabolism, shivering, panting, cutaneous shunting, sudomotor activity, sweating, pilomotor activity, and vasomotor activity.

Therefore the temperature occupies an important role within the interactions that occur inside tissues, such as coagulation, tissue necrosis, hyperthermia, and vaporization, with several specific temperature ranges as shown in Fig. 3.1 (Huang and Horng, 2015).

Consequently, several research groups all over the world submitted mathematical models to precisely represent the fluid flow and heat transfer in the biological processes through a finite, vascularized, heated tissue by assuming some simplifying assumptions as described in Penne's bioheat and Chen and Holmes models in the following subsections (Afrin et al., 2011).

3.2.2 Penne's bioheat equation

Penne's bioheat transfer equation (PBHTE) is considered the typical model used for forecasting the distribution of temperature inside tissues. It covers a special descriptive term for the heat exchange which occurs between tissues and blood flow. It is deduced by measuring the temperature of the arterial blood and tissues in the human forearm, and the equation is described as follows (Fang et al., 2018):

$$q_m + \nabla(K\nabla T) + J.E - w_b c_b \rho_b (T - T_b) = c\rho \frac{\partial T}{\partial t} \tag{3.57}$$

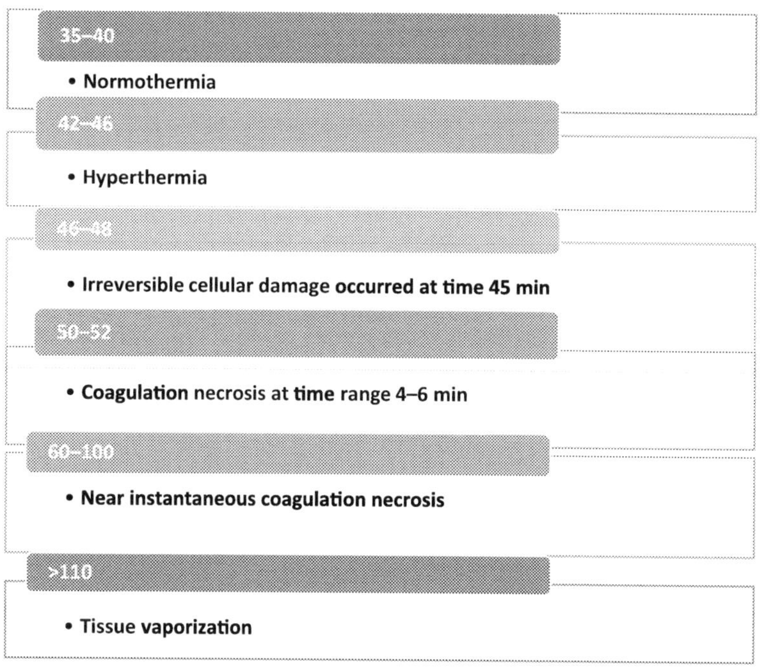

Figure 3.1 Relationship between the temperature ranges in degree Celsius and generated tissue interactions.

where q_m is the energy generated from the metabolism process (w/m^3), which can be neglected due to its smaller value concerning the externally applied energy density (Zorbas and Samaras, 2014), J is the current density (A/m^2), E is the electric field (v/m), w_b is the blood perfusion rate (1/s), c_b is the blood specific heat (J/kg K), ρ_b is the density of blood (kg/m^3), T is the time-dependent temperature (°C), T_b is the baseline temperature of arterial blood, that is constant with a standard value 37°C, c is the specific heat of tissue (J/kg K), t is the time, and K is the thermal conductivity of tissue (w/m K), which may be constant for a simple approximation of a dependent temperature as described in the following equation (Singh and Repaka, 2018b):

$$K(T) = K_b + (13*10^{-4})(T - T_b) \tag{3.58}$$

where K_b is the constant tissue thermal conductivity at the baseline temperature. Moreover, the blood perfusion rate is time and damaged degree dependent, as observed in the following expression (Singh and Repaka, 2018a):

$$w_b = \left\{ \begin{array}{ll} w_{bo} & at\ \Omega(t) \leq 1 \\ w_{bo}(-260\Omega(t)^2 + 25\Omega(t) + 1) & at\ 0 < \Omega(t) \leq 0.1 \\ w_{bo}\exp^{-\Omega(t)} & \Omega(t) > 0.1 \end{array} \right\} \tag{3.59}$$

where w_{bo} is the constant blood perfusion rate at the baseline temperature, and $\Omega(t)$ is the thermal induced damaged degree. The blood acts as a heat sink in the hyperthermia treatment procedure because it moves the energy away from the operating region (Kai et al., 2013). Nevertheless, PBHTE has been used in several applications as hyperthermia models, which is simple and has a best ability to predict the temperature parameter in these applications. It has considerable drawbacks, including not having a mechanism for the heat transfer, and the effect of the blood flow directionality. The Chen and Holmes model is explained in the next subsection (Huang and Horng, 2015).

3.2.3 The Chen and Holmes equations

Several researchers have developed an alternative heat transfer formulation to calculate the temperature in all points inside tissues. Chen and Holmes (CH) concluded one formula that included a strong physiological and physical basis as described in the following equation (Nakayama and Kuwahara, 2008):

$$q_m + \nabla(K\nabla T) + \nabla(K_p\nabla T) + J.E - uc_b\rho_b\nabla T - w_b c_b \rho_b (T - T_b) = c\rho\frac{\partial T}{\partial t} \quad (3.60)$$

where $uc_b\rho_b\nabla T$ is the term concerned with the convective heat transfer, which expresses the thermal interaction generated between the tissue and blood vessels, $\nabla(K_p\nabla T)$ is the term concerned with the conductive heat transfer of enhanced tissues due to blood perfusion in tissues, and K_p is the perfusion conductivity, depending on the blood perfusion rate. Hence, the CH model contains a stronger physical basis than Penne's model. As well, it requires details, and knowledge for the flow pattern, and vascular anatomy for solving it, which increase the model accuracy (Huang and Horng, 2015).

3.2.4 The Weinbaum and Jiji equation

Furthermore, Weinbaum and Jiji (WJ) suggested a different mathematical model for BHTE (Nakayama et al., 2010). This model depends on the impacts observed from the vascular network, where the blood vessels, which are responsible for the heat transfer inside the biological tissues, occur in countercurrent pairs (Nakayama et al., 2009). The main heat transfer between the blood and tissues is due to an imperfect countercurrent heat exchanger, which exist between the thermally significant veins and arteries. Their model uses a tensor notation, as described in the following equation (Huang and Horng, 2015):

$$c\rho\frac{\partial T}{\partial t} - \frac{\partial}{\partial x_i}\left(K_{ij_{eff}}\frac{\partial T}{\partial x_j}\right) = Q_{m'} - \frac{{K_b}^2 a^2 \pi^2 n}{4k\mathfrak{z}} L_j {N_{pe}}^2 \frac{\partial L_i}{\partial K_j}\frac{\partial T}{\partial x_j} \quad (3.61)$$

where K_b is the thermal conductivity of the blood, a is the radius of the local blood vessel, n is the blood vessel number density, $\mathfrak{z}$ is the shape factor related to the thermal conduction resistance existing between the neighboring countercurrent blood vessels, L_i is the cosine direction for the ith pair related to the countercurrent vessels. Also, N_{pe} is the localized Peclet number, which is expressed using the following expression:

$$N_{pe} = \frac{2acnu\rho_b}{k_b} \quad (3.62)$$

where u is the blood velocity inside the vessels, and $K_{ij_{eff}}$ is the tensor element of effective conductivity, which is expressed as follows (Huang and Horng, 2015):

$$K_{ij_{eff}} = K\left[¥_{ij} + \frac{{K_b}^2 a^2 \pi^2 n}{4k^2 \mathfrak{z}} L_i L_j {N_{pe}}^2 \right] \tag{3.63}$$

where $¥_{ij}$ is the known Kronecker delta function. Accordingly, the WJ equation is considered one of the most suitable additions to the bioheat transfer equation. Nonetheless, in some applied cases, the WJ equation requires comprehensive information about the velocities of blood flow inside the countercurrent blood vessels, their orientations, and sizes (Singh et al., 2015a). Moreover, there are multiple issues with the WJ model, including the following (Mahmood et al., 2005):

- Comparisons for macroscopic experiments and predicted temperatures are required.
- The model formulation is advanced for normal superficial tissues, in which the heat is countercurrently transferred.

Furthermore, the vascular anatomy differs from the normal superficial tissues, hence another model can be applied for tumors. Wissler investigated two main assumptions related to the WJ model, including the following (O'Neill et al., 2009):

- The arithmetic mean of the blood temperature for the venue and arteriole is approximated by the mean temperature of the biological tissue.
- The heat transfer between the surrounding tissue and the thermally considerable venule–arteriole pairs can be neglected.

3.2.5 The Weinbaum, Jiji, and Lemons model

This WJL model designates the blood flow effect through the heat transfer operation inside small blood vessels. Moreover, Keller and Seiler employed the specified effective conductivity of the nonisothermal region using several blood flow conditions (Nakayama et al., 2011). The WJL model resembles the Keller and Seiler approach using the same three mathematical equations. However, it relies on completely different vascular generations (Brinck and Werner, 1995).

3.2.6 Baish et al. model

Based on Baish et al., one of the main suppositions for deducing the WJ model was done to the following reasons (Huang and Horng, 2015):

- The vessel's proximity to the countercurrent pair;

- The conducted heat in the arterial wall extends to the venous wall, where the tissue temperature between the blood vessels is unaffected.

However, the previous hypothesis was criticized, where some of the heat that leaves the small arteriole wall remains inside the tissue. The model assumed that the heat transfer generated between the countercurrent blood vessels is largely based on two main factors, including (Yuan, 2009):

- The variance between the arterial temperature (T_a) and the venous temperature (T_v), which is as considered ($T_a - T_v$);
- The variance between the biological tissue temperature (T_{tissue}) and the average temperature of the blood ($\frac{T_a + T_v}{2}$).

Furthermore, many studies have been proposed to estimate the validity of these approaches, where some approximations and arguments are considered to examine the effect of the blood flow on the biothermal modeling. Wissler assessed the WJ model due to the hypothesis taken for the blood temperature within the venous and arterial vessels, and the temperature of the nearby biological tissues. Wissler suggested another model, which describes the heat exchange between the tissue and vessels and added a perfusion term rather than that existing in the WJL model. This model refused the WJL equation hypothesis to derive the thermal conductivity tensor (Cognata and Durrant, 2014). Moreover, Charny et al. modified the WJL model concerning blood vessels by replacing the governed equation with the energy conservation equation of the biological tissue. According to their analysis of the transient and steady-state temperature fields within a limb under the normothermic and hyperthermic conditions, the temperature profiles of the biological tissues are considered with models similar to those described by Penne's model through the organ regions with the great vessels having diameters greater than 0.4 mm (Okajima et al., 2009).

3.2.7 O'Neill et al. model

Furthermore, the cell's condition is affected by the generated heat, and can be expressed using a cell death model of the temperature gradient using the ordinary differential equations (ODEs), where the cells within the target region are divided into killed, denatured, or destroyed (Pearce, 2013). Additionally, a vulnerable state, which exists between the dead and living states, is included due to the direct observation of the cells responding to the thermal treatments. Hence, the process is described using the following equation (Tehrani et al., 2020):

$$S_A \underset{\leftarrow T_b}{\overset{\rightarrow T_f}{}} S_V \overset{T_f}{\rightarrow} S_D \tag{3.64}$$

where T_f is the forward rate constant describing a transition operation from the living to the vulnerable cells, and is considered as the initial damage that caused the injured state, T_b is the backward rate constant, that defines the self-healing process in which the vulnerable state is transitioned into the alive state with full function, and S_D, S_A, and S_V are the fractions of the dead, alive, and vulnerable cells, respectively, that are described using the following equations (Hall et al., 2015):

$$\frac{dS_A}{dt} = -T_f S_A + T_b(1 - S_A - S_D) \tag{3.65}$$

$$\frac{dS_D}{dt} = T_f(1 - S_A - S_D) \tag{3.66}$$

$$T_f = \overline{T_f} e^{T/P_k}(1 - S_A) \tag{3.67}$$

where P_k is a parameter setting concerning the exponential rate increase with temperature, and $\overline{T_f}$ is the scaling factor $(1/s)$. This model tolerates data fitting along with a greater thermotherapeutic temperature range, containing the initial stage in the case of the cells' exposure to the heat for the first time. In addition, $(S_V + S_D)$, which is known as the cell viability, is used to compute the lesion size. The model was first applied to investigate the MWA to obtain the shape and size of lesions, and to observe their side effects (O'Neill et al., 2011).

3.3 Tissue contraction during thermal ablation

The tissue contraction and shrinkage, which can be described with a globular protein model, occur due to the tissue exposure to the elevated temperatures, collagen contraction, and protein denaturation (Amabile et al., 2017). It is considered a problem related to specifying the ablated region during the obtaining of postoperative images. Generally, the transitions that occur between the protein states are attached to a change in the molecule length, which causes a change in the tissue length (Brace et al., 2010). Therefore a globular protein model with three states can be used to describe the contraction tissues due to the elevated temperature exposure (Iancu et al., 2011). Moreover, the protein thermal denaturation occur to the biological tissue in the case of the unfolding of the protein molecules of the helical structure to the coiled structure (Qin et al., 2014). The operation can be irreversible or reversible due to the refolding and unfolding that occur within the different intermediate transition states (Erxleben et al., 2020). Several experiments have been performed for computing the tissue shrinkage, where each one is categorized into two protocols, including the controlled heating rate and the temperature jump (Benjwal et al., 2006). A specimen is quickly heated at a constant rate to a target threshold temperature and is kept at this target temperature through the testing, hence time and temperature are kept as independent variables (Park et al., 2016). This process is distinguished by a minimum required number of specimens (Park et al., 2016).

Consequently, the mathematical framework of the tissue contraction behavior due to thermal ablation is realized using a model with three states containing the reaction kinetics with first order to describe the molecule transition between several states (Lopresto et al., 2018). The values of the used parameters to predict the tissue contraction are exposed to the elevated temperature and the respective sensitivities. The transition occurs between the different states, which change the length of the constituent molecules, causing residual stress within the tissue and presenting as a gross tissue contraction on the macroscale (Chae et al., 2001). Moreover, the model with three states is preferred rather than that with two states because the two-state model does not have the ability of partial recovery capture for the tissue shrinkage (Singh and Melnik, 2020). The three-step model for the treatment plan has mathematical models to improve the accuracy, and correct the thermal ablation predictions for the shrinkage (Payne et al., 2011).

The mathematical model gathers several complex effects to obtain an easily implementable and simple model. It considers the globular protein response to heating, and contains three main intermediate states, including denatured (D), unfolded (U), and native (N), which are gathered into an overall state as described in the following reaction equation (Wright and Humphrey, 2002):

$$N \underset{\leftarrow K_{r_3}}{\overset{\rightarrow K_{r_1}}{}} U \overset{K_{r_2}}{\rightarrow} D \tag{3.68}$$

where R_i is the reaction rates that are controlled using the following Arrhenius equation (Park et al., 2016):

$$K_r(T) = Ae^{E_a/RT} \tag{3.69}$$

where A is the frequency factor $(1/s), E_a$ is the activation energy (J/mol), that is required for transforming the normal tissue state to the damaged state and different according to the tissue type, and R is the universal gas constant (J/mol K). The described equations of the system are expressed as follows (Singh and Melnik, 2020):

$$\frac{dN}{dt} = -K_{r_1}N + K_{r_3}U \tag{3.70}$$

$$\frac{dU}{dt} = K_{r_1}N - K_{r_2}U - K_{r_3}U \tag{3.71}$$

$$\frac{dD}{dt} = K_{r_2}U \tag{3.72}$$

where $(D + U + N = 1)$. To specify the length change of the tissue, a model, which is the linear sum for the different protein states, is described using the following equation (Park et al., 2016):

$$L_{tot} = D\,L_D + U\,L_U + N\,L_N \tag{3.73}$$

where L_{tot} is the total tissue length and L_i is the total tissue length for the protein state i. The protein state variables $i \in \{D, U, N\}$ should be interpreted with the protein proportion within each state as explained in the following expression (Park et al., 2016):

$$i = \frac{N_x}{N_o} \tag{3.74}$$

where N_o is the number of total proteins, and N_x is the protein number within a state i, that is, to include all the proteins within the unfolded state. The total protein length is obtained as explained in the following equation (Dueck et al., 2011):

$$L_{tot} = U\,L_U = \frac{N_o}{N_o}L_U = L_U \tag{3.75}$$

The relative shrinkage (ℶ), which describes the experimental obtained data, is a function of the time resulting from the isothermal conditions as expressed in the following equation (Chen et al., 1997):

$$\beth = \frac{L_o - L_{tot}}{L_o} = \frac{L_o - D\,L_D - U\,L_U - N\,L_N}{L_o} \tag{3.76}$$

In the beginning, all the proteins are supposed to be within the native state, hence the relative shrinkage is simplified as follows (Park et al., 2016):

$$\daleth = 1 - N - D\frac{L_D}{L_N} - U\frac{L_U}{L_N} \tag{3.77}$$

The proportion of proteins and their length ratios L_D/L_N and L_U/L_N affect the length of the tissue and its relative shrinkage. This model introduces more parameters than are generated from the mass kinetics model because the protein state observations are related and indirect instead to L_{tot}. Additionally, the parameters set belong to $\{A_1, E_{a_1}, A_2, E_{a_2}, A_3, E_{a_3}, L_D/L_N, \text{and } L_U/L_N\}$, and are experimentally determined (Park et al., 2016). The nonlinear fitting algorithm is used and adjusted to have values greater than zero. Additionally, the Monte Carlo approach can be used to fit all target parameters (Oyeyemi et al., 2015). During the curve-fitting operation, 200 of the initial random parameter values are taken to obtain the values with a good curve fit. Therefore the model sensitivity related to each parameter included within the reaction rates is determined using the sensitivity analysis, in which a greater value range of the parameters is considered. In addition, it can be determined using the Morris technique, which is considered a one-at-a-time global sensitivity analysis technique (Singh et al., 2019).

3.4 First-order Arrhenius rate equation

Tissues are damaged due to heating that results from temperature elevation, which occurs at a specific time according to the tissue type. Several studies have been performed to quantify the tissue damage resulting from temperature elevation, where Moritz, and Henriques were the first to provide kinetic models for the tissue damage by temperature increase (Prakash, 2010). Hereafter, the induced thermal damage $\Omega(t)$ is calculated using the first-order Arrhenius rate equation as follows (Singh et al., 2015b):

$$\Omega(t) = \int_0^t A\exp^{\left(-\frac{E_a}{R\,T(t)}\right)} dt = \ln\left(\frac{C_o}{C(t)}\right) \tag{3.78}$$

where C_o is the original undamaged cell concentration before the heating, and $C(t)$ is the remaining undamaged cell concentration after the heating. A thermal damage value of 1 corresponds to a cell death probability of 63%, and 4.6 corresponds to a cell death probability of 99% (Chang and Nguyen, 2004). $\Omega(t) = 1$ at each point inside the tumor is considered enough for tumor ablation (Singh and Repaka, 2015). To apply these equations and obtain the results, the finite element method (FEM) should be built as described in the following subsection.

3.5 Cooling effect of large blood vessels

Tissues of the organs inside the body may have neighboring blood vessels of different sizes, locations, and shapes. For example, the liver is a perfused complex organ, which circulates blood between the portal vein and hepatic artery (Kai et al., 2013), where the liver and portal veins are of various sizes through the liver. It was deduced that during thermal ablation, great blood vessels cause considerable cooling through the tissue, hence they limit the coagulation region extent (Nakai et al., 2007). Accordingly, to contemplate the effect of the discrete blood vessels adjacent to the target tumor, blood vessels of variable sizes and located variable distances from the tumor have been studied (Solovchuk et al., 2012). This procedure is accomplished using a donut-shaped blood vessel in three dimensions. Therefore the generated heat sink can be expressed with boundary conditional parameterization of the convective heat flux, as described in the following equation (Deng and Liu, 2006):

$$h = \frac{k_b N_{UD}}{D} \tag{3.79}$$

where h is the coefficient of heat transfer between the blood and tissue (watt/m$^{2\circ}C$), K_b is the thermal conductivity of the blood (watt/m $^\circ C$), D is the diameter of the blood vessel (m), and N_{UD} is the Nusselt number, which depends on the flow rate and blood vessel length, and is calculated as described in the following equations according to the surface (Becker and Kuznetsov, 2007):

$$N_{UD} = \begin{array}{ll} 0.825 + \dfrac{0.387\ Ra^{1/6}}{(1+(0.492/Pr)^{9/16})^{8/27}} & \text{for vertical surfaces} \\ 0.54\ Ra^{1/4} & \text{for horizontal surfaces cooling} \\ 0.27\ Ra^{1/4} & \text{for horizontal surfaces heating} \end{array} \tag{3.80}$$

where Pr is the common Prandtl number and Ra is the Rayleigh number, which are expressed using the following expression (Aminossadati and Ghasemi, 2009):

$$Ra = \frac{\text{₵} g {L_c}^3 (T_s - T_\infty)}{\chi_\upsilon \upsilon_K} \tag{3.81}$$

where ₵ is the volumetric thermal expansion coefficient (K^{-1}), g is the gravity, L_c is the characteristic length scale (m), T_s is the surface temperature (K), T_∞ is the surrounding air temperature (K), $\mathfrak{X}_v$ is the thermal diffusivity (m^2/s), and υ_K is the kinematic viscosity (m^2/s). Also, υ_K, $\mathfrak{X}_v$, ₵, and k_b are calculated at the temperature of the air film (T_f), that is given by (Becker and Kuznetsov, 2007):

$$T_f = \frac{T_\infty + T_s}{2} \tag{3.82}$$

The agreement existing between the cooled blood vessel and the simulated baseline (B_s) is summarized using the Dice coefficient (Q_{DC}) as follows (Deshazer, 2016):

$$Q_{DC} = \frac{2|B_{S_{BV}} \cap B_s|}{|B_{S_{BV}}| + |B_s|} \tag{3.83}$$

where $B_{S_{BV}}$ is the boundary volume of the vessel-cooled treatment effect regions. Furthermore, the average temperature in the blood vessel is an incessant function $T_{av}(S_A, t)$ of the time t, and the axial position S_A. Accordingly, the heat transfer equation, that contains the large vessel parameters, such as the thermal exchange with target surrounding tissue, axial diffusion, and convection, is expressed using the one-dimensional following equation (Chen and Saidel, 2009):

$$\frac{\partial T_{av}}{\partial t} = V_{av}\frac{\partial T_{av}}{\partial S_A} + \mathfrak{X}_v \frac{\partial^2 T_{av}}{\partial {S_A}^2} + h(T - T_{av}) \tag{3.84}$$

where V_{av} is the average velocity (m/s).

3.6 Finite element method

The FEM is the most commonly used technique to solve the problems related to mathematical models and engineering, such as fluid flow, heat transfer, electromagnetic potential, mass transport, and structural analysis (Huebner et al., 2001). It is considered a numerical technique to solve the partial differential equations (PDEs) with two or three variables, such as the Poisson equation in the elliptic shape, the wave equation in the hyperbolic shape, and the Fourier law in the parabolic shape (Liu et al., 2014). The FEM divides the great geometric model, whether gas, liquid, or solid, into simpler, smaller, and interconnected parts with node-specified joints, which are called the finite elements (FEs) to obtain results with minimum acceptable modifications and/or distortion of their main shape, where the accuracy of the results increases, the obtained computational error decreases, and the processing time increases with increasing number of finite elements. For example, a parasol system, which is built to compute the temperature of beverage cans in a cooler, that is exposed to the sun as an external radiative source all the time, is explained in Fig. 3.2.

It is difficult to obtain the exact system response, like temperature, under the effect of solar radiation. Hence, this parasol structure can be approximated to multiple smaller subregions using the FEM technique shown in Fig. 3.3.

An approximate appropriate solution is estimated in each element, and the overall structure response is derived, where the FEM is built using the following sequential steps (Rao, 2017):

- Discretization, wherein the geometric structure is separated into discrete small elements, or subdivisions, where these elements are decided according to their arrangement, size, type, and number.

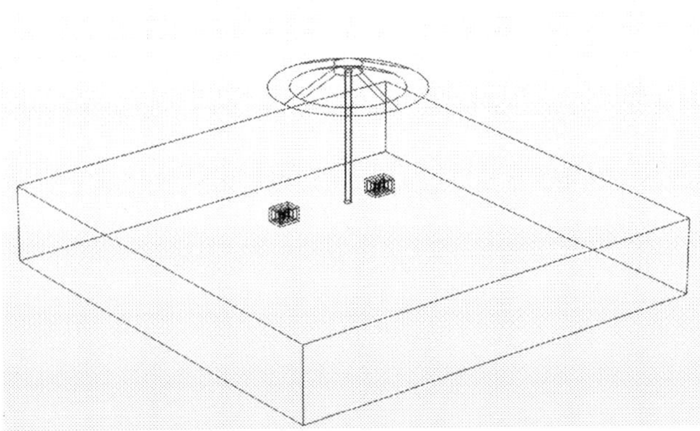

Figure 3.2 Parasol system for two Styrofoam coolers.

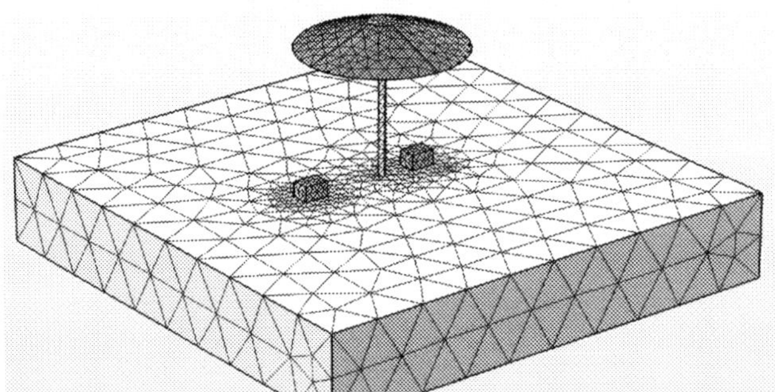

Figure 3.3 Mesh structure for the parasol system.

- The displacement model or interpolation must be properly selected, simple from the computational standpoint, and achieve the determined convergence requirements.
- The load vectors and stiffness metrics of the elements are derived using suitable equilibrium conditions, a weighted residual approach, or variational principle.
- The elements equations are assembled to obtain the equilibrium equations. Accordingly, the geometric structure, which contains multiple finite elements, individual load vector, and stiffness metrices is assembled in an appropriate method to generate the overall equilibrium responding equations as follows:

$$\underset{\sim}{\vec{P}} = [\underset{\sim}{K}]\underset{\sim}{\vec{\text{Ø}}} \tag{3.85}$$

where $\underset{\sim}{\vec{P}}$ is the nodal forces vector for the complete geometric structure, $\underset{\sim}{K}$ is the stiffness matrix, and $\underset{\sim}{\vec{\text{Ø}}}$ is the nodal displacements vector.

- The nodal displacements are solved where the overall equilibrium obtained equations are modified with the problem boundary conditions as follows:

$$\vec{P} = [K]\vec{\text{Ø}} \tag{3.86}$$

Two solvers are available, including the linear problems solver, in which the vector $\vec{\text{Ø}}$ is easily solved, and the nonlinear problems solver, in which the vector $\vec{\text{Ø}}$ is solved using sequential steps, and each step includes a modification of the load vector $\vec{P}$ or/and the stiffness matrix $[K]$.

- Finally, the stresses and strains of the elements are computed using the equations concerning structural and solid mechanics.

Any engineering problem solution uses the FE software and involves three main steps as follows (Papazafeiropoulos et al., 2017):

- The preprocessing, in which the geometry, the materials' characteristics, the boundary conditions, and load are provided as input data, the in-built modules of automatic mesh generation involve the element mesh using a lower input obtained from the analyst with the mesh density, and element types and the analyst display the data, and the FE mesh produced for verification, and visual inspection for correctness.
- Numerical analysis, in which the software can automatically create the characteristics metrics, and vectors of the elements, gather them to produce the system equations, execute the determined boundary conditions, solve the generated equations to find the value of the nodal displacement, and calculate the resultants of the elements, such as the strains and stresses.

- The postprocessing, in which the problem solution and element stresses are displayed with two methods, including graphically in two- or three-dimensional plotting windows of the stress variation, or the deformed shape, and numerically in a tabular shape according to the wishes of the analyst.

These previous steps can be executed as a graphical user interface using ANSYS and COMSOL Multiphysics software (Madenci and Guven, 2015; Pepper and Heinrich, 2017), or writing a script file using the MATLAB® software (Kwon and Bang, 2018). Hence, the obtained approximations and computational time amounts differ according to the used discretization method. An efficient FEM requires some knowledge and experience of the simple guidelines. Nevertheless, for the best systems, including complex geometric models, the manual FEM discretization needs a lot of time and effort from the analyst. Furthermore, some programs of the automatic mesh generation are advanced for the effective discretization of complex problems and require minimum interface with the existing analyst (Rao, 2017).

Generally, the finite element shape is one of the most important parameters, which must be carefully chosen to properly simulate the original domain without the need for a great computational effort for the solution. It is chosen based on the domain geometry, and the number of the independent coordinate representing the system (Adjerid et al., 1993). The available one-dimensional elements are described in Fig. 3.4.

Fig. 3.4 displays the one-dimensional mesh elements that contain only two nodes that exist at the end of each element. For two spatial coordinates representing the geometry, field variables, and materials properties, a two-dimensional mesh is used (Beckett et al., 2001). Henceforth, the available two-dimensional elements are described as shown in Fig. 3.5.

Fig. 3.5 displays the two-dimensional mesh elements, which contain four nodes that exist at the corner of each element. For three spatial coordinates representing the geometry, field variables, and materials properties, a three-dimensional mesh is used (Gain et al., 2014). Thus, the available three-dimensional elements are described as shown in Fig. 3.6.

Fig. 3.6 displays the three-dimensional mesh elements, which are specified according to their shape, where the tetrahedron element contains four faces, six edges, and four vertices, the pyramid element contains five faces, eight edges, and five vertices, each element of the rectangular prism and hexahedron contains six faces, twelve edges, and eight vertices, and the triangular prism element contains five faces, nine edges, and six vertices (Devloo et al., 2019).

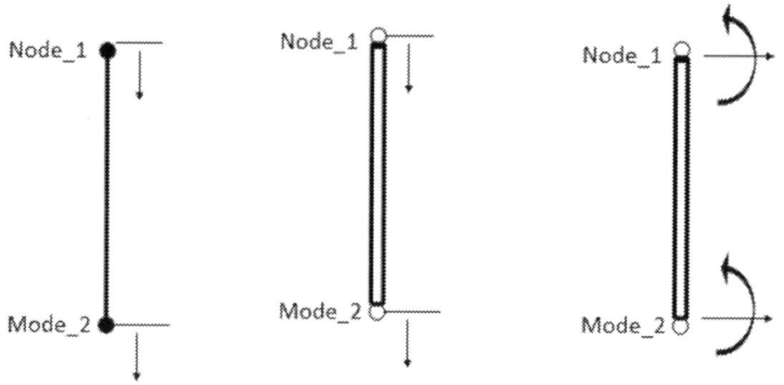

Figure 3.4 One-dimensional mesh elements.

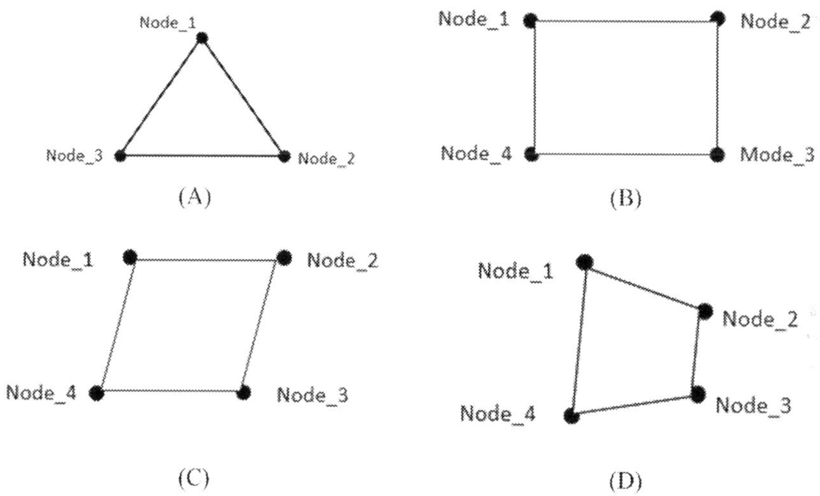

Figure 3.5 Two-dimensional mesh elements, including (A) triangle, (B) rectangle, (C) parallelogram, and (D) quadrilateral.

Moreover, the element size is the second most important parameter for building the geometric mesh, as it directly affects the solution convergence. However, the smaller element size achieves more accurate results, and consumes a large computational time. Hence, a tradeoff is taken into account to generate a minimum acceptable error and small processing time (Shi et al., 2008). In addition, some geometric cases are divided into finite elements with different sizes, including a finer size in the operating target region, and a coarse size in the other secondary regions as shown in the inductive heating model in Fig. 3.7.

The node location, which is determined according to the percentage continuity existing in the model, mainly affects the

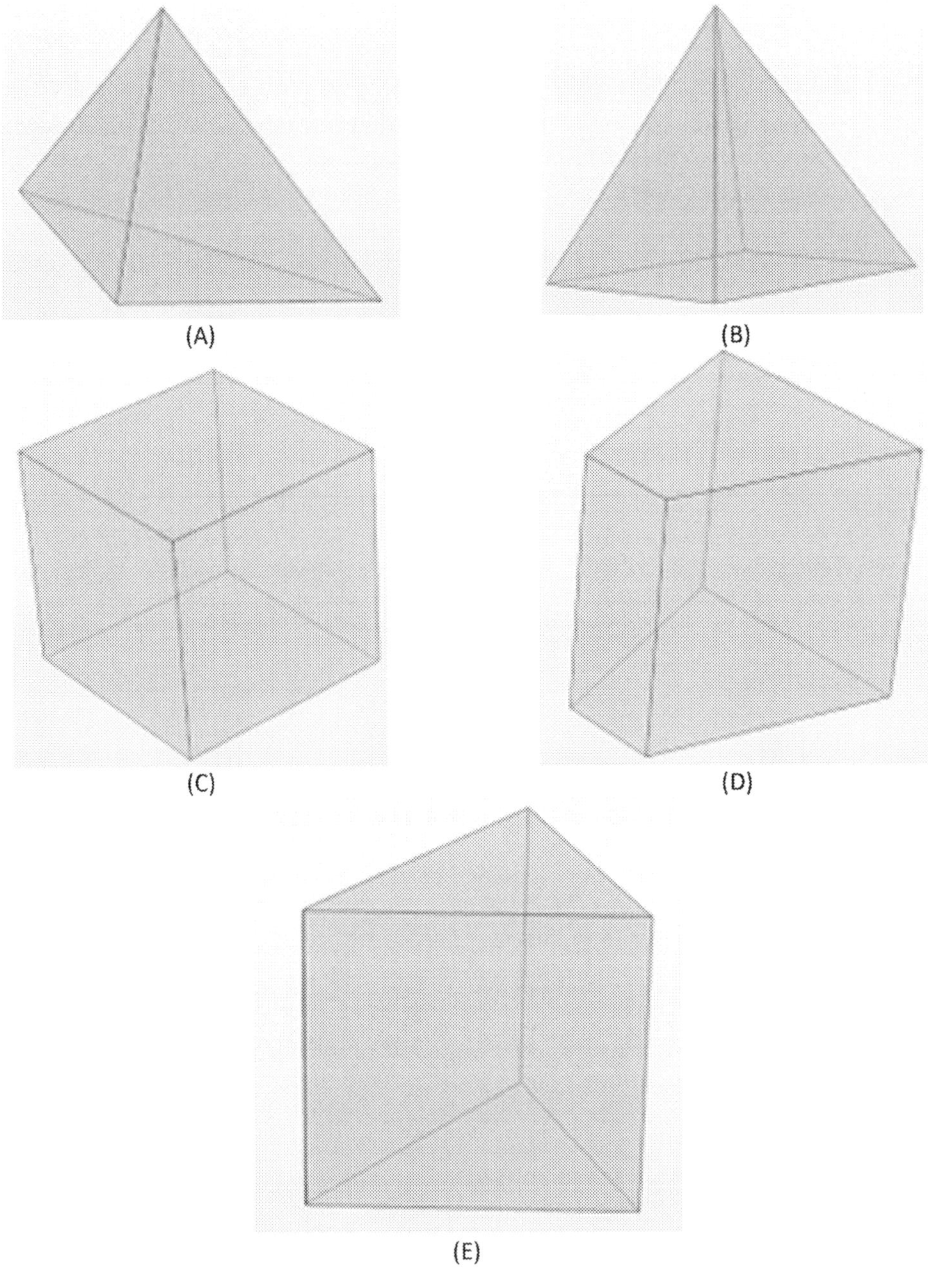

Figure 3.6 Three-dimensional mesh elements, including (A) tetrahedron, (B) pyramid, (C) rectangular prism, (D) hexahedron, and (E) triangular prism.

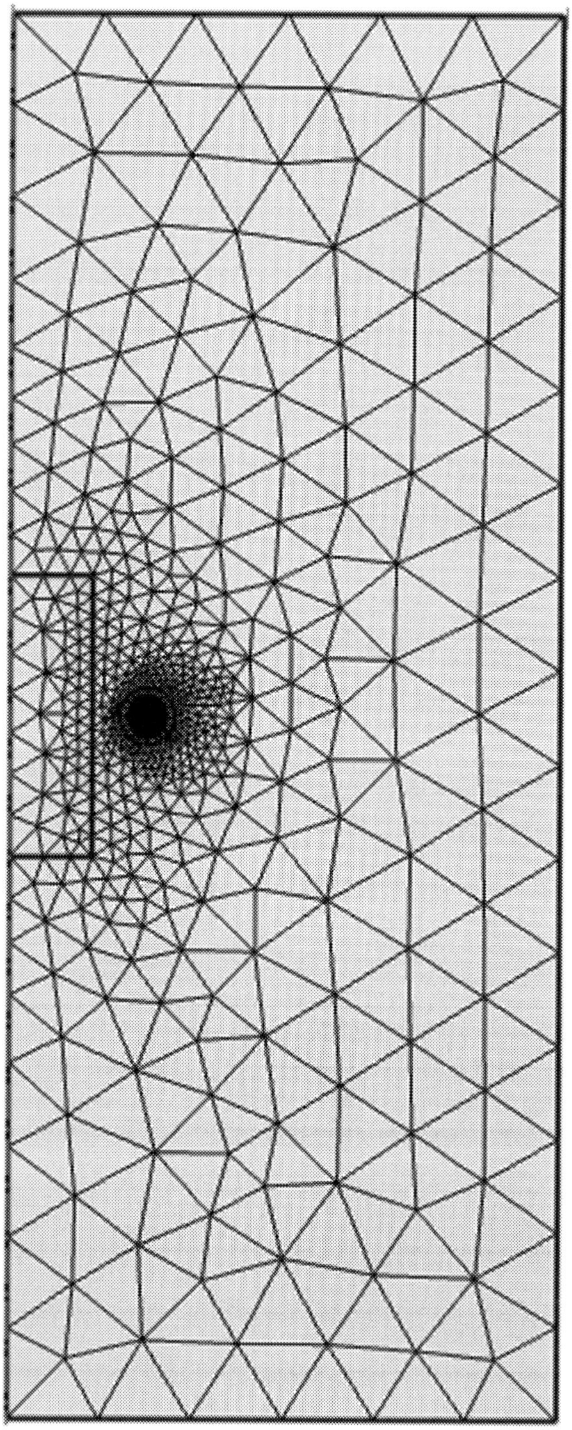

Figure 3.7 Geometric mesh for the inductive heating model with different size elements.

solution. The model that does not has abrupt changes in its external conditions, material properties, and/or geometry; is simply divided into similar subdivisions with uniform spacing between the nodes. In contrast, nodes of the model that contains discontinuities (Engel et al., 2002) are set at the existing discontinuities, as presented in Fig. 3.8.

The element number, which is related to the size of the element, the involved degree of freedom (DOF) number, and the desired accuracy, are precisely determined to be available for storing the computer memory (Kim et al., 2007). Furthermore, the infinite and semi-infinite systems, such as the foundations, dams, and soil, whose boundaries are difficult to determine, can be represented by taking a unit slice for the idealization and analysis for the strain computation. However, the mesh of the simple domain is manually generated, and most of the practical complex systems, such as the construction, automobile, and aerospace industries, require thousands and millions of finite elements for mesh generation. These complex systems of mesh generation are automatically accomplished using the solid modeling package or computer-aided design (CAD) systems (Neves et al., 2011). Many schemes of automatic mesh generation use the bottom-up technique to produce the sequential domain nodes, including the corners, boundaries (curves), surfaces, and solids, where the problem geometric region is built using some determined steps as follows (Engvall and Evans, 2016):

- In the beginning, the nodes are set on the corner points inside the domain.
- Then, the nodes are spread into the geometric curves representing the boundaries.
- This is followed by developing the surface nodes using the boundary nodes.
- Finally, the nodes through a given domain are developed using the nodes on several surfaces.

The allowable minimum sizes of the elements in the different domain regions, and several schemes of mesh generation, create all the possible nodes that are determined. Then, the elements of the triangle are generated by connecting these nodes in the plane region. In some cases, the Delaunay triangulation scheme is done, in which the triangular mesh elements are produced through the sum maximizing the triangles' smallest angle, and hence thin element generation can be avoided (Lee and Schachter, 1980). The octree and tessellation method is the most commonly used technique for the generation of automatic mesh, in which node points set, and a spot starting specified node is provided to create the first simple element from the neighboring

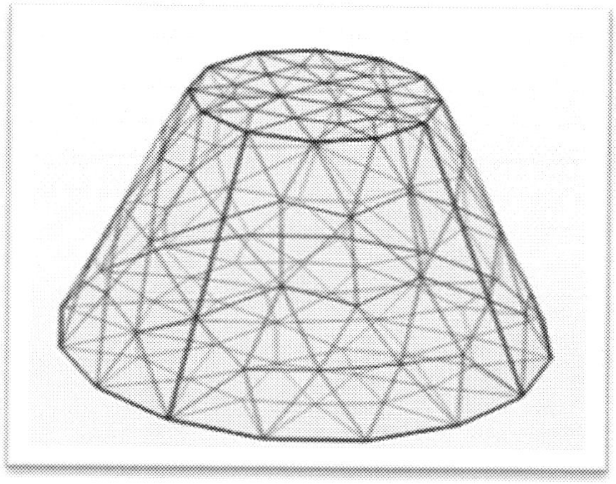

(A)

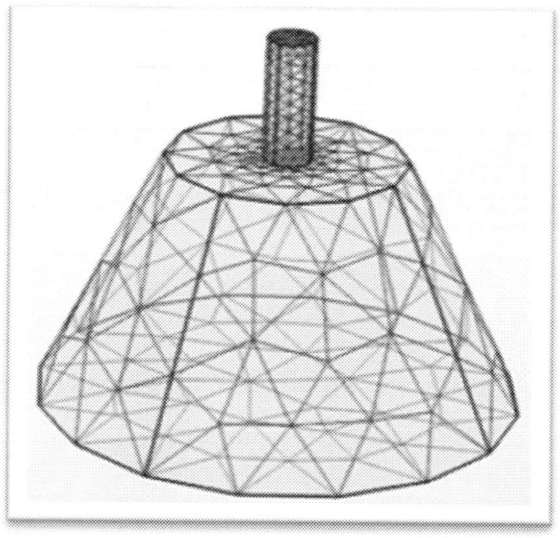

(B)

Figure 3.8 Geometric mesh for two models, including (A) continuous model, and (B) discontinuous model with an abrupt change in the geometry.

nodes. Accordingly, the neighboring or subsequent element is produced by specifying the node points, which achieves minimum distortion of the element shape (Fritzen et al., 2009). The procedure is repeated for the generation of all the elements using a step-by-step operation as shown in Fig. 3.9.

The octree technique, which is called the tree structure method, is used in the computer graphics display and solid

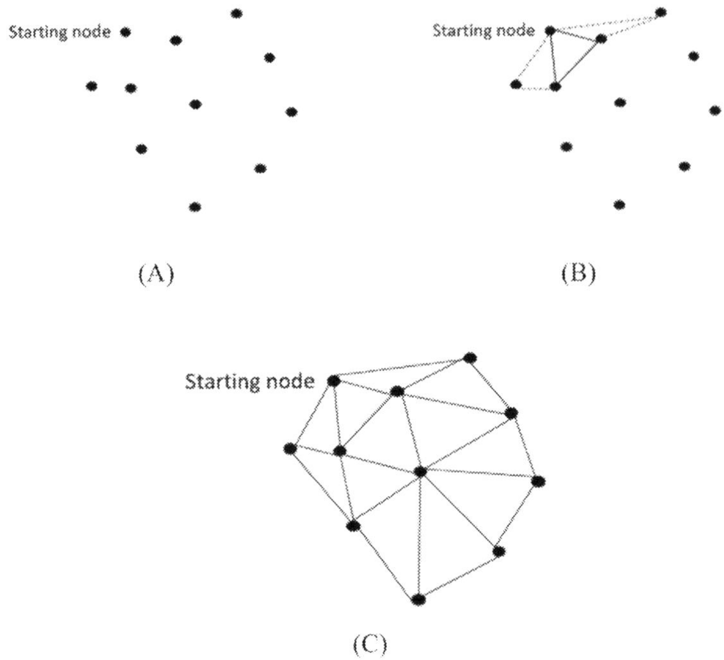

Figure 3.9 Tessellation method for mesh generation, including (A) nodes inside a region, (B) simple elements generation, and (C) complete elements collection.

modeling as it encloses the object in a 3D cube, with the cube divided into four similar quadrants in the case of object disability to uniformly cover the 3D cube. Likewise, in the 2D octree technique (quadtree method), first, the object is enclosed inside a determined square region. Hence, the square domain is divided into four similar quadrants in the case of object disability of completely covering the square, if the generated quadrant is empty, and cannot cover the object, or full, and completely covers the object. Then, the subdivision procedure is stopped, and divided into four similar subquadrants in the case of partial object occupation (Huo et al., 2020). The subdivision procedure is repeated until a predetermined resolution level or all the generated regions are achieved.

For thermal ablation, the FEM can mode complex geometries that are insolvable by finite difference models or analytical methods. Several researchers have controlled the power used during the ablation process manually using trial-and-error to preserve the constant temperature of the probe tip. Other studies have designed programmable controllers to monitor the tip's temperature after each time instant and accordingly control the used voltage keeping the temperature constant. Hence, closed-loop systems entailing an FE model, and controlling software can be designed (Haemmerich and Webster, 2005) to reduce the processing time. Using the FEM for modeling temperature-based ablation to destroy tissue by thermal energy produced from converting the electric energy to thermal temperature allows the temperature circulation within the targeted tissues. Tissue heating during the ablation process is enabled by the bioheat equation with consideration of the tissue properties. Fig. 3.10 illustrates the main stages of FE modeling during the ablation process.

The three key stages to FE modeling generation include preprocessing to generate the model geometry with assignment of the fitting material properties of the applicator's different parts and the tissue characteristics (electric conductivity, thermal conductivity, density), and the boundary/initial conditions. Typically, the boundary conditions include the temperature and applied energy (current/voltage) at the model's edges. Accordingly, the FE mesh model is generated using software, such as COMSOL Biophysics, MATLAB, or any commercial software. For transient analysis, after the mesh generation stage, the second stage starts

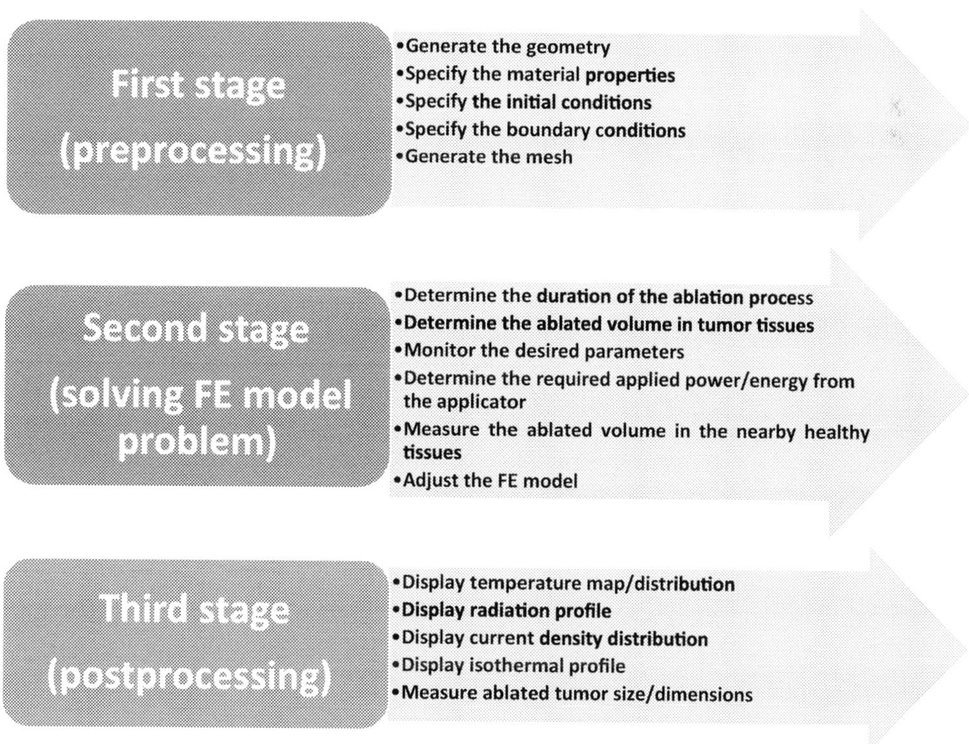

Figure 3.10 Typical stages of FE modeling of any ablation system.

to solve the specific FE modeling ablation problem under study to reveal the time duration of the ablation process, the required power for complete ablation, and to measure the temperature at the applicator's tip. Finally, in the third stage, the temperature distribution, radiation profile, current density distribution, and isothermal profile, are displayed along with measuring the finally ablated volume dimensions. In Chapters 6–8, a detailed explanation of the use of COMSOL software which applies the governing equations, bioheat transfer equations, and FE model generation of the mesh to explain the stages, case study for modeling an RFA system, and case study for modeling an MWA system, respectively, is given.

3.7 Conclusion

Numerical modeling to solve and analyze electro-thermal ablation problems has a noteworthy effect on a proper ablation procedure before clinical trials. For a mathematical description of the heat transfer, the tissue model is designed to predict the changes in temperature during the hypothermia or hyperthermia treatment methods. Penne's equation is considered the main bioheat transfer governing equation and includes the blood perfusion effect. It is a Fourier heat conduction equation integrated with source elements that represent the global heat flow effects between the blood and nearby tissues. The main concept of Penne's equation is that if the blood temperature is equal to the normal temperature of the blood in the aortic flow over the tissues at a specific speed, a thermal balance occurs, and afterward, the temperature collects in the veins. Thereafter, the blood perfusion rate becomes a key factor for calculating the bioheat transfer in several ablation systems and simulations. The mathematical explanation of the electric field of the electromagnetic waves in RFA and MWA, respectively, are used as a governing equation to mathematically represent the generated energy or wave from the applicator based on the used ablation procedure, as described in this chapter. By coupling the bioheat transfer equations and the generated wave governing equation, the FE model can be implemented as a result of interacting with the generated energy source in the bioheat transfer equation. This describes heat transfer in living tissues. In addition, determination of the temperature profile/distribution in the tissues is influenced by the type of tumor and the external thermal actions. These calculations are used to estimate the damage level in the tissues due to the increase or decrease in temperature. The destruction level in the biological tissues can be calculated using the first-order Arrhenius rate equation, where the perfusion coefficient is considered the main parameter to measure the tissue destruction. Several studies have derived formulations to describe the relationship between tissue destruction and perfusion, as well as between perfusion and temperature. In addition, this chapter explained the FEM and main stages of the model and solved an ablation-based problem by linking the generated wave equations, the bioheat transfer equations, and the damage degree rate with the mesh structure of the geometry under study, which also includes the material and tissue properties and characteristics.

References

Adjerid, S., Flaherty, J.E., Wang, Y.J., 1993. A posteriori error estimation with finite element methods of lines for one-dimensional parabolic systems. Numerische Mathematik 65, 1–21.

Afrin, N., Zhang, Y., Chen, J., 2011. Thermal lagging in living biological tissue based on nonequilibrium heat transfer between tissue, arterial and venous bloods. Int. J. Heat. Mass. Transf. 54, 2419–2426.

Ahmed, M., Goldberg, S.N., 2002. Thermal ablation therapy for hepatocellular carcinoma. J. Vasc. Interv. Radiol. 13, S231–S243.

Alba-Martínez, J., Trujillo, M., Blasco-Gimenez, R., Berjano, E., 2011. Could it be advantageous to tune the temperature controller during radiofrequency ablation? A feasibility study using theoretical models. Int. J. Hyperth. 27, 539–548.

Amabile, C., Farina, L., Lopresto, V., Pinto, R., Cassarino, S., Tosoratti, N., et al., 2017. Tissue shrinkage in microwave ablation of liver: an ex vivo predictive model. Int. J. Hyperth. 33, 101–109.

Aminossadati, S., Ghasemi, B., 2009. Natural convection cooling of a localised heat source at the bottom of a nanofluid-filled enclosure. Eur. J. Mech. B Fluids 28, 630–640.

Audigier, C., Mansi, T., Delingette, H., Rapaka, S., Mihalef, V., Carnegie, D., et al., 2015. Efficient lattice boltzmann solver for patient-specific radiofrequency ablation of hepatic tumors. IEEE Trans. Med. Imaging 34, 1576–1589.

Audigier, C., Mansi, T., Delingette, H., Rapaka, S., Passerini, T., Mihalef, V., et al., 2017. Comprehensive preclinical evaluation of a multi-physics model of liver tumor radiofrequency ablation. Int. J. Computer Assist. Radiol. Surg. 12, 1543–1559.

Becker, S., Kuznetsov, A., 2007. Thermal damage reduction associated with in vivo skin electroporation: a numerical investigation justifying aggressive pre-cooling. Int. J. Heat. Mass. Transf. 50, 105–116.

Beckett, G., Mackenzie, J.A., Robertson, M., 2001. A moving mesh finite element method for the solution of two-dimensional Stefan problems. J. Comput. Phys. 168, 500–518.

Benjwal, S., Verma, S., Röhm, K.H., Gursky, O., 2006. Monitoring protein aggregation during thermal unfolding in circular dichroism experiments. Protein Sci. 15, 635–639.

Berjano, E.J., 2006. Theoretical modeling for radiofrequency ablation: state-of-the-art and challenges for the future. Biomed. Eng. Online 5, 24.

Berjano, E.J., Hornero, F., 2004. Thermal-electrical modeling for epicardial atrial radiofrequency ablation. IEEE Trans. Biomed. Eng. 51, 1348–1357.
Berjano, E.J., Hornero, F., 2005. A cooled intraesophageal balloon to prevent thermal injury during endocardial surgical radiofrequency ablation of the left atrium: a finite element study. Phys. Med. Biol. 50, N269.
Berjano, E.J., Alió, J.L., Saiz, J., 2005. Modeling for radio-frequency conductive keratoplasty: implications for the maximum temperature reached in the cornea. Physiol. Meas. 26, 157.
Blazejczyk, K., Epstein, Y., Jendritzky, G., Staiger, H., Tinz, B., 2012. Comparison of UTCI to selected thermal indices. Int. J. Biometeorol. 56, 515–535.
Borisov, O.V., Mao, X., Russo, R.E., 2000. Effects of crater development on fractionation and signal intensity during laser ablation inductively coupled plasma mass spectrometry. Spectrochim. Acta Part B: At. Spectrosc. 55, 1693–1704.
Brace, C.L., 2008. Temperature-dependent dielectric properties of liver tissue measured during thermal ablation: toward an improved numerical model. 2008 30th Annual International Conference of the IEEE Engineering in Medicine and Biology Society. IEEE, pp. 230–233.
Brace, C.L., Diaz, T.A., Hinshaw, J.L., Lee JR, F.T., 2010. Tissue contraction caused by radiofrequency and microwave ablation: a laboratory study in liver and lung. J. Vasc. Interv. Radiol. 21, 1280–1286.
Brinck, H., Werner, J., 1995. Use of vascular and non-vascular models for the assessment of temperature distribution during induced hyperthermia. Int. J. Hyperth. 11, 615–626.
Chae, Y., Diaz-Valdes, S.H., Lavernia, E.J., Wong, B.J.-F., 2001. Finite element analysis of thermal residual stress and temperature changes in cartilage during laser radiation. Laser-Tissue Interaction XII: Photochemical, Photothermal, and Photomechanical. International Society for Optics and Photonics, pp. 255–268.
Chakravarty, P., Marches, R., Zimmerman, N.S., Swafford, A.D.-E., Bajaj, P., Musselman, I.H., et al., 2008. Thermal ablation of tumor cells with antibody-functionalized single-walled carbon nanotubes. Proc. Natl. Acad. Sci. 105, 8697–8702.
Chan, J.Y., Ooi, E.H., 2016. Sensitivity of thermophysiological models of cryoablation to the thermal and biophysical properties of tissues. Cryobiology 73, 304–315.
Chang, I.A., Nguyen, U.D., 2004. Thermal modeling of lesion growth with radiofrequency ablation devices. Biomed. Eng. Online 3, 27.
Chen, X., Saidel, G.M., 2009. Mathematical modeling of thermal ablation in tissue surrounding a large vessel. J. Biomech. Eng. 131, 011001-1–011001-5.
Chen, S., Wright, N., Humphrey, J., 1997. Heat-induced changes in the mechanics of a collagenous tissue: isothermal free shrinkage. J. Biomech. Eng.
Chen, J.S., Agnissey, K., Shaffer, T., Philips, C., Wolfson, M., 2002. Transient heat transfer in human endometrium during cryoablation. Proceedings of the IEEE 28th Annual Northeast Bioengineering Conference (IEEE Cat. No. 02CH37342). IEEE, pp. 63–64.
Cheng, C.-W., Wang, S., Chang, K., Chen, J., 2016. Femtosecond laser ablation of copper at high laser fluence: modeling and experimental comparison. Appl. Surf. Sci. 361, 41–48.
Chiang, J., Wang, P., Brace, C.L., 2013. Computational modelling of microwave tumour ablations. Int. J. Hyperth. 29, 308–317.
Cho, J., Yoon, J., Cho, S., Kwon, K., Lim, S., Kim, D., et al., 2006. In-vivo measurements of the dielectric properties of breast carcinoma xenografted on nude mice. Int. J. Cancer 119, 593–598.

Cognata, T., Durrant, T., 2014. 3D modeling of human thermal interaction in complex environments using the Wissler human thermal model. In: 44th International Conference on Environmental Systems.

Deng, Z.-S., Liu, J., 2006. Numerical study of the effects of large blood vessels on three-dimensional tissue temperature profiles during cryosurgery. Numer. Heat. Transfer Part A: Appl. 49, 47–67.

Deshazer, G., 2016. Improved Modeling of Image-Guided Thermal Ablation Procedures Towards Patient-Specific Treatment Planning Applications.

Devloo, P.R., Duran, O., Gomes, S.M., Ainsworth, M., 2019. High-order composite finite element exact sequences based on tetrahedral–hexahedral–prismatic–pyramidal partitions. Computer Methods Appl. Mech. Eng. 355, 952–975.

Dodi, R., Ferraguti, F., Ristolainen, A., Secchi, C., Sanna, A., 2014. Planning and simulation of percutaneous cryoablation. AASRI Procedia 6, 118–122.

Dueck, J., Marashdeh, M., Breiter, R., 2011. Experimental investigation and mathematical modeling of the thermal shrinkage of bovine pericardium. J. Med. Biol. Eng. 31, 193–200.

Ekstrand, V., Wiksell, H., Schultz, I., Sandstedt, B., Rotstein, S., Eriksson, A., 2005. Influence of electrical and thermal properties on RF ablation of breast cancer: is the tumour preferentially heated? Biomed. Eng. Online 4, 1–16.

Engel, G., Garikipati, K., Hughes, T.J., Larson, M.G., Mazzei, L., Taylor, R.L., 2002. Continuous/discontinuous finite element approximations of fourth-order elliptic problems in structural and continuum mechanics with applications to thin beams and plates, and strain gradient elasticity. Computer Methods Appl. Mech. Eng. 191, 3669–3750.

Engvall, L., Evans, J.A., 2016. Isogeometric triangular Bernstein–Bézier discretizations: automatic mesh generation and geometrically exact finite element analysis. Computer Methods Appl. Mech. Eng. 304, 378–407.

Erxleben, C., Niehues, S.M., Geyer, B., Poch, F., Bressem, K.K., Lehmann, K.S., et al., 2020. CT-based quantification of short-term tissue shrinkage following hepatic microwave ablation in an in vivo porcine liver model. Acta Radiol. 0284185120914452.

Ewertowska, E., Quesada, R., Radosevic, A., Andaluz, A., Moll, X., Arnas, F.G., et al., 2018. A clinically oriented computer model for radiofrequency ablation of hepatic tissue with internally cooled wet electrode. Int. J. Hyperth. 35, 194–204.

Fang, Z., Zhang, B., Moser, M., Zhang, E., Zhang, W., 2018. Design of a novel electrode of radiofrequency ablation for large tumors: a finite element study. J. Eng. Sci. Med. Diagn. Ther. 1.

Fritzen, F., Böhlke, T., Schnack, E., 2009. Periodic three-dimensional mesh generation for crystalline aggregates based on Voronoi tessellations. Comput. Mech. 43, 701–713.

Gain, A.L., Talischi, C., Paulino, G.H., 2014. On the virtual element method for three-dimensional linear elasticity problems on arbitrary polyhedral meshes. Computer Methods Appl. Mech. Eng. 282, 132–160.

Gallinato, O., De Senneville, B.D., Seror, O., Poignard, C., 2019. Numerical workflow of irreversible electroporation for deep-seated tumor. Phys. Med. Biol. 64, 055016.

Gallinato, O., De Senneville, B.D., Seror, O., Poignard, C., 2020. Numerical modelling challenges for clinical electroporation ablation technique of liver tumors. Math. Model. Nat. Phenom. 15, 11.

Garrean, S., Hering, J., Saied, A., Hoopes, P., Helton, W., Ryan, T., et al., 2009. Ultrasound monitoring of a novel microwave ablation (MWA) device in porcine liver: lessons learned and phenomena observed on ablative effects near major intrahepatic vessels. J. Gastrointest. Surg. 13, 334–340.

Georgii, J., Pätz, T., Rieder, C., Ballhausen, H., Schwenke, M., Schwen, L.O., et al., 2019. Efficient GPU-based numerical simulation of cryoablation of the kidney. Computational Biomechanics for Medicine. Springer.

Gilbert, D., Stoesslein, M., Axinte, D., Butler-Smith, P., Kell, J., 2014. A time based method for predicting the workpiece surface micro-topography under pulsed laser ablation. J. Mater. Process. Technol. 214, 3077–3088.

Guerrero López, G.D., Cepeda Rubio, M.F.J., Hernández Jácquez, J.I., Vera Hernandez, A., Leija Salas, L., Valdes Perezgasga, F., et al., 2017. Computational FEM model, phantom and ex vivo swine breast validation of an optimized double-slot microcoaxial antenna designed for minimally invasive breast tumor ablation: theoretical and experimental comparison of temperature, size of lesion, and SWR, preliminary data. Comput. Math. Methods Med. 2017.

Haemmerich, D., Webster, J.G., 2005. Automatic control of finite element models for temperature-controlled radiofrequency ablation. Biomed. Eng. Online 4, 42.

Haemmerich, D., Chachati, L., Wright, A.S., Mahvi, D.M., Lee, F.T., Webster, J.G., 2003. Hepatic radiofrequency ablation with internally cooled probes: effect of coolant temperature on lesion size. IEEE Trans. Biomed. Eng. 50, 493–500.

Hall, S.K., Ooi, E.H., Payne, S.J., 2015. Cell death, perfusion and electrical parameters are critical in models of hepatic radiofrequency ablation. Int. J. Hyperth. 31, 538–550.

Hijnen, N.M., Heijman, E., Köhler, M.O., Ylihautala, M., Ehnholm, G.J., Simonetti, A.W., et al., 2012. Tumour hyperthermia and ablation in rats using a clinical MR-HIFU system equipped with a dedicated small animal set-up. Int. J. Hyperth. 28, 141–155.

Hocquelet, A., Aube, C., Rode, A., Cartier, V., Sutter, O., Manichon, A.F., et al., 2017. Comparison of no-touch multi-bipolar vs. monopolar radiofrequency ablation for small HCC. J. Hepatol. 66, 67–74.

Hou, G.Y., Luo, J., Marquet, F., Maleke, C., Vappou, J., Konofagou, E.E., 2011. Performance assessment of HIFU lesion detection by harmonic motion imaging for focused ultrasound (HMIFU): a 3-D finite-element-based framework with experimental validation. Ultrasound Med. Biol. 37, 2013–2027.

Huang, H.-W., Horng, T.-L., 2015. Bioheat transfer and thermal heating for tumor treatment. Heat Transfer and Fluid Flow in Biological Processes. Elsevier.

Huebner, K.H., Dewhirst, D.L., Smith, D.E., Byrom, T.G., 2001. The Finite Element Method for Engineers. John Wiley & Sons.

Huo, S., Liu, G., Zhang, J., Song, C., 2020. A smoothed finite element method for octree-based polyhedral meshes with large number of hanging nodes and irregular elements. Computer Methods Appl. Mech. Eng. 359, 112646.

Iancu, C., Mocan, L., Bele, C., Orza, A.I., Tabaran, F.A., Catoi, C., et al., 2011. Enhanced laser thermal ablation for the in vitro treatment of liver cancer by specific delivery of multiwalled carbon nanotubes functionalized with human serum albumin. Int. J. Nanomed. 6, 129.

Jiang, Y., Possebon, R., Mulier, S., Wang, C., Chen, F., Feng, Y., et al., 2017. A methodology for constraining power in finite element modeling of radiofrequency ablation. Int. J. Numer. Methods Biomed. Eng. 33, e2834.

Kai, J., Ming, S., Yang, L., Wen-Zhi, Z., Xiang-Qian, Z., Zhe, L., et al., 2013. Complete radio frequency ablation of hepatocellular carcinoma adjacent to the main bile duct and blood vessels between the first and the second hepatic portal. Cell Biochem. Biophys. 66, 397–402.

Kaur, S., Maini, S., 2014. Microwave ablation therapy for the treatment of hepatocellular carcinoma using double slot interstitial antenna. Int. J. Res. Computer Appl. Robot. 2, 56–61.
Keangin, P., Rattanadecho, P., 2018. A numerical investigation of microwave ablation on porous liver tissue. Adv. Mech. Eng. 10, 1687814017734133.
Keangin, P., Rattanadecho, P., Wessapan, T., 2011. An analysis of heat transfer in liver tissue during microwave ablation using single and double slot antenna. Int. Commun. Heat. Mass. Transf. 38, 757–766.
Kim, J., Yoon, J.-C., Kang, B.-S., 2007. Finite element analysis and modeling of structure with bolted joints. Appl. Math. Model. 31, 895–911.
Kwon, Y.W., Bang, H., 2018. The Finite Element Method Using MATLAB. CRC press.
Laquiere, A., Boustiere, C., Leblanc, S., Penaranda, G., Desilets, E., Prat, F., 2016. Safety and feasibility of endoscopic biliary radiofrequency ablation treatment of extrahepatic cholangiocarcinoma. Surg. Endosc. 30, 1242–1248.
Lee, D.-T., Schachter, B.J., 1980. Two algorithms for constructing a Delaunay triangulation. Int. J. Computer Inf. Sci. 9, 219–242.
Liu, D., Brace, C.L., 2017. Numerical simulation of microwave ablation incorporating tissue contraction based on thermal dose. Phys. Med. Biol. 62, 2070.
Liu, H.L., Chang, H., Chen, W.S., Shih, T.C., Hsiao, J.K., Lin, W.L., 2007. Feasibility of transrib focused ultrasound thermal ablation for liver tumors using a spherically curved 2D array: a numerical study. Med. Phys. 34, 3436–3448.
Liu, Y., Fang, Z., Li, H., He, S., 2014. A mixed finite element method for a time-fractional fourth-order partial differential equation. Appl. Math. Comput. 243, 703–717.
Lopresto, V., Strigari, L., Farina, L., Minosse, S., Pinto, R., D'alessio, D., et al., 2018. CT-based investigation of the contraction of ex vivo tissue undergoing microwave thermal ablation. Phys. Med. Biol. 63, 055019.
Machi, J., Uchida, S., Sumida, K., Limm, W.M., Hundahl, S.A., Oishi, A.J., et al., 2001. Ultrasound-guided radiofrequency thermal ablation of liver tumors: percutaneous, laparoscopic, and open surgical approaches. J. Gastrointest. Surg. 5, 477–489.
Madenci, E., Guven, I., 2015. The Finite Element Method and Applications in Engineering Using ANSYS®. Springer.
Mahmood, N., Feng, Y., Browne, J.C., 2005. A case study in application family development by automated component composition: h-p adaptive finite element codes. International Conference on Computational Science. Springer, pp. 347–354.
Majumdar, P., Xia, H., 2007. A Green's function model for the analysis of laser heating of materials. Appl. Math. Model. 31, 1186–1200.
Nakai, M., Sato, M., Sahara, S., Kawai, N., Tanihata, H., Kimura, M., et al., 2007. Radiofrequency ablation in a porcine liver model: effects of transcatheter arterial embolization with iodized oil on ablation time, maximum output, and coagulation diameter as well as angiographic characteristics. World J. Gastroenterol. WJG 13, 2841.
Nakatsuka, S., Yashiro, H., Inoue, M., Kuribayashi, S., Kawamura, M., Izumi, Y., et al., 2010. On freeze-thaw sequence of vital organ of assuming the cryoablation for malignant lung tumors by using cryoprobe as heat source. Cryobiology 61, 317–326.
Nakayama, A., Kuwahara, F., 2008. A general bioheat transfer model based on the theory of porous media. Int. J. Heat. Mass. Transf. 51, 3190–3199.
Nakayama, A., Kuwahara, Y., Iwata, K., Kawamura, M., 2008. The limiting radius for freezing a tumor during percutaneous cryoablation. J. Heat. Transf. 130.

Nakayama, A., Kuwahara, F., Liu, W., 2009. A macroscopic model for countercurrent bioheat transfer in a circulatory system. J. Porous Media 12.

Nakayama, A., Sano, Y., Yoshikawa, K., 2010. A rigorous derivation of the bioheat equation for local tissue heat transfer based on a volume averaging theory. Heat. Mass. Transf. 46, 739–746.

Nakayama, A., Kuwahara, F., Liu, W., 2011. A general set of bioheat transfer equations based on the volume averaging theory. Porous Media: Appl. Biol. Syst. Biotechnol. 1–44.

Neal II, R.E., Garcia, P.A., Robertson, J.L., Davalos, R.V., 2012. Experimental characterization and numerical modeling of tissue electrical conductivity during pulsed electric fields for irreversible electroporation treatment planning. IEEE Trans. Biomed. Eng. 59, 1076–1085.

Neves, L.A., Machado, J.M., Shiyou, Y., 2011. Integrating open source codes for solid modeling and automatic mesh generation in FEM applications. Proceedings of 2011 Cross Strait Quad-Regional Radio Science and Wireless Technology Conference. IEEE, pp. 36–39.

O'Neill, D., Peng, T., Payne, S., 2009. A two-equation coupled system model for determination of liver tissue temperature during radio frequency ablation. 2009 Annual International Conference of the IEEE Engineering in Medicine and Biology Society. IEEE, pp. 3893–3896.

O'Neill, D.P., Peng, T., Stiegler, P., Mayrhauser, U., Koestenbauer, S., Tscheliessnigg, K., et al., 2011. A three-state mathematical model of hyperthermic cell death. Ann. Biomed. Eng. 39, 570–579.

Okajima, J., Maruyama, S., Takeda, H., Komiya, A., 2009. Dimensionless solutions and general characteristics of bioheat transfer during thermal therapy. J. Therm. Biol. 34, 377–384.

Orr, C.-S., Eberhart, R.C., 1995. Overview of bioheat transfer. Optical-Thermal Response of Laser-Irradiated Tissue. Springer.

Oyeyemi, G.M., Ogunjobi, E.O., Folorunsho, A.I., 2015. On performance of shrinkage methods – a Monte Carlo Study. Int. J. Stat. Appl. 5, 72–76.

Papazafeiropoulos, G., Muñiz-Calvente, M., Martínez-Pañeda, E., 2017. Abaqus2Matlab: a suitable tool for finite element post-processing. Adv. Eng. Softw. 105, 9–16.

Park, C.S., Hall, S.K., Liu, C., Payne, S.J., 2016. A model of tissue contraction during thermal ablation. Physiol. Meas. 37, 1474.

Payne, S., Flanagan, R., Pollari, M., Alhonnoro, T., Bost, C., O'Neill, D., et al., 2011. Image-based multi-scale modelling and validation of radio-frequency ablation in liver tumours. Philos. Trans. Royal Soc. A Math. Phys. Eng. Sci. 369, 4233–4254.

Pearce, J.A., 2013. Comparative analysis of mathematical models of cell death and thermal damage processes. Int. J. Hyperth. 29, 262–280.

Pech, M., Janitzky, A., Wendler, J.J., Strang, C., Blaschke, S., Dudeck, O., et al., 2011. Irreversible electroporation of renal cell carcinoma: a first-in-man phase I clinical study. Cardiovasc. Interv. Radiol. 34, 132–138.

Pepper, D.W., Heinrich, J.C., 2017. The Finite Element Method: Basic Concepts and Applications With MATLAB, MAPLE, and COMSOL. CRC press.

Phipps, C., 2007. Laser Ablation and Its Applications. Springer.

Pillai, K., Akhter, J., Chua, T.C., Shehata, M., Alzahrani, N., Al-Alem, I., et al., 2015. Heat sink effect on tumor ablation characteristics as observed in monopolar radiofrequency, bipolar radiofrequency, and microwave, using ex vivo calf liver model. Medicine 94, e580.

Prakash, P., 2010. Theoretical modeling for hepatic microwave ablation. Open. Biomed. Eng. J. 4, 27.
Prakash, P., Diederich, C.J., 2012. Considerations for theoretical modelling of thermal ablation with catheter-based ultrasonic sources: implications for treatment planning, monitoring and control. Int. J. Hyperth. 28, 69–86.
Qin, J., Chen, J.-Y., Zhao, W.-P., Hu, L., Chen, W.-Z., Wang, Z.-B., 2012. Outcome of unintended pregnancy after ultrasound-guided high-intensity focused ultrasound ablation of uterine fibroids. Int. J. Gynecol. Obstet. 117, 273–277.
Qin, Z., Balasubramanian, S.K., Wolkers, W.F., Pearce, J.A., Bischof, J.C., 2014. Correlated parameter fit of arrhenius model for thermal denaturation of proteins and cells. Ann. Biomed. Eng. 42, 2392–2404.
Rao, S.S., 2017. The Finite Element Method in Engineering. Butterworth-heinemann.
Rombouts, S.J., Van Dijck, W.P., Nijkamp, M.W., Derksen, T.C., Brosens, L.A., Hoogwater, F.J., et al., 2017. Clinical and pathological outcomes after irreversible electroporation of the pancreas using two parallel plate electrodes: a porcine model. HPB 19, 1058–1065.
Rubio, M.F.J.C., López, G.D.G., Perezgasga, F.V., García, F.F., Hernández, A.V., Salas, L.L., 2015. Computer modeling for microwave ablation in breast cancer using a coaxial slot antenna. Int. J. Thermophys. 36, 2687–2704.
Sedaghatkish, A., Rezaeian, M., Heydari, H., Ranjbar, A.M., Soltani, M., 2020. Acoustic streaming and thermosensitive liposomes for drug delivery into hepatocellular carcinoma tumor adjacent to major hepatic veins; an acoustics–thermal–fluid-mass transport coupling model. Int. J. Therm. Sci. 158, 106540.
Shi, Y., Li, Z., Hao, H., 2008. Mesh size effect in numerical simulation of blast wave propagation and interaction with structures. Trans. Tianjin Univ. 14, 396–402.
Singh, S., Melnik, R., 2020. Thermal ablation of biological tissues in disease treatment: a review of computational models and future directions. Electromagn. Biol. Med. 1–40.
Singh, S., Repaka, R., 2015. Pre-clinical modelling and simulation of hepatic radiofrequency ablation. Proc COMSOL Conference 2015. pp. 29–30.
Singh, S., Repaka, R., 2017a. Effect of different breast density compositions on thermal damage of breast tumor during radiofrequency ablation. Appl. Therm. Eng. 125, 443–451.
Singh, S., Repaka, R., 2017b. Temperature-controlled radiofrequency ablation of different tissues using two-compartment models. Int. J. Hyperth. 33, 122–134.
Singh, S., Repaka, R., 2018a. Numerical investigation of convective cooling in minimizing skin burns during radiofrequency ablation of breast tumor. Sādhanā 43, 90.
Singh, S., Repaka, R., 2018b. Parametric sensitivity analysis of critical factors affecting the thermal damage during RFA of breast tumor. Int. J. Therm. Sci. 124, 366–374.
Singh, R., Das, K., Okajima, J., Maruyama, S., Mishra, S.C., 2015a. Modeling skin cooling using optical windows and cryogens during laser induced hyperthermia in a multilayer vascularized tissue. Appl. Therm. Eng. 89, 28–35.
Singh, S., Bhowmik, A., Repaka, R., 2015b. Radiofrequency ablation of malignant breast tumor: a numerical study. Proc. 23rd National Heat and Mass Transfer Conference and 1st International ISHMT-ASTFE Heat and Mass Transfer Conference. pp. 17–20.

Singh, S., Repaka, R., Al-Jumaily, A., 2019. Sensitivity analysis of critical parameters affecting the efficacy of microwave ablation using Taguchi method. Int. J. RF Microw. Computer-Aided Eng. 29, e21581.

Solovchuk, M.A., Sheu, T.W., Lin, W.-L., Kuo, I., Thiriet, M., 2012. Simulation study on acoustic streaming and convective cooling in blood vessels during a high-intensity focused ultrasound thermal ablation. Int. J. Heat. Mass. Transf. 55, 1261–1270.

Solovchuk, M.A., Sheu, T.W., Thiriet, M., Lin, W.-L., 2013. On a computational study for investigating acoustic streaming and heating during focused ultrasound ablation of liver tumor. Appl. Therm. Eng. 56, 62–76.

Soltani, M., Tehrani, M.H., Kashkooli, F.M., Rezaeian, M., 2020. Effects of magnetic nanoparticle diffusion on microwave ablation treatment: a numerical approach. J. Magn. Magn. Mater. 514, 167196.

Soneson, J.E., Myers, M.R., 2010. Thresholds for nonlinear effects in high-intensity focused ultrasound propagation and tissue heating. IEEE Trans. Ultrason. Ferroelectr. Freq. Control. 57, 2450–2459.

Stephens, D.N., Truong, U.T., Nikoozadeh, A., Oralkan, Ö., Seo, C.H., Cannata, J., et al., 2012. First in vivo use of a capacitive micromachined ultrasound transducer array–based imaging and ablation catheter. J. Ultrasound Med. 31, 247–256.

Sweeney, D.C., Reberšek, M., Dermol, J., Rems, L., Miklavčič, D., Davalos, R.V., 2016. Quantification of cell membrane permeability induced by monopolar and high-frequency bipolar bursts of electrical pulses. Biochim. Biophys. Acta (BBA)-Biomembranes 1858, 2689–2698.

Tehrani, M.H., Soltani, M., Kashkooli, F.M., Raahemifar, K., 2020. Use of microwave ablation for thermal treatment of solid tumors with different shapes and sizes—a computational approach. PLoS One 15, e0233219.

Thomson, K.R., Kavnoudias, H., Neal II, R.E., 2015. Introduction to irreversible electroporation—principles and techniques. Tech. Vasc. Interv. Radiol. 18, 128–134.

Tungjitkusolmun, S., Staelin, S.T., Haemmerich, D., Tsai, J.-Z., Cao, H., Webster, J.G., et al., 2002. Three-dimensional finite-element analyses for radio-frequency hepatic tumor ablation. IEEE Trans. Biomed. Eng. 49, 3–9.

Vasantgadkar, N.A., Bhandarkar, U.V., Joshi, S.S., 2010. A finite element model to predict the ablation depth in pulsed laser ablation. Thin Solid. Films 519, 1421–1430.

Vogl, T.J., Basten, L.M., Nour-Eldin, N.-E.A., Kaltenbach, B., Bodelle, B., Wichmann, J.L., et al., 2018. Evaluation of microwave ablation of liver malignancy with enabled constant spatial energy control to achieve a predictable spherical ablation zone. Int. J. Hyperth. 34, 492–500.

Wani, M.A., Kant, N., 2016. Nonlinear propagation of Gaussian laser beam in an inhomogeneous plasma under plasma density ramp. Optik 127, 6710–6714.

Wright, N., Humphrey, J., 2002. Denaturation of collagen via heating: an irreversible rate process. Annu. Rev. Biomed. Eng. 4, 109–128.

Wu, X., Liu, B., Xu, B., 2016. Theoretical evaluation of high frequency microwave ablation applied in cancer therapy. Appl. Therm. Eng. 107, 501–507.

Yamashita, Y., Yokomine, T., Ebara, S., Shimizu, A., 2006. Heat transport analysis for femtosecond laser ablation with molecular dynamics-two temperature model method. Fusion. Eng. Des. 81, 1695–1700.

Yin, X., Epstein, L.M., Hynynen, K., 2006. Noninvasive transesophageal cardiac thermal ablation using a 2-D focused, ultrasound phased array: a simulation study. IEEE Trans. Ultrason. Ferroelectr. Freq. Control. 53, 1138–1149.

Yuan, P., 2009. Numerical analysis of an equivalent heat transfer coefficient in a porous model for simulating a biological tissue in a hyperthermia therapy. Int. J. Heat. Mass. Transf. 52, 1734–1740.

Zanus, G., Boetto, R., Gringeri, E., Vitale, A., D'amico, F., Carraro, A., et al., 2011. Microwave thermal ablation for hepatocarcinoma: six liver transplantation cases. Transplantation Proceedings. Elsevier, pp. 1091–1094.

Zhang, C., Zhang, L., Jiang, X., Jia, B., Liao, W., Dai, R., et al., 2020. Influence of pulse length on heat affected zones of evaporatively-mitigated damages of fused silica optics by CO2 laser. Opt. Lasers Eng. 125, 105857.

Zhao, X., Shin, Y.C., 2013. Coulomb explosion and early plasma generation during femtosecond laser ablation of silicon at high laser fluence. J. Phys. D: Appl. Phys. 46, 335501.

Zhou, W., Arellano, R.S., 2018. Thermal ablation of T1c renal cell carcinoma: a comparative assessment of technical performance, procedural outcome, and safety of microwave ablation, radiofrequency ablation, and cryoablation. J. Vasc. Interv. Radiol. 29, 943–951.

Zorbas, G., Samaras, T., 2014. Simulation of radiofrequency ablation in real human anatomy. Int. J. Hyperth. 30, 570–578.

4

Clinical applications of thermal ablation

Chapter outline

4.1 Introduction

Ablation technologies have a remarkable impact on thermal therapy of cancers and are still undergoing progress. Since ablation therapy has various techniques gaining clinical acceptance, manufacturers/instruments, ablation operators, and settings, trials are currently not comparable (Mcnicholas, 2012). Both theoretical studies and clinical practice have established an optimistic future for thermal ablation procedures. From the clinical standpoint, randomized controlled trials with meta-analyses can be applied to compare thermal ablation methods with accepted therapeutic treatment studies (Feng et al., 2012; Ljungberg et al., 2015; Lencioni and Crocetti, 2013; Bertolotti et al., 2020; Munday et al., 2020). Numerous studies have assessed the practicability and safety of thermal ablation procedures by focusing on the treatment of several types of tumors in different organs (Shi et al., 2020; Tehrani et al., 2020; Pacella and Mauri, 2020; Testoni et al., 2020; Ashour et al., 2020a,b).

Thermal Ablation Therapy. DOI: https://doi.org/10.1016/B978-0-12-819544-4.00009-5
© 2021 Elsevier Inc. All rights reserved.

Thermal ablative procedures provide the ability for treatment of patients while avoiding surgical risks as they offer entire cancer ablation with superior cosmetic results, shorter hospital stay, and lower morbidity rates. Nevertheless, the thermal ablation disadvantages include growth of remaining tumor at the ablation margin, inconsistency in the entire ablation based on tumor location, the inability to efficiently ablate large tumors in a single ablation session, and incomplete destruction of the tumor (Parhi and Mandru, 2020). In most clinical cases, the thermal ablation role requires more studies in wide prospective trials using uniform ablation protocols, accurate outcome measurements over acceptable follow-up durations, and rigorous setting criteria.

Thermal ablation has a key role in the treatment of malignant and benign tumors. In clinical practice, observations of the ablation of each tumor type and various ablation techniques have been carried out (Rivero et al., 2018; Wilkins, 2019). Clinically, thermal ablation is considered the main therapeutic strategy for cancer treatment based on the patient case, where some patients have an increased risk from surgery if tumors exist in several positions in different organs in the same patient, or organ resection (Wilkins, 2019). Replacing the traditional surgical interventions with nominally invasive ablation procedures has great financial and physical impacts. Different steps are followed to implement any ablation process, namely, (1) assess accurately the location and size of the tumor, (2) damage the tumorous tissues completely, including in situ cancer, and (3) monitor the ablation treatment procedure in real time.

Accordingly, the ablation process can be carried out by radiologists, gastro-oncologists, or surgeons. It is performed using nominally invasive percutaneously/laparoscopy procedures by inserting one or more applicators through <1 cm incisions in the patient's skin under the guidance of monitoring systems, such as ultrasonography (US), computed tomography (CT), and magnetic resonance imaging (MRI) (Southwell et al., 2018; Yao et al., 2018; Cornud et al., 2018; Dewitt et al., 2019). These imaging-guidance devices are used for thermal mapping to direct the placement of the ablation applicator into the targeted tumor position, monitor the targeting/localizing of the thermal influence on the tumor for energy deposition control, treatment response assessment, and treatment. Nevertheless, some problems may be faced during the ablation procedure, including:

- Lack of accuracy during the probe insertion;
- Shortage of consistent real-time temperature assessment within the targeted tissues; and

- Failure of control during the heat generation of the energy deposition.

For small tumors, thermal ablation procedures can be performed under light general anesthesia, light sedation, full consciousness, in a single session (Prud'homme et al., 2019). For tumors larger than 3 cm, multiple sessions may be required, and multiple overlapping ablation zones may be created. Since most types of ablation devices have a single needle, sequential ablation procedures may be required.

Many ablation techniques have been used for accurate, complete, and efficient tumor tissue ablation. After designing an ablation system, the approval of the US Food and Drug Administration (FDA) or the European Medicines Agency (EMA) is essential for regulatory clearance before being marketed for commercial sale (Runge, 2018; Chen et al., 2018). During the manufacturing of ablation devices several designs have been produced, including embedded systems for heat deposition, prevailing generators, and upgraded electrode designs for ablating large tissue necrosis volumes.

Recently, widespread attention has been directed to clinical trials and settings for nominally invasive ablation practices using several thermal energy sources, including radiofrequency (RF), microwave (MW), high-intensity focused ultrasound (HIFU), and laser to damage tumors in different organ sites. Such ablation therapeutic techniques enable the avoidance of surgery-based tumor removal and allow nonsurgical treatment candidates for the eradication of solid tumors in different organs, including the adrenals, lung, liver, bone, prostate, neck, head, and kidney (Tanaka et al., 2020).

The current clinical thermal ablation methods follow paradigm protocols that raise the tissue temperature to >50°C for >5 minutes to induce coagulation for any tissue type. Recent trials have proved that 50°C isotherm can be considered as the key surrogate treatment instead of surgery in tumors. Since 50°C is considered the coagulation zone's margin at which the real critical thermal dosimetry occurs, however, it varies based on the tissue characteristics as this temperature margin is tissue specific.

This increases the necessity for empathetic heat–tissue interactions for each tissue type. In addition, based on the varying high-temperature effects of thermal ablation, it is not necessary to follow a standard hyperthermia procedure as they depend on the prolonged paradigm, with uniform heating of the complete tumorous tissue area after realizing the thermal steady state (Spirou et al., 2018). In addition, the thermal dosimetry of the ablation electrode depends on the tissue type as well as the critical ablation margin distance from the ablation needle, and also the amount of energy delivered to the tumor tissues (Ewertowska et al., 2018).

Several clinical trials have studied hyperthermia with chemotherapy and/or radiation therapy for treating different cancer types, such as head and neck, breast, lung, liver, melanoma of the skin, brain, and cervix (Davatgaran-Taghipour et al., 2017; Testori et al., 2017; Potenza et al., 2018). A noteworthy reduction in tumor size is achieved by combining hyperthermia with other treatment methods (Vankayala and Hwang, 2018; Chang et al., 2018; Elming et al., 2019). Numerous hyperthermia approaches are under development, such as local, regional, and whole-body hyperthermia, as illustrated in Fig. 4.1.

Owing to the effectiveness of hyperthermia in cancer treatment, it becomes an effectual counterpart compared to the traditional treatment methods of cancer, including chemo- and radiation therapy. By increasing the temperature above 60°C in the thermal ablation, cell destruction occurs, where the hyperthermia success in treatment is linked to the achieved temperature and the treatment duration. However, whole-body hyperthermia procedures may cause vomiting, nausea, and vascular/cardiac disorders. Also, using regional hyperthermia may cause high-temperature spots because of the variances in tissue characteristics in the different regions (Behrouzkia et al., 2016). The resultant temperature spots may cause wounds and burns. In addition, using the regional perfusion procedures may result in bleeding, and blood clots in nearby tissues.

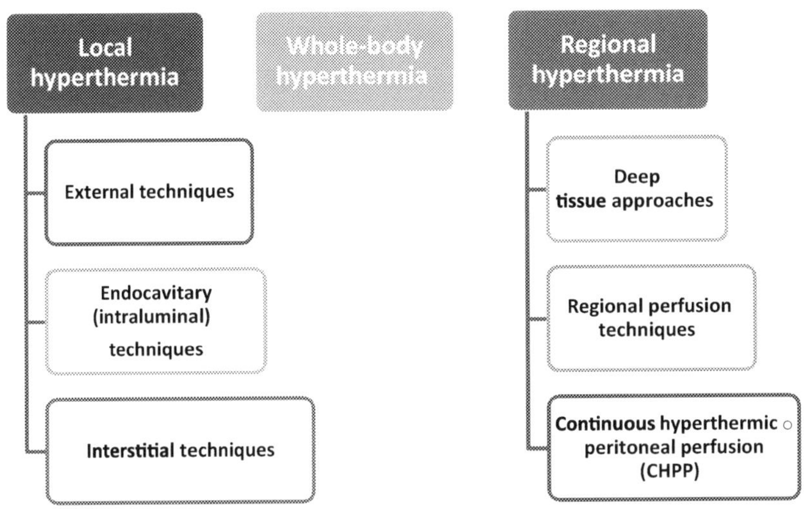

Figure 4.1 Hyperthermia treatment techniques.

4.2 Clinical trials of thermal ablation therapy of tumors in different organs

Resection surgery has several criteria along with the oncology guidelines (Fleshman et al., 2015), including:

- The appropriateness of the patient for anesthesia; and
- The expected reduction in organ functionality.

These criteria and restrictions have triggered an urgent need for using minimally invasive procedures using thermal ablation methods. Percutaneous energy-/thermal-based ablation therapy depends on locally applied extreme temperature levels using energy different sources to cause permanent cell damage and eventually tumor coagulative necrosis and apoptosis. Such thermal ablation techniques have been employed to treat several tumor types, such as adrenal gland tumor, rectal cancer, breast cancer, kidney cancer, pancreatic cancer, lung tumor, liver tumor, prostate cancer, brain cancer, bone cancer, esophageal cancer, head and neck cancer, skin cancer, and thyroid cancer (Kok et al., 2020).

The advancement of up-to-date imaging-guidance modalities increases the engagement of the thermo-ablative techniques as minimally invasive for treatment of tumors in different organs. Cryoablation (CA), radiofrequency ablation (RFA), laser ablation (LA), and microwave ablation (MWA) have a growing role in tumor treatment. In particular, these procedures are upgraded to resolve the limitations of the ablation process in clinical trials. For example, MWA proved its effectiveness with large tumors with complete ablation rate, while RFA failed to deal with large tumors (Meloni et al., 2017). Before approving the selection of a specific ablation technique, controlled trials are required to evaluate the efficiency of the tumor control, complication/survival rates, and patient health treatment using the ablation procedure under investigation to provide conclusive confirmation of the impact of the interstitial practices.

Typically, from a clinical point of view, compared to surgical resection, thermal ablation methods provide numerous advantages, including (Sidoff and Dupuy, 2017; Poulou et al., 2015):

- Reduced cost;
- Less injury;
- Protection of nearby healthy tissues;
- Efficient;
- Reduced mortality rate based on the used ablation method;
- Functional outcomes;

- Visual intrapractical monitoring; and
- Short hospitalization period after ablation.

Nevertheless, the most common shortcomings in some cases of thermal ablation methods during the treatment of different organs include deficient ablation and cancer recurrence.

Among the thermal ablation techniques, CA, which is a decreased-temperature-based ablation method, and RFA and MWA, which are increased-temperature-based ablation methods, are the most frequently used ablation procedures. In addition, LA and HIFU are increased-temperature-based methods which are still under development (Waseem et al., 2015). Typically, each method has particular advantages and disadvantages, as reported in Figs. 4.2 and 4.3, resulting in the favorability of each in specific organ treatments.

In addition, Fig. 4.4 summarizes the main advantages and disadvantages of the main thermal ablation therapy techniques.

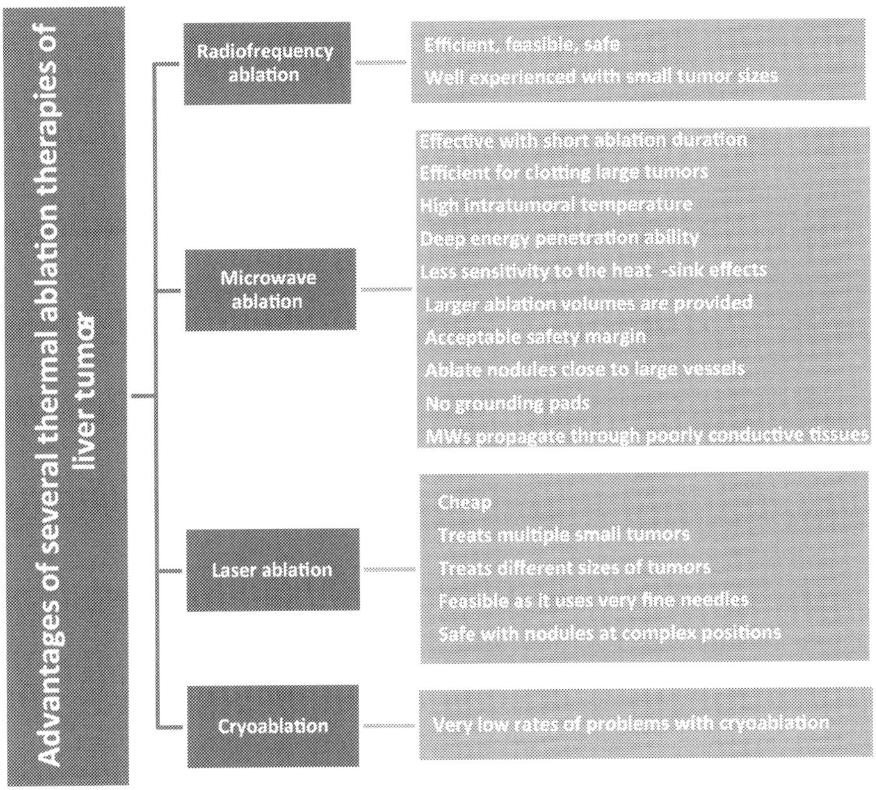

Figure 4.2 Advantages of different thermal ablation therapies.

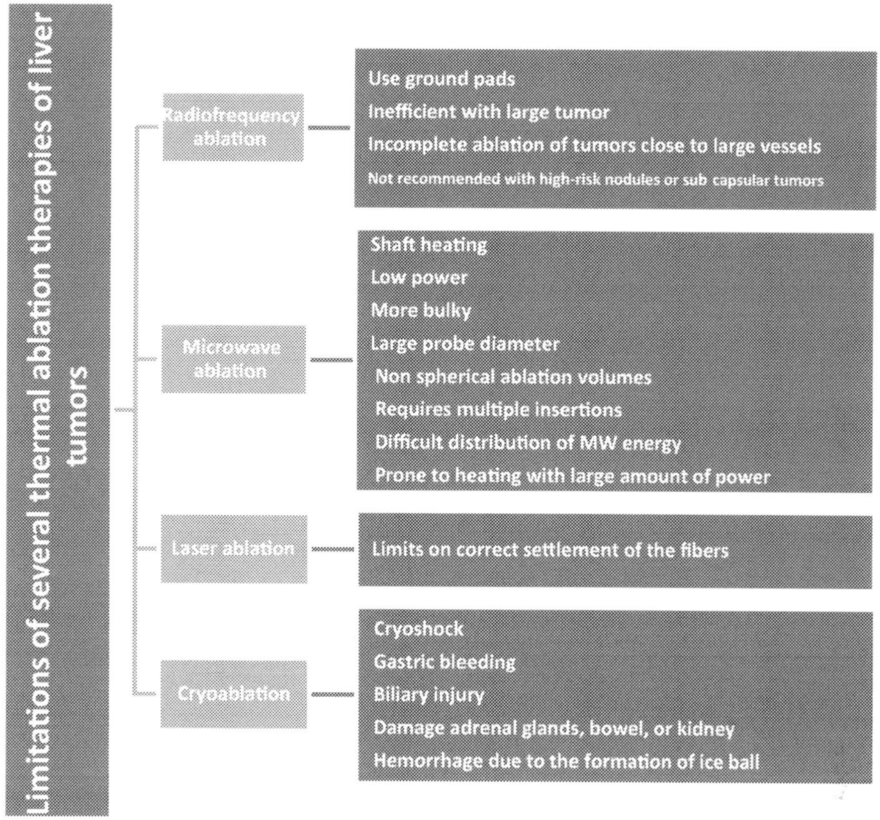

Figure 4.3 Limitations of different thermal ablation therapies.

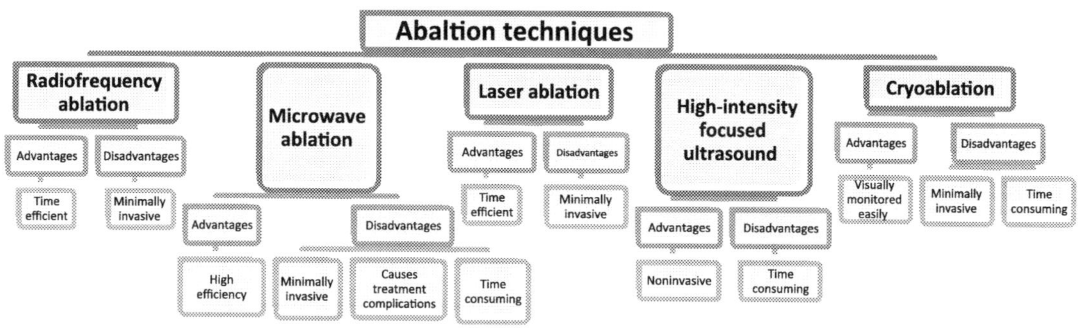

Figure 4.4 General advantages and disadvantages of the main ablation methods.

4.2.1 Thermal ablation therapy of liver tumors

The liver is a solid vascular organ that has a large number of large blood vessels causing vascular cooling, which acts as a heat sink (Huang, 2013; Fang et al., 2017; Jiang et al., 2015).

It may be affected by metastatic or primary malignancy diseases, including hepatocellular carcinoma (HCC) which is widespread worldwide. Advanced ablation methods are seen as a treatment method for patients with single metastases or early-phase tumors. Conventionally, surgery or transplantation for liver tumors were considered the original gold-standard therapy (Izzo et al., 2019; Petrowsky et al., 2020). However, several patients are not surgery candidates, leading thermal ablation becoming the main alternative.

Based on tumor number and size, the thermal ablation techniques can be selected, and can also be used jointly with other systemic treatments for the best therapy outcomes (Wong et al., 2010; Cucchetti et al., 2013). Although RFA is a well-known effective ablation method, numerous clinical studies have recommended LA and LTA (Lu et al., 2005). Generally, LTA, RFA, and MWA have common technical concepts, where they depend on measured thermal energy generation/transfer to increase the temperature of the targeted tissues to between 60°C and 100°C for coagulating necrosis of the tumor. In addition, they use the same procedure by placing a needle tip, where LA inserts a fiber using a tiny needle, RFA uses RF electrodes, and MWA uses antennas, in the targeted tumor tissues. The selection of the most suitable ablation modality is based on a match between the advantages/limitations with the individual patient and tumor characteristics (Wang et al., 2012; Vitale et al., 2017). The individual patient scenario may add some limitations to a specific ablation method, for example, RFA is discounted in patients with pacemakers. Also, the tumor characteristics include the location, size, and number.

Typically, MWA can better overcome large heat sinks and perfusion compared with other high-temperature-based ablation methods. Also, it can treat large hepatic zones of up to 10 mm in high-perfusion regions, such as hepatic vessels larger than 3 mm, while RFA has limited performance to ablate such cases (Stewart et al., 2018). For hepatic carcinoma (HCC) treatment, thermal ablation may be combined with chemotherapy/systemic resection to improve the survival of patients with liver metastases. Both LA and MWA are as safe and efficient as RFA. Consequently, thus far, the selection of the used ablation method is based mainly on physician experience, along with the previously mentioned considerations (Fatima et al., 2017).

The tumor size and location affect the RFA effectiveness, where complete ablation occurs using RFA with tumors that have diameters of less than 3 cm, while RFA of patients with tumors with diameters greater than 5 cm leads to very poor results. On the contrary, MWA is better recommended to ablate large hepatic tumors than those ablated by RFA. Unlike RFA, which uses only one probe, multiple MWA antennas in proximity permit thermal and electrical interaction, which can be powered concurrently (Poggi et al., 2015; Swan et al., 2012). MWA can produce a very high temperature of over 150°C by using water cooling systems, while in RFA, the temperature cannot exceed 100°C to avoid water vaporization from the cells. It can ablate large zones, where MW can generate heat in high impedance, low thermal conductivity, or low electrical conductivity desiccated and charred tissues. Since the liver has an irritating vascular source, MWA is recommended as it is not prone to heat sink like RFA. Even with the several advantages of MWA and its fast process in about 10 minutes, its drawbacks have limited its use. Several studies have been conducted to overcome the MWA limitations by reducing the back-heating effects and also reducing the thickness of the MW antenna (Ashour et al., 2020b; Orsi and Varano, 2015; Poggi et al., 2015). In addition, CA is not recommended in the case of liver ablation of tumors due to the heat sink effect of the vascular structures in the liver.

Compared to RFA and MWA, LA which uses a bare-tip fiber laser probe to convert electrical energy into light energy to heat tissue by coagulative necrosis can provide spherical ablated large tumors (Tombesi et al., 2015; Klass, 2010; Stewart et al., 2017). LA can use multisource multifibers for use in the bare tip fibers and inserted through a needle into the tumor to treat nodules of different diameters. In terms of cost, both LA and RFA are cheaper than MWA. However, MWA is used to treat liver tumors near large blood vessels. In terms of safety, LA is safer for ablating nodules in risky positions due to its fine needle tip compared to MWA and RFA. As a remedy to the small HCCs of diameter less than 3 cm, RFA can be used with overlapping insertions or three fibers in the LA needle. For multiple small tumors, LA is preferred using 1–4 fibers. In contrast, for large HCC tumors and tumors near large vessels, MWA is applied (Shady et al., 2016; Wang et al., 2013; Dubay et al., 2011).

4.2.2 Thermal ablation therapy of pancreatic cancer

The pancreas is a glandular organ in the abdomen behind the stomach, which regulates blood sugar and assists the digestion process (Tang et al., 2020). It is divided anatomically into four main parts: the head, neck, body, and tail, where the head contains blood vessels and veins. The most common pancreatic cancer is pancreatic adenocarcinoma which leads to several symptoms, such as back/abdominal pain, weight loss, dark urine, and yellow skin (Amin et al., 2020). Promising research studies are on-going based on mathematical modeling, and preclinical or clinical trials, as an attempt to find standard selection criteria based on investigating the cutting-edge thermal ablation methods for ablating pancreatic cancer.

In some cases, surgical resection is considered for treating pancreatic cancer with increased life expectancy. To date, pancreatectomy, which is an operator-dependent method, is considered the exclusively accepted treatment method although it is based on an invasive surgical process.

Chemoradiation/chemotherapy is considered another therapeutic method for patients with pancreatic cancer; however, they provide poor outcomes (Van Veldhuisen et al., 2019). These methods cause systemic complications and reduce the chances of long-term survival due to their various negative effects, such as lack of appetite and fatigue, leukopenia, neutropenia, and anemia. Alternatively, several minimally invasive therapeutic options using ablation therapy modalities have been recommended for pancreatic cancer treatment (Paiella et al., 2018). These energy-based thermal therapy techniques insert one or more needles into the targeted tumor target to induce a precise change in the tumor tissue conditions, causing tumor apoptosis and necrosis (Saccomandi et al., 2018). However, the diffused margins of pancreatic cancer complicate complete ablation using one thermal ablation method. Clinically, for pancreatic cancer treatment, RFA suffers from several limitations due to:

- Adverse trials, including thermal injury to structures surrounding the pancreas, such as the bile duct, portal vein, stomach, superior mesenteric vessels, and duodenum;
- Thermal damage in the pancreas;
- Risk of pancreatic parenchyma;
- The heat-sink effect of the heat dissipation due to the closeness of the tumor to blood vessels; and
- Other clinical/technical restrictions.

Using RFA, the RF electrode is inserted into the target tumor through the GI tract, where a current is created using an external RF generator, and the temperature sensors attached to the elector are used to monitor the temperature threshold. The RFA process may occur by selecting the appropriate treatment time, power, and temperature threshold according to the size of the pancreatic tumor under real-time monitoring of an ultrasound imaging modality (Kallis et al., 2015; Song et al., 2016; Paiella et al., 2016; D'Onofrio et al., 2017; Scopelliti et al., 2018; Bang et al., 2019; Barthet et al., 2019; Sodergren et al., 2020).

Several developments have been conducted toward finding an efficient pancreatic cancer ablation by combining different therapies, such as radiotherapy with chemotherapy, multiple chemotherapeutical systems, and stereotactic radiation therapy (Hirata et al., 2015). Generally, systemic toxicity is enhanced by integrating chemotherapy with radiation (Lowery et al., 2011; Kowalewski et al., 2018).

Targeting long-term outcomes, an enhancement of local pancreatic tumor control is considered using ablative practices. In preclinical/clinical trials, RFA, MWA, LA, CA, and HIFU have proved their role in ablating pancreatic tumors (Marrocchio et al., 2018; Zhang et al., 2021). Ex vivo and preclinical studies have demonstrated the significance of real-time temperature control based on the induced thermal injury of the tumors. A novel trend has been directed to using nanoparticles for the enhancement of the thermal properties of the ablation process (Navya et al., 2019; Canavese et al., 2018). Nevertheless, the long-term effects and biocompatibility of nanoparticles are under examination.

For pancreatic tumor ablation, several limitations (Oleinikov et al., 2019; Duc and Keserci, 2019; Ruarus et al., 2020) have been raised, as follows:

- The interpositioning of bowel gases obstructs the ablation process, leading to imperfect ablation of the pancreatic tumor;
- The gas–tissue interface leads to heat deposition that affects the thermal ablation of the interposed bowel;
- The existing possible interposition of fibrotic, vascular, and fatty tissues; and
- The dependency of RFA on the impedance of the necrotic tissues, which limits RFA to MWA and HIFU ablation methods.

To overcome such drawbacks, it is recommended to undertake the following procedures:

- Evacuate the existing gases in the colon and stomach to avoid ablation-related difficulties;
- Make a cautious pretechnical plan;
- Make direct contact during RFA, MWA, and CA with the neoplastic tissues using the percutaneous/endoscopic method; and
- Consider the postprocedural difficulties during the mechanical insertion of the antennas/electrodes/cryoprobes into the pancreatic tumor tissues and the maximum local temperature.

Experimentally, ultrasound ablation (HIFU) of the pancreatic tumor ablation provided a low complication occurrence rate compared to RFA, MWA, LA, and CA (Maiettini et al., 2019; Xu et al., 2019). In addition, HIFU is recommended for the cancer-related treatment of patients with pancreatic tumors for palliation. HIFU can be used alone to treat local pancreatic cancer, or alongside chemotherapy (Bhosale et al., 2018).

4.2.3 Thermal ablation therapy of renal cancer

Another vascular-solid organ in the abdomen is the kidney, which has high dominant perfusion rates producing noteworthy heat sink effects (Singh and Melnik, 2019). Tumor ablation techniques are used for the treatment of solid tumors, including RFA, MWA, HIFU, LA, and CA. For renal cancer ablation, both RFA and CA, using needle-type probes, are recognized as the best ablation techniques for the ablation of small renal tumors (O'Neal et al., 2018; Johnson et al., 2019; Monfardini et al., 2019).

During the ablation of renal tumors, the produced local heat using RFA results in clotting necrosis of the tumor tissues. By using high-frequency 400–460 kHz of alternating current, the ions oscillate in the targeted tumor tissues, leading to an elevation in the temperature of up to 60°C–90°C. The thermally applied energy destroys the vessels that supply the renal tumor leading to a cytotoxicity effect, which is based on the achieved maximum temperature and the ablation duration. To evaluate the success of RFA, increasing the impedance value illustrates the dryness of the ablated renal tissues. To ensure complete tumor coagulation necrosis, the ablation procedure is repeated if the impedance threshold is obtained (Ridouani and Srimathveeravalli, 2019). In addition, MWA is like RFA, where it applies a high-frequency 0.9–2.5 GHz alternating current using an ablation antenna for quick agitation of the water molecules in the renal tumor tissues in a short time.

Conversely, CA freezes the targeted renal tumor tissues by inserting a cryoprobe into the tumor to decrease the temperature to −40°C to −60°C, causing irreversible hydropic cell injury. The tumor cell destruction occurs by repeated freeze–thaw cycles (Kawabata et al., 2020).

The American and European urological societies guidelines approved CA or RFA as a thermo-ablation choice for small renal tumor treatment. T1a renal tumors with diameters less than 3–4 cm can be preserved efficiently using RFA (Ljungberg et al., 2015; Finelli et al., 2017; Ljungberg et al., 2019). However, for large renal tumors, MWA is used. These ablation methods do not achieve undamaged renal functionality.

During kidney thermal ablation, care to avoid damaging the ureter and renal pelvis, which are located close to the tumor, complicates the ablation process, especially RFA, while the use of MWA is supported by the cooling mechanism. Furthermore, based on a number of clinical studies, MWA is endorsed for ablating larger zones in kidney tumors using higher temperatures (more than 100°C) compared to RFA (Hao et al., 2018; Yu et al., 2020). It is considered an efficient treatment method for renal tumors as it achieves complete ablation in one session without tumor recurrence.

4.2.4 Thermal ablation therapy of lung tumor

The lung consists of two springy parts located in the chest taking in oxygen when inhaling and releasing carbon dioxide during exhaling (Cascio et al., 2012). The most common lung cancer is nonsmall cell lung cancer. The typical care for patients with stage I or II of this cancer is surgical resection (Onishi et al., 2011). However, in some cases, surgery is impossible due to the patient's status and health. Accordingly, an alternative thermal ablation-based therapy has been considered. However, the lung tissues have poor thermal conduction and low electrical conductivity, restricting the efficiency of using RFA. Clinical studies have proved that an imperfect ablation rate is related to tumor size, which also affects the required ablation margin (key to the success of the ablation process). Conversely, MWA has substantial advantages in lung ablation. The high impedance and low conductivity of the lung lead to complete MW heating effect on the lung volume. The inherent lower conductivity and permittivity in the lung permit deeper penetration of the MW into the lung compared to into the liver.

Clinical studies have proved that no there is effect on the local efficiency of the ablation process based on the tissue type. However, lung tumor size shows a relevant effect on the ablation process selection and success (Ye et al., 2015). Local returns of lung cancer are extremely common with lung tumor diameters larger than 3 cm.

The most common complications associated with lung tumor ablation are:

- Postinterventional pneumothorax;
- Damage to the small blood vessels;
- Hemothorax necessitating thoracic surgery;
- Peripheral tumor ablation that may cause pleural irritation;
- Hemorrhage of the hemothorax/pleural effusion after ablation;
- Skin burns may occur around the ground pad during RFA;
- Hemorrhage of the small alveolar due to the insertion of the ablation needle;

- Massive lethal hemothorax may occur after RFA of the recurrent tumor; and/or
- Cutaneous metastases may occur at the insertion site of the ablation electrode.

For renal tumor treatment, thermal ablation methods are operative, mostly for small tumors of size less than 4 cm. RFA is a low-cost, effective, and simple procedure, while CA is more multifaceted and favored for treatment of large tumors. Also, MWA can be used to treat large tumors. In conclusion, the following points are summarized for lung tumor ablation:

- Low injury rate when applying the safety rules;
- RFA is recommended for small lung tumors;
- MWA ablates large lung tumor zones;
- CA visualizes the ablation zone without damaging the supporting urothelial tissues;
- Focused ultrasound remains under study for use in lung tumor ablation;
- Consistent follow-up, for about a year, is essential after the ablation process.

4.2.5 Thermal ablation therapy of cancers in other body organs

The thermal damage caused by heating has several characteristics that are based on the attainable temperature of the tissue and the required duration to reach this temperature level. In addition, the sensitivity of the tumorous cells to any systemic radiation or chemotherapy increases with increasing the cell temperature to 42°C. By heating tissues to 45°C for several hours, permanent cellular injury occurs. Elevating the temperature to 55°C reduces the required ablation time for irreversible cell destruction. Almost instant cell and tissue coagulation is realized at temperature levels of 60°C–100°C. At temperatures higher than 100°C–110°C, carbonization/vaporization of tissues occurs, leading to an increase in impedance, which stops the RFA process. The coagulation necrosis diameter is reached using the thermal ablation bioheat transfer equation, which corresponds to the local tissue interactions multiplied by the deposited energy in the tissue minus the heat loss (convective and conductive) (Ni et al., 2005). Typically, solid tumors have greater blood flow values compared to their corresponding healthy tissues, leading to a chaotic/primitive/highly acidic nature, and greater sensitivity to the heat's cytotoxic effect on solid tumors. Since heating tissues is easier with a lower blood flow rate, nervous tissues are particularly sensitive to heat with more possibility of damage.

This requires any ablation process, even in solid tumors, to be careful while ablating organs close to nervous tissues to avoid any negative effects and complications that may affect the patient's general health. The most common thermal ablation modalities introduced in the previous sections have proved their impact on successful ablative results in tumors of other body organs. The thermal ablation therapy of cancers in other body organs are discussed next.

1. *Bone cancer*: Bones have high impedance and low thermal conductivity, which limits the efficiency of RFA. However, MW energy is comparatively unresponsive to low thermal conductivity and thus penetrates deeper into bones (Nakatsuka et al., 2004). Moreover, bone tissue heating and dryness are unaffected by MW due to the relative permittivity of bones (Ringe et al., 2016). Osteoid osteoma (OO) is one of the common benign osteoblastic bone tumors with a small radiolucent nidus less than 15 mm. For OO treatment, different thermal ablation therapy methods under the guidance of CT have been established, where RFA has been predominant (Paladini et al., 2018).
2. *Thyroid cancer*: The thyroid gland has a butterfly shape and is found low in the front of the neck. Papillary thyroid carcinoma is one of the most common thyroid carcinomas, and is less than 1.0 cm in size. Comparatively, the thyroid nodules are elliptical and large. Surgical operations are the first-line treatment method, but are unsuitable in some cases (Hasan, 2016). Accordingly, nonsurgical treatment methods using thermal therapy ablation, including RFA, LA, and ethanol ablation (EA), can be used for the treatment of thyroid nodules. LA is recommended for ablating solid nodules, which EA fails to treat efficiently. At the same time, EA is used to ablate cystic nodules effectively. Based on clinical trials, RFA is used efficiently to treat thyroid nodules, including lymph nodes, and benign/recurrent thyroid nodules. Recently, MWA and HIFU ablation have been introduced to treat thyroid nodules (Mauri et al., 2017).

   Since RFA has been recommended, several studies were conducted in developing RF probes by shortening the straight-type RF probe for accurate control during the ablation process or perfusing chilled water using a peristaltic pump to improve the pattern of RF energy deposition and prevent tissue charring. The main outcomes of the clinical studies on thyroid cancer ablation shared the limitations while using the standard ablation modalities. These limitations include disappearance of the thyroid nodule, the return of the nodule's tumor, and the reduction rate of the nodule volume (Mainini et al., 2017; Guo et al., 2020).

3. *Breast cancer*: Breast cancer can be treated using thermal approaches based on freezing/heating the cancer tissues. In clinical trials on breast cancer treatment, CA, RFA, LA, MWA, and HIFU ablation are introduced to reduce the requirement for surgical excision, surgical difficulties, and hospital stay (Habrawi et al., 2021). For breast cancer ablation, LA has a short ablation time and produces a paler zone surrounding the central cavity. Using RFA has several benefits, such as short ablation time and good results with small tumors. However, patients need general anesthesia and there is an increased complication rate. Like RFA, HIFU is chosen for smaller breast cancer ablation, however it has a longer ablation time. Also, HIFU is considered the only ablative method that does not require skin incisions. MWA is also preferred for breast cancer ablation providing high thermal efficiency, with the risk of ablating surrounding healthy tissues (Li et al., 2017; Zhu et al., 2018). On the other hand, CA requires reduced anesthesia compared to other ablative methods. The main drawback of all the ablation methods is the lack of a histopathological assessment prior to treatment and in the follow-up stages. According to the recommendations of clinical studies, histological validation and evaluation are essential to confirm the treatment response.
4. *Brain tumors*: Brain tumors are destructive, with a short survival outcome even if standard therapies are used, including radiation therapy, chemotherapy, or surgery (Jain, 2018). Therefore thermal ablative therapies are currently being explored in clinical trials. There are numerous continuing efforts to develop new thermal therapy methods for brain tumor ablation (Calabrese et al., 2018). Generally, the most well-known thermal ablation methods, except for LA and CA, in brain tumors are considered safe when using a slow heating procedure, low power, and trifling injury to the nearby normal brain tissues.

4.3 Clinical studies using radiofrequency ablation

Soft-tissue tumors and osseous metastasis can be treated using ablation techniques, which are categorized into thermal ablation techniques, such as radiofrequency, laser, microwave, radiofrequency ionization, HIFU, and cryo ablations. Also, chemical ablation techniques can be applied, such as acetic acid or ethanol injection, and the ablation procedure can be used alone or combined with other surgical percutaneous techniques.

The preceding sections have concluded the effectiveness of the RFA procedure for ablating different organs. This section reports several clinical studies which have been conducted using RFA. For example, the efficiency of the ablation techniques in the treatment of malignant and benign lesions having diameters up to 30 mm, was reported in (Filippiadis et al., 2014). The results have indicated that the treatment is palliative, achieves pain reduction, local disease control, reduction of tumor burden, and its mass effect on the organs. However, these ablation techniques may lead to some weakening of the bones. Thus, when stabilization undermining occurs, augmentation of bone must follow the ablation procedure, depending on the location and size of the lesion. The results have demonstrated that thermal ablation of soft tissues and bones has low complication and high success rates. However, iatrogenic damage to the sensitive surrounding structures is considered one of the most common complications. It is commonly known that the nervous structures, such as skin, cartilage, and nerves are highly sensitive to quite low and high temperatures that can cause permanent neurological thermal damage or resultant transience. The thermal damage may cause osteonecrosis of the normal bones at the periphery of the lesion, skin burns, and scarring and atrophy in the surrounding muscular areas. Consequently, thermal ablation success requires complete ablated tumor volume as well as protection of the surrounding vulnerable structures from thermal damage, using all kinds of neural and thermal monitoring, and skin cooling must be performed.

Another study conducted by Hoffmann et al. (2010) used a clinical trial to treat 39 patients, including 16 females and 23 males, with ages ranging from 7 to 53 years, with OO, within 61 months of using the RFA procedure, the lesion distribution included 20 lesions inside the femur, 10 lesions in the tibia, 5 lesions in the spine, 1 lesion in the radius, 1 in the talus, and 1 in the pelvis. In the case of children, the RFA procedure was done using general anesthesia. However, for adults, the procedure was done under conscious sedation. The treatment procedure included skin cooling in the OOs that are located inside bones, with a minor covering of soft tissues, such as the radius and tibia, and flushing using saline through another needle in the case of nerve structures close to the OO. The results demonstrated that 38 patients were efficiently ablated, with complaints in only three cases, including minor complications, such as prolonged pain and hematoma. Also, other major complications occurred, such as infections and broken drills. Hence, RFA is a safe and efficient treatment method of OOs with a low relapse rate, and patients who suffer from a relapse can be ablated with a second session of the RFA procedure rather than surgical resection. The RFA treatment reduces the need for patient rehabilitation, hospital stay periods, and costs as compared with other therapy techniques. Nevertheless, surgical resection is only used in some rare cases in patients who have suffered from two or more relapses after the RFA session, or with percutaneously inaccessible lesions.

Koelblinger et al. (2014) conducted a retrospective study to treat a number of patients who suffered from sarcoma lung metastases in the period between 2007 and 2012 using the RFA therapy technique. The study included 15 women whose ages ranged from 10 to 78 years, with 55 sarcoma lung metastases with size ranging from 0.5 to 2 cm, which was ablated in 30 sessions. The treatment sessions were performed using a cool-tip device under general conscious/anesthesia sedation. The results indicated that the control rate was 95%, and the progression-free survival rate was 53% and 23% over 1 and 2 years, respectively. The overall survival rate was 94% and 85% over 2 and 3 years, respectively. Consequently, the clinical results demonstrated that RFA is an effective and safe treatment technique for patients suffering from small sarcoma metastases.

Furthermore, Gregory et al. (2013) studied the RFA feasibility of giant angiomyolipomas (AMLs) using a saline solution-cooled probe under CT guidance. The clinical study included four female patients with ages ranging from 33 to 67 years who suffered from giant AMLs with a maximal axis of range 61–324 mm, which were not suitable for surgery or embolotherapy. The results established that all four patients were successfully treated without any intraprocedural complications, over 48 months as a minimum period. There were no new specific renal symptoms or hemorrhages, and the renal function was normal, however, during the radiological imaging follow-up, the tumors became fattier with the effect of soft-tissue involution.

Moreover, Fowler et al. (2012) used pulsed RFA under US guidance to treat a female having aged 23 years suffering from consistent pain with meralgia paresthetica and paresthesias in the left anterolateral thigh over 18 months. The treatment procedure included:

- Determining the femoral lateral cutaneous nerve location using the ultrasound wave;
- Reproducing the dysthesia of the patient with the simulation; and

- Injecting the pulsed RF waves for 2 minutes at a temperature of 42°C using a needle that was rotated by an angle of 180 degrees for an additional period of pulsed RFA treatment.

Then, the ablation procedure was followed by 4 mg dexamethasone injection with 0.25% ropivacaine. After a follow-up period lasting for 3 months, the patient indicated perfect pain relief and enhanced ability for performing her tasks as a member of an active duty service.

Also, RFA was used in another study, where Shaheen et al. (2011) investigated RFA durability in a patient suffering from Barrett's esophagus (BE) with dysplasia as well as reduced esophageal adenocarcinoma rates. The long-term eradication rates, neosquamous epithelium durability, and disease progression were considered. The study included 127 randomized trial subjects with dysplastic BE, a follow-up time of 3.05 years, and was expanded to 5 years with patients suffering from intestinal metaplasia eradication at 2 years. The results showed that RFA treatment has an acceptable and durable safety profile with disease progression of low rate over up to 3 years.

Pouw et al. (2010) studied the effectiveness and safety of RFA combined with resection by baseline endoscopic for treatment of early cancer and high-grade intraepithelial neoplasia (HGIN). The study involved 24 patients with a mean age of 65 years, suffering from Barrett's esophagus with a size of less than or equal to 12 cm, including early cancer or HGIN. The visible lesions were resected using an endoscope, followed by serial RFA. Endoscopic resection of the focal escape was used in the case of Barrett tissue persistence. The results illustrated that the intestinal metaplasia and neoplasia in patients were eradicated with percentages of 88% and 95%, respectively, while the escape endoscopic resection of two patients improved the rates to 96% and 100%, respectively. However, the complications resulting after the RFA procedure were dysphagia and melena, after an additional follow-up period with a median value of 22 months, the results showed no neoplasia recurrence.

Patients suffering from an obstruction in the malignant bile duct were treated using self-expandable metal stents (SEMSs) as a standard technique for ensuring the continuity of biliary drainage. Unfortunately, 50% of these patients were exposed to an occlusion in the stent within 6 months, also, the changes made to the design, concomitant therapy, and composition of the stent did not result in perfect stent patency. Hence, Steel et al. (2011) found an alternative method, called bipolar RFA, for safely prolonging stent patency. RFA catheter deployment succeeded in 22 patients, and SEMSs were placed in all successful deployment cases of the RFA catheter. The results showed that after the placement of SEMS, only one case failed to achieve successful decompression of the biliary and died in 90 days. Moreover, all patient cases preserved the patency of the stent at 30 days, one case had biochemical pancreatitis, one case developed rigors, and two cases required percutaneous gallbladder drainage. Also, the results illustrated that, at a follow-up period of 90 days, three cases had occluded bile ducts stents, and one additional case died due to a patent stent.

Song et al. (2016) illustrated the safety and feasibility of using EUS-RFA in the treatment of unresectable pancreatic cancer with diameter ranging from 3 to 9 cm in six patients with an age range of between 43 and 73 years, two of whom had the stage 4 disease, and four had stage 3 disease. In addition, two pancreatic cancers were located in the body, and four in the head. The clinical study comprised an RF generator and an 18-gauge endoscopic RF probe with an exposed tip length of 1 cm. After insertion of the RF probe into the tumor, activation of the RF generator was adjusted to deliver power within the range of 20–50 watts for 10 seconds. The procedure was repeated as required by the size of the tumor. The results showed that the EUS-RFA treatment was successfully performed in all six patients, with two patients having mild abdominal pain. However, there were no side effects, such as bleeding or pancreatitis.

It is known that cardiac arrhythmia treatment using RFA therapy may be greatly affected by the contact force (CF) between the tissue and the catheter tip. Excessive CF causes complications, where insufficient CF causes an ineffective target lesion. Hence, Kuck et al. (2012) studied the ability of a novel force-sensing RFA catcher to measure real-time CF, and the possible implications for patient safety. The study included 77 patients, 34 of which suffered from atrial fibrillation (AF) and 43 cases with supraventricular tachycardia (SVT) in the right side, who were percutaneously ablated with the novel catheter containing the CF measuring. The study indicated that real-time CF catheter ablation is a safe technology for AF and SVT treatment. High CF measurement must be performed during catheter manipulation, including during the ablation procedure to obtain useful information for the operator to achieve the safety of catheter manipulation.

The local tumor progression rate, which is considered an important prognostic agent in patients suffering from HCC tumors ablated with RFA, rapidly increases when the diameter of the target tumor increases to more than or equal to 3 cm. Hence, Kim et al. (2011) compared the effectiveness of RFA alone and RFA combined with transarterial chemoembolization (TACE).

This study involved 57 patients suffering from HCC with size ranging from 3.1 to 5 cm who were ablated with RFA combined with TACE. Also, 66 cases were ablated with RFA only. The results illustrated that a blood flow decrease resulted from the TACE and decreased the heat loss, increased the ablated tumor volume, decreased the progression rate, improved the recurrence, and achieved overall survival rates of HCC tumors with medium size, during a follow-up period ranging from 2.6 to 126.2 months. The local progression rates of tumors were 40% and 70% using RFA combined with TACE and RFA only, respectively. The major complication rates were 0% and 3% for RFA combined with TACE and RFA only, respectively.

Joo et al. (2013) compared repeated thermal RFA and alcohol ablation (AA) in the treatment of recurrent facet joint thoracolumbar pain. The clinical study involved 40 patients divided evenly into the two ablation procedures. The results showed that during a follow-up period of 24 months after the AA and RFA, 1 patient treated with RFA and 17 patients treated with AA were ablated without recurrence of facet joint thoracolumbar pain. The average efficient periods were 24 and 10.7 months in the AA and RFA groups, respectively. Furthermore, there were no considerable complications resulting from either groups except for injection site pain.

The efficacy of RFA combined with lyso-thermosensitive liposomal doxorubicin (LTLD) was studied by Yang et al. (2019). This study considered an encapsulated doxorubicin inside a heat-sensitive liposome in treatment of patients with unresectable HCC ranging from 3 to 7 cm. Eighteen patients were included in this clinical study. They were divided into two groups: RFA and RFA with LTLD, with an age range of between 40 and 73 years, and included 4 females and 14 males. The results showed only one case in the combination group that was exposed to a major complication of empyema. The RFA and RFA with LTLD groups produced:

- Primary success rates of 77.8% and 93.3%, respectively;
- Coagulation volume differences between post- and pre-RFA of 37.3 ± 8.5 and 105.7 ± 73.8 cm^3, respectively;
- Local progression rates over an average follow up 49.1 ± 24.8 months of 22.2% and 6.7%, respectively; and
- Overall survival of 46 ± 10.6 and 68.5 ± 7.2 months, respectively.

Wood et al. (2012) combined LTLD with RFA to define the safe maximum tolerated dose (MTD), dose-limiting toxicity (DLT), and pharmacokinetic properties. The study included 9 patients with primary tumors and 15 patients with metastatic tumors in the liver, with 4 or fewer lesions and 70 mm diameter. The RFA procedure was started 15 minutes after starting the infusion of intravenous LTLD, which lasted for 30 minutes. The results showed that the MTD and DLT criteria were 50 and 60 mg/m^2, respectively.

In addition, Cucchetti et al. (2013) compared the cost-effectiveness of RFA and hepatic resection for early HCC. The study included 4424 patients treated with the RFA and 3996 patients treated with the resection. The costs were estimated from the healthcare provider perspective. The results demonstrated that for single-nodule HCC smaller than 2 cm or bi- or tri nodules with a size smaller than or equal to 3 cm, the RFA technique achieved a similar quality-adjusted life expectancy at a minimal cost compared with the resection. In addition, for single-nodule HCC with size ranging from 3 to 5 cm, the resection achieved better cost-effectiveness and life expectancy than RFA.

Comparing the performance of RFA with the other ablation techniques is important with respect to treatment time, applied energy, survival, recurrence, progression rates, complications, technical success, and reduction in the volumetric percentage. Hence, Mauri et al. (2017) compared RFA and percutaneous laser ablation (PLA) using the same equipment to treat benign thyroid nodules. The study comprised 59 patients aged 55.8 ± 14.1 years, treated with RFA, and 31 patients aged 55.2 ± 14.2 years treated with PLA under ultrasound guidance. The results proved a technical success, and no major complications occurred, also, they explained that PLA and RFA produced:

- Mean ablation times of 30.1 ± 13.8 and 13.9 ± 5.9 minutes, respectively;
- Mean energy deployments of 5.4223 ± 2.4845 and 34.6627 ± 15.8123 kJ, respectively;
- Reductions in the mean volume by 42% ± 17% in the first month, 60% ± 15% in the sixth month, and 70% ± 16% in the twelfth month, and 51% ± 15% in the first month, 64% ± 14% in the sixth month, and 74% ± 14% in the twelfth month, respectively.

Hence, the RFA technique is faster than PLA, however, it needs significantly greater energy.

Moreover, Yu et al. (2011) compared the effectiveness of RFA and MWA to treat porcine hepatic tumor using an internally cooled RF probe. The clinical study included an in vivo experiment, in which 30 patients were treated with both MWA using needle antennas operating at 915 and 2450 MHz at a power of 60 watts, and RFA using a bipolar probe with an active tip length of 30 mm (T30) at the power of 30 watts. Also, an ex vivo experiment was conducted, in which 40 cases were treated with MWA using antennas operating at 915 and 2450 MHz at powers of 70 and 60 watts, respectively, and

20 cases were treated with RFA using a bipolar probe with active tip length of 30 and 40 mm (T40) at powers of 30 and 40 watts, respectively. According to the automatic power system setting, all the treatment sessions were for 10 minutes. The results achieved in both in vivo and ex vivo models, showed the diameters of the long (DL) and short axis (DS) after MWA were significantly greater than those obtained from all the RFA trials, where they were 5.79 ± 1.03, 3.86 ± 0.81, and 3.21 ± 0.51, and 2.95 ± 0.32, 2.35 ± 0.75, and 1.61 ± 0.33, for the 915, 2450 MHz MWA, and bipolar RFA, respectively. Moreover, T30 RFA and 2450 MHz MWA rarely produced elliptical-shaped ablation zones. Additionally, the 915, 2450 MHz MWA, and bipolar RFA produced a maximum tissue temperature at 0.5 cm from the probe of $89.07°C \pm 3.52°C$, $80.07°C \pm 12.82°C$, and $65.56°C \pm 15.31°C$, respectively, and the time required for temperature reaching 54°C were24.50 ± 4.09, 37.50 ± 7.62, and 57.29 ± 23.24 seconds, respectively.

The RF probes used in the RFA procedure represented one of the main parts of the RFA system. Accordingly, Ihara et al. (2016) studied the effect of the deployed electrode array diameter on tumor progression after treating lung tissue with RFA. The clinical study contained 651 tumors with the long axis diameter ranging from 0.2 to 4.2 cm, which were treated with multi-tined expandable electrodes. The results illustrated that, to avoid significant risk of local progression, the electrode-deployed array diameter should be greater than the diameter of the tumor by a value that is larger than or equal to 1 cm.

Tumors have an effect on the ablation procedure with respect to the closeness of vessels, hence, Zorbas and Samaras (2015) studied the heat-sink effect of blood vessels near a probe in a numerical model composed of a spherical tumor inside liver tissue during RFA. The mathematical study included several blood vessels that were different in size and distance from the probe. The results illustrated that the minimum distance of a straight, cylindrical, and parallel blood vessel for an acceptable drop of only 10% in the injured volume compared to the nonblood vessel model varied from 0.45 cm for a vessel with a diameter of 0.2 cm, to 2 cm for a vessel with a diameter of 2 cm. However, the worst case scenario, in which a large blood vessel of diameter 2 cm was located at a distance 0.2 cm from the probe, produced a drop in the necrotic volume that reached 21.23%.

When the tumorous tissues are located close to blood vessels or a bile duct, it is of concern that a complete ablation is difficult to achieve due to the vessel heat-sink effect, and damage to the blood vessels and ducts. Therefore, Lin et al. (2012) combined RFA with a radioactive seed implantation of iodine-125 (I-125) to treat HCC close to great blood vessels under 1.5T MRI guidance. This clinical study included 16 patients who had a total of 24 HCC tumors with an average diameter of 2.35 ± 1.03 cm, and the implantation process was performed in lesion portions close to the blood vessels. The results clarified that combined RFA with radioactive seed implantation of I-125 is a successful treatment for HCC tumors adjacent to great blood vessels. Likewise, Kai et al. (2013) achieved complete ablation of HCC tumor in a location adjacent to the main blood vessels and a bile duct between the first and second hepatic portal using careful designed pretreatment and RFA treatment monitoring.

Colorectal liver metastases (CRLM) resection is sometimes hindered due to their locations adjacent to the major hepatic vessels. Therefore Snoeren et al. (2015) examined the effectiveness and safety of RFA with multipolar probes for the treatment of CRLM close to the major hepatic vessels. The clinical study included 52 patients who suffered from CRLM with 144 tumors, who were consecutively treated with RFA using three simultaneously placed electrodes. The experimental results indicated that, through a follow-up period of at least 3 years after the RFA treatment procedure, the patients with CRLM adjacent to the great hepatic vessels were effectively and safely treated with the multipolar RFA system.

Some complications may have occurred from the RFA treatment. Hence, Baek et al. (2012) evaluated the imaging features and clinical aspects of complications resulting from benign thyroid nodule treatment using RFA. This retrospective study, which was performed in the period between June 2002 and September 2009, involved 1543 patients who suffered from thyroid nodules treated with RFA using internally cooled electrodes. The treatment sessions were done at 13 thyroid centers, which were members of the Korean Thyroid Radiology Society. The results included 48 complications with a percentage of 3.3%, including 28 minor and 20 major complications. The minor complications included hematoma in 15 cases, skin burns in 4 cases, and vomiting in 9 cases. Additionally, the major complications included voice changes in 15 cases, permanent hypothyroidism in 1 case, tumor rupture in 3 cases, and brachial plexus injury in 1 case. All cases spontaneously recovered except for one case who was exposed to surgery and one case with permanent hypothyroidism.

4.4 Discussion

Thermal ablation uses a wide variety of energy sources to increase the temperature of the targeted tumor tissues by monitoring the produced thermal doses using temperature profiles. This enables the applied energy to be for shorter time intervals compared to the time required during standard hyperthermia. The goal of thermal ablation is to damage the tumor/cancer tissues by creating a cytotoxic temperature for a short period in a nominally invasive nature without destroying neighboring vigorous structures out of the required safety margin. RF, LA, ultrasound, and MW are frequently used ablative techniques which elevate the temperature locally to above 55°C to damage the tumor/cancer tissues. On the other hand, the CA methods use subzero temperature values to freeze and damage the tumor tissues. These methods deliver a noninvasive or minimal (e.g., laparoscopically/percutaneously) method for cancer healing, which has gained increased attention as an alternate to the typical surgical approaches, particularly for patients who refuse surgery, with contraindications, or for nonoperable cases. Several research studies have assessed the viability and safety of thermal ablation methods for numerous types of tumors. Clinical studies and practice have uncovered constructive outcomes of different ablative applications with some limitations affecting the ablation procedure efficiency being reported, such as the inability to efficiently ablate large tumors in a single treatment setting and the constant growth of residual tumors close to the ablation margin.

Generally, the greatest impact of thermal ablation techniques is the reduced mortality and morbidity compared with an ordinary invasive resection. Currently, thermal ablative techniques are applied in several clinical applications to treat liver, kidney, and prostate tumors, with an increased number of applications to other organ sites, such as the brain, bone, thyroid, pancreas, lung, and breast. From the previously reported clinical studies on using RFA, it is concluded that finding the most suitable modality is essential to the success of an ablation procedure. The type of tumor tissues to be ablated and the tumor size are two of the most significant factors in the decision of which modality to use. Commonly, RF ablation fits the treatment of tumors in different organs, such as kidney and liver, with diameters less than 2 cm. For larger tumors, RFA efficacy decreases and requires multiple series.

Regarding MWA, clinical studies have proved its applicability to a wide-ranging spectrum of tumor tissues, such as bone, kidney, liver, and lung. In addition, generations of MW systems have been developed to be more active for ablating larger tumors, nonetheless, longer term clinical information is required for evaluating the role of tumor size on the efficacy of MWA. CA was contradicted for use in primary liver tumors, especially in patients with severe cirrhosis. However, it was used fruitfully in tumors smaller than 2 cm in the bone, prostate, liver, lung, and breast. For tumors in superficial/stationary regions, the high-accuracy HIFU noninvasive method has become an attractive treatment in the uterus or prostate. Nonetheless, HIFU has limited efficiency in ablating other organs. Finally, even though clinical studies have supported LA, RFA, MWA, and CA showing their robustness, supplementary clinical studies are required to support the use of HIFU in clinical trials on humans. In addition, based on clinical studies, it is concluded that:

- All ablative methods have variability in the clinical outcomes in terms of treatment duration, complications, and ablation completeness.
- The presented clinical studies have missed standard protocols and collected pathological information to recognize the best ablative method for treatment.
- Qualified studies are compulsory for comparing ablative procedures to adopt the most capable technique.
- Analysis of cost-effectiveness and standard of care methods is mandatory.
- Assessment of the human exposure risk to electromagnetic and thermal energy/heat must be considered and studied.
- The temperature elevation has negative effects on the circulatory system and heart as well as other effects, including some pain, nausea, disorientation, weakness, dizziness, an increase in blood pressure, and cataracts.
- There is a necessity for treatments that can precisely localize tissue destruction.

4.5 Conclusion

Numerous clinical trials have established the impact of thermal ablation therapy in a variety of cancers in different organs, such as the pancreas, liver, lung, uterine, breast, brain, and bones as well as pelvic tumors, melanoma, and superficial tumors.

Preclinical and clinical studies have reported that RFA can treat both palliative and curative indicators as well as solid small tumors in the breast, lung (nodules), liver, kidney, and bones. Since in RFA, the heat is not delivered by the probe itself, the ablated volume is limited. This adds a restriction to the ablated area by RFA, which differs from the other ablative methods. To resolve this limitation, RFA can be used to ablate large tumors using multiple probes or by increasing the duration of the ablation process.

Regarding MWA, it has several advantages compared to other thermal ablation therapy modalities, which are confirmed by the preclinical and initial clinical studies as it can:

- Generate, in a short time, very high temperature to ablate large ablation zones with less susceptibility to the effects of a heat sink;
- Ablate high-impedance tissues, including bones and lungs.

Furthermore, other studies have been carried out to develop robust MWA clinical systems.

Currently, HIFU has proved its efficiency in the treatment of both malignant and benign tumors in the prostate, kidney, breast, and liver, as well as being used in bone metastasis for pain relief and the treatment of osteosarcoma.

This chapter concludes with key messages based on the preclinical and clinical practices that can be highlighted as follows:

- The interstitial ablation methods have key advantages of being minimally invasive, have low complication rates, and minor loss of the treated organ's functionality.
- Both CA and RFA are the most frequently used ablative techniques for tumors in the kidney and other solid organs, while MWA remains experimental.
- MWA has a great impact on the treatment of large tumors after undertaking all safety procedures.
- Modalities of image-guided thermal ablation are essential to monitoring the ablation process and temperature level.
- The used image-guidance modality affects the monitoring process during the ablation process.
  - CT is cheap and provides superior spatial resolution, but it cannot visualize a temperature map or blood flow in the vessels.
  - US is cheap and offers real-time imaging, but it cannot visualize a temperature map or blood flow in the vessels.
  - MRI offers temperature mapping and superior resolution; but it is expensive and labor intensive.
- After evaluating all the obtainable treatment choices, the decision of which thermal ablation method to use is taken after considering the individual patient and to guarantee nonrecurrence of the tumor after weighing up the recommended treatment options.
- Clinical studies should be carried out on potential randomized trials associating different local methods, such as thermal ablation, stereotactic radiotherapy, and limited resection.
- However, since the evidence level for the ablative methods remains incomplete, thermal ablation must only be applied as a part of clinical studies.

Researchers and clinicians should study the tradeoff between the risk of cancer and the EM exposure risks. The safety issues related to the thermal effects of the transferred energy from generated radiation to the tissues depend on the operating frequency governed by the dielectric loss.

References

Amin, S., Meza, J., Song, B., Narang, A., Kitchen, H., Hodgin, M., 2020. Advances in pancreatic cancer diagnosis and treatment. Adv. Pancreat. Cancer Diagn. Treat. 126.

Ashour, A.S., Asran, M., Fotiadis, D.I., 2020a. Optimal power for microwave slotted probes in ablating different hepatocellular carcinoma sizes. Comput. Biol. Med. 127, 104101.

Ashour, A.S., Asran, M.I., Mohamed, W.S., Fotiadis, D.I., 2020b. Optimal localization of a novel shifted 1T-ring based microwave ablation probe in hepatocellular carcinoma. IEEE Trans. Biomed. Eng. 68, 505–514.

Baek, J.H., Lee, J.H., Sung, J.Y., Bae, J.-I., Kim, K.T., Sim, J., et al., 2012. Complications encountered in the treatment of benign thyroid nodules with US-guided radiofrequency ablation: a multicenter study. Radiology 262, 335–342.

Bang, J.Y., Sutton, B., Hawes, R.H., Varadarajulu, S., 2019. EUS-guided celiac ganglion radiofrequency ablation versus celiac plexus neurolysis for palliation of pain in pancreatic cancer: a randomized controlled trial (with videos). Gastrointest. Endosc. 89, 58–66. e3.

Barthet, M., Giovannini, M., Lesavre, N., Boustiere, C., Napoleon, B., Koch, S., et al., 2019. Endoscopic ultrasound-guided radiofrequency ablation for pancreatic neuroendocrine tumors and pancreatic cystic neoplasms: a prospective multicenter study. Endoscopy 51, 836–842.

Behrouzkia, Z., Joveini, Z., Keshavarzi, B., Eyvazzadeh, N., Aghdam, R.Z., 2016. Hyperthermia: how can it be used? Oman Med. J. 31, 89.

Bertolotti, A., Ferdynus, C., Milpied, B., Dupin, N., Huiart, L., Derancourt, C., 2020. Local management of anogenital warts in non-immunocompromised adults: a network meta-analysis of randomized controlled trials. Dermatol. Ther. 10, 249–262.

Bhosale, R.R., Gangadharappa, H.V., Gowda, D.V., Osmani, R.A.M.A., Vaghela, R., Kulkarni, P.K., et al., 2018. Current perspectives on novel drug carrier systems and therapies for management of pancreatic cancer: an updated inclusive review. Crit. Rev. Ther. Drug Carrier Syst. 35.

Calabrese, G., Daou, A., Barbu, E., Tsibouklis, J., 2018. Towards carborane-functionalised structures for the treatment of brain cancer. Drug Discov. Today 23, 63–75.

Canavese, G., Ancona, A., Racca, L., Canta, M., Dumontel, B., Barbaresco, F., et al., 2018. Nanoparticle-assisted ultrasound: a special focus on sonodynamic therapy against cancer. Chem. Eng. J. 340, 155–172.

Cascio, D., Magro, R., Fauci, F., Iacomi, M., Raso, G., 2012. Automatic detection of lung nodules in CT datasets based on stable 3D mass–spring models. Comput. Biol. Med. 42, 1098–1109.

Chang, D., Lim, M., Goos, J.A., Qiao, R., Ng, Y.Y., Mansfeld, F.M., et al., 2018. Biologically targeted magnetic hyperthermia: potential and limitations. Front. Pharmacol. 9, 831.

Chen, Y.-J., Chiou, C.-M., Huang, Y.-W., Tu, P.-W., Lee, Y.-C., Chien, C.-H., 2018. A comparative study of medical device regulations: US, Europe, Canada, and Taiwan. Ther. Innov. Regul. Sci. 52, 62–69.

Cornud, F., Bomers, J., Futterer, J., Ghai, S., Reijnen, J., Tempany, C., 2018. MR imaging-guided prostate interventional imaging: ready for a clinical use? Diagn. Interv. Imaging 99, 743–753.

Cucchetti, A., Piscaglia, F., Cescon, M., Colecchia, A., Ercolani, G., Bolondi, L., et al., 2013. Cost-effectiveness of hepatic resection versus percutaneous radiofrequency ablation for early hepatocellular carcinoma. J. Hepatol. 59, 300–307.

D'Onofrio, M., Crosara, S., Robertis, D.E., Butturini, R., Salvia, G., Paiella, R., et al., 2017. Percutaneous radiofrequency ablation of unresectable locally advanced pancreatic cancer: preliminary results. Technol. Cancer Res. Treat. 16, 285–294.

Davatgaran-Taghipour, Y., Masoomzadeh, S., Farzaei, M.H., Bahramsoltani, R., Karimi-Soureh, Z., Rahimi, R., et al., 2017. Polyphenol nanoformulations for cancer therapy: experimental evidence and clinical perspective. Int. J. Nanomed. 12, 2689.

Dewitt, M.R., Lattouche, E., Kaufman, J.D., Fesmire, C.C., Swet, J., Kirks, R., et al., 2019. Simplified non-thermal tissue ablation with a single insertion device enabled by bipolar high-frequency pulses. IEEE Trans. Biomed. Eng. 67, 2043–2051.

Dubay, D.A., Sandroussi, C., Kachura, J.R., Ho, C.S., Beecroft, J.R., Vollmer, C.M., et al., 2011. Radiofrequency ablation of hepatocellular carcinoma as a bridge to liver transplantation. HPB 13, 24–32.

Duc, N.M., Keserci, B., 2019. Emerging clinical applications of high-intensity focused ultrasound. Diagn. Interv. Radiol. 25, 398.

Elming, P.B., Sørensen, B.S., Oei, A.L., Franken, N.A., Crezee, J., Overgaard, J., et al., 2019. Hyperthermia: the optimal treatment to overcome radiation resistant hypoxia. Cancers 11, 60.

Ewertowska, E., Quesada, R., Radosevic, A., Andaluz, A., Moll, X., Arnas, F.G., et al., 2018. A clinically oriented computer model for radiofrequency ablation of hepatic tissue with internally cooled wet electrode. Int. J. Hyperth. 35, 194–204.

Fang, Z., Zhang, B., Zhang, W., 2017. Current solutions for the Heat-Sink effect of blood vessels with radiofrequency ablation: a review and future work. Advanced Computational Methods in Life System Modeling and Simulation. Springer.

Fatima, S., Lee, N.P., Bian, Z.X., 2017. Current and emerging cancer therapies for treatment of hepatocellular carcinoma. Front. Anti-Cancer Drug Discov. 8, 72.

Feng, K., Yan, J., Li, X., Xia, F., Ma, K., Wang, S., et al., 2012. A randomized controlled trial of radiofrequency ablation and surgical resection in the treatment of small hepatocellular carcinoma. J. Hepatol. 57, 794–802.

Filippiadis, D.K., Tutton, S., Mazioti, A., Kelekis, A., 2014. Percutaneous image-guided ablation of bone and soft tissue tumours: a review of available techniques and protective measures. Insights Imaging 5, 339–346.

Finelli, A., Ismaila, N., Bro, B., Durack, J., Eggener, S., Evans, A., et al., 2017. Management of small renal masses: American Society of Clinical Oncology clinical practice guideline. J. Clin. Oncol 35, 668–680.

Fleshman, J., Branda, M., Sargent, D.J., Boller, A.M., George, V., Abbas, M., et al., 2015. Effect of laparoscopic-assisted resection vs open resection of stage II or III rectal cancer on pathologic outcomes: the ACOSOG Z6051 randomized clinical trial. JAMA 314, 1346–1355.

Fowler, I.M., Tucker, A.A., Mendez, R.J., 2012. Treatment of meralgia paresthetica with ultrasound-guided pulsed radiofrequency ablation of the lateral femoral cutaneous nerve. Pain. Pract. 12, 394–398.

Gregory, S., Anderson, C., Patel, U., 2013. Radiofrequency ablation of large renal angiomyolipoma: median-term follow-up. Cardiovasc. Interv. Radiol. 36, 682–689.
Guo, Y., Li, Z., Wang, S., Liao, X., Li, C., 2020. Single-fiber laser ablation in treating selected metastatic lymph nodes of papillary thyroid carcinoma and benign cold thyroid nodules—preliminary results. Lasers Surg. Med. 52, 408–418.
Habrawi, Z., Melkus, M.W., Khan, S., Henderson, J., Brandi, L., Chu, V., et al., 2021. Cryoablation: a promising non-operative therapy for low-risk breast cancer. Am. J. Surg. 221, 127–133.
Hao, G., Hao, Y., Cheng, Z., Zhang, X., Cao, F., Yu, X., et al., 2018. Local tumor progression after ultrasound-guided percutaneous microwave ablation of stage T1a renal cell carcinoma: risk factors analysis of 171 tumors. Int. J. Hyperth. 35, 62–70.
Hasan, M.N., 2016. A Prescription Based Survey on Prevalence and Management of Thyroid Cencer. East West University.
Hirata, T., Teshima, T., Nishiyama, K., Ogawa, K., Otani, K., Kawaguchi, Y., et al., 2015. Histopathological effects of preoperative chemoradiotherapy for pancreatic cancer: an analysis for the impact of radiation and gemcitabine doses. Radiother. Oncol. 114, 122–127.
Hoffmann, R.-T., Jakobs, T.F., Kubisch, C.H., Trumm, C.G., Weber, C., Duerr, H.-R., et al., 2010. Radiofrequency ablation in the treatment of osteoid osteoma—5-year experience. Eur. J. Radiol. 73, 374–379.
Huang, H.W., 2013. Influence of blood vessel on the thermal lesion formation during radiofrequency ablation for liver tumors. Med. Phys. 40, 073303.
Ihara, H., Gobara, H., Hiraki, T., Mitsuhashi, T., Iguchi, T., Fujiwara, H., et al., 2016. Radiofrequency ablation of lung tumors using a multitined expandable electrode: impact of the electrode array diameter on local tumor progression. J. Vasc. Interv. Radiol. 27, 87–95.
Izzo, F., Granata, V., Grassi, R., Fusco, R., Palaia, R., Delrio, P., et al., 2019. Radiofrequency ablation and microwave ablation in liver tumors: an update. Oncologist 24, e990.
Jain, K.K., 2018. A critical overview of targeted therapies for glioblastoma. Front. Oncol. 8, 419.
Jiang, K., Chen, J., Liu, Y., Liu, J., Liu, A., Dong, J., et al., 2015. Heat-irrigate effect'of radiofrequency ablation on relevant regional hepatocyte in living swine liver-initial study on pathology. Cell Biochem. Biophys. 72, 37–41.
Johnson, B.A., Sorokin, I., Cadeddu, J.A., 2019. Ten-year outcomes of renal tumor radio frequency ablation. J. Urol. 201, 251–258.
Joo, Y.-C., Park, J.-Y., Kim, K.-H., 2013. Comparison of alcohol ablation with repeated thermal radiofrequency ablation in medial branch neurotomy for the treatment of recurrent thoracolumbar facet joint pain. J. Anesthesia 27, 390–395.
Kai, J., Ming, S., Yang, L., Wen-Zhi, Z., Xiang-Qian, Z., Zhe, L., et al., 2013. Complete radio frequency ablation of hepatocellular carcinoma adjacent to the main bile duct and blood vessels between the first and the second hepatic portal. Cell Biochem. Biophys. 66, 397–402.
Kallis, Y., Phillips, N., Steel, A., Kaltsidis, H., Vlavianos, P., Habib, N., et al., 2015. Analysis of endoscopic radiofrequency ablation of biliary malignant strictures in pancreatic cancer suggests potential survival benefit. Dig. Dis. Sci. 60, 3449–3455.
Kawabata, T., Hiraki, T., Iguchi, T., Matsui, Y., Uka, M., Masaoka, Y., et al., 2020. Post-ablation syndrome after percutaneous cryoablation of small renal tumors: a prospective study of incidence, severity, duration, and effect on lifestyle. Eur. J. Radiol. 122, 108750.

Kim, J.H., Won, H.J., Shin, Y.M., Kim, S.H., Yoon, H.-K., Sung, K.-B., et al., 2011. Medium-sized (3.1–5.0 cm) hepatocellular carcinoma: transarterial chemoembolization plus radiofrequency ablation versus radiofrequency ablation alone. Ann. Surg. Oncol. 18, 1624–1629.

Klass, D., 2010. Bimodal Electric Tissue Ablation (BETA): An Investigation of the Ablative Potential of Combining Alternating and Direct Current in the Liver. University of East Anglia.

Koelblinger, C., Strauss, S., Gillams, A., 2014. Outcome after radiofrequency ablation of sarcoma lung metastases. Cardiovasc. Interv. Radiol. 37, 147–153.

Kok, H.P., Cressman, E.N., Ceelen, W., Brace, C.L., Ivkov, R., Grull, H., et al., 2020. Heating technology for malignant tumors: a review. Int. J. Hyperth. 37, 711–741.

Kowalewski, A., Szylberg, Ł., Saganek, M., Napiontek, W., Antosik, P., Grzanka, D., 2018. Emerging strategies in BRCA-positive pancreatic cancer. J. Cancer Res. Clin. Oncol. 144, 1503–1507.

Kuck, K.-H., Reddy, V.Y., Schmidt, B., Natale, A., Neuzil, P., Saoudi, N., et al., 2012. A novel radiofrequency ablation catheter using contact force sensing: Toccata study. Heart Rhythm. 9, 18–23.

Lencioni, R., Crocetti, L., 2013. Image-guided ablation for hepatocellular carcinoma. Multidisciplinary Treatment of Hepatocellular Carcinoma. Springer.

Li, L., Wang, W., Pan, H., Ma, G., Shi, X., Xie, H., et al., 2017. Microwave ablation combined with OK-432 induces Th1-type response and specific antitumor immunity in a murine model of breast cancer. J. Transl. Med. 15, 1–10.

Lin, Z.-Y., Chen, J., Deng, X.-F., 2012. Treatment of hepatocellular carcinoma adjacent to large blood vessels using 1.5T MRI-guided percutaneous radiofrequency ablation combined with iodine-125 radioactive seed implantation. Eur. J. Radiol. 81, 3079–3083.

Ljungberg, B., Bensalah, K., Canfield, S., Dabestani, S., Hofmann, F., Hora, M., et al., 2015. EAU guidelines on renal cell carcinoma: 2014 update. Eur. Urol. 67, 913–924.

Ljungberg, B., Albiges, L., Abu-Ghanem, Y., Bensalah, K., Dabestani, S., Fernández-Pello, S., et al., 2019. European association of urology guidelines on renal cell carcinoma: the 2019 update. Eur. Urol. 75, 799–810.

Lowery, M.A., Kelsen, D.P., Stadler, Z.K., Yu, K.H., Janjigian, Y.Y., Ludwig, E., et al., 2011. An emerging entity: pancreatic adenocarcinoma associated with a known BRCA mutation: clinical descriptors, treatment implications, and future directions. Oncologist 16, 1397.

Lu, M.-D., Xu, H.-X., Xie, X.-Y., Yin, X.-Y., Chen, J.-W., Kuang, M., et al., 2005. Percutaneous microwave and radiofrequency ablation for hepatocellular carcinoma: a retrospective comparative study. J. Gastroenterol. 40, 1054.

Maiettini, D., Mauri, G., Varano, G., Bonomo, G., Della Vigna, P., Rebonato, A., et al., 2019. Pancreatic ablation: minimally invasive treatment options. Int. J. Hyperth. 36, 53–58.

Mainini, A.P., Monaco, C., Pescatori, L.C., Angelis, D.E., Sardanelli, C., Sconfienza, F., et al., 2017. Image-guided thermal ablation of benign thyroid nodules. J. Ultrasound 20, 11–22.

Marrocchio, C., Dababou, S., Catalano, C., Napoli, A., 2018. Nonoperative ablation of pancreatic neoplasms. Surg. Clin. 98, 127–140.

Mauri, G., Cova, L., Monaco, C.G., Sconfienza, L.M., Corbetta, S., Benedini, S., et al., 2017. Benign thyroid nodules treatment using percutaneous laser ablation (PLA) and radiofrequency ablation (RFA). Int. J. Hyperth. 33, 295–299.

Mcnicholas, T.A., 2012. Lasers in Urology: Principles and Practice. Springer Science & Business Media.

Meloni, M.F., Chiang, J., Laeseke, P.F., Dietrich, C.F., Sannino, A., Solbiati, M., et al., 2017. Microwave ablation in primary and secondary liver tumours: technical and clinical approaches. Int. J. Hyperth. 33, 15–24.
Monfardini, L., Gennaro, N., Della Vigna, P., Bonomo, G., Varano, G., Maiettini, D., et al., 2019. Cone-beam CT-assisted ablation of renal tumors: preliminary results. Cardiovasc. Interv. Radiol. 42, 1718–1725.
Munday, J., Higgins, N., Mathew, S., Dalgleish, L., Batterbury, A.S., Burgess, L., et al., 2020. Nurse-led randomized controlled trials in the perioperative setting: a scoping review. J. Multidiscip. Healthc. 13, 647.
Nakatsuka, A., Yamakado, K., Maeda, M., Yasuda, M., Akeboshi, M., Takaki, H., et al., 2004. Radiofrequency ablation combined with bone cement injection for the treatment of bone malignancies. J. Vasc. Interv. Radiol. 15, 707–712.
Navya, P., Kaphle, A., Srinivas, S., Bhargava, S.K., Rotello, V.M., Daima, H.K., 2019. Current trends and challenges in cancer management and therapy using designer nanomaterials. Nano Convergence 6, 23.
Ni, Y., Mulier, S., Miao, Y., Michel, L., Marchal, G., 2005. A review of the general aspects of radiofrequency ablation. Abdom. Imaging 30, 381–400.
O'Neal, D., Cohen, T., Peterson, C., Barr, R.G., 2018. Contrast-enhanced ultrasound-guided radiofrequency ablation of renal tumors. J. Kidney Cancer VHL 5, 7.
Oleinikov, K., Dancour, A., Epshtein, J., Benson, A., Mazeh, H., Tal, I., et al., 2019. Endoscopic ultrasound-guided radiofrequency ablation: a new therapeutic approach for pancreatic neuroendocrine tumors. J. Clin. Endocrinol. Metab. 104, 2637–2647.
Onishi, H., Shirato, H., Nagata, Y., Hiraoka, M., Fujino, M., Gomi, K., et al., 2011. Stereotactic body radiotherapy (SBRT) for operable stage I non–small-cell lung cancer: can SBRT be comparable to surgery? Int. J. Radiat. Oncol. Biol. Phys. 81, 1352–1358.
Orsi, F., Varano, G., 2015. Minimal invasive treatments for liver malignancies. Ultrason. Sonochem. 27, 659–667.
Pacella, C.M., Mauri, G., 2020. History of Laser Ablation. Image-Guided Laser Ablation. Springer.
Paiella, S., Salvia, R., Ramera, M., Girelli, R., Frigerio, I., Giardino, A., et al., 2016. Local ablative strategies for ductal pancreatic cancer (radiofrequency ablation, irreversible electroporation): a review. Gastroenterol. Res. Pract. 2016.
Paiella, S., De Pastena, M., D'Onofrio, M., Crinò, S.F., Pan, T.L., De Robertis, R., et al., 2018. Palliative therapy in pancreatic cancer—interventional treatment with radiofrequency ablation/irreversible electroporation. Transl. Gastroenterol. Hepatol. 3.
Paladini, A., Lucatelli, P., Cappelli, F., Pizzi, G., Anelli, V., Amodeo, E., et al., 2018. Osteoid osteoma treated with radiofrequency ablation in non-operating room anesthesia. A different way of approaching ablative therapy on osteoid osteoma. Eur. Rev. Med. Pharmacol. Sci. 22, 5438–5446.
Parhi, R., Mandru, A., 2020. Enhancement of skin permeability with thermal ablation techniques: concept to commercial products. Drug Deliv. Transl. Res. 1–25.
Petrowsky, H., Fritsch, R., Guckenberger, M., De Oliveira, M.L., Dutkowski, P., Clavien, P.-A., 2020. Modern therapeutic approaches for the treatment of malignant liver tumours. Nat. Rev. Gastroenterol. Hepatol. 17, 755–772.
Poggi, G., Tosoratti, N., Montagna, B., Picchi, C., 2015. Microwave ablation of hepatocellular carcinoma. World J. Hepatol. 7, 2578.
Potenza, C., Bernardini, N., Balduzzi, V., Losco, L., Mambrin, A., Marchesiello, A., et al., 2018. A review of the literature of surgical and nonsurgical treatments of invasive squamous cells carcinoma. BioMed. Res. Int. 2018.

Poulou, L.S., Botsa, E., Thanou, I., Ziakas, P.D., Thanos, L., 2015. Percutaneous microwave ablation vs radiofrequency ablation in the treatment of hepatocellular carcinoma. World J. Hepatol. 7, 1054.

Pouw, R.E., Wirths, K., Eisendrath, P., Sondermeijer, C.M., Ten Kate, F.J., Fockens, P., et al., 2010. Efficacy of radiofrequency ablation combined with endoscopic resection for Barrett's esophagus with early neoplasia. Clin. Gastroenterol. Hepatol. 8, 23–29.

Prud'homme, C., Deschamps, F., Moulin, B., Hakime, A., Al-Ahmar, M., Moalla, S., et al., 2019. Image-guided lung metastasis ablation: a literature review. Int. J. Hyperth. 36, 37–45.

Ridouani, F., Srimathveeravalli, G., 2019. Percutaneous image-guided ablation: from techniques to treatments. La Presse Médicale 48, e219–e231.

Ringe, K., Panzica, M., Von Falck, C., 2016. Thermoablation of bone tumors. RöFo-Fortschritte auf dem Gebiet der Röntgenstrahlen und der bildgebenden Verfahren. © Georg Thieme Verlag KG, pp. 539–550.

Rivero, J.R., De La Cerda III, J., Wang, H., Liss, M.A., Farrell, A.M., Rodriguez, R., et al., 2018. Partial nephrectomy versus thermal ablation for clinical stage T1 renal masses: systematic review and meta-analysis of more than 3,900 patients. J. Vasc. Interv. Radiol. 29, 18–29.

Ruarus, A.H., Vroomen, L.G., Geboers, B., Van Veldhuisen, E., Puijk, R.S., Nieuwenhuizen, S., et al., 2020. Percutaneous irreversible electroporation in locally advanced and recurrent pancreatic cancer (PANFIRE-2): a multicenter, prospective, single-arm, Phase II study. Radiology 294, 212–220.

Runge, V.M., 2018. Dechelation (transmetalation): consequences and safety concerns with the linear gadolinium-based contrast agents, in view of recent health care Rulings by the EMA (Europe), FDA (United States), and PMDA (Japan). Investig. Radiol. 53, 571–578.

Saccomandi, P., Lapergola, A., Longo, F., Schena, E., Quero, G., 2018. Thermal ablation of pancreatic cancer: a systematic literature review of clinical practice and pre-clinical studies. Int. J. Hyperth. 35, 398–418.

Scopelliti, F., Pea, A., Conigliaro, R., Butturini, G., Frigerio, I., Regi, P., et al., 2018. Technique, safety, and feasibility of EUS-guided radiofrequency ablation in unresectable pancreatic cancer. Surg. Endosc. 32, 4022–4028.

Shady, W., Petre, E.N., Gonen, M., Erinjeri, J.P., Brown, K.T., Covey, A.M., et al., 2016. Percutaneous radiofrequency ablation of colorectal cancer liver metastases: factors affecting outcomes—a 10-year experience at a single center. Radiology 278, 601–611.

Shaheen, N.J., Overholt, B.F., Sampliner, R.E., Wolfsen, H.C., Wang, K.K., Fleischer, D.E., et al., 2011. Durability of radiofrequency ablation in Barrett's esophagus with dysplasia. Gastroenterology 141, 460–468.

Shi, S., Vissapragada, R., Abi Jaoude, J., Huang, C., Mittal, A., Liu, E., et al., 2020. Evolving role of biomaterials in diagnostic and therapeutic radiation oncology. Bioact. Mater. 5, 233–240.

Sidoff, L., Dupuy, D.E., 2017. Clinical experiences with microwave thermal ablation of lung malignancies. Int. J. Hyperth. 33, 25–33.

Singh, S., Melnik, R., 2019. Coupled thermo-electro-mechanical models for thermal ablation of biological tissues and heat relaxation time effects. Phys. Med. Biol. 64, 245008.

Snoeren, N., Nijkamp, M.W., Berendsen, T., Govaert, K.M., Van Kessel, C.S., Rinkes, I.H.B., et al., 2015. Multipolar radiofrequency ablation for colorectal liver metastases close to major hepatic vessels. Surgeon 13, 77–82.

Sodergren, M., Mangal, N., Wasan, H., Sadanandam, A., Balachandran, V., Jiao, L., et al., 2020. Immunological combination treatment holds the key to improving survival in pancreatic cancer. J. Cancer Res. Clin. Oncol. 146, 2897–2911.

Song, T.J., Seo, D.W., Lakhtakia, S., Reddy, N., Oh, D.W., Park, D.H., et al., 2016. Initial experience of EUS-guided radiofrequency ablation of unresectable pancreatic cancer. Gastrointest. Endosc. 83, 440–443.

Southwell, D.G., Birk, H.S., Larson, P.S., Starr, P.A., Sugrue, L.P., Auguste, K.I., 2018. Laser ablative therapy of sessile hypothalamic hamartomas in children using interventional MRI: report of 5 cases. J. Neurosurg. Pediatr. 21, 460–465.

Spirou, S.V., Costa Lima, S.A., Bouziotis, P., Vranješ-Djurić, S., Efthimiadou, E.K., Laurenzana, A., et al., 2018. Recommendations for in vitro and in vivo testing of magnetic nanoparticle hyperthermia combined with radiation therapy. Nanomaterials 8, 306.

Steel, A.W., Postgate, A.J., Khorsandi, S., Nicholls, J., Jiao, L., Vlavianos, P., et al., 2011. Endoscopically applied radiofrequency ablation appears to be safe in the treatment of malignant biliary obstruction. Gastrointest. Endosc. 73, 149–153.

Stewart, C.L., Edil, B.H., Ryu, R.K., Rajebi, M.R., 2017. Principles of thermal ablation. Radiation Therapy for Liver Tumors. Springer.

Stewart, C.L., Warner, S., Ito, K., Raoof, M., Wu, G.X., Lu, W.-P., et al., 2018. Cytoreduction for colorectal metastases: liver, lung, peritoneum, lymph nodes, bone, brain. When does it palliate, prolong survival, and potentially cure? Curr. Probl. Surg. 55, 330.

Swan, R.-Z., Tsirline, V., Sindram, D., Martinie, J.-B., Iannitti, D.-A., 2012. Fundamentals of microwave physics: application to hepatic ablation. J. Microw. Surg. 30, 25–40.

Tanaka, T., Yang, M., Froemming, A.T., Bryce, A.H., Inai, R., Kanazawa, S., et al., 2020. Current imaging techniques for and imaging spectrum of prostate cancer recurrence and metastasis: a pictorial review. RadioGraphics. 40, 709–726.

Tang, Z.-C., Chu, Y., Tan, Y.-Y., Li, J., Gao, S., 2020. Pancreatic and duodenal homeobox-1 in pancreatic ductal adenocarcinoma and diabetes mellitus. Chin. Med. J. 133, 344.

Tehrani, M.H., Soltani, M., Kashkooli, F.M., Raahemifar, K., 2020. Use of microwave ablation for thermal treatment of solid tumors with different shapes and sizes—a computational approach. PLoS One 15, e0233219.

Testoni, S.G.G., Healey, A.J., Dietrich, C.F., Arcidiacono, P.G., 2020. Systematic review of endoscopy ultrasound-guided thermal ablation treatment for pancreatic cancer. Endosc. Ultrasound 9, 83.

Testori, A., Ribero, S., Bataille, V., 2017. Diagnosis and treatment of in-transit melanoma metastases. Eur. J. Surg. Oncol. (EJSO) 43, 544–560.

Tombesi, P., Vece, F.D., Sartori, S., 2015. Radiofrequency, microwave, and laser ablation of liver tumors: time to move toward a tailored ablation technique? Hepatoma Res. 1, 52–57.

Van Veldhuisen, E., Van Den Oord, C., Brada, L.J., Walma, M.S., Vogel, J.A., Wilmink, J.W., et al., 2019. Locally advanced pancreatic cancer: work-up, staging, and local intervention strategies. Cancers 11, 976.

Vankayala, R., Hwang, K.C., 2018. Near-infrared-light-activatable nanomaterial-mediated phototheranostic nanomedicines: an emerging paradigm for cancer treatment. Adv. Mater. 30, 1706320.

Vitale, A., Peck-Radosavljevic, M., Giannini, E.G., Vibert, E., Sieghart, W., Van Poucke, S., et al., 2017. Personalized treatment of patients with very early hepatocellular carcinoma. J. Hepatol. 66, 412–423.

Wang, J.-H., Wang, C.-C., Hung, C.-H., Chen, C.-L., Lu, S.-N., 2012. Survival comparison between surgical resection and radiofrequency ablation for patients in BCLC very early/early stage hepatocellular carcinoma. J. Hepatol. 56, 412–418.

Wang, X., Sofocleous, C.T., Erinjeri, J.P., Petre, E.N., Gonen, M., Do, K.G., et al., 2013. Margin size is an independent predictor of local tumor progression after ablation of colon cancer liver metastases. Cardiovasc. Interv. Radiol. 36, 166–175.

Waseem, H., Osborn, K.E., Schoenberg, M.R., Kelley, V., Bozorg, A., Cabello, D., et al., 2015. Laser ablation therapy: an alternative treatment for medically resistant mesial temporal lobe epilepsy after age 50. Epilepsy Behav. 51, 152–157.

Wilkins, L., 2019. Monitoring liver stiffness changes during controlled hyperthermia and microwave ablation using point shear-wave elastography. Radiol. Imaging Cancer 1, e194013.

Wong, S.L., Mangu, P.B., Choti, M.A., Crocenzi, T.S., Dodd Iii, G.D., Dorfman, G. S., et al., 2010. American Society of Clinical Oncology 2009 clinical evidence review on radiofrequency ablation of hepatic metastases from colorectal cancer. J. Clin. Oncol. 28, 493–508.

Wood, B.J., Poon, R.T., Locklin, J.K., Dreher, M.R., Ng, K., Eugeni, M., et al., 2012. Phase I study of heat-deployed liposomal doxorubicin during radiofrequency ablation for hepatic malignancies. J. Vasc. Interv. Radiol. 23, 248–255. e7.

Xu, Z., Xie, H., Zhou, L., Chen, X., Zheng, S., 2019. The combination strategy of transarterial chemoembolization and radiofrequency ablation or microwave ablation against hepatocellular carcinoma. Anal. Cell. Pathol. 2019.

Yang, W., Lee, J.-C., Chen, M.-H., Zhang, Z.-Y., Bai, X.-M., Yin, S.-S., et al., 2019. Thermosensitive liposomal doxorubicin plus radiofrequency ablation increased tumor destruction and improved survival in patients with medium and large hepatocellular carcinoma: a randomized, double-blinded, dummy-controlled clinical trial in a single center. J. Cancer Res. Ther. 15, 773.

Yao, W., Lu, M., Fan, W., Huang, J., Gu, Y., Gao, F., et al., 2018. Comparison between microwave ablation and lobectomy for stage I non-small cell lung cancer: a propensity score analysis. Int. J. Hyperth. 34, 1329–1336.

Ye, X., Fan, W., Chen, J.H., Feng, W.J., Gu, S.Z., Han, Y., et al., 2015. C hinese expert consensus workshop report: guidelines for thermal ablation of primary and metastatic lung tumors. Thorac. Cancer 6, 112–121.

Yu, J., Liang, P., Yu, X., Liu, F., Chen, L., Wang, Y., 2011. A comparison of microwave ablation and bipolar radiofrequency ablation both with an internally cooled probe: results in ex vivo and in vivo porcine livers. Eur. J. Radiol. 79, 124–130.

Yu, J., Zhang, X., Liu, H., Zhang, R., Yu, X., Cheng, Z., et al., 2020. Percutaneous microwave ablation versus laparoscopic partial nephrectomy for cT1a renal cell carcinoma: a propensity-matched Cohort study of 1955 patients. Radiology 294, 698–706.

Zhang, H., Zhu, X., Zeng, Z., Gao, X., 2021. Interventional therapy combined with radiotherapy for pancreatic carcinoma. Integrative Pancreatic Intervention Therapy. Elsevier.

Zhu, J., Yu, M., Chen, L., Kong, P., Li, L., Ma, G., et al., 2018. Enhanced antitumor efficacy through microwave ablation in combination with immune checkpoints blockade in breast cancer: a pre-clinical study in a murine model. Diagn. Interv. Imaging 99, 135–142.

Zorbas, G., Samaras, T., 2015. A study of the sink effect by blood vessels in radiofrequency ablation. Comput. Biol. Med. 57, 182–186.

5

Ablation probes

Chapter outline

5.1 Radiofrequency ablation probe

Radiofrequency ablation (RFA) is a minimally invasive treatment technique to achieve tumorous tissue necrosis by protein denaturation of the cell. It is based on heating the target tumorous tissues to above 45°C to treat various focal secondary and primary malignancies in tissues, such as the liver, kidney, bone, breast, lung, thyroid, adrenal gland, and pancreas (Armellini et al., 2015; Fang et al., 2018). It has been recently used in preference to other traditional treatment techniques, such as hormonal therapy, chemotherapy, radiation therapy, and surgery to overcome their side effects and drawbacks, such as time consuming, bad cosmetic results, and expensive costs, as well as achieving minimal scarring, a reduced recovery time, reduction in morbidity, mortality, cost, hospital stay, complication risks from anesthesia, less pain, improved life quality of cancer patients, and potential reduction of financial, emotional, and physical impacts (Singh et al., 2015, 2016; Singh and Repaka, 2015, 2018a). According to the DNA Advanced Pain Treatment Center, RFA is considered an effective treatment technique for tumors, bone spurs, and cervical and lumbar facet joint pain. RFA can be used to treat low back pain resulting from sacroiliac joint complex, pain arising in the

Thermal Ablation Therapy. DOI: https://doi.org/10.1016/B978-0-12-819544-4.00003-4
© 2021 Elsevier Inc. All rights reserved.

sphenopalatine ganglion that causes a headache, pain in the dorsal root ganglion resulting from disc herniation, trigeminal neuralgia, sympathetic ganglia, and arthritis pain, as well as atypical facial pain. However, it is unsuitable for patients with the following conditions: (1) weigh above 250 pounds, (2) pregnancy, (3) having blood clotting disorders, (4) having an active infection, or (5) diabetes. Patients who suffer from heart conditions, diabetes, or take blood-thinning drugs, such as aspirin and Coumadin, should stop their drugs for 1 week before the RFA procedure. Medications, such as nonsteroidal antiinflammatory drugs (NSAIDS) and ibuprofen, need to be stopped for at least 24 hours, and preferably 72 hours before the RFA procedure. According to the Mayo Clinic, RFA, which is also called radiofrequency (RF) neurotomy, is the main preparation before conducting the RFA procedure. During the procedure, the patient wears a hospital gown, lies on an X-ray table in the prone position, with an intravenous (IV) line placed in the patient's hand or arm for patient comfort during the procedure, numbing medication is injected into the patient's skin before the RF needle insertion, and energy is supplied from an RFA generator by a connected RF needle which is inserted inside the tumor with the aid of image-guidance modalities, such as magnetic resonance imaging (MRI), ultrasound, or computed tomography (CT) to ensure the needle is properly placed (Singh and Repaka, 2015). The ablated lesion volume mainly depends on the external current intensity, the ablation duration, density composition in the heterogeneous organs, tumor location from the core of the body, needle location inside the tumor, the length of the needle's active tip, tissue conductivity, blood vessel size, location, and distance from the needle's active tip (Singh et al., 2015, 2016; Singh and Repaka, 2018b).

After the RFA, several procedures are followed, including rest, the soreness resulting from the needle insertion disappears after a day or two, with an ice pack not a heating pad placed on the injection region for 20 minutes, three or four times per day during the first recovery day, and baths are avoided for 2 days. However, the patient may take a warm shower with 24 hours after the RFA procedure. The main limitation related to the standard RFA is tissue charring at the needle interface, where the extended durations of ionic heating lead to desiccation of the tissues adjacent to the needle. Hence, these charred tissues act as an insulator, which prevents any energy transfer through the tumorous charred tissues and limits the lesion size and shape, from both engineering and medical points of view. There are other factors responsible for RFA failure, including inaccuracy of the imaging system and RF needle placement (Zhang et al., 2016), as shown in Fig. 5.1.

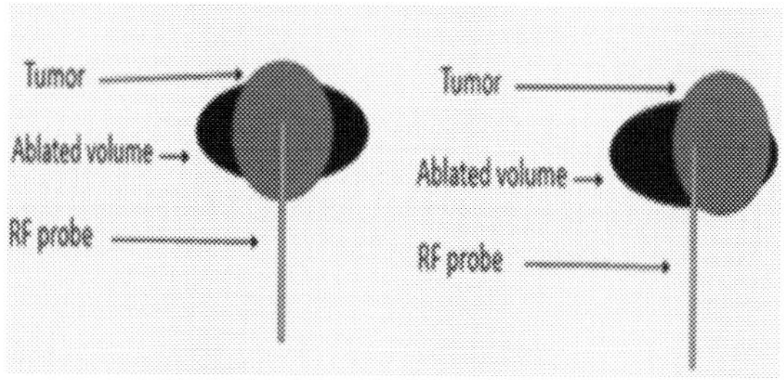

Figure 5.1 Factor causing the failure of RFA.

Continuous RFA is considered the initial power supply method, and uses continuous radio waves. These waves are increased until the tumorous tissue temperature reaches a critical value (Erdogan et al., 2000). Generally, the benefits of pain relief using the continuous RFA procedure are longer than those using the pulsed procedure. From previous studies, it was deduced that clinical RFA efficiency sharply decreases while treating tumors with diameters larger than 3 cm. Consequently, RFA is restricted to tumors with diameters less than 3 cm using a single needle due to the tissue-charring problem. To overcome this problem, three aspects are considered, including:

- The design of the RF needle, in which the contact between the needle and the tissue is required to be expanded to spread the power into a wide region as in the Christmas tree needle, the umbrella needle, the bicomponent conformal needle, the origami needle, the InCircle Monarch needle, and perfusion needle, produces a round and larger ablation zone.
- The design of the energy source in a pulsed RFA using short duration (20 ms) of high current or voltage level followed by low current or voltage level for about 489 ms. During the currents' alternating pattern, the temperature changes between 40°C and 42°C for 2 or 8 minutes depending on the injection place. Thus, the charring problem and the tissues around a needle can be cooled during the periods of low levels of current or voltage. Thus, more energy is delivered to the tumorous tissue, and a large necrosis area is achieved. The pulsed method was developed to reduce the adverse reaction risk, including deafferentation pain and nerve damage. Also, the loss of temperature in the off phase of the duty cycles with a following rapid temperature increase during the energy application in the on phase in the duty cycles produces more homogeneous and deeper lesions (Li et al., 2012; Erdogan et al., 2000). However, this pulsed power supply method is still capable of avoiding the tissue-charring problem and as a result fails to

treat tumors with diameters larger than 3 cm (Singh and Melnik, 2019; Pangarkar and Miedema, 2014; Zhang et al., 2016). Consequently, the power supply method is used, which includes a control system, during the RFA. It is possible to control the power delivered to the tissues according to three main modes, including:

- Impedance-controlled mode, in which the impedance between the ground pad and RF needle is monitored until reaching a preset value. Then, the RF power delivery is halted for a time period, once the dehydrated charred tissues are cooled, the impedance decreases again to be lower than the prohibited preset value, and the power generator also resumes the power delivery. This method is repeated multiple times to obtain a large necrosis size (Zhang et al., 2016), and
- Temperature-controlled mode, in which the RF power produced from the generator is adjusted manually using a trial-and-error method or using programming such as proportional integral derivative (PID) embedded within the power generator, on the target threshold temperature, which is observed by an embedded temperature sensor at the RF needle tip, which sends temperature information as feedback to the controller to be adjusted as shown in Fig. 5.2, to avert biological tissue overheating and charring, and
- The power-controlled mode (Singh and Repaka, 2017; Zhang et al., 2016).

Finally, combining RFA with conventional adjuvant therapies, such as radiotherapy or chemotherapy, is used in the treatment of tumors with diameters larger than 3 cm. In RFA combined with radiotherapy, two mechanisms are available to

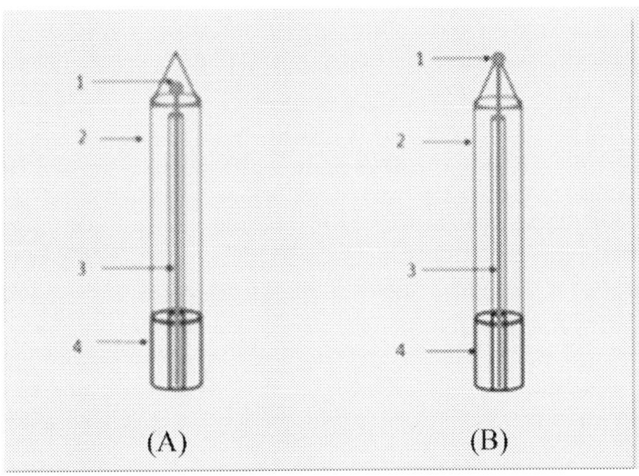

Figure 5.2 RF needle with an embedded thermal sensor, which is present at the electrode tip: (A) sensor is not contact with the tumorous tissue, and (B) sensor is contact with it. (1) Thermal sensor, (2), nickel titanium, (3) wire, and (4) polyurethane.

increase the target necrosis size, including (1) tumor sensitization to subsequent radiation resulting from increased oxygenation due to hyperthermia-induced area, and (2) inhibition of recovery and repair in the radiation-induced area, and increasing in the free radical formation. In RFA combined with chemotherapy, three different mechanisms exist to increase the area of target tumor necrosis, including (1) the effect of two hits on the susceptible tumorous tissues, which contains the initial reversible injury to cells resulting from the RFA procedure in the more peripheral target ablation zone, which is followed by an irreversible injury through chemotherapy agents, (2) the effect of increasing cell stress on the tumor necrosis, and (3) temporary blood supply occlusion of the tumor, including the hepatic artery or portal vein, to avoid the heat-sink effect, where, during the laparoscopy or open RFA, the inflow portal occlusion causes a temperature increase of 10°C around the RF needle and the larger necrosis tumor zone. For percutaneous RFA, the hepatic or celiac artery can be blocked with a balloon catheter and the feeding arteries can be embolized with particles made from gelatin sponge (Zhang et al., 2016; Widmann et al., 2009; Yamashiki et al., 2005; Rossi et al., 2000). Nevertheless, these additional treatment therapies have known disadvantages, such as epithelial surface damage, gastrointestinal distress, and immune system depression (Iwamoto, 2013; Burdelya et al., 2012).

5.1.1 Working operation of the radiofrequency probe

The electromagnetic spectrum (EMS) contains various waves at different frequency ranges, which includes radio waves that range from 3 Hz to 300 GHz, infrared, visible, ultraviolet, X-rays, and gamma-rays (Hong and Georgiades, 2010).

The RF waves can be used in thermal ablation to burn a defined tumorous tissue volume. The RF needle is considered the cathode of the electric circuit that is closed by grounded pads, which act as an anode, on the thighs of the patient as observed in Fig. 5.3 (Zhang et al., 2016).

Fig. 5.3 illustrates that the needle tip in a small section area causes a large energy flux to exist around it. Otherwise, the grounded pad has a large section area, which causes energy dispersion and reduces the energy flux, while the distance from the needle tip increases. Consequently, the damage to the tumorous tissues is restricted to the electric circuit part that surrounds the needle tip. Hence, the water molecules

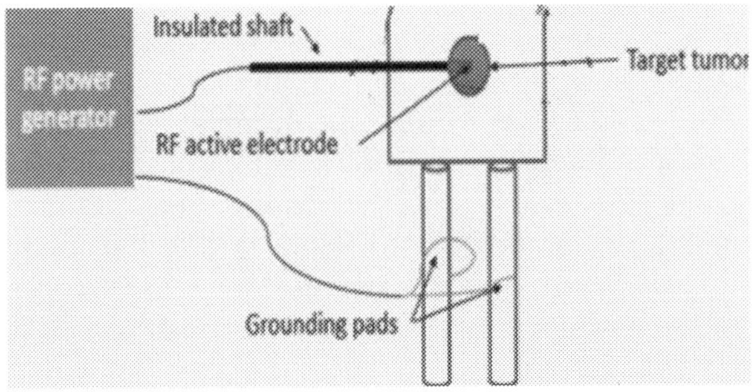

Figure 5.3 Equivalent electric circuit for a tumor inside a patient's body, and the RF needle.

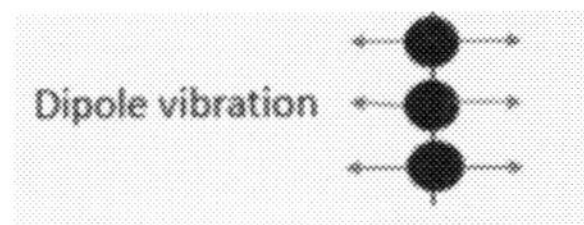

Figure 5.4 Vibration of tissue molecules with the changing direction of alternating current.

inside the tumorous tissue next to the RF needle tip try to align in the alternative current direction and are obliged to rapidly vibrate according to the externally applied RF current. The molecules, which are far from the needle, also vibrate due to the effect of the adjacent vibrating molecules. The vibration effect of the tissue molecules is illustrated in Fig. 5.4 (Hong and Georgiades, 2010).

Fig. 5.4 demonstrates that the alternative current causes water molecules to be moved with the change in direction. Hence, the friction between adjacent water molecules causes energy loss and a temperature increase (Singh et al., 2016; Singh and Repaka, 2018a,b). Generally, the temperature and energy exponentially decrease with increasing distance from the needle until the constant core temperature is reached and zero tissue necrosis at a larger distance (Hong and Georgiades, 2010; Chang and Nguyen, 2004). The RF needle itself is not the heat source, but it does produce an RF field, which forces the adjacent tissue molecules to move and cause intense agitation. These water molecules near the needle are considered the heat source that is transported further by the effect of tissue conductivity (Hong and Georgiades, 2010).

One of the main RFA limitations is the tissue-charring problem, which heavily depends on the good thermal and electrical conductivities of the tissue. It is considered one of the main reasons for the small necrosis area due to RFA (Zhang et al., 2016). In the case of pressing the power button of the RF generator too quickly and causing a faster energy increase, the tissues around the needle become desiccated before the ability to achieve a maximum ablation tumor region. A larger ablation zone occurs due to the rapid increase in temperature inside the tissue around the needle. Also, the tissue impedance increases causing tissue charring and limiting the required heat transmission for maximum ablation tumor (Hong and Georgiades, 2010). Also, the electrical impedance increases, where the temperature increases due to the occurrence of roll-off (Fang et al., 2019). Thus, these desiccated tissues near the electrode act as an insulator that limits the energy and temperature transmission to the remainder of the tumor region. This phenomenon is known as roll-off (Romero-Méndez and Berjano, 2017; Trujillo et al., 2012; Yap et al., 2019, 2020; Hiyoshi et al., 2019).

5.1.2 RF electrode types and design

The design of the RF electrode plays an important role in the ablation therapy procedure, where the first traditional one is called a plain electrode. The plain RF electrode consists of an active tip and an insulated shaft, as shown in Fig. 5.5 (Mulier et al., 2005). It was used in the first RFA experiments to treat limited ablated tumor volumes, and tiny lesions of size less than 5 mm, such as a nidus of osteoid osteomas due to the rapid increase in the electric impedance and charring around the RF needle (Mulier et al., 2005; Widmann et al., 2009).

To overcome the limitations in the ablated size, several approaches were used to develop an advanced RF electrode, such as expandable electrodes, cooled electrodes, wet electrodes, and bipolar electrodes. Accordingly, the expandable electrode consists of curved needles inside a hollow-shaft cannula, and are deployed for diameters greater than 7 cm in the proper position inside the tumor (Aubé et al., 2004; Bruners et al., 2008; Dupuy, 2001). This expandable electrode type can produce large conical or spherical shapes according to the surface size of the electrode (Wood et al., 2007). It has two well-known types:

- The multitined expandable electrode contains of an array consisting of 4–12 curved prongs that are expanded from the cannula in a Christmas tree-like or umbrella-like style. Thus, tissue charring decreases due to the power distribution over a broader surface area (Mulier et al., 2005).
- The coiled expandable electrode contains a spring that allows the hollow needle tip out is deployed vertically to the shaft (Mulier et al., 2003).

Figure 5.5 Plain RFA electrode.

Unfortunately, during tines expansion, the expandable electrodes are a potential damage risk to the pleura, adjacent vessels, liver capsule, surgical staples, and bile ducts (Lin et al., 2007; Chen et al., 2006; Rhim et al., 2001). Therefore cooled electrodes were developed consisting of an inner chamber that divides the internal electrode space into a concentric inner and outer lumen. The internal lumen first pumps a chilled saline solution with temperature ranging from 0°C to 8°C into the electrode tip, and the outer lumen returns the saline solution to an outer collection unit (Wagata et al., 2016; Mulier et al., 2005; Widmann et al., 2009; Kho et al., 2020).

To prevent the tissue-charring problem, great deposition of the energy into the target tissue is achieved, and a larger necrosis tumor volume than achieved with the plain electrode, a single cooled electrode can be used (Goldberg et al., 1996). Furthermore, a cluster cooled tip electrode with a specific number of electrodes and distance between them can produce a higher intensity of RF current. This causes less tissue charring around the electrode tip, and achieves larger necrosis volumes than a single electrode (Goldberg et al., 1998; Choi et al., 2016; Zhou et al., 2019). In Fig. 5.6, a cluster electrode is shown, which consists of three cooled probes with the distance between (Hong and Georgiades, 2010).

The distance between the electrodes of the cluster cooled RFA tip in the same cluster system should be studied. For example, if the electrodes are separated by a distance larger than 1 cm, the shape produced from overlapping will change to obtain a coagulation area with the shape of a dumbbell or dog bone. Also, if the distance between the cluster electrodes is increased, the shape will consist of separate and distinct areas

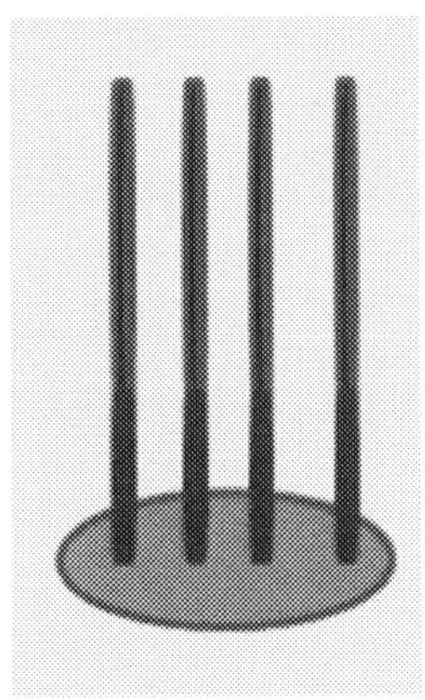

Figure 5.6 Cluster cooled RFA tip electrode.

around each single electrode, producing viable tumorous tissues in the area between (Hong and Georgiades, 2010; Burdío et al., 2008; Haemmerich et al., 2005; Yoon et al., 2014).

Additionally, the wet electrode (Fig. 5.7) is a hollow electrode, which contains a number of holes at the uninsulated end to inject a hypertonic or isotonic saline solution into the target tissue region to increase the electrical and thermal conductivities (Mulier et al., 2005). This allows an increase in the power deposition by values that are more than 10-fold as compared with a plain electrode. Hence, the necrosis area increases (Leveillee and Hoey, 2003; Ni et al., 2000; Miao et al., 2000; Livraghi et al., 2000; Bruners et al., 2008).

The perfusion of the saline solution inside the tissue is done at a percentage ranging from 0.9% to 36% at a rate of 0.5–2 mL/minute (Widmann et al., 2009). However, using saline solution perfusion has a drawback related to the obtained irregular shape of the necrosis volume due to the irregular distribution of infused saline (Lee et al., 2003, 2005b,c). In addition, hot saline diffusion along the liver capsule, needle track, and vessels is correlated with thermal injury to the neighboring organs, or portal vein thrombosis (Lee et al., 2005a; Frich et al., 2006; Park et al., 2010).

Commonly, for RFA, monopolar probes are used, in which the electric current passes through the target tumorous tissue from an active tip embedded in the probe to a great dispersive grounding pad on the patient's skin, which leads to a skin burn problem (Al-Alem et al., 2014; Kröger et al., 2006). Then, the bipolar probes were developed, in which the electric current is transferred through the tumorous tissue between two electrodes on the same probe and separated by a variable distance inside the tumor without the need for grounding pads. The ground pad is used to disperse equal energy and heat amounts, and avoid skin burns in its position (Vorländer et al., 2018; Osaki et al., 2013).

The electrodes embedded in the bipolar probe inside the target tumorous tissue are active with minimal loss of energy. Therefore

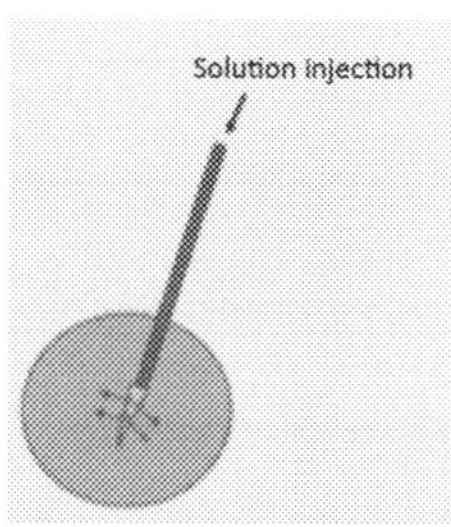

Figure 5.7 Wet RFA electrode with saline solution infusion using two methods.

it efficiently can achieve greater ablated tumor volumes gaining the approval of the US Food and Drug Administration (FDA), the InCircle bipolar probe (RFA Medical, Freemont, California), which is compatible with several available RF generators manufactured at different corporations. The InCircle probe contains two electrodes that are put on the tumor sides and surrounded without penetrating the lesion.

In addition, a multiple electrode system is a combination of several available single-shaft electrodes of any type, such as a cooled-wet electrode, expandable wet electrode, bipolar wet electrode, bipolar cooled electrode, bipolar expandable electrode, and bipolar cooled wet electrode (Francica et al., 2019; Tan et al., 2018; Mulier et al., 2005; Ben-David et al., 2019). The electrode system is designed according to several parameters, including:

- The electrode number embedded in the system, such as double, triple, or quadruple;
- The electric current mode, such as monopolar or bipolar;
- The activation mode of electrodes, such as simultaneous, consecutive, or switching;
- The tumor size; and
- The type of insertion site, such as a monofocal tumor, which can be ablated by insertion of electrodes at the same tumor region, or a multifocal tumor, such as trifocal or bifocal, which can be ablated at the same time by insertion of the electrodes at different sites inside the tumor.

The switching box in the RFA system is programmed to alternately dispense the electric current to distinct locations. It can be also programmed in any electric mode as well as any activation mode (Tomasian and Jennings, 2018; Shiina et al., 2018; Mulier et al., 2020; Shao et al., 2017). There is inaccuracy in the bipolar RFA term, that may cause confusion, and so the meaning should be explained more to determine whether there is a bipolar mode between more than or exactly two electrodes embedded parallel to the multiple electrode system or if it is meant that there is a bipolar single-shaft probe (Tomasian and Jennings, 2018; Shao et al., 2017; Khiabani et al., 2020; Petit et al., 2020).

The embedded electrodes in the multiple system are activated by three modes, namely:

- The simultaneous mode, in which a single RF power generator is connected to all electrodes in the multiple system and supplies them with power at the same time (Xu et al., 2016; Brace et al., 2008). However, according to the Faraday cage, the electrostatic shielding phenomenon occurs, leading to scattered electric flux lines in the region between electrodes and an unheated portion in this region (Yoon et al., 2013). Hence, the level of damage and the electrostatic shielding are affected by the distance between the electrodes, which should be relatively short to avoid the resulting previous phenomenon and to obtain complete tumor ablation. Therefore the simultaneous activation mode is appropriate for small tumors and can save time (Haemmerich et al., 2005; Huang et al., 2020).

- The consecutive mode, in which the electric current flow is changed to the second electrode to be activated after the completion of the ablation session for the first electrode (Xu et al., 2016; Kim et al., 2017). The time between the consecutive RF ablation sessions should be reduced to 5 seconds or less to avoid tissue cooling between them. This duration is required to change the sequential connection in the RF generator as well as one electrode from the existing group (Lee et al., 2007; Yoon et al., 2015). During the consecutive process, the first electrode in the group of electrodes is supplied from the RF generator until a specific time, then this first electrode stops the ablation process, and the second RF needle begins to ablate. The temperature around the second electrode increases, while the temperature around the first electrode is gradually decreased. It has been proved that, though the temperature can decrease through the ablation procedure under the effect of consecutive mode, the tumorous tissue damage by hyperthermia is irreversible (Xu et al., 2016; Korkusuz et al., 2017; Park et al., 2014).
- The switching mode, in which the energy delivery switches between the electrodes in the system and is automatically adjusted by a microprocessor embedded with the RF generator according to the impedance, and is continuously measured (Widmann et al., 2012; Yoon et al., 2015; Bale et al., 2012; Tan et al., 2018).

Generally, like any treatment technique, RFA has several advantages and disadvantages. In terms of cost, RFA is relatively cheaper compared with other ablation techniques, such as high-intensity ultrasound ablation (HIUA) and microwave ablation (MWA). In addition, RFA is considered a typical single procedure, with oncologist awareness of the ablation technique, the small size of the RF needle with gauges from 14 to 17, long record of accumulated data, wide availability, availability of compatible MRI, accepted safety profile, gentle burns ingrained in the RFA technique, the effect of hemostasis, which is useful for periprocedural bleeding, multiorgan experience, and other RFA roles used in medicine, including electrocautery and cardiac EP ablations, (Seror, 2015; Chen et al., 2016; Kim et al., 2011), are all counted as the main advantages of RFA. On the contrary, RFA disadvantages are due to the generated skin pad burns, its operator-dependent procedure, relatively slower speed, limitations in ablated size, with volumes from 4 to 7 cm, concerns and difficulty of needle track seeding, heat sink resulting from blood vessels, limited visualization as a result of ultrasound imaging guidance during the monitoring of RFA therapy, and dependence on additional imaging guidance (Hong and Georgiades, 2010). Moreover, other side effects caused by RFA include bruising at the injection site, skin discoloration, and bleeding, which subside in a few days (Fischer et al., 2015; McCormick and Walega, 2018; Guttler, 2020).

5.1.3 Blood vessel effect on RF ablation

Blood vessels close to tumorous tissues cause heat sink, which limits the ablated tumor volumes and the temperature distribution (Singh and Repaka, 2015; Zorbas and Samaras, 2015; Lim et al., 2010; Jiang et al., 2014; Kai et al., 2013). The blood vessel cooling effect causes the necrosis shape to be slightly asymmetric, and the distribution of temperature to be highly nonuniform. This phenomenon occurs due to the higher electrical conductivity of the blood passing through the vessels, which diverts larger RF current toward them, and dissipates the thermal heat resulting under the ablation therapy from the surrounding tumorous tissues. In addition, the bifurcated blood vessel and complex vascular system have a greater effect on incomplete tumor destruction, such as the liver and its inhomogeneous nature, which contains three different types of blood vessels, called hepatic veins, portal veins, and hepatic arteries, with different flow velocities, diameters, hepatic tumors, liver parenchyma, stroma, and bile ducts, each of which has unique thermal and electrical properties (Tungjitkusolmun et al., 2002; Berber and Siperstein, 2008).

The blood flow inside the vessels is one of the most important parameters affecting the formation of RF lesions. There can be normal flow, high flow, and reduced flow in patients who suffer from cirrhosis due to portal hypertension (Iwakiri, 2012). For example, it is known that the normal velocity of the blood flow in the portal vein has a value of 22.9 cm/s, however, its velocity reduces to 16.9 cm/s due to portal hypertension (Haemmerich et al., 2003). Subsequently, the heat transfer coefficient h (W/(m^{2}K)), which is between the tissue and vessel is calculated for reduced ($h = 639$ W/(m^{2}K)) and normal

($h = 810\ \mathrm{W/(m^2K)}$) blood velocities of the portal venous using the following formula (Nazemian and Nabaei, 2020):

$$h = \frac{k_b Nu_D}{D} \tag{5.1}$$

where k_b is the blood's thermal conductivity, D is the diameter of the blood vessel, and Nu_D is the average local Nusselt number, which is approximated by the following expression (Haemmerich et al., 2003):

$$Nu_D = 4 + 0.48624 \ln^2 (Pr \times Re \times D/(18 \times L)) \tag{5.2}$$

where Pr is the Prandtl number, which is equal to 25 for blood, L is the length of the blood vessel, and Re is the Reynolds number, which is computed using the following equation (Nazemian and Nabaei, 2020):

$$Re = \frac{v_b \rho_b D}{\mu} \tag{5.3}$$

where μ is the blood viscosity, ρ_b is the blood's mass density, and v_b is the average velocity of blood, with a portal vein length of 70 mm, and the Nusselt number is calculated to be 16.2 and 18.8 for normal blood flow and reduced flow due to portal hypertension from cirrhosis, respectively (Haemmerich et al., 2003).

In terms of the RF electrode, there are some parameters, such as the length, diameter, orientation, and location of blood vessels that have dominant effects on the lesion formation, as follows.

- The tumor necrosis volume decreases with the increase in the perfusion rate of lesion blood.
- The great blood vessels that are orthogonal with respect to the electrode lead to less ablated volume than other parallel blood vessels.
- During RFA, the convective RF energy can easily destroy the parallel blood vessels and their neighboring normal tissues.
- The small vessels or large vessels containing blood with a small flow rate can form thermal tail-like lesions, which can damage the downstream areas of the vessel (Huang, 2013; Kai et al., 2013). The higher flow velocity of blood has a great influence on the heat transfer. Therefore the aorta inside the brain tissue leads to larger ablated volume and temperature distribution with a greater temperature than obtained in the case of the terminal and main veins due to the higher velocity of blood flow inside the aorta and its greater diameter (Keangin and Rattanadecho, 2020).

Generally, the worst case is considered to be a great blood vessel which is very close to the probe (Zorbas and Samaras, 2015; Lin et al., 2012; Iguchi et al., 2007; Huang, 2013). Generally, the blood vessel effect varies according to the ablated probe, such as a microwave (MW) probe, bipolar (BP) probe, and monopolar (MP) probe. The BP is slightly affected by heat sink rather than the MW and MP probes, whether the heat sink effect exists or not. It produces larger ablated volumes and masses as compared with those generated from the MW and MP ablation probes. Also, it is known that RF devices can produce ablated volumes with a spherical shape rather than the MW devices that produce ellipsoidal shapes (Pillai et al., 2015).

Generally, during RFA treatment planning, it is very important to successfully determine the location of blood vessels to apply additional heating to be able to compensate for the cooling effect resulting from them. A semiautomated algorithm has been developed to detect small veins and arteries in the liver using magnetic resonance venography (MRV) and angiography (MRA). This skeletonization algorithm is called ordered region growing, in which the paths of veins and arteries are extracted. It requires the identification of all distal and proximal endpoints inside all blood vessels. The path extraction of the vessel is done interactively, and the wrong paths are manually canceled. The 3D paths of vessels are smoothed, and meshes of tubular surfaces are then created from all available paths of the vessel. The tubular mesh diameter is tapered for its radius at any point to be proportional to the determined distance from the location of the distal endpoint (Yim et al., 2001). During the RFA procedure on the liver using a Pringle maneuver, the vascular occlusion of the liver reduces the cooling effects from the vascular perfusion and the larger blood vessels. It is used in a minimally invasive mode, such as laparoscopic and percutaneous, also, another method, using vaso-active pharmacological agents, is used to minimize the flow of blood inside the liver, and hence vessel cooling is effected (Haemmerich et al., 2003; Haase et al., 2012). Likewise, RFA is combined with percutaneous ethanol injection (PEI) to improve tumor treatment in high-risk locations, which contain great blood vessels of diameter more than or equal to 3 mm, subcapsular (tumors with 10 mm of capsule), or a vital structure, or hepatocellular carcinomas (HCCs) tumors with high-risk locations near hepatic blood vessels or bile ducts. It is difficult to realize complete tumor ablation using only RFA to avoid bile duct stenosis, bilioma, or postoperative bile leak (Jiang et al., 2014). Commonly, a specific amount of ethanol is injected into the tumorous region close to the vital structure or blood vessel while the RF needle is set at a distance ranging between 5 and 10 mm from the ethanol injection needle, and is immediately activated after the PEI (Wong et al., 2008).

5.1.4 Practical RF generator and commercial probes

The fabricated RF generators consist of a voltage acquisition unit, input voltage unit, power amplifier unit, load unit, current acquisition unit, smart power unit, processor unit, input control unit, drive unit, and signal transmission unit. They can produce alternating current with power ranging between 200 and 250 watts, and frequencies of 460–500 kHz through RF needles with gauge usually of 14–17 of different configurations. According to the FDA approval, there are three main manufacturers of monopolar RF electrodes, and one multipolar/bipolar manufacturer in the United States and Germany (Armaiz Flores et al., 2018; Rathke et al., 2014), including the Boston Scientific probe, which produced RF generators, called RF 3000 (Boston Scientific, Natick, Massachusetts, United States), which generates an RF signal of frequency 480 kHz, and power with a maximum value of 200 watts, starting with 20–80 watts, and increasing at a rate of 5–10 watt per minute according to the target tumor size until reaching the maximum value of 55–200 watts (Hong and Georgiades, 2010). If the roll off (increase in the impedance feedback) occurs before reaching the target ablated power, its delivery is automatically stopped for a specific duration and then resumes (Rathke et al., 2014). Moreover, the transmitted energy is in the monopolar mode, also, the Boston Scientific probes have active tips of 3/5 cm, and the user control mechanism of induced energy is impedance controlled (Rathke et al., 2014).

The first manufacturer, Boston Scientific RF, which made a monopolar probe with an umbrella = like array, called LeVeen, has deployment diameters ranging from 2 to 5 cm. During the RFA procedure, all the RF electrode tips return the impedance feedback to avoid the charring problem (Hiraki et al., 2010; Hong and Georgiades, 2010). The single-step expansion by the LeVeen probe is available except for the 5 cm LeVeen electrode that has a staged deployment. Boston Scientific also provides an introducer coaxial system, which permits CT scanning related to placement of the RF needle without the need for bulky hubs (Mahnken et al., 2011).

The second manufacturer is called Covidien, and produced the Cool tip by Radionics/Valley Lab/Covidien (Mansfield, Massachusetts, United States), which can generate a signal frequency of 480 kHz, and maximum power value of 200 watts. It operates in a multipolar mode using a switching controller, and has an impedance-controlled technique, which monitors and regulates the impedance of tissue to enable the most suitable ablation power (Lanuti et al., 2012; Skonieczki et al., 2011; Krokidis and Ahmed, 2013). The device is attached to an infusion pump during the ablation procedure to inject a chilled saline solution to avoid vaporization and carbonization at and around the needle tip (Rathke et al., 2014; Lee et al., 2018). The Covidien system uses the RF Cool tip probe (MedWrench.com) with electrodes of different lengths to ablate different tumor volumes ranging to 3 cm ablated volume (Rachbauer et al., 2003). The Covidien offers a cluster needle containing three electrodes mounted on a handle to ablate larger tumor volumes up to 5 cm in diameter (Hong and Georgiades, 2010).

The third major manufacturer of RF devices is Angio Dynamics, which produced Model 1500X (RITA medical devices, Mountain View, California, United States; DOTmed.com). Its commercial RF device has the ability to generate signals with a frequency of 460 kHz, producing power with a maximum value of 250 watts, and has a temperature-controlled algorithm with thermocouples at all electrode tines that monitor the maximum temperature inside the ablation region to prevent reaching unwanted temperatures (Krokidis and Ahmed, 2013; Ankory et al., 2019). The RITA produces Starburst XL electrodes to ablate tumors with volume ranging between 3 and 5 cm. According to the Angio Dynamics protocol, tumors with a size of 3 cm in diameter can be ablated with 150 watts applied power and 9 minutes ablating time. In addition, the same power can ablate tumors of 5 cm in diameter with an increased ablation time of 15 minutes (Rathke et al., 2014). The RITA electrode has nine deployable curved tines, which are configured in a Christmas tree shape, and expanded to different sizes to ablate various tumors.

Furthermore, the Angio Dynamics manufacturer produces a versatile collection of flexible RF electrodes, such as Starburst Flex and Starburst Semi Flex, that are made of flexible (polymer) and rigid (metal) trocar sections, and have the ability to bend up to an angle of 90 degrees in all possible directions to mitigate the gantry constraints of CT and MRI systems (Hong and Georgiades, 2010).

The FDA approved only the smaller SB SDE and SB MRI (Hong and Georgiades, 2010; Zakeri et al., 2017). The SB SDE device is used for small tumor ablation of 2 cm. It has arrays with a side deployed that allows physicians to accurately place the trocar tip of the electrode and easily view its position.

The RITA produces a saline infusion system, called the intelliflow pump, which infuses a saline solution into the target ablation region to decrease tissue charring and enhance the ablation. For example, the Starburst Xli-enhanced electrode is able to ablate large tumors with size up to 7 cm (Hong and Georgiades, 2010). Also, the Starburst Xli- enhanced semi-flex RFA device has flexibility features (Gregory et al., 2013).

In the Starburst categories, the Starburst Talon is deployed and designed for lesions with difficult access, which are close to critical structures, and may move in the soft tissues. Also, the Talon electrode is used in intraoperative, percutaneous, and laparoscopic procedures (Mukund et al., 2020; De Filippo et al., 2020).

5.2 Microwave ablation

MWA is efficient for clinical handling of several tumor types in different organs, including lung, bone, liver, and kidney. The MW warms the targeted tissues by generating an alternating electromagnetic (EM) field by MW antennas (Shiomi et al., 2008; Chan et al., 2017). Temperatures higher than 60°C cause instantaneous cell destruction, which requires prediction and precise control of the tissue temperature in the tumor tissues to avoid damaging surrounding healthy tissues.

5.2.1 Working operation of a microwave probe

The EM energy propagation depends on the magnetic permeability and dielectric permittivity of the medium through which the EM waves travel. For biological tissues, the relative magnetic permeability equals 1, while, the dielectric permittivity differs to a high degree based on the tissue type. Dielectric heating occurs inside tissues due to the generated heat by dipole rotation (Prakash, 2010). Both effective conductivity and relative permittivity depend on the temperature and frequency. The wavelength, calculated from the relative permittivity, is a significant parameter for the antenna design. It restricts the minimum antenna size for practical MWA applications.

The system of the MWA includes four main parts are: (1) the MW source to control the power and MW frequency, (2) conductor, (3) connector (coaxial cable), which joins and transmits the

MW generator to the conductor, and then, to the MW probe, and (4) the MW probe (applicator), which is an antenna to transfer and radiate MW power into the neighboring tissues, leading to raising the temperature of the targeted tissues to be ablated through the tip.

In the MWA system, the power is passed from the high-power MW generator to the targeted tissues via an applicator. In clinical use, most of MWA systems operate at 915 MHz and 2.45 GHz, which are the most commonly used frequency bands in the scientific, medical, and industrial applications. These two frequencies exhibit less cable heating and large penetration depth. Nevertheless, higher frequencies such as 9.2 GHz, 10 GHz, 14.5 GHz, 900 MHz–18 GHz, and 24 GHz are sometimes used (Luyen et al., 2014).

5.2.2 Microwave applicator types and designs

The different parts of the MWA system are still under study to enable effective ablation. The efforts are directed at design applicators equipped to generate large spherical ablation patterns. The applicator (probe) is considered the main part of the MWA system as it has direct contact with the targeted tissues. Since tumor tissues in different body organs have different thermal and conductivity properties, researchers have been devoted to designing new MWA probes, which are different in their dimensions and designs, aiming to implement an efficient MW probe in terms of the radiation pattern, the transferred MW power, and minimum time of the ablation process, which match the targeted ablated region's size/shape without damaging large areas of the surrounding healthy tissues.

MW applicators comprise a transmission line for feeding, which ends with a radiating antenna. The antenna radiates the electric field, which is absorbed in the adjacent tissue regions causing heating. The transmission line heats up once the MW power is provided from the generator, where the commercial transmission lines are fabricated from dielectrics and conductors with significant attenuation. The applicator's efficacy at transmitting the power from the feeding transmission line to the antenna is based on several design parameters and settings, including the operating frequency and radiating antenna's geometry, as well as the EM properties of the tissues. These parameters also control the spatial contour of the radiated electric field into targeted tissues. Generally, at the operating frequency, the antenna geometry in relation to the EM wavelength also determines the radiation pattern. It is required to choose the antenna geometry which produces a matched electric field "E" radiation pattern with the targeted tissue regions to be ablated. Though the absorbed EM power is relative to $|E|^2$, the passive conduction of cooling/heating also affects the thermal contours in the tissues owing to the blood flow. Such effects must be measured while designing the MW antenna.

- Most antennas applied in the MWA of near-spherical shaped tumors yield elongated ablation regions. These elongated ablated regions damage the nearby healthy tissues and result in complicated burns. Researchers have developed numerous antennas for MWA of near-spherical tumors, some of which are reported in Fig. 5.8, including floating sleeve antenna (Yang et al., 2006), choke antenna (Prakash et al., 2009), and cap-choke antenna (Cavagnaro et al., 2010).

Other studies established small-MW antennas to create spherical ablation zones (Wang et al., 2015; Luyen et al., 2016) as reported in Fig. 5.9.

In MWA, an antenna which is connected to an MW generator is inserted in the targeted tumor tissues for ablation, where the generated energy produces heat and friction. The main categories

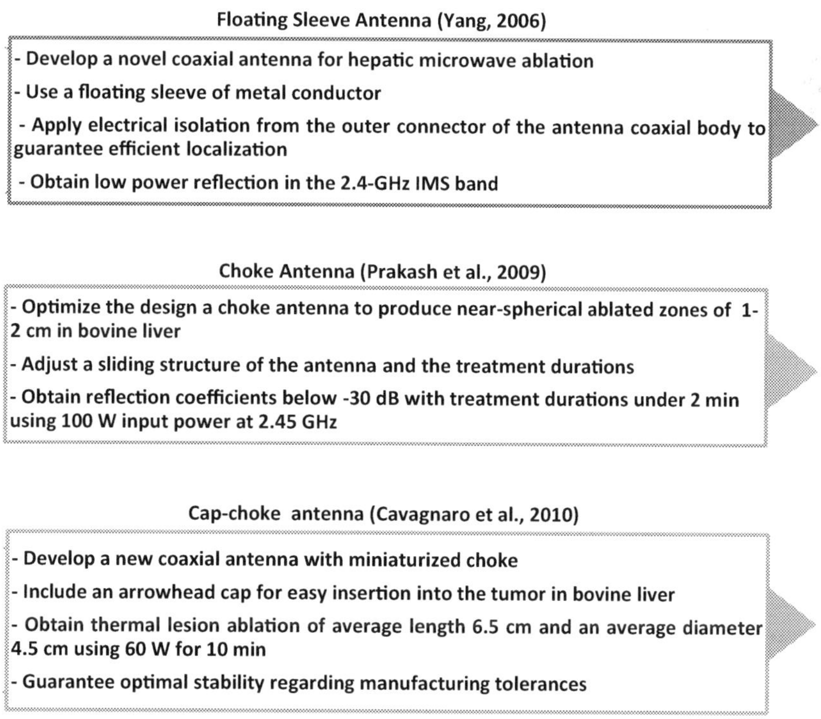

Figure 5.8 Studies on MWA antennas of near-spherical tumors.

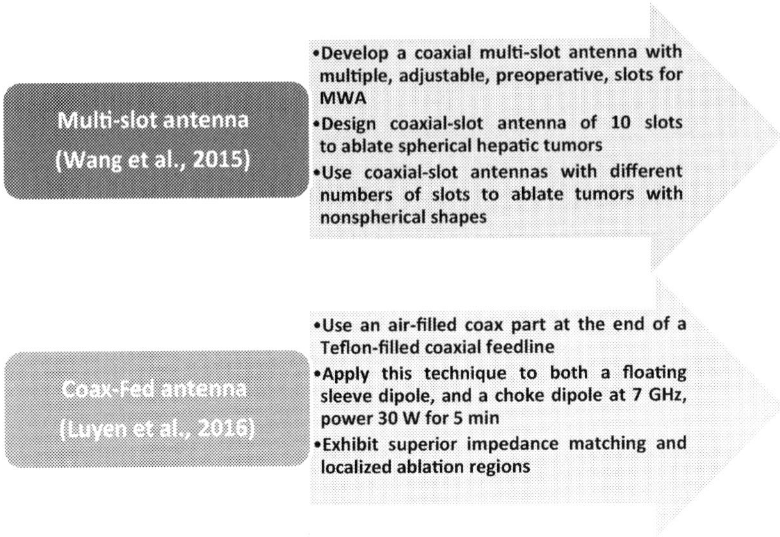

Figure 5.9 Studies on a small antenna for spherical tumor ablation.

of the MW probes are classified based on their main structure into slotted probes (single-slot, multislot, single slot with 1T ring), choked probes (choked, cap choke, double-slot cap choke), pole probes (monopole, dipole), axial probes (triaxial, floating sleeve), and helix and spiral probes as illustrated in Fig. 5.10.

The most common feeding transmission line inside MWA applicators is the semirigid coaxial cable due to its cylindrical geometry. The cylindrical transmission line enables easy inset of the MW probe within the rigid catheters. Also, the MW antenna can be combined by adapting the distal tip geometry of the cable. The MW applicator/antenna is an essential element of the ablation device due to its noteworthy influence on the ablation pattern. It is designed for fitting inside an insulating material to simplify the insertion of the MW probe in the human body, offer mechanical stability and protection, and enhance the power coupling with the tissues. The design of the MW antenna is considered a tradeoff between tissue heating pattern, power competence, and antenna diameter to ablate a specific tissue region. Commonly, the antennas are designed from coaxial cables, thus, the small-diameter antennas may have trouble handling high power levels without undesirable thermal harm to the surrounding tissues around the proximal antenna tip (Brace et al., 2005; Wolf et al., 2008). Most MW probes use a needle-based antenna causing an ovoid ablation pattern. In practice, the probe is positioned in the midpoint of the targeted tumorous tissues, where the ablation region grows

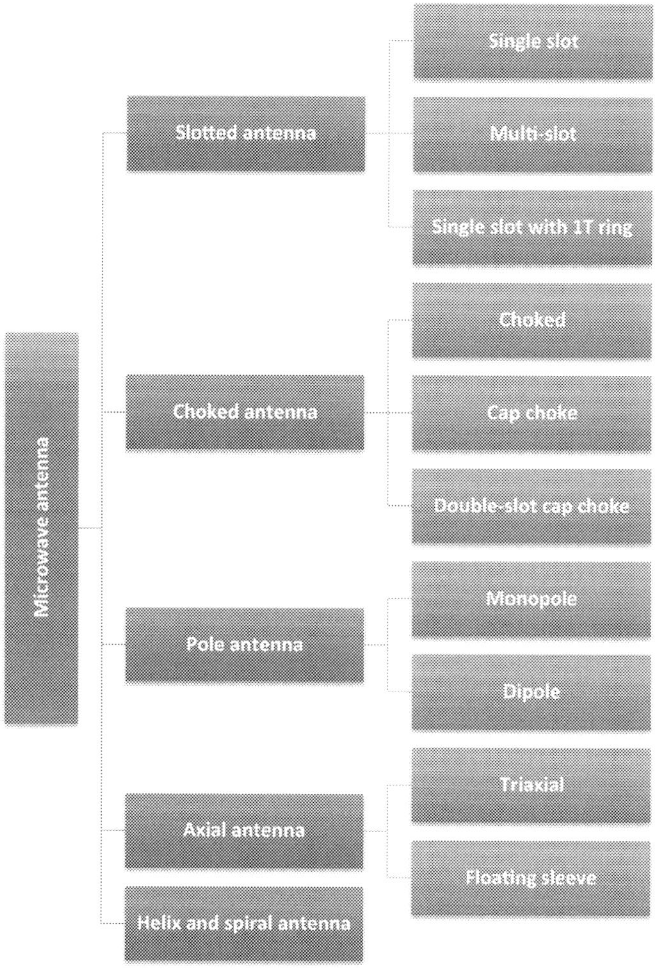

Figure 5.10 Categories of MWA antennas.

radially outward. Accordingly, different MWA probes are designed to realize an affective ablation with consideration of the cases at which the tumor exists at critical regions in the body. Thus, the ablation process should be conducted cautiously to avoid any harm to nontargeted tissues. Different MWA antenna designs are reported below that can be used in clinical applications for various applications.

5.2.2.1 Slotted MW antennas

Slotted antennas are classified into single-slot (SS) antenna, multislot (MS) antenna, and single slot with 1T ring antenna. The single-slot (SS) antennas have proved their superiority compared to using monopole antenna, a dipole antenna, triaxial antenna, choked antenna, and RF probes. To obtain a larger ablated diameter that covers the whole tumor zone, a greater number of SS probes can be used with power 30 watts for each probe, which is more efficient than to one probe with 90 watts. This is owing to enhancement of the energy spatial distribution rather than directing the power using one probe with a great value in one position (Brace, 2010). Recently, several studies have been conducted to compare the SS antenna to the monopole, dipole, dual slot, and sleeved antennas (Fallahi and Prakash, 2018). At 0.5–4 GHz, a dipole with a sleeve and more symmetric power absorption profile was achieved with a good distribution of the heat around the antenna, where the backward heating decreased. However, the reflection coefficient was superior using the traditional dipole compared to the one with a sleeve, while the SS antenna provided good results.

To enhance the radiation location while attaining a more spherical ablation profile a multislot (MS) coaxial antenna can be used (Brace, 2011). The MS antenna is like the SS antenna with more scattered slots along the coaxial cable. In the case of a multislot antenna, it is found that the number of slots affected greatly the MWA efficiency. A comparison between using one air slot, two slots, five slots, and fifteen slots proved that increasing the number of slots changed the isothermal contour and ablation profile. Also, changing the length of the slots using the same power showed that decreasing the slot length provided a more spherical shape of the ablated zone (Wang et al., 2015). The reflection coefficient (SAR) also changes with a change in the number of slots (Gas, 2017). At 2.45 GHz, the reflection coefficient values were −21.55, −29.55, and −19.40, using one slot (SS), two slots, and three-slots, respectively, where the best value was obtained using the two-slot antenna. In addition, the area of the ablated region increases with increasing number of slots and adding the inner copper hollows (Ge et al., 2018). Other studies measured the reflection coefficient to evaluate the number of slots at different power values (Gas and Szymanik, 2018).

To reduce the problem of backward heating and gain the highest power distribution behind the existing slot, a single slot with a 1T ring was developed by adding external T-rings (Acikgoz and Turer, 2014). By comparing the performance of the traditional SS antenna, SS with one T-ring, three T-rings, five T-rings, seven T-rings, and nine T-rings, it was found that using five T-rings was superior to one T-ring in terms of the SAR, even with increasing temperature.

5.2.2.2 Choke applicators

The three main types of coaxial antennas are choked antennas, including (1) traditional choked antenna, (2) cap choke antenna, and (3) double-slot cap choke antenna. The choke antennas are usually used with coaxial dipole antennas, where adding a choke element enhances the impedance matching of the antenna (Longo et al., 2003). Also, adding a cap to the antenna's tip localizes the antenna's radiation to the distal end of the probe, and increases the capacitance. The traditional MW choke antenna was simulated numerically by Nevels et al. (1998) for catheter cardiac tissue ablation compared to a practical monopolar applicator covered with Teflon and holding a choke and disk at the end of the antenna. The disk impulses the MW radiation toward the probe tip and reduces the operating frequency to its halved value. Accordingly, the length of the probe is reduced to half at 2.45 GHz, leading to a small heart cavity area.

The choke on the feeding line reduces the probe length, and the backward heating is reduced by directing the currents inside the choke. It stops the returning wave, which prevents any undesirable heating of the blood in the vein. Later, the design of the choke MW antenna was enhanced by Prakash et al. (2009) to ablate nearly spherical tumors with radius 1–2 cm. The size of the ablated region is based on the probe structure and the ablation treatment duration. Probe simulation was performed and mathematical results were confirmed by experiments using ex vivo bovine liver.

To enhance the spherical pattern of the ablated region, a cap is added to the antenna radiating part with dielectric materials, such as ceramic, alumina, or Teflon. The cap choke antenna was developed by Lin and Wang (1996) by adding a cap choke to guarantee the essential distribution of the stated absorption rate with good reflection coefficient. The choke acts as an open circuit to prevent the returned current, and the cap increases the probe tip capacitance, which improves the generated radiation by the MWA probe. The cap can be designed using variability of shapes, including arrow, ring, or disk. The reduction of the reflex current causes the cap choke MWA antenna to be unaffected by the insertion depth.

To reduce the coaxial cable heating and minimize the returning current, Pisa et al. (2001) considered a cap choke antenna working at 915 MHz, which is unaffected by the insertion depth of the MWA applicator in the tissues. Later, for MWA at 2.45 GHz, an axial antenna with choke in the outer conductor including a cap and an arrowhead cover for the tip was designed (Wakaki et al., 2017). The cap and choke are assisted in localizing the heating locations along the probe and provided uniform temperature distribution for ablating spherical tumors using a power of 10 watts in 110 seconds. To increase the operating frequency from 0 to 20 GHz, a double-slot cap choke antenna can be used (Bertram et al., 2006).

5.2.2.3 Pole applicators

The main categories of pole applicators are (1) monopole applicator and (2) dipole applicator, which are considered to be directional applicators. Such probe designs can restrict MW radiation toward the ideal direction to ablate targets in the proximity of critical structures (Chaichanyut and Tungjitkusolmun, 2015).

Typically, the monopole antenna is designed by shedding a $\lambda_{effective}/4$ sector of the outer conductor in the coaxial cable. The reflection coefficient and absorption rate are considered the main factors to compare the characteristics of the metal-tip monopole antennas (Labonte et al., 1995). Changing the antenna design in terms of the dielectric length and metal tip length has a significant effect on the ablation process. Three monopole applicators were designed using a metal tip, a dielectric tip, and an open tip, respectively (Labonte et al., 1996), for MWA of endocardial catheterization. The monopole antenna with a metal tip proved its superiority compared to the others. However, in terms of the shape of the ablated zone, the dual-slot antenna provided more a spherical ablated region compared to the monopole antenna (Chiang et al., 2013). To achieve steering of heating during MWA, a modified monopole was designed by adding a slot on one side of the monopole antenna with a reflector backed (Mohtashami et al., 2017). The added portion of the external conductor works as a reflector, which focuses the radiation in a direction away from the reflector. The reflector decreases the undesirable returned currents along the antenna, which diminishes the required antenna's diameter. In addition, the monopole antenna with a reflector acts as a drill (Fallahi and Prakash, 2018). The monopole applicator can have a ring shape to be located around the tumor for ablating it from all directions for rapid ablation. Using the monopole probe with high power at between 110 and 120 watts does not require a cooling unit, while a low power less than 50 watts necessitates using a cooling unit to diminish the ablated zone, and reduce the backward heating along with the applicator (Adewumi et al., 2019).

The common dipole probe consists of two equal-length conductors detached by a small gap, while the transmission line of feeding is linked. The coaxial dipole integrates a conductor length of $\lambda_{effective}/4$ (effective wavelength) parted from the cable. Additionally, a dipole can include a sleeve.

To increase the temperature, the conventional dipole probe is modified by connecting a $\lambda/4$ sleeve positioned on the feed line to design a dipole antenna with a sleeve (Hurter et al., 1991). The generated heating patterns are independent of the depth of penetration into the tissues. This dipole with a sleeve produced highly concentrated energy in the targeted tissues with reduced damage in the nearby healthy tissues, where the sleeve acts as with high resistance (open end) leading to zero return current on the outer conductor. To control the MWA process, the dielectric properties of biological tissues are measured using an open-ended coaxial antenna and a dipole antenna (Wang and Brace, 2011) by calculating the reflection coefficient, and the complex permittivity of the nearby tissues to the probe. However, compared to the SS antenna, the dipole antenna provides more backward heating, causing tissue burning along the axis of the antenna insertion using 10 watts at 2.45 GHz (Maini and Marwaha, 2012).

The operating frequency has a significant role during the ablation process. In MWA using a dipole antenna at a frequency of 915 MHz, the ablated zone diameter was almost 25% smaller than the diameter when using 2.45 GHz (Taj-Eldin and Prakash, 2014). Also, the ablated region is more spherical. Consequently, to ablate large tumors, and more spherical ablated zones, the 2.45 GHz frequency is recommended with a dipole antenna. To design an MWA device, which destroys tumors in a predefined direction while avoiding any damage to neighboring nervous structures, a dipole antenna with the addition of a catheter metal was implemented (Alnassan et al., 2014). The added catheter metal was used to direct the EM waves in a specific direction. The produced radiation from the active part of the dipole antenna increases the rooted area's size by increasing the used power.

The chief limitations of the applicators during MWA are because of the nonspherical ablation profile, backward heating, reliance of the ablation pattern, and operating frequency on the insertion depth of the applicator. Typically, the current passes through the antenna within the coaxial cable on the inner conductor surface and within the outer conductor's surface. A portion of the current would pass back in the direction of the MW source on the outer conductor's outer surface owing to an unbalanced coaxial cable structure. This current produces undesirable heat lengthwise of the feeding cable and shrinks the ablation pattern sphericity. Accordingly, an effective key to overcoming these limitations, by overpowering the backward currents and preserving the direction of the power deposition to the antenna's radiating part, is to add a sleeve or choke to the outer conductor of the cable (Yang et al., 2006). Also, the solution to recover the ablation profile sphericity is to add a cap and dielectric loading materials, such as ceramic, alumina, or Teflon, on the antenna radiating part (Chiu et al., 2003).

5.2.2.4 *Axial applicator*

Generally, a coaxial cable with transverse electromagnetic propagation is used as a feed line to the MW ablation probe. However, the unstable design of the probe causes a flow of returned current into the probe's outer connector. The returned current leads to an impedance mismatch and unwanted damage close to the probe's insertion point. To solve the impedance mismatch problem between the probe and feed line, other applicators were designed using a choked, monopole, or dipole probe. Nevertheless, these probes are unpractical due to their large diameter, which impairs the insertion process of the MWA probe through the skin. Hence, to prevent the return currents (loss) using small-diameter probes while improving the reflection coefficient, axial applicators are recommended, where the ablation needle (active part) is located at one-fourth of the wavelength from the coaxial monopole antenna's base forming a triaxial structure (Brace et al., 2004).

The triaxial MW probe is used mainly to enhance both the reflection coefficient and the impedance match, where the standing waves caused by the reflected power cause overheating of the coaxial cable, which in turn may lead to cable failure and surface burns (Brace et al., 2007). Compared to the RFA probe, the triaxial MW probe with high power provides a larger ablation zone (Laeseke et al., 2009, Andreano et al., 2010). Also, the triaxial MW probe provides an elliptical ablation zone as stated by Santos et al. (2010). Furthermore, cooling during the ablation process can be applied to solve the problems of backward heating and make the ablated tissues more spherical shaped (Knavel et al., 2012).

For therapeutic applications using MWA, the most commonly used frequencies are 433 MHz, 915 MHz, and 2.45 GHz, even though the International Telecommunication Union (ITU) and the Federal Communications Commission (FCC) permitted several unlimited frequency bands in medical applications.

Accordingly, Brace (2010) defined the MW frequencies in MWA to be from 300 MHz to 300 GHz.

In MWA, to realize a high precise absorption rate with a low reflection coefficient, another axial applicator is the floating sleeve probe that can be used. It is an isolated metal conductor from the outer conductor of the antenna, where the metal sleeve reduces the background thermal emission along with the antenna and increases the spherical emission zone around the probe (Yang et al., 2006). It can be improved by adding multiple sections along its divided edge as designed by Maini and Marwaha (2013), who implemented a modified shape of the traditional floating sleeve probe. Also, a modification in the dual-slot antenna for MWA was performed by adding a metal sleeve, which controls the specific absorption rate (Ibitoye et al., 2015). Another modification is considered in the floating sleeve probe, called the tapered cap floating sleeve (Maini, 2016). Also, the sleeve can be in a single-slot dipole probe to reduce the return current at the probe's outer conductor (Fallahi and Prakash, 2018).

5.2.2.5 Helix and spiral applicators

Spiral, helical, and lobed antennas are of a compact size, which provides homogeneous distribution of the absorption rate, uniform heat pattern along the helical axis, and constant heat pattern irrespective of the insertion depth. The significant gain in using the helical probe compared to the traditional dipole probe is the stability of the heat pattern. In addition, the spiral antenna can be used inside a filled balloon with a low-insulation liquid to control the generated heat in the tissues. The structure of the antenna supports MW energy propagation, where the helix antenna configuration includes a helix dielectric-core antenna and a helix air-core antenna.

The helix probe is designed by removing a chunk of the outer conductor from the coaxial cable end. The common structure of the spiral probe includes a helical wire with a helix shape (Fallahi and Prakash, 2018). The maximum MW radiation intensity is obtained along the helix axis when the helix diameter and distance between the helix rolls are equal. Different probe designs were settled for MWA with the objective of competently transferring the MW power to the targeted tissues with a radiation pattern well matched to the shape and size of the tumor tissues. For thermal treatment, the most common spiral antenna designs are as follows.

- Helical coil-opened for cardiac eradication, which is designed by shortening the proximal end of the helix antenna;

- Helical coil-separated for atrioventricular ablation, where the distal end of the coil is linked to the inner conductor of the coaxial cable;
- Helical coil-connected for esophagus treatment, where the distal end is electrically linked to the inner conductor of the coaxial cable with a shortened proximal end to the outer conductor.

The main factors affecting the produced heat during the MWA process are a proper determination of the radiating part length and the number of coil turns. Several studies have been conducted to improve the design of the spiral probe for a more realistic pattern of the ablated regions such as those by Ryan (2013), and Lee and Son (2019). For cardiac ablation, spiral and helix antennas can be combined in the interior catheter-based devices (Fallahi and Prakash, 2018; Maini and Shekhawat, 2018).

5.2.3 General considerations in the antenna design

For effective ablation, attentions should be focused while designing the used applicator in the following (Hulsey et al., 2015; Lyons and Pua, 2019; Luyen et al., 2015; Chen et al., 2017):

- Confine the heating, increase the energy absorption within tumor zones, and increase the size compactness of the probe, frequencies higher than 0.9, 1.5, and 2.45 GHz are used with MWA to achieve high thermal conductivity.
- Decrease the cancer recurrence possibility by realizing adequate safety margins during the ablation, where for example, 50°C can ablate a 5–10 mm radial distance surrounding the tumor in healthy tissues to guarantee no expansion of the cancerous cells via the blood vessels.
- Create electromagnetic radiation matched to the tumor shape by using Maxwell equations to calculate the current distribution on the antenna surface while considering the boundary conditions.
- Generate heat suitable to the tumor shape with minimized redundant radiation exposure by synthesizing the essential electric field.
- Maintaining reflection less than −10 dB is crucial to overcoming the increase in reflection throughout the ablation process.
- Create a narrowed homogeneous heat pattern by using sufficient temperature levels essential for effective ablation, which produces end-fire (highly directed) radiation within the tumor.
- Control the heating inside the tumor by applying directed radiation using low input power.

Generally, the power-effective designs of MWA applicators can be classified as a coaxial slot, linear element, looped, or helical-based designs. Such designs provide good patterns of broadside radiation. However, they are comparatively narrow-band, and the radiation can be disposed of in elongated patterns or hot spots. The slot or coaxial applicators are considered to yield relatively low reflections with satisfactory radiation patterns, which are required for only small-volume ablation. Nevertheless, the slot antennas suffer from extreme backward heating along the length of the probe shaft. To reduce such backward heating, coaxial chokes were designed by eradicating the current on the coaxial cable's outer conductor. Still, the chokes are undesirable in percutaneous applications. Constantly improving the antenna's performance includes achieving size compactness with focused energy radiation, wider bandwidth, and minimum reflection. These requirements are challenging and indispensable for designing efficient antennas to guarantee an optimal ablation treatment process. Ultimately, the ablated region shape and size in live tissues are based on the antenna design, change in the thermal conduction, tissue type/properties, and thermal sinks due to neighboring structures like blood vessels. The interplay between these factors is intricate, leading to different system designs.

From the preceding studies of MWA probes, there are differing inconsistencies with their design or the added modifications. Each study compares the presented probe under the same conditions with other varieties to determine its efficiency and features. The thermal distribution of the triaxial probe was studied in Brace et al. (2007) and Laeseke et al. (2009). Also, Brace (2010) compared the single-slot, triaxial, monopole, choked, and dipole MWA probes and the RFA probe, showing that the greatest efficiency was achieved by the choked probe and triaxial probe using 90 watts of power. To increase the spherical shape of the ablated area, and reduce the backward heating, a metal casing was used to modify the dual-slot antenna (Ibitoye et al., 2015). Another addition was used with the unipolar probe by adding water for cooling and a reflector (Fallahi and Prakash, 2018). The position of the slot in a single-slot probe was studied by stirring the slot in three different locations with regard to the center of the tumor (Razib et al., 2016). Also, changing the number of slots and the length of one slot affected the thermal distribution during the ablation process (Wang et al., 2015). Adding a T-ring to the single-slot probe and changing its position affects the thermal distribution, as studied in Ashour et al. (2020). A change of the tip's type, including dielectric, metal, and open, in monopole probes has an effect on the heat distribution (Labonte et al., 1996). In addition, it was found that higher temperature results from a two-pronged monopole probe compared with a one-pronged probe (Chaichanyut and Tungjitkusolmun, 2015). The used frequency and power affect the thermal profile greatly during the ablation procedure (Ryan, 2013). Adding a choke and cap produces a spherical breast tumor area (Wakaki et al., 2017), and adding a spiral portion affects the measured temperature (Lee and Son, 2019).

5.2.4 Practical microwave applicators

The MWA commercial systems consists of a generator, antennas, a power distribution system, and a cooling system to reduce the effect of the shaft heating due to the reflected power. The generated power is produced from solid-state sources or a magnetron, which can be controlled independently of the impedance spikes and tissue types. The generator frequencies are usually 915 MHz or 2.45 GHz, as permitted by the Federal Communications Commission. The EM energy is distributed from the generator to the antenna through a coaxial transmission line, which has excellent propagation characteristics. However, with decreasing cable diameter, the power loss increases, which is considered the main tradeoff between the cable diameter and cable heating. Fig. 5.11 reports several MWA systems in commercially use.

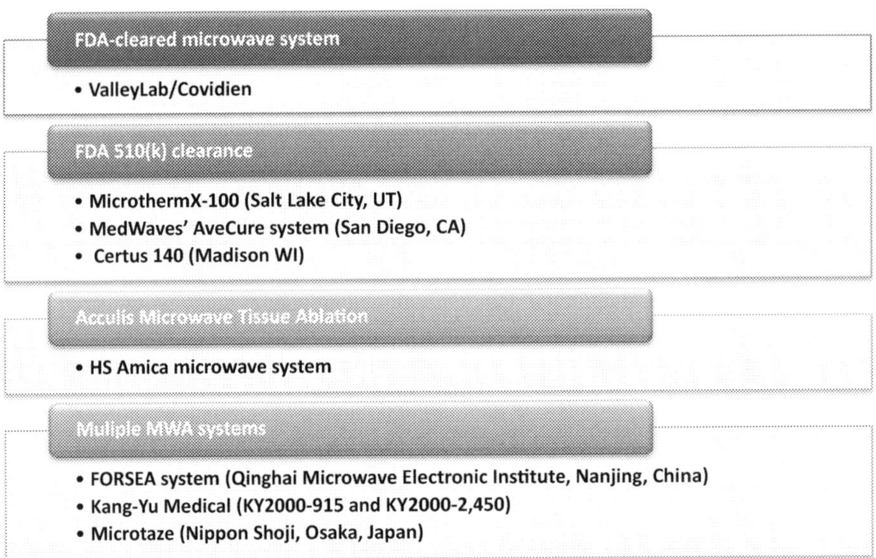

Figure 5.11 Examples of commercial MWA systems.

5.3 Conclusion

MWA of tumors is an exciting procedure with several theoretical advantages compared to other existing ablation systems as it generates larger ablation zones, faster, using very high temperatures, with less predisposition to heat-sink effects. It is recommended to ablate high-impedance tissues like bone and lung using MWA. Preclinical and clinical trials have confirmed these advantages; nonetheless, more studies are required to be conducted to develop more robust clinical systems. By overviewing the most common ablation techniques, both RFA and MJWA are the most powerful methods for ablating small and large tumor volumes, respectively. This chapter has introduced several RF and MW probe designs. To compare the success and performance of the different probe designs, the same conditions should be applied. Such comparisons are conducted between the different probes in terms of the reflection coefficient, heat transfer profile, the ablated zone size/shape, and the required power/temperature for complete ablation of the tumor zones. Previous studies have been carried out using different probe types to improve the efficiency during the ablation process by studying several relations, including (1) the relationship between the emitted maximum temperature and the used frequency and (2) the relationship between the ablation period and the required power for complete ablation. Typically, decreasing the probe frequency, increased the required power to eradicate the tumor using the same time period, leading to an increase in the maximum temperature.

In clinical use, RF and MW ablation devices are the most extensively used in various ablation technologies, such as tumor ablation and treatment of cardiac arrhythmias. MW energy was reported in the late 1970s for thermal soft-tissue ablation using monopolar MW antennas for surgical clotting. However, most of the progress in MWA technology is intensive to thermal ablation of large unresectable tumors in vascular organs like the liver. Compared to RFA, MWA enables the heating of larger tissue volumes. Researchers have focused on augmenting the needle-based ablation applicators to offer large spherical ablation patterns while suppressing the backward current on the outer conductor surface of the coaxial cable by adding sleeves and chokes in the MWA probe along with keeping the electric fields focused around the antenna tip. Also, by loading the probe with a dielectric material with electrical permittivity approximately equal to that of the tissues can be used in antenna design for creating better coupling between the antenna and the surrounding tissue. Although different modifications can be performed to improve the ablation pattern, there remains room to design less invasive and thinner applicators. An air-filled coaxial slice can be added at the end of a Teflon-filled coaxial line with a balloon assimilated on the outer conductor of the coaxial cable to obtain thinner applicators. Although antennas using rigid coaxial feedlines are industrialized for tumor ablation, spiral and helix antennas are joined within flexible catheter-based devices for cardiac ablation. For endometrial ablation, bone drilling, and collagen reforming on the cornea, localized and small ablation patterns using transmission-line-based applicators are used. Along with the RFA and MWA introduced in this chapter, other ablation probes with different structures and concepts exist for laser ablation, cryosurgical ablation, and surgical ablation.

References

Acikgoz, H., Turer, İ., 2014. A novel microwave coaxial slot antenna for liver tumor ablation. Adv. Electromagn. 3, 20–25.

Adewumi, O., Agbasi, O., Fatile, O., 2019. Effect of water-cooled loop monopole antenna on microwave ablation efficiency. Mehmet Akif Ersoy Üniversitesi Sağlık Bilimleri Enstitüsü Dergisi 7, 41–54.

Al-Alem, I., Pillai, K., Akhter, J., Chua, T.C., Morris, D.L., 2014. Heat sink phenomenon of bipolar and monopolar radiofrequency ablation observed using polypropylene tubes for vessel simulation. Surg. Innov. 21, 269–276.

Alnassan, H., Kastler, A., Wang, X., Kastler, B., 2014. Modified dipole antenna for directional microwave ablation using 3D numerical simulation. The 7th 2014 Biomedical Engineering International Conference. IEEE, pp. 1–4.

Andreano, A., Huang, Y., Meloni, M.F., Lee Jr, F.T., Brace, C., 2010. Microwaves create larger ablations than radiofrequency when controlled for power in ex vivo tissue. Med. Phys. 37, 2967–2973.

Ankory, R., Kadar, A., Netzer, D., Schermann, H., Gortzak, Y., Dadia, S., et al., 2019. 3D imaging and stealth navigation instead of CT guidance for radiofrequency ablation of osteoid osteomas: a series of 52 patients. BMC Musculoskelet. Disord. 20, 579.

Armaiz Flores, A., Oppenheimer Velez, M.L., Thompson, S.M., Windebank, A.J., Greenberg-Worisek, A.J., 2018. Navigating the clinical translation of medical devices: the case of radiofrequency ablation. Clin. Transl. Sci. 11, 8–10.

Armellini, E., Crinò, S.F., Ballare, M., Occhipinti, P., 2015. Endoscopic ultrasound-guided radiofrequency ablation of a pancreatic neuroendocrine tumor. Endoscopy 47, E600–E601.

Ashour, A.S., Asran, M.I., Mohamed, W.S., Fotiadis, D.I., 2020. Optimal localization of a novel shifted 1T-ring based microwave ablation probe in hepatocellular carcinoma. IEEE Trans. Biomed. Eng. 68, 505–514.

Aubé, C., Schmidt, D., Brieger, J., Schenk, M., Helmberger, T., Koenig, C.W., et al., 2004. Magnetic resonance imaging characteristics of six radiofrequency electrodes in a phantom study. J. Vasc. Interv. Radiol. 15, 385–392.

Bale, R., Widmann, G., Schullian, P., Haidu, M., Pall, G., Klaus, A., et al., 2012. Percutaneous stereotactic radiofrequency ablation of colorectal liver metastases. Eur. Radiol. 22, 930–937.
Ben-David, E., Nissenbaum, I., Gurevich, S., Cosman, E.R., Goldberg, S.N., 2019. Optimization and characterization of a novel internally-cooled radiofrequency ablation system with optimized pulsing algorithm in an ex-vivo bovine liver. Int. J. Hyperth. 36, 81–88.
Berber, E., Siperstein, A., 2008. Local recurrence after laparoscopic radiofrequency ablation of liver tumors: an analysis of 1032 tumors. Ann. Surg. Oncol. 15, 2757–2764.
Bertram, J.M., Yang, D., Converse, M.C., Webster, J.G., Mahvi, D.M., 2006. Antenna design for microwave hepatic ablation using an axisymmetric electromagnetic model. Biomed. Eng. Online 5, 15.
Brace, C.L., 2010. Microwave tissue ablation: biophysics, technology, and applications. Crit. Rev. Biomed. Eng. 38, 65–78.
Brace, C.L., 2011. Dual-slot antennas for microwave tissue heating: parametric design analysis and experimental validation. Med. Phys. 38, 4232–4240.
Brace, C.L., Van Der Weide, D.W., Lee, F.T., Laeseke, P.F., Sampson, L., 2004. Analysis and experimental validation of a triaxial antenna for microwave tumor ablation. 2004 IEEE MTT-S International Microwave Symposium Digest (IEEE Cat. No. 04CH37535). IEEE, pp. 1437–1440.
Brace, C.L., Laeseke, P.F., Van Der Weide, D.W., Lee, F.T., 2005. Microwave ablation with a triaxial antenna: results in ex vivo bovine liver. IEEE Trans. Microw. Theory Tech. 53, 215–220.
Brace, C.L., Laeseke, P.F., Sampson, L.A., Frey, T.M., Van Der Weide, D.W., Lee Jr, F.T., 2007. Microwave ablation with a single small-gauge triaxial antenna: in vivo porcine liver model. Radiology 242, 435–440.
Brace, C., Sampson, L., Hinshaw, J., Sandhu, N., Lee, F., 2008. Radiofrequency ablation: simultaneous application of multiple electrodes via switching creates larger, more confluent ablations than sequential application in a large animal model. J. Vasc. Interv. Radiol.: JVIR 20, 118–124.
Bruners, P., Schmitz-Rode, T., Gunther, R., Mahnken, A., 2008. Multipolar hepatic radiofrequency ablation using up to six applicators: preliminary results. RöFo-Fortschritte auf dem Gebiet der Röntgenstrahlen und der bildgebenden Verfahren. © Georg Thieme Verlag KG Stuttgart, New York, pp. 216–222.
Burdelya, L.G., Gleiberman, A.S., Toshkov, I., Aygun-Sunar, S., Bapardekar, M., Manderscheid-Kern, P., et al., 2012. Toll-like receptor 5 agonist protects mice from dermatitis and oral mucositis caused by local radiation: implications for head-and-neck cancer radiotherapy. Int. J. Radiat. Oncol. Biol. Phys. 83, 228–234.
Burdío, F., Navarro, A., Berjano, E., Burdio, J., Gonzalez, A., Guemes, A., et al., 2008. Radiofrequency hepatic ablation with internally cooled electrodes and hybrid applicators with distant saline infusion using an in vivo porcine model. Eur. J. Surg. Oncol. (EJSO) 34, 822–830.
Cavagnaro, M., Amabile, C., Bernardi, P., Pisa, S., Tosoratti, N., 2010. A minimally invasive antenna for microwave ablation therapies: design, performances, and experimental assessment. IEEE Trans. Biomed. Eng. 58, 949–959.
Chaichanyut, M., Tungjitkusolmun, S., 2015. FEM modeling for performance evaluation of microwave ablation applicator when using T prong monopole antennas. 7th WACBE World Congress on Bioengineering 2015. Springer, pp. 114–117.
Chan, P., Velasco, S., Vesselle, G., Boucebci, S., Herpe, G., Debaene, B., et al., 2017. Percutaneous microwave ablation of renal cancers under CT guidance: safety and efficacy with a 2-year follow-up. Clin. Radiol. 72, 786–792.

Chang, I.A., Nguyen, U.D., 2004. Thermal modeling of lesion growth with radiofrequency ablation devices. Biomed. Eng. Online 3, 27.
Chen, M.H., Wei, Y., Yan, K., Gao, W., Dai, Y., Huo, L., et al., 2006. Treatment strategy to optimize radiofrequency ablation for liver malignancies. J. Vasc. Interv. Radiol. 17, 671–683.
Chen, Q.-W., Ying, H.-F., Gao, S., Shen, Y.-H., Meng, Z.-Q., Chen, H., et al., 2016. Radiofrequency ablation plus chemoembolization versus radiofrequency ablation alone for hepatocellular carcinoma: a systematic review and meta-analysis. Clin. Res. Hepatol. Gastroenterol. 40, 309–314.
Chen, Q.-F., Jia, Z.-Y., Yang, Z.-Q., Fan, W.-L., Shi, H.-B., 2017. Transarterial chemoembolization monotherapy versus combined transarterial chemoembolization–microwave ablation therapy for hepatocellular carcinoma tumors ≤ 5 cm: a propensity analysis at a single center, Cardiovasc. Interv. Radiol., 40. pp. 1748–1755.
Chiang, J., Hynes, K.A., Bedoya, M., Brace, C.L., 2013. A dual-slot microwave antenna for more spherical ablation zones: ex vivo and in vivo validation. Radiology 268, 382–389.
Chiu, H.-M., Mohan, A.S., Weily, A.R., Guy, D.J., Ross, D.L., 2003. Analysis of a novel expanded tip wire (ETW) antenna for microwave ablation of cardiac arrhythmias. IEEE Trans. Biomed. Eng. 50, 890–899.
Choi, J.W., Lee, J.M., Lee, D.H., Yoon, J.-H., Suh, K.-S., Yoon, J.-H., et al., 2016. Switching monopolar radiofrequency ablation using a separable cluster electrode in patients with hepatocellular carcinoma: a prospective study. PLoS One 11, e0161980.
De Filippo, M., Ziglioli, F., Russo, U., Pagano, P., Brunese, L., Bertelli, E., et al., 2020. Radiofrequency ablation (RFA) of T1a renal cancer with externally cooled multitined expandable electrodes. Radiol. Med. 1–8.
Dupuy, D., 2001. Image-guided radiofrequency tumor ablation: challenges and opportunities – Part II. J. Vasc. Interv. Radiol. 12, 1035–1048.
Erdogan, A., Grumbrecht, S., Carlsson, J., Roederich, H., Schulte, B., Sperzel, J., et al., 2000. Homogeneity and diameter of linear lesions induced with multipolar ablation catheters: in vitro and in vivo comparison of pulsed versus continuous radiofrequency energy delivery. J. Interv. Card. Electrophysiol. 4, 655–661.
Fallahi, H., Prakash, P., 2018. Antenna designs for microwave tissue ablation. Crit. Rev. Biomed. Eng. 46, 495–521.
Fang, Z., Zhang, B., Moser, M., Zhang, E., Zhang, W., 2018. Design of a novel electrode of radiofrequency ablation for large tumors: a finite element study. J. Eng. Sci. Med. Diagn. Ther. 1.
Fang, Z., Moser, M.A., Zhang, E., Zhang, W., Zhang, B., 2019. Design of a novel electrode of radiofrequency ablation for large tumors: in vitro validation and evaluation. J. Biomech. Eng. 141, 031007.
Fischer, F., Klapdor, R., Gruessner, S., Ziert, Y., Hillemanns, P., Hertel, H., 2015. Radiofrequency endometrial ablation for the treatment of heavy menstrual bleeding among women at high surgical risk. Int. J. Gynecol. Obstet. 131, 123–128.
Francica, G., Altiero, M., Laccetti, E., Pezzullo, F., Tanga, M., Avitabile, G., et al., 2019. Long-term follow-up of unresectable medium-large hepatocellular carcinoma nodules treated with radiofrequency ablation using a multiple-electrode switching system. Br. J. Radiol. 92, 20180625.
Frich, L., Hol, P.K., Roy, S., Mala, T., Edwin, B., Clausen, O.P.F., et al., 2006. Experimental hepatic radiofrequency ablation using wet electrodes:

electrode-to-vessel distance is a significant predictor for delayed portal vein thrombosis. Eur. Radiol. 16, 1990–1999.
Gas, P., 2017. Optimization of multi-slot coaxial antennas for microwave thermotherapy based on the S11-parameter analysis. Biocybern. Biomed. Eng. 37, 78–93.
Gas, P., Szymanik, B., 2018. Shape optimization of the multi-slot coaxial antenna for local hepatic heating during microwave ablation. 2018 International Interdisciplinary PhD Workshop (IIPhDW). IEEE, pp. 319–322.
Ge, M., Jiang, H., Huang, X., Zhou, Y., Zhi, D., Zhao, G., et al., 2018. A multi-slot coaxial microwave antenna for liver tumor ablation. Phys. Med. Biol. 63, 175011.
Goldberg, S.N., Gazelle, G.S., Halpern, E.F., Rittman, W.J., Mueller, P.R., Rosenthal, D.I., 1996. Radiofrequency tissue ablation: importance of local temperature along the electrode tip exposure in determining lesion shape and size. Acad. Radiol. 3, 212–218.
Goldberg, S.N., Solbiati, L., Hahn, P.F., Cosman, E., Conrad, J.E., Fogle, R., et al., 1998. Large-volume tissue ablation with radio frequency by using a clustered, internally cooled electrode technique: laboratory and clinical experience in liver metastases. Radiology 209, 371–379.
Gregory, S., Anderson, C., Patel, U., 2013. Radiofrequency ablation of large renal angiomyolipoma: median-term follow-up. Cardiovasc. Interv. Radiol. 36, 682–689.
Guttler, R.B., 2020. SAT-LB77 thyroid radiofrequency ablation: a new office based procedure for endocrinologists. J. Endocr. Soc. 4, SAT-LB77.
Haase, S., Suss, P., Schwientek, J., Teichert, K., Preusser, T., 2012. Radiofrequency ablation planning: an application of semi-infinite modelling techniques. Eur. J. Oper. Res. 218, 856–864.
Haemmerich, D., Wright, A., Mahvi, D., Lee, F., Webster, J., 2003. Hepatic bipolar radiofrequency ablation creates coagulation zones close to blood vessels: a finite element study. Med. Biol. Eng. Comput. 41, 317–323.
Haemmerich, D., Lee Jr, F.T., Schutt, D.J., Sampson, L.A., Webster, J.G., Fine, J.P., et al., 2005. Large-volume radiofrequency ablation of ex vivo bovine liver with multiple cooled cluster electrodes. Radiology 234, 563–568.
Hiraki, T., Gobara, H., Mimura, H., Sano, Y., Toyooka, S., Shibamoto, K., et al., 2010. Brachial nerve injury caused by percutaneous radiofrequency ablation of apical lung cancer: a report of four cases. J. Vasc. Interv. Radiol. 21, 1129–1133.
Hiyoshi, Y., Miyamoto, Y., Kiyozumi, Y., Sawayama, H., Eto, K., Nagai, Y., et al., 2019. CT-guided percutaneous radiofrequency ablation for lung metastases from colorectal cancer. Int. J. Clin. Oncol. 24, 288–295.
Hong, K., Georgiades, C., 2010. Radiofrequency ablation: mechanism of action and devices. J. Vasc. Interv. Radiol. 21, S179–S186.
Huang, H.W., 2013. Influence of blood vessel on the thermal lesion formation during radiofrequency ablation for liver tumors. Med. Phys. 40, 073303.
Huang, Z., Gu, Y., Wu, S., Lai, C., Wang, X., Huang, J., 2020. Computed tomography-guided radiofrequency ablation of the retained iodized oil after simultaneous combination with transarterial embolization in small recurrent or residual hepatocellular carcinoma. J. Interv. Med. 3, 49–54.
Hulsey, R., Asili, M., Topsakal, E., 2015. Adjustable zone microwave ablation. 2015 USNC-URSI Radio Science Meeting (Joint with AP-S Symposium). IEEE, 329–329.
Hurter, W., Reinbold, F., Lorenz, W., 1991. A dipole antenna for interstitial microwave hyperthermia. IEEE Trans. Microw. Theory Tech. 39, 1048–1054.

Ibitoye, Z., Nwoye, E., Aweda, M., Oremosu, A., Annunobi, C., Akanmu, O., 2015. Optimization of dual slot antenna using floating metallic sleeve for microwave ablation. Med. Eng. Phys. 37, 384–391.

Iguchi, T., Hiraki, T., Gobara, H., Mimura, H., Fujiwara, H., Tajiri, N., et al., 2007. Percutaneous radiofrequency ablation of lung tumors close to the heart or aorta: evaluation of safety and effectiveness. J. Vasc. Interv. Radiol. 18, 733–740.

Iwakiri, Y., 2012. Endothelial dysfunction in the regulation of cirrhosis and portal hypertension. Liver Int. 32, 199–213.

Iwamoto, T., 2013. Clinical application of drug delivery systems in cancer chemotherapy: review of the efficacy and side effects of approved drugs. Biol. Pharm. Bull. 36, 715–718.

Jiang, K., Zhang, W.-Z., Liu, Y., Su, M., Zhao, X.-Q., Dong, J.-H., et al., 2014. One-off" complete radiofrequency ablation for hepatocellular carcinoma in a "high-risk location" adjacent to the major bile duct and hepatic blood vessel. Cell Biochem. Biophys. 69, 605–617.

Kai, J., Ming, S., Yang, L., Wen-Zhi, Z., Xiang-Qian, Z., Zhe, L., et al., 2013. Complete radio frequency ablation of hepatocellular carcinoma adjacent to the main bile duct and blood vessels between the first and the second hepatic portal. Cell Biochem. Biophys. 66, 397–402.

Keangin, P., Rattanadecho, P., 2020. An analysis of temperature distribution and ablated volume in the 3-D FEM tissue model with blood vessel during radio-frequency ablation. IOP Conference Series: Materials Science and Engineering. IOP Publishing, p. 012048.

Khiabani, A.J., Macgregor, R.M., Manghelli, J.L., Ruaengsri, C., Carter, D.A., Melby, S.J., et al., 2020. Bipolar radiofrequency ablation on explanted human hearts: how to ensure transmural lesions. Ann. Thorac. Surg. 110, 1933–1939.

Kho, A.S., Foo, J.J., Ooi, E.T., Ooi, E.H., 2020. Shape-shifting thermal coagulation zone during saline-infused radiofrequency ablation: a computational study on the effects of different infusion location. Comput. Methods Prog. Biomed. 184, 105289.

Kim, J.H., Won, H.J., Shin, Y.M., Kim, S.H., Yoon, H.-K., Sung, K.-B., et al., 2011. Medium-sized (3.1–5.0 cm) hepatocellular carcinoma: transarterial chemoembolization plus radiofrequency ablation versus radiofrequency ablation alone. Ann. Surg. Oncol. 18, 1624–1629.

Kim, S.M., Shin, S.S., Lee, B.C., Kim, J.W., Heo, S.H., Lim, H.S., et al., 2017. Imaging evaluation of ablative margin and index tumor immediately after radiofrequency ablation for hepatocellular carcinoma: comparison between multidetector-row CT and MR imaging. Abdom. Radiol. 42, 2527–2537.

Knavel, E.M., Hinshaw, J.L., Lubner, M.G., Andreano, A., Warner, T.F., Lee Jr, F.T., et al., 2012. High-powered gas-cooled microwave ablation: shaft cooling creates an effective stick function without altering the ablation zone. Am. J. Roentgenol. 198, W260–W265.

Korkusuz, Y., Mader, A., Gröner, D., Ahmad, S., Mader, O.M., Grunwald, F., et al., 2017. Comparison of mono- and bipolar radiofrequency ablation in benign thyroid disease. World J. Surg. 41, 2530–2537.

Kröger, T., Altrogge, I., Preusser, T., Pereira, P.L., Schmidt, D., Weihusen, A., et al., 2006. Numerical Simulation of Radio Frequency Ablation With State Dependent Material Parameters in Three Space Dimensions. Springer, Berlin, Heidelberg, pp. 380–388.

Krokidis, M., Ahmed, I., 2013. Overview of thermal ablation devices: radiofrequency ablation. Interventional Radiology Techniques in Ablation. Springer.

Labonte, S., Blais, A., Roy, L., 1995. Metal-tip monopoles for microwave ablation. Proceedings of 17th International Conference of the Engineering in Medicine and Biology Society. IEEE, pp. 279–280.
Labonte, S., Blais, A., Legault, S.R., Ali, H.O., Roy, L., 1996. Monopole antennas for microwave catheter ablation. IEEE Trans. Microw. Theory Tech. 44, 1832–1840.
Laeseke, P.F., Lee Jr, F.T., Sampson, L.A., Van Der Weide, D.W., Brace, C.L., 2009. Microwave ablation versus radiofrequency ablation in the kidney: high-power triaxial antennas create larger ablation zones than similarly sized internally cooled electrodes. J. Vasc. Interv. Radiol. 20, 1224–1229.
Lanuti, M., Sharma, A., Willers, H., Digumarthy, S.R., Mathisen, D.J., Shepard, J.-A.O., 2012. Radiofrequency ablation for stage I non-small cell lung cancer: management of locoregional recurrence. Ann. Thorac. Surg. 93, 921–928.
Lee, M., Son, T., 2019. Helical slot antenna for the microwave ablation. Int. J. Antennas Propag. 2019.
Lee, J.M., Han, J.K., Kim, S.H., Sohn, K.L., Lee, K.H., Ah, S.K., et al., 2003. A comparative experimental study of the in-vitro efficiency of hypertonic saline-enhanced hepatic bipolar and monopolar radiofrequency ablation. Korean J. Radiol. 4, 163–169.
Lee, J.M., Han, J.K., Kim, S.H., Han, C.J., An, S.K., Lee, J.Y., et al., 2005a. Wet radio-frequency ablation using multiple electrodes: comparative study of bipolar versus monopolar modes in the bovine liver. Eur. J. Radiol. 54, 408–417.
Lee, J.M., Han, J.K., Kim, S.H., Sohn, K.L., Choi, S.H., Choi, B.I., 2005b. Bipolar radiofrequency ablation in ex vivo bovine liver with the open-perfused system versus the cooled-wet system. Eur. Radiol. 15, 759–764.
Lee, J.M., Kim, S.H., Han, J.K., Sohn, K.L., Choi, B.I., 2005c. Ex vivo experiment of saline-enhanced hepatic bipolar radiofrequency ablation with a perfused needle electrode: comparison with conventional monopolar and simultaneous monopolar modes. Cardiovasc. Interv. Radiol. 28, 338–345.
Lee, J.M., Han, J.K., Kim, H.C., Kim, S.H., Kim, K.W., Joo, S.M., et al., 2007. Multiple-electrode radiofrequency ablation of in vivo porcine liver: comparative studies of consecutive monopolar, switching monopolar versus multipolar modes. Investig. Radiol. 42, 676–683.
Lee, D., Park, S., Ang, M.J.C., Park, J.-G., Yoon, S., Kim, C., et al., 2018. Evaluation of liver lesions by use of shear wave elastography and computed tomography perfusion imaging after radiofrequency ablation in clinically normal dogs. Am. J. Vet. Res. 79, 1140–1149.
Leveillee, R.J., Hoey, M.F., 2003. Radiofrequency interstitial tissue ablation: wet electrode. J. Endourol. 17, 563–577.
Li, X., Ni, J., Yang, L., Wu, B., He, M., Zhang, X., et al., 2012. A prospective study of Gasserian ganglion pulsed radiofrequency combined with continuous radiofrequency for the treatment of trigeminal neuralgia. J. Clin. Neurosci. 19, 824–828.
Lim, D., Namgung, B., Woo, D.G., Choi, J.S., Kim, H.S., Tack, G.R., 2010. Effect of input waveform pattern and large blood vessel existence on destruction of liver tumor using radiofrequency ablation: finite element analysis. J. Biomech. Eng. 132, 061003.
Lin, J.C., Wang, Y.-J., 1996. The cap-choke catheter antenna for microwave ablation treatment. IEEE Trans. Biomed. Eng. 43, 657–660.
Lin, S.-M., Lin, C.-C., Chen, W.-T., Chen, Y.-C., Hsu, C.-W., 2007. Radiofrequency ablation for hepatocellular carcinoma: a prospective comparison of four radiofrequency devices. J. Vasc. Interv. Radiol. 18, 1118–1125.
Lin, Z.-Y., Chen, J., Deng, X.-F., 2012. Treatment of hepatocellular carcinoma adjacent to large blood vessels using 1.5 T MRI-guided percutaneous

radiofrequency ablation combined with iodine-125 radioactive seed implantation. Eur. J. Radiol. 81, 3079–3083.
Livraghi, T., Goldberg, S.N., Lazzaroni, S., Meloni, F., Ierace, T., Solbiati, L., et al., 2000. Hepatocellular carcinoma: radio-frequency ablation of medium and large lesions. Radiology 214, 761–768.
Longo, I., Gentili, G.B., Cerretelli, M., Tosoratti, N., 2003. A coaxial antenna with miniaturized choke for minimally invasive interstitial heating. IEEE Trans. Biomed. Eng. 50, 82–88.
Luyen, H., Gao, F., Hagness, S.C., Behdad, N., 2014. Microwave ablation at 10.0 GHz achieves comparable ablation zones to 1.9 GHz in ex vivo bovine liver. IEEE Trans. Biomed. Eng. 61, 1702–1710.
Luyen, H., Hagness, S.C., Behdad, N., 2015. A balun-free helical antenna for minimally invasive microwave ablation. IEEE Trans. Antennas Propag. 63, 959–965.
Luyen, H., Hagness, S.C., Behdad, N., 2016. Reduced-diameter designs of coax-fed microwave ablation antennas equipped with baluns. IEEE Antennas Wirel. Propag. Lett. 16, 1385–1388.
Lyons, G.R., Pua, B.B., 2019. Ablation planning software for optimizing treatment: challenges, techniques, and applications. Tech. Vasc. Interv. Radiol. 22, 21–25.
Mahnken, A.H., Bruners, P., Delbruck, H., Gunther, R.W., 2011. Radiofrequency ablation of osteoid osteoma: initial experience with a new monopolar ablation device. Cardiovasc. Interv. Radiol. 34, 579–584.
Maini, S., 2016. Design optimization of tapered cap floating sleeve antenna for interstitial microwave ablation for liver tumor. 2016 IEEE International Conference on Consumer Electronics (ICCE). IEEE, pp. 305–308.
Maini, S., Marwaha, A., 2012. Development and analysis of coaxial slot antenna in comparision with coaxial dipole antenna for interstitial microwave ablation. Proceedings of the 6th International Conference on Communications and Information Technology, and Proceedings of the 3rd World Conference on Education and Educational Technologies. World Scientific and Engineering Academy and Society (WSEAS), pp. 192–196.
Maini, S., Marwaha, A., 2013. Design and performance analysis of multisection floating sleeve antenna using FEM for interstitial microwave ablation for HCC. 2013 IEEE Point-of-Care Healthcare Technologies (PHT). IEEE, pp. 256–259.
Maini, S., Shekhawat, D.S., 2018. Analysis of copper tube sleeve coaxial spiral antenna for interstitial hepatic microwave ablation. 2018 Eighth International Conference on Information Science and Technology (ICIST). IEEE, pp. 31–34.
McCormick, Z.L., Walega, D.R., 2018. Third-degree skin burn from conventional radiofrequency ablation of the inferiomedial genicular nerve. Pain. Med. 19, 1095–1097.
Miao, Y., Ni, Y., Yu, J., Marchal, G., 2000. A comparative study on validation of a novel cooled-wet electrode for radiofrequency liver ablation. Investig. Radiol. 35, 438–444.
Mohtashami, Y., Hagness, S.C., Behdad, N., 2017. A hybrid slot/monopole antenna with directional heating patterns for microwave ablation. IEEE Trans. Antennas Propag. 65, 3889–3896.
Mukund, A., Vats, P., Jindal, A., Patidar, Y., Sarin, S.K., 2020. Early Hepatocellular carcinoma treated by Radiofrequency ablation–mid and long term outcomes. J. Clin. Exp. Hepatol. 10, 563–573
Mulier, S., Ni, Y., Miao, Y., Rosiere, A., Khoury, A., Marchal, G., et al., 2003. Size and geometry of hepatic radiofrequency lesions. Eur. J. Surg. Oncol. (EJSO) 29, 867–878.

Mulier, S., Miao, Y., Mulier, P., Dupas, B., Pereira, P., De Baere, T., et al., 2005. Electrodes and multiple electrode systems for radiofrequency ablation: a proposal for updated terminology. Eur. Radiol. 15, 798–808.
Mulier, S., Possebon, R., Jiang, Y., Jamart, J., Wang, C., Miao, Y., et al., 2020. Radiofrequency ablation with four electrodes as a building block for matrix radiofrequency ablation: ex vivo liver experiments and finite element method modelling. Influence of electric and activation mode on coagulation size and geometry. Surg. Oncol. 33, 145–157.
Nazemian, S., Nabaei, M., 2020. Numerical investigation of the effect of vascular network complexity on the efficiency of cryosurgery of a liver tumor. Int. J. Therm. Sci. 158, 106569.
Nevels, R.D., Arndt, G.D., Raffoul, G.W., Carl, J.R., Pacifico, A., 1998. Microwave catheter design. IEEE Trans. Biomed. Eng. 45, 885–890.
Ni, Y., Miao, Y., Mulier, S., Yu, J., Baert, A., Marchal, G., 2000. A novel "cooled-wet" electrode for radiofrequency ablation. Eur. Radiol. 10, 852–854.
Osaki, Y., Ikeda, K., Izumi, N., Yamashita, S., Kumada, H., Hatta, S., et al., 2013. Clinical effectiveness of bipolar radiofrequency ablation for small liver cancers. J. Gastroenterol. 48, 874–883.
Pangarkar, S., Miedema, M.L., 2014. Pulsed versus conventional radio frequency ablation for lumbar facet joint dysfunction. Curr. Phys. Med. Rehabil. Rep. 2, 61–65.
Park, S.Y., Tak, W.Y., Jeon, S.W., Cho, C.M., Kweon, Y.O., Kim, S.K., et al., 2010. The efficacy of intraperitoneal saline infusion for percutaneous radiofrequency ablation for hepatocellular carcinoma. Eur. J. Radiol. 74, 536–540.
Park, E.K., Kim, H.J., Kim, C.Y., Hur, Y.H., Koh, Y.S., Kim, J.C., et al., 2014. A comparison between surgical resection and radiofrequency ablation in the treatment of hepatocellular carcinoma. Ann. Surg. Treat. Res. 87, 72–80.
Petit, A., Hocquelet, A., N'Kontchou, G., Varin, E., Sellier, N., Seror, O., et al., 2020. No-touch multi-bipolar radiofrequency ablation for the treatment of subcapsular hepatocellular carcinoma ≤ 5 cm not puncturable via the non-tumorous liver parenchyma. Cardiovasc. Interv. Radiol. 43, 273–283.
Pillai, K., Akhter, J., Chua, T.C., Shehata, M., Alzahrani, N., Al-Alem, I., et al., 2015. Heat sink effect on tumor ablation characteristics as observed in monopolar radiofrequency, bipolar radiofrequency, and microwave, using ex vivo calf liver model. Medicine 94.
Pisa, S., Cavagnaro, M., Bernardi, P., Lin, J.C., 2001. A 915-MHz antenna for microwave thermal ablation treatment: physical design, computer modeling and experimental measurement. IEEE Trans. Biomed. Eng. 48, 599–601.
Prakash, P., 2010. Theoretical modeling for hepatic microwave ablation. Open. Biomed. Eng. J. 4, 27.
Prakash, P., Converse, M.C., Webster, J.G., Mahvi, D.M., 2009. An optimal sliding choke antenna for hepatic microwave ablation. IEEE Trans. Biomed. Eng. 56, 2470–2476.
Rachbauer, F., Mangat, J., Bodner, G., et al., 2003. Heat distribution and heat transport in bone during radiofrequency catheter ablation. Arch. Orthop. Trauma. Surg. 123, 86–90.
Rathke, H., Hamm, B., Guttler, F., Rathke, J., Rump, J., Teichgräber, U., et al., 2014. Comparison of four radiofrequency ablation systems at two target volumes in an ex vivo bovine liver model. Diagn. Interv. Radiol. (Ankara, Turk.) 20, 251–258.
Razib, A., Hossain, K., Hossain, S., 2016. Microwave ablation technique (MWA) for cancer treatment: simulation of single Slot MCA for different slot position. 2016 International Conference on Medical Engineering, Health Informatics and Technology (MediTec). IEEE, pp. 1–6.

Rhim, H., Goldberg, S.N., Dodd, G.D., Solbiati, L., Lim, H.K., Tonolini, M., et al., 2001. Essential techniques for successful radio-frequency thermal ablation of malignant hepatic tumors. Radiographics 21, S17–S35.

Romero-Méndez, R., Berjano, E., 2017. An analytical solution for radiofrequency ablation with a cooled cylindrical electrode. Math. Probl. Eng. 2017.

Rossi, S., Garbagnati, F., Lencioni, R., Allgaier, H.-P., Marchianò, A., Fornari, F., et al., 2000. Percutaneous radio-frequency thermal ablation of nonresectable hepatocellular carcinoma after occlusion of tumor blood supply. Radiology 217, 119–126.

Ryan, P.T., 2013. Microwave ablation for cancer: physics, performance, innovation, and the future. Image-Guided Cancer Therapy: A Multidisciplinary Approach. Springer-Verlag, New York, pp. 37–59.

Santos, R.S., Gan, J., Ohara, C.J., Daly, B., Ebright, M.I., Desimone, M., et al., 2010. Microwave ablation of lung tissue: impact of single-lung ventilation on ablation size. Ann. Thorac. Surg. 90, 1116–1119.

Seror, O., 2015. Ablative therapies: advantages and disadvantages of radiofrequency, cryotherapy, microwave and electroporation methods, or how to choose the right method for an individual patient? Diagn. Interv. Imaging 96, 617–624.

Shao, Y.L., Leo, H.L., Chua, K.J., 2017. Studying the thermal performance of a bipolar radiofrequency ablation with an improved electrode matrix system: in vitro experiments and modelling. Appl. Therm. Eng. 116, 623–635.

Shiina, S., Sato, K., Tateishi, R., Shimizu, M., Ohama, H., Hatanaka, T., et al., 2018. Percutaneous ablation for hepatocellular carcinoma: comparison of various ablation techniques and surgery. Can. J. Gastroenterol. Hepatol. 2018, 4756147.

Shiomi, H., Naka, S., Sato, K., Demura, K., Murakami, K., Shimizu, T., et al., 2008. Thoracoscopy-assisted magnetic resonance guided microwave coagulation therapy for hepatic tumors. Am. J. Surg. 195, 854–860.

Singh, S., Repaka, R., 2015. Pre-clinical modelling and simulation of hepatic radiofrequency ablation. In: Proc COMSOL Conference 2015, pp. 29–30.

Singh, S., Repaka, R., 2017. Temperature-controlled radiofrequency ablation of different tissues using two-compartment models. Int. J. Hyperth. 33, 122–134.

Singh, S., Repaka, R., 2018a. Numerical investigation of convective cooling in minimizing skin burns during radiofrequency ablation of breast tumor. Sādhanā 43, 90.

Singh, S., Repaka, R., 2018b. Parametric sensitivity analysis of critical factors affecting the thermal damage during RFA of breast tumor. Int. J. Therm. Sci. 124, 366–374.

Singh, S., Melnik, R., 2019. Computational analysis of pulsed radiofrequency ablation in treating chronic pain. International Conference on Computational Science. Springer, pp. 436–450.

Singh, S., Bhowmik, A., Repaka, R., 2015. Radiofrequency ablation of malignant breast tumor: a numerical study. In: Proc. 23rd National Heat and Mass Transfer Conference and 1st International ISHMT-ASTFE Heat and Mass Transfer Conference, pp. 17–20.

Singh, S., Bhowmik, A., Repaka, R., 2016. Thermal analysis of induced damage to the healthy cell during RFA of breast tumor. J. Therm. Biol. 58, 80–90.

Skonieczki, B.D., Wells, C., Wasser, E.J., Dupuy, D.E., 2011. Radiofrequency and microwave tumor ablation in patients with implanted cardiac devices: is it safe? Eur. J. Radiol. 79, 343–346.

Taj-Eldin, M., Prakash, P., 2014. Microwave ablation at 915 MHz vs. 2.45 GHz: single and multiple-antenna considerations. 2014 IEEE Antennas and Propagation Society International Symposium (APSURSI). IEEE, pp. 1139–1140.

Tan, Y., Jiang, J., Wang, Q., Guo, S., Ma, K., Bie, P., 2018. Radiofrequency ablation using a multiple-electrode switching system for hepatocellular carcinoma within the Milan criteria: long-term results. Int. J. Hyperth. 34, 298–305.

Tomasian, A., Jennings, J.W., 2018. Spinal osteoid osteoma: percutaneous radiofrequency ablation using a navigational bipolar electrode system. Am. J. Roentgenol. 211, 856–860.

Trujillo, M., Alba, J., Berjano, E., 2012. Relationship between roll-off occurrence and spatial distribution of dehydrated tissue during RF ablation with cooled electrodes. Int. J. Hyperth. 28, 62–68.

Tungjitkusolmun, S., Staelin, S.T., Haemmerich, D., Tsai, J.-Z., Cao, H., Webster, J.G., et al., 2002. Three-dimensional finite-element analyses for radio-frequency hepatic tumor ablation. IEEE Trans. Biomed. Eng. 49, 3–9.

Vorländer, C., David Kohlhase, K., Korkusuz, Y., Erbelding, C., Luboldt, W., Baser, I., et al., 2018. Comparison between microwave ablation and bipolar radiofrequency ablation in benign thyroid nodules: differences in energy transmission, duration of application and applied shots. Int. J. Hyperth. 35, 216–225.

Wagata, M., Murakoshi, T., Ishii, K., Muromoto, J., Sasahara, J., Murotsuki, J., 2016. Radiofrequency ablation with an internally cooled electrode for twin reversed arterial perfusion sequence. Fetal Diagn. Ther. 40, 110–115.

Wakaki, T., Michiyama, T., Kuwano, S., 2017. A new interstitial choke embedded antenna for microwave ablation. IEICE Commun. Express 6, 435–438.

Wang, P., Brace, C.L., 2011. Tissue dielectric measurement using an interstitial dipole antenna. IEEE Trans. Biomed. Eng. 59, 115–121.

Wang, T., Zhao, G., Qiu, B., 2015. Theoretical evaluation of the treatment effectiveness of a novel coaxial multi-slot antenna for conformal microwave ablation of tumors. Int. J. Heat. Mass. Transf. 90, 81–91.

Widmann, G., Bodner, G., Bale, R., 2009. Tumour ablation: technical aspects. Cancer Imaging 9, S63.

Widmann, G., Schullian, P., Haidu, M., Bale, R., 2012. Stereotactic radiofrequency ablation (SRFA) of liver lesions: technique effectiveness, safety, and interoperator performance. Cardiovasc. Interv. Radiol. 35, 570–580.

Wolf, F.J., Grand, D.J., Machan, J.T., Dipetrillo, T.A., Mayo-Smith, W.W., Dupuy, D. E., 2008. Microwave ablation of lung malignancies: effectiveness, CT findings, and safety in 50 patients. Radiology 247, 871–879.

Wong, S.N., Lin, C.-J., Lin, C.-C., Chen, W.-T., Cua, I.H.Y., Lin, S.-M., 2008. Combined percutaneous radiofrequency ablation and ethanol injection for hepatocellular carcinoma in high-risk locations. Am. J. Roentgenol. 190, W187–W195.

Wood, B.J., Locklin, J.K., Viswanathan, A., Kruecker, J., Haemmerich, D., Cebral, J., et al., 2007. Technologies for guidance of radiofrequency ablation in the multimodality interventional suite of the future. J. Vasc. Interv. Radiol. 18, 9–24.

Xu, L., Cai, K., Yang, R., Lin, Q., Yue, H., Liu, F., 2016. Simulation of multi-probe radiofrequency ablation guided by optical surgery navigation system under different active modes. Comput. Assist. Surg. 21, 107–116.

Yamashiki, N., Tateishi, R., Yoshida, H., Shiina, S., Teratani, T., Sato, S., et al., 2005. Ablation therapy in containing extension of hepatocellular carcinoma: a simulative analysis of dropout from the waiting list for liver transplantation. Liver Transplant. 11, 508–514.

Yang, D., Bertram, J.M., Converse, M.C., O'Rourke, A.P., Webster, J.G., Hagness, S.C., et al., 2006. A floating sleeve antenna yields localized hepatic microwave ablation. IEEE Trans. Biomed. Eng. 53, 533–537.

Yap, S., Cheong, J.K., Ooi, E.H., Liao, I.Y., Foo, J.J., Nair, S.R., et al., 2019. A numerical study on the no-touch bipolar radiofrequency ablation. 2019 41st Annual International Conference of the IEEE Engineering in Medicine and Biology Society (EMBC). IEEE, pp. 2887–2890.

Yap, S., Cheong, J.K., Foo, J.J., Ooi, E.T., Ooi, E.H., 2020. The effects of the no-touch gap on the no-touch bipolar radiofrequency ablation treatment of liver cancer: a numerical study using a two compartment model. Appl. Math. Model. 78, 134–147.

Yim, P.J., Marcos, H.B., Choyke, P.L., Hvizda, J.L., Libutti, S.K., Wood, B.J., 2001. Detection of blood vessels for radio-frequency ablation treatment planning. Proceedings 14th IEEE Symposium on Computer-Based Medical Systems. CBMS 2001. IEEE, pp. 123–126.

Yoon, J.-H., Lee, J.M., Han, J.K., Choi, B.I., 2013. Dual switching monopolar radiofrequency ablation using a separable clustered electrode: comparison with consecutive and switching monopolar modes in ex vivo bovine livers. Korean J. Radiol. 14, 403–411.

Yoon, J.H., Lee, J.M., Hwang, E.J., Hwang, I.P., Baek, J., Han, J.K., et al., 2014. Monopolar radiofrequency ablation using a dual-switching system and a separable clustered electrode: evaluation of the in vivo efficiency. Korean J. Radiol. 15, 235–244.

Yoon, J.H., Lee, J.M., Woo, S., Hwang, E.J., Hwang, I., Choi, W., et al., 2015. Switching bipolar hepatic radiofrequency ablation using internally cooled wet electrodes: comparison with consecutive monopolar and switching monopolar modes. Br. J. Radiol. 88, 20140468.

Zakeri, S.A., Ngo, A., Kibriya, N., 2017. Psoas abscess and spondylodiscitis following radiofrequency ablation of a renal tumour: unexpected complications. Cardiovascular Interv. Radiol. 40, 1484–1486.

Zhang, B., Moser, M.A., Zhang, E.M., Luo, Y., Liu, C., Zhang, W., 2016. A review of radiofrequency ablation: large target tissue necrosis and mathematical modelling. Phys. Med. 32, 961–971.

Zhou, W., Herwald, S.E., Mccarthy, C., Uppot, R.N., Arellano, R.S., 2019. Radiofrequency ablation, cryoablation, and microwave ablation for T1a renal cell carcinoma: a comparative evaluation of therapeutic and renal function outcomes. J. Vasc. Interv. Radiol. 30, 1035–1042.

Zorbas, G., Samaras, T., 2015. A study of the sink effect by blood vessels in radiofrequency ablation. Comput. Biol. Med. 57, 182–186.

6

COMSOL Multiphysics software for ablation system simulation

Chapter outline

6.1 Introduction

Simulation technology and computer modeling provide powerful tools to assist clinicians in finding the most suitable ablation treatment instead of surgery and monitor temperature changes in the ablation zone (Berjano, 2006; Prakash and Diederich, 2012; Primavesi et al., 2018). Several researchers have implemented mathematical models by using computer electromagnetic (EM) models for ablation probe design. The main computer EM models are the method of moments, finite element method (FEM), and finite difference method, which are used to solve the EM fundamental equations, such as the Maxwell equations. The FEM provides complex geometries modeling by representing them in small elements in a mesh system with edges or nodes signified by partial differential equations, which are unsolved by finite difference models or analytical techniques (Haemmerich et al., 2001; Tungjitkusolmun et al., 2002; Chang and Nguyen, 2004; Mohammed and Verhey, 2005; Huang and Liauh, 2011). It is a numerical method which can be formulated as a function minimization. The FEM provides solutions to multiple differential equations, thus it solves heat-transfer problems (O'Hara et al., 2009; Bazyar and Talebi, 2015).

Commercial FEM packages are unable to support automated temperature control. Most researchers control the power of the applicators manually using trial and error to maintain a constant electrode tip temperature. Using the FEM for modeling temperature-controlled ablation therapy, the applied voltage should

Thermal Ablation Therapy. DOI: https://doi.org/10.1016/B978-0-12-819544-4.00007-1
© 2021 Elsevier Inc. All rights reserved.

be adopted to maintain a constant tip temperature to achieve equivalent results to clinical ablation devices. To provide accurate models for the ablation therapy process, the tissue EM properties are required. These properties include the conductivity and permittivity of tumor tissues and nearby healthy tissues based on the tissue type of the affected organ, and also the initial and boundary conditions are defined in the finite element (FE) model.

Simulation models can be employed to establish the temperature distribution models in terms of the hyperbolic and Penne's heat transfer analysis to find the effects of different heat transfer models on the temperature map during the thermal ablation procedure (Zhang et al., 2008; Liu and Chen, 2009; Liu et al., 2012; Kengne et al., 2013; Salimpour and Shirani, 2017). Also, other models can be used to adjust and design the used electrode needle and probe structure using simulation tools for a successful ablation therapy process (Woo et al., 2000; Villard et al., 2005; Kuzhagaliyev et al., 2018; Ziegle et al., 2018; Ashour et al., 2021). Another role for simulation modeling is to find the optimal parameter settings and their changes over time in the ablation process to achieve complete tumor ablation (Chen et al., 2004; Ahmed et al., 2008; Hall et al., 2015). Such parameters include adjusting the center temperature, temperature-raising rate, and the applicator insertion positioning to determine the output power and attain the essential thermal coagulation results. Furthermore, the temperature dependency of the tissue electrothermal properties, which considers the temperature dependence of thermal and electrical conductivities, should be taken into account to determine the EM energy absorption and estimate the temperature profile (Wang et al., 2018). Considering the effect of the changes in the tissue parameters on the tumor volume is essential to improving the simulation model accuracy.

The prior planning and predictability of the proper ablation applicator and its settings related to the tumor location and characteristics have a significant role in assisting the physicians. The increased predictability rate increases the ablation performance and success rate. The simulation increases the physician's predictability to provide a safer and effective treatment plan that is less time consuming. To create a precise ablation zone with real-time monitoring and feedback using radiometers, ablation device (applicator, energy source, required power) optimization is essential. Typically, the radiometers measure the EM radiation in the ablation zone, which enables the description of the EM field spatial distribution. Such monitoring enables physicians to fine-tune the ablated region during the ablation process by ensuring the ability of the generated radiation to damage the targeted tumor tissues with minimal destruction of the nearby surrounding healthy tissue (Habash et al., 2007; Horton et al., 2014). Moreover, the simulation modeling aims to optimize the design of the applicator for optimal emitting/radiating energy and monitoring/receiving properties. Several simulators are available to apply the FEM numerically to model the ablation process. The main goal is to provide directional complete ablation of the targeted tumor tissues while saving the surrounding healthy tissues. Accordingly, the necrotic tissue fraction is evaluated at different points in the tumor zone and at several points in time.

The most efficient simulation tool for ablation therapy is the COMSOL. COMSOL Multiphysics includes several features and methods, such as boundary element methods, meshing techniques, time stepping methods, and different FEMs. Consequently, it covers an extensive range of different numerical methods appropriate for solving an inclusive range of problems (Duan et al., 2016; Perera-Bel et al., 2020). The FE model using COMSOL Multiphysics has an axial symmetry structure. Such a model is used in the case of symmetric geometry about an axis (Zampolli et al., 2007; Ghose et al., 2011; Bonomo and Isakson, 2016). Two-dimensional (2D) axisymmetric models denote a slice of the real three-dimensional (3D) model, where it has the following characteristics:

- The geometry must lie in the $X \geq 0$ part of the XY plane;
- The geometry parts must lie in the XY plane;
- Displacements are defined in the XY plane.

If a 2D axisymmetric model rotates around the Y-axis as a reference of the Cartesian coordinate system, it will provide the original 3D structure. This 2D axisymmetric modeling reduces the computational time by representing the tumor in a 2D plan and repeats the calculations in the 3D domain (Pryor, 2009; Mi et al., 2017; de Oliveira et al., 2018). The FEM using COMSOL is applied to simulate different ablation systems and the temperature field which is sensitive to the tissue parameters, such as the resistance (R), density (ρ), electrical conductivity (σ), thermal conductivity (k), thermal specific capacity (Cp), and dielectric constant (ε). Additionally, the simulation procedures depend on the values of several parameters to be accomplished with simulation software (Jiang et al., 2017). The following section reports different simulation software tools that can be used to build the FEM of the ablation therapy system.

6.2 Simulation software programs for ablation therapy modeling

The experimental justification for the different types of thermal ablation therapy applicators is that it is a widespread tedious task due to the expensive hardware trials which are insufficiently powerful to evaluate the proposed ablation system. To overcome this limitation, computational numerical models are used to solve the complex physical and EM application for component designs. There are several commercial software programs for FE modeling, such as ABAQUS, SolidWorks, COMSOL, and ANSYS, which can be used for FE model to simulate a specific phenomenon by modeling a specific geometry suitable for the desired system as shown in Figs. 6.1 and 6.2.

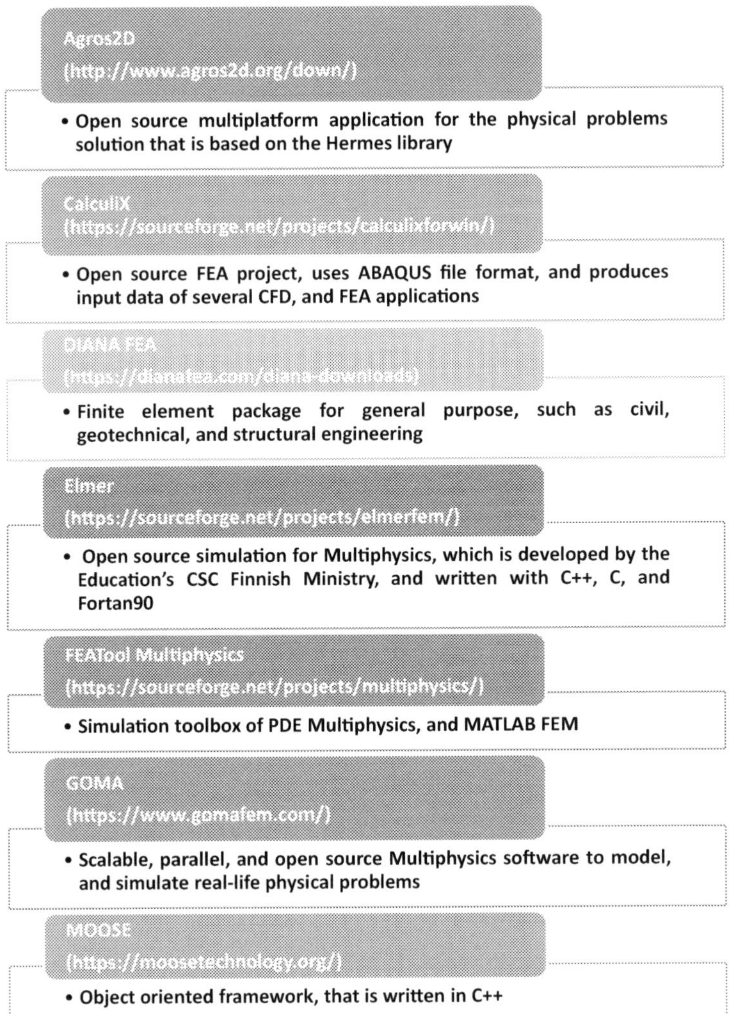

Figure 6.1 Simulation programming using the finite element method.

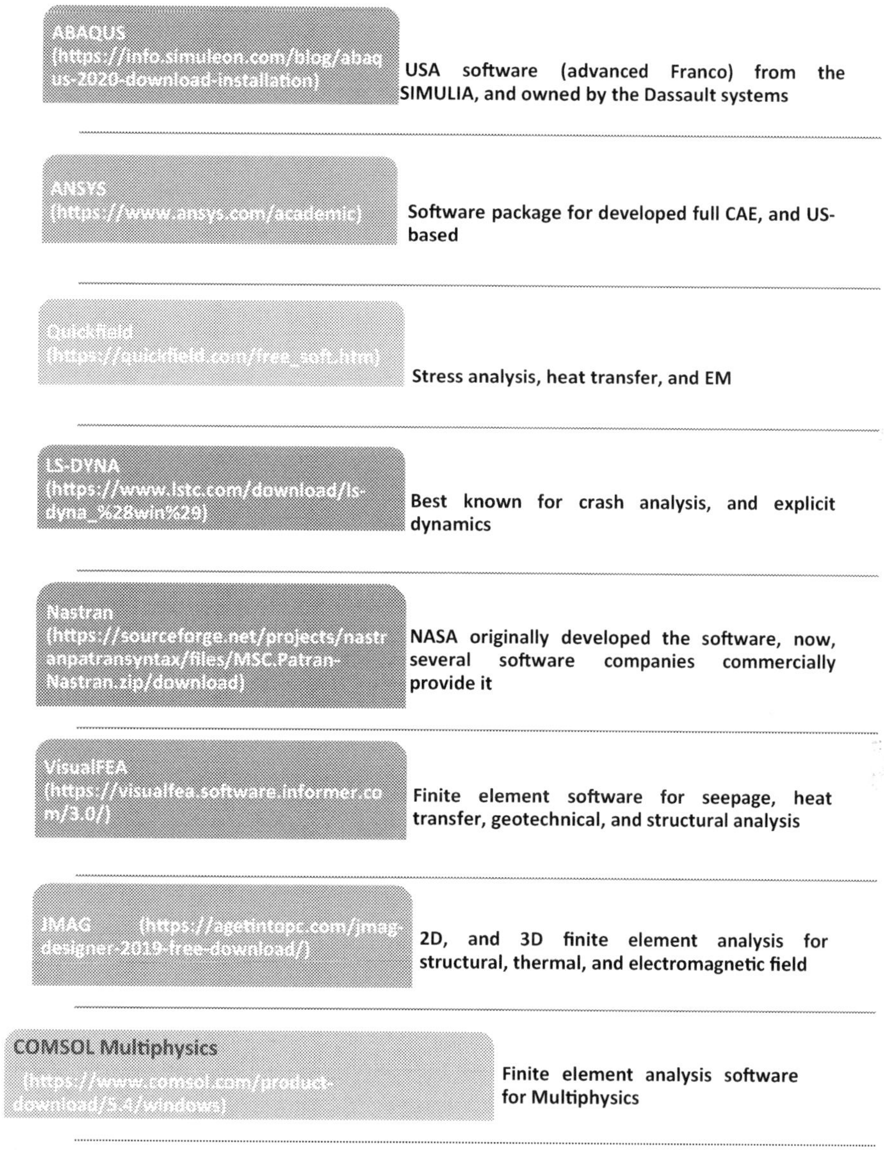

Figure 6.2 Extra simulation programming using the finite element method.

These simulation software tools can be applied using Windows or Linux platforms. To model physical problems for thermal therapy ablation systems modeling, the COMSOL software uses EM and thermal libraries which allow users to solve multiphysics problems by simulating the EM effect of the applicator and the thermal effect on the tumor produced in a single computation. The following section focuses on explaining the COMSOL menus during the simulation of any ablation system.

6.3 COMSOL for ablation system simulation

When using COMSOL software for ablation system modeling, a model for the used antenna including the characteristics and dimensions of the material of each part is included. Also, the model will include the tumor zone, that is, its position, dimensions, and tissue properties as well as the same for the surrounding healthy tissues. Moreover, with the z-axis as axial symmetry, the boundary conditions for the EM simulation are considered for the exterior boundaries. Generally, for thermal simulation, the boundary conditions are a constant temperature of 25°C. The FEM is used through COMSOL to simulate the applicator performance to achieve complete ablation of the tumor zones with minimal damage to the surrounding healthy tissues using the minimum required power and ablation processing time (duration) as the optimal case. Accordingly, the mesh settings which represent the FEM model are determined by providing the minimum and maximum mesh element size, where the maximum size of the elements is selected by considering the minimum wavelength presented in the simulation experiment. Accordingly, the number of boundary elements, vertex elements, and degrees of freedom are solved. The traditional simulation methods' accuracy and performance of the ablation system are affected by the number of the dynamic factors raised concurrently in multiple physics domains. However, COMSOL can achieve the appropriate complex modeling easily and rapidly, which improves the simulation design. A very significant feature in COMSOL software is its concurrence with MATLAB® software. This integration allows COMSOL to optimize easily the composite models with extremely sophisticated algorithms. Using the module in COMSOL, which is used in simulating the ablation system, several illustrations and plots can be obtained, such as:

- The predicted ablated volume, and the predicted tissue damage in both the tumor and the surrounding tissue, which can be used to adapt the perfusion conditions in the tissues;
- The heat pattern and profile, which illustrate the changes over time, where the tissue temperature increases with time during the ablation process;
- The heat distribution in the model;

- The power dissipation density, which is determined by the calculated thermal damage calculation;
- The variation in the heat capacity in the tissues surrounding the probe;
- The weighted error plot, which indicates the reflected power regions.

Generally, the COMSOL Multiphysics application builder provides creation simulation applications with an interface, as shown in Fig. 6.3. It deploys them outside or within their organizations with COMSOL Compiler and COMSOL Server. It has several advantages, such as a seamless and easy interface with heat transfer physics field and other fields. Also, COMSOL has the ability to modify the governing existing equations, the available flexibility concerned in the solver selection in iterative, segregated, direct, mixed solvers, and fully coupled solutions. The COMSOL Multiphysics program is integrated with other programs through the LiveLink for AutoCAD, Excel, Inventor, PTC_Creo_Parametric, PTC_ProENGINEER, Revit, Solid_Edge, SOLIDWORKS, and MATLAB. It allows generating a MATLAB code from a COMSOL file, in addition, the file can be modified, extended, and run within MATLAB.

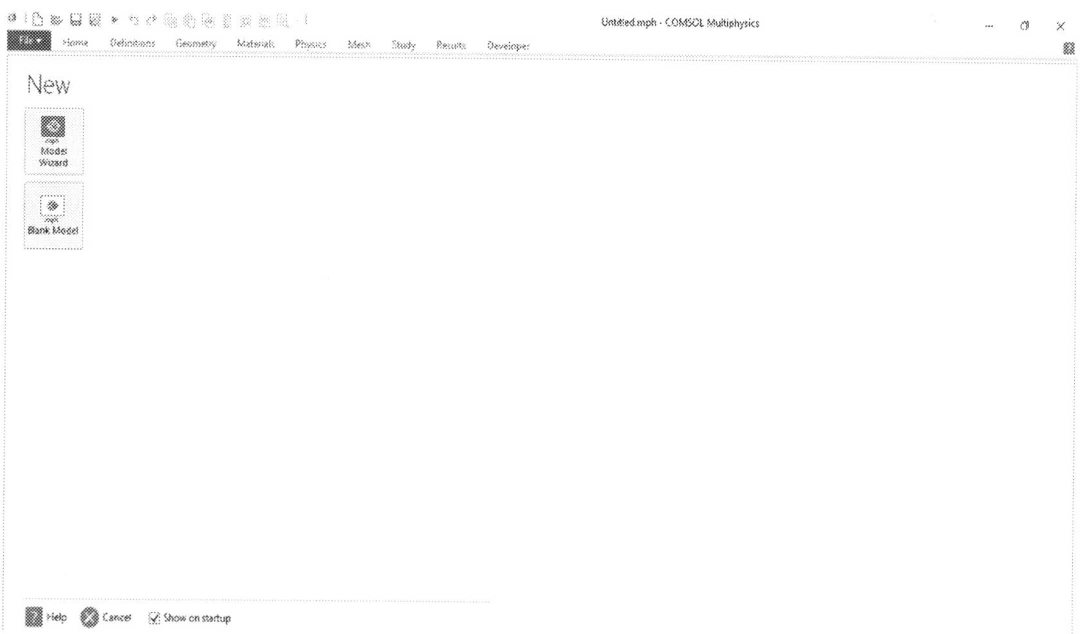

Figure 6.3 COMSOL Multiphysics interface.

In the heat transfer module using COMSOL, analysis of the heat transfer in the tissues is considered based on the bioheat equations, which can consider the cessation of blood flow once the tissues are coagulated. Such consideration occurs by including the perfusion term in the calculations to represent the heat transfer to cells surrounding the tumor and determine the temperature distribution for ensuring the efficiency of the energy delivery. Fig. 6.3 illustrates the main menu interface of COMSOL.

Fig. 6.3 displays two available wizards in COMSOL Multiphysics, which are:

- Model wizard, which sets the model step by step before entering the integrated development environment (IDE); and
- Blank model, which enters an empty model. Furthermore, the model space dimension is determined according to the target work and selected by pressing the done icon as shown in Fig. 6.4.

Fig. 6.4 displays the available space dimensions inside COMSOL software including 0D, which means no space dimension, 1D, which means asymmetric one-dimensional, 1D axisymmetric, 2D, which means asymmetric two-dimensional, 2D axisymmetric, and 3D, which means three-dimensional model. Moreover, COMSOL Multiphysics provides simulation for AC/DC interfaces, which are used to model the electric current conduction, electrical circuits, and electric and magnetic fields with orbits of charged particles, rotating machinery, such as generator motors, and full vector formulations, hence, it computes them in the static and low-frequencies system, in addition to acoustics interface, chemical species transport, electrochemistry, fluid flow, heat transfer, optics, plasma, radiofrequency (RF), EM, semiconductor, structural mechanics, and mathematics as observed in Fig. 6.5.

Fig. 6.5 displays the existing available physics, especially the heat transfer interface in different media, such as fluids, solids, and pipes to compute radiative intensity fields, temperature,

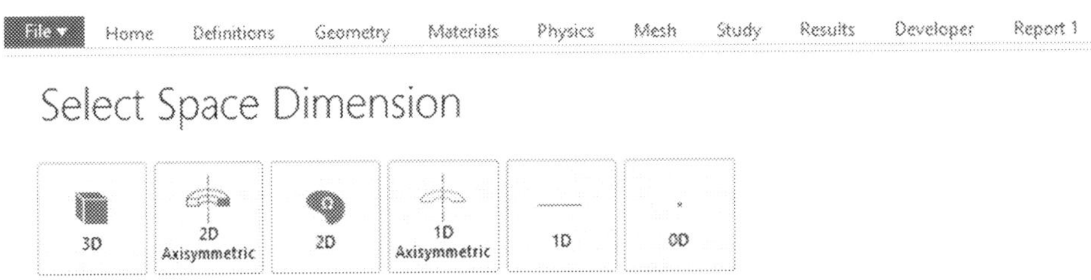

Figure 6.4 Space dimension in COMSOL Multiphysics.

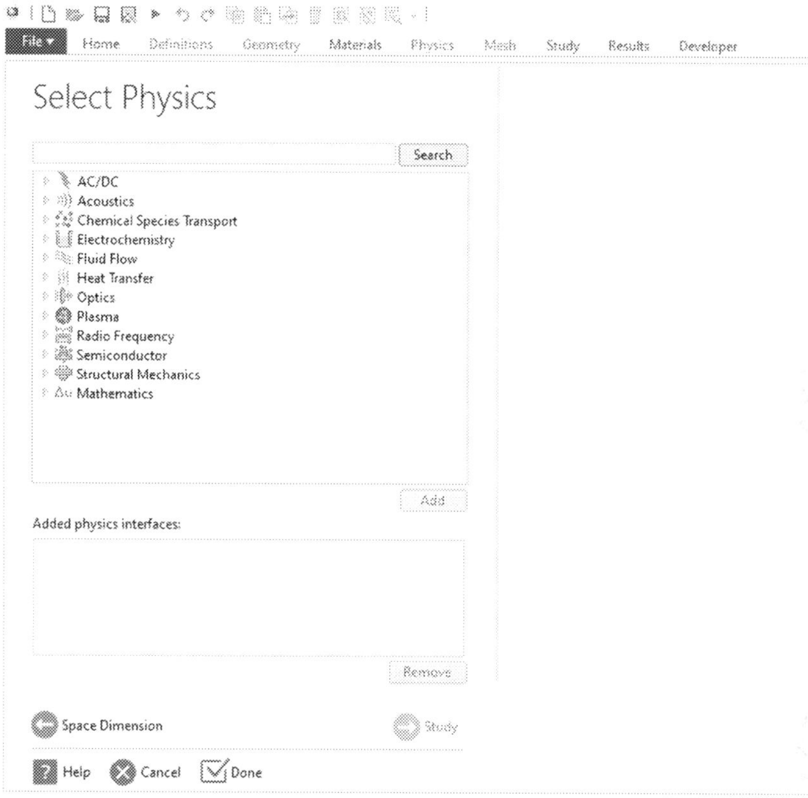

Figure 6.5 Multiple physics in COMSOL Multiphysics.

and other physical fields, such as EM fields, pressure, and velocity as shown in Fig. 6.6.

Fig. 6.6 guides the selection of suitable physics and adds them to the model by pressing the study icon as shown in Fig. 6.7.

Fig. 6.7 displays the selected physics, which are chosen according to the studied work. It is essential to determine the study type with respect to the time and frequency, including the dependent or independent time and frequency domain as shown in Fig. 6.8.

Fig. 6.8 illustrates the frequency domain study, which is used to compute the model response with the harmonic excitation for one or multiple frequencies. Moreover, the small-signal analysis, frequency-domain study is used to study small oscillations, which is done through two steps, including a stationary study step for computing biased solution, and a frequency-domain,

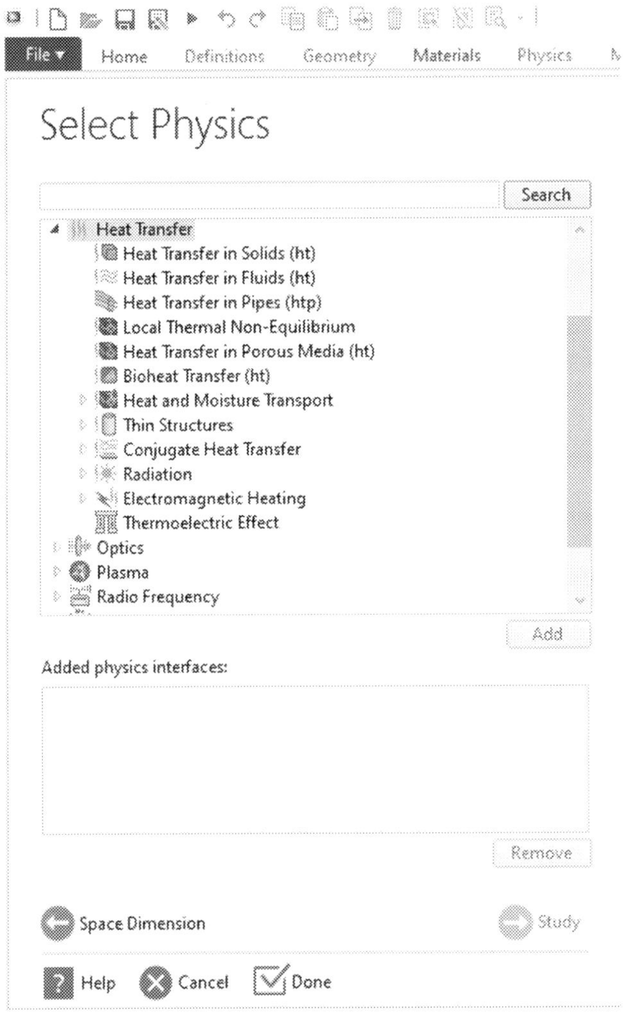

Figure 6.6 Heat transfer interface in a different medium.

perturbation study step for computing the frequency response about the biased solution. The stationary study is used in the case of field variables stability with time as existing while computing the temperature, stresses, strains, deformations, and chemical composition at the equilibrium. The stationary source sweep study is used for extracting automatically lumped matrices, such as a resistance matrix in Electric Currents or a capacitance matrix in Electrostatics. Finally, the time-dependent study is used in the case of field variables change with time, for example, the propagation of the EM wave in the time domain.

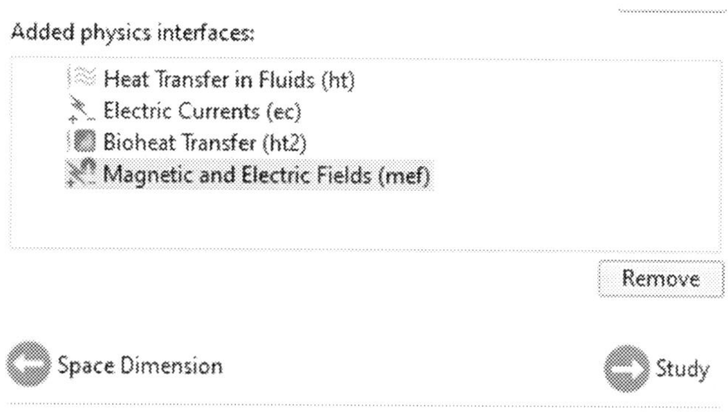

Figure 6.7 Selected physics added to the model.

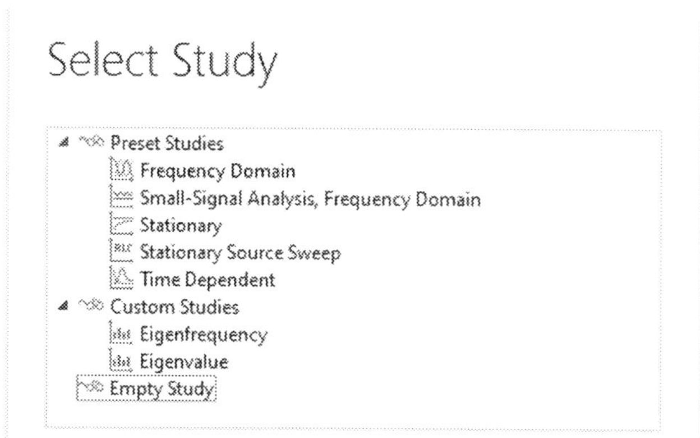

Figure 6.8 Study concerning time and frequency.

Accordingly, the COMSOL Multiphysics IDE contains six main sections, namely the customize quick access toolbar, ribbon taps section, model builder that contains any additional parts, settings that adjust the selected option, graphics that display the geometry, and results and message/tables section that displays the final tables as shown in Fig. 6.9.

Fig. 6.9 displays a general overview of the options inside the program. Consequently, the customize quick access toolbar, like other programs, includes some general icons, such as new, open, save, save as, undo, redo, delete, select, clear, find, file name at the toolbar center, minimize, maximize, restore down, and close as shown in Fig. 6.10.

Fig. 6.10 displays the most commonly used icons in the COMSOL Multiphysics module. For building the model, adding another physics, materials, building the mesh, adding the study,

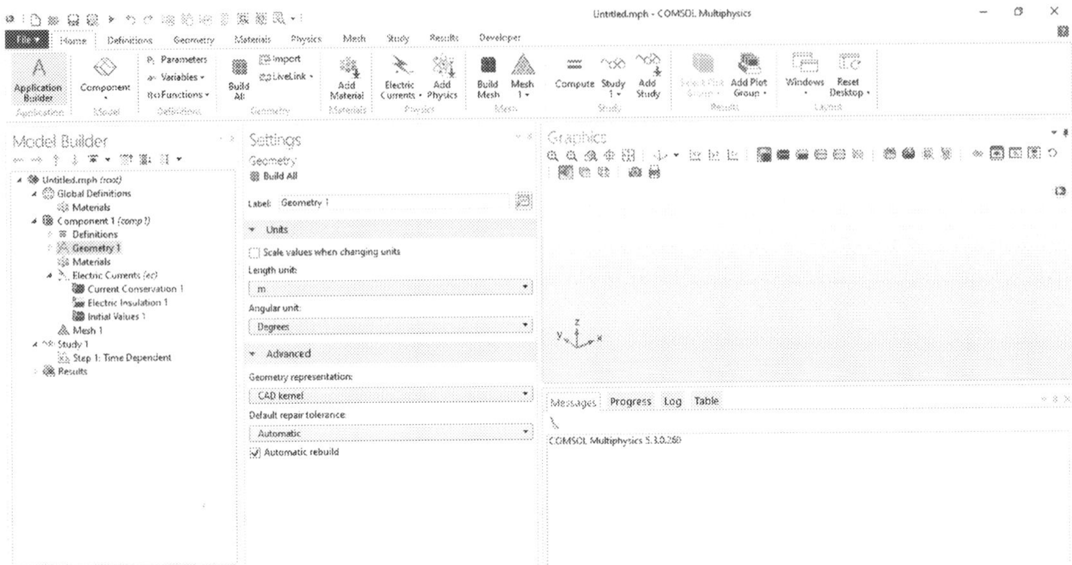

Figure 6.9 Integrated development environment of COMSOL Multiphysics.

Figure 6.10 Customize quick access toolbar.

Figure 6.11 Customize quick access toolbar.

obtaining the results, and help icon, the ribbon taps section is used as observed in Fig. 6.11.

Fig. 6.11 displays the icons inside the ribbon taps section, where the home icon presents some essential options, such as defining the parameters, variables, and functions, building one selected or all objects, importing an external geometry, LiveLink with other programs, adding materials for each part inside the geometry, adding additional physics, building a standard mesh, adding a study, plot groups, such as 1D, 2D, 3D, polar plot, and smith plot as shown in Fig. 6.12.

Fig. 6.12 clarifies the available options inside the home icon. In addition, there is an application builder icon, which opens

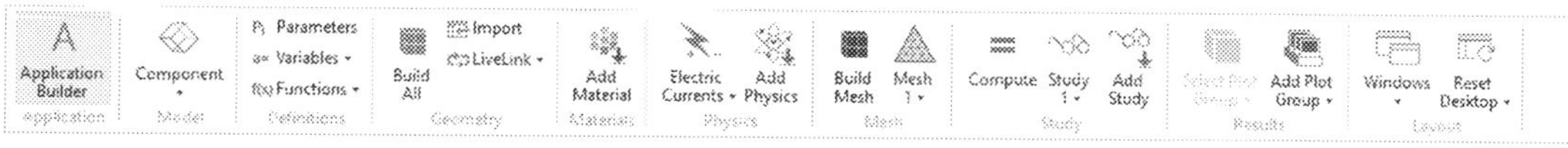

Figure 6.12 Home icon in the ribbon tap section.

Figure 6.13 Application builder in the Home icon.

the application builder to modify the application user interface, create, and edit application codes as shown in Fig. 6.13.

COMSOL Multiphysics provides a lot of main global and local functions, such as:

- Waveform, which can generate different waves, sine, cosine, sawtooth, square, and triangle;
- Elevation function, which can import geospatial elevation data from the digital elevation model, and map the elevation data to a function of X and Y;
- Switch function, which can switch between the global functions during a solver sweep; and
- External function, which has interfaces to other external functions that are written in C language programming as shown in Fig. 6.14.

To search about any tool inside COMSOL Multiphysics, the help icon in the ribbon toolbar and find icon in the quick access toolbar can be used, as shown in Fig. 6.15.

To integrate COMSOL Multiphysics with other programs to send and receive data between them, the LiveLink can be used as shown in Fig. 6.16.

The COMSOL layout can be modified to widescreen layout or regular screen layout from the reset desktop icon, as shown in Fig. 6.17.

Furthermore, multiadditional options, which divide all the geometry into subsections, can be used as shown in Fig. 6.18.

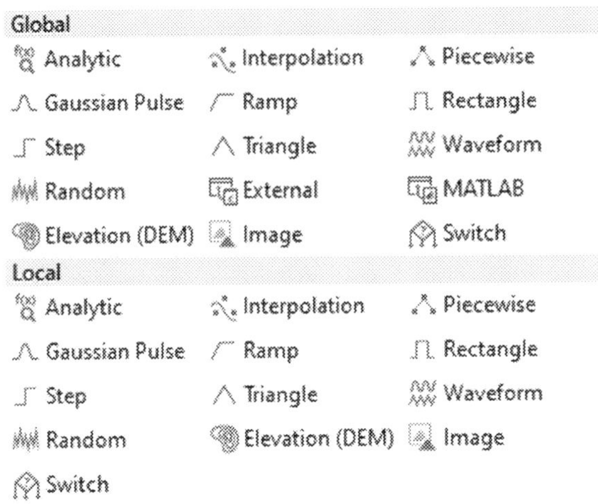

Figure 6.14 Global and locally available functions.

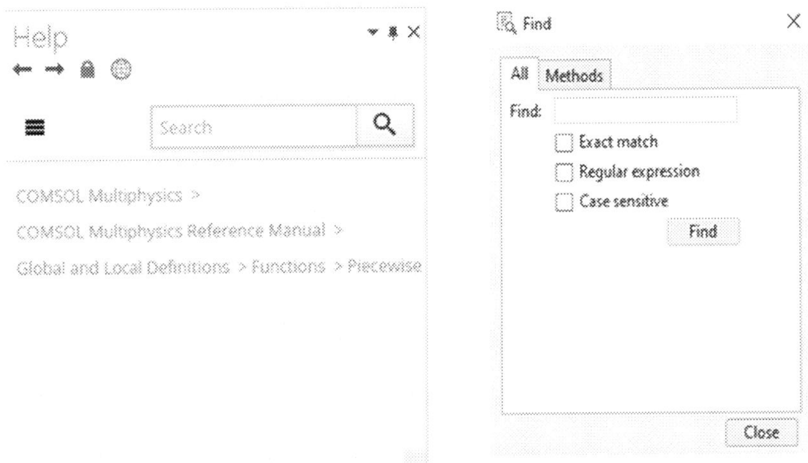

Figure 6.15 Search icons in COMSOL Multiphysics, including help and find.

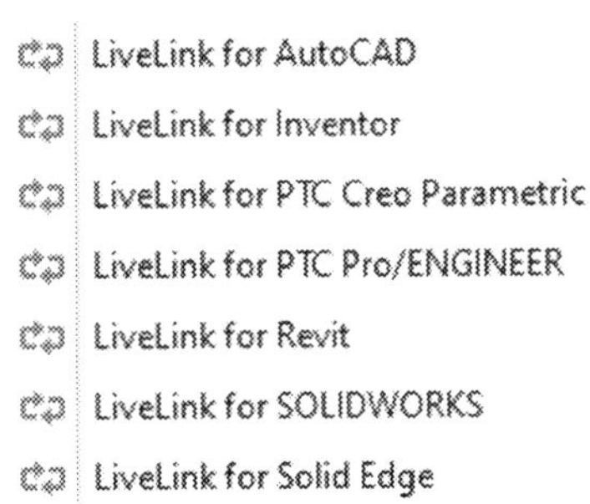

Figure 6.16 COMSOL Multiphysics LiveLink with several programs.

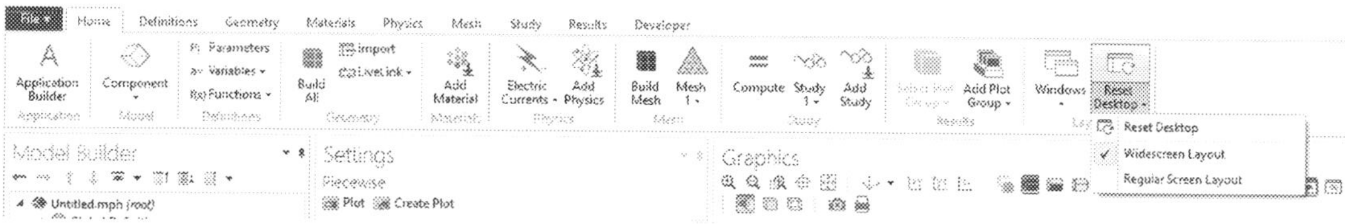

Figure 6.17 COMSOL screen layouts.

Figure 6.18 COMSOL definitions layouts.

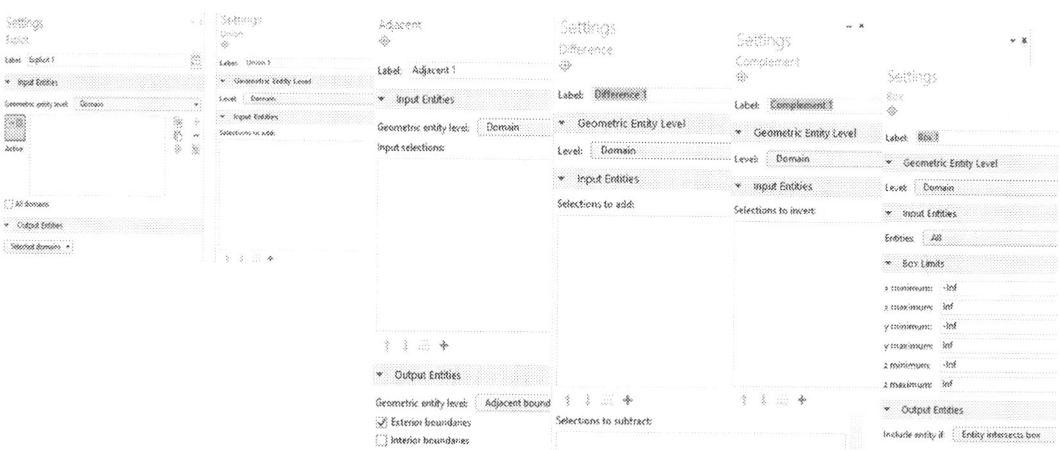

Figure 6.19 Some examples of selections in COMSOL Multiphysics.

Fig. 6.18 provides many definitions that facilitate all operations for the user, such as local variable icon, which can add a variable node to the current model component to define variables with a local scope, also the selections part, such as explicit, adjacent, union, difference, box, cylinder, and complement, which determine a specific part inside the geometry to simply deal with it. Accordingly, section selection occurs through the setting window for each option, as shown in Fig. 6.19.

In Fig. 6.19, the adjacent icon can select entities that are adjacent to the entities in other selections, union, which can gather a group of selections, difference, which can generate the difference between two selections set, box, which can select the entities that intersect or be inside a box, and complement, which can produce the complement of a group of selections. In addition, an explicit icon can select the geometric entities set, such as boundaries or domains. Moreover, the values at different locations inside the model are measured using the probes in the definitions section as shown in Fig. 6.20.

Fig. 6.20 displays a group of probes that can identify the value of any parameter at any point and any time, such as the domain probe, which defines values like the maximum, minimum, integral, and average over the domain, boundary probe, which defines the previous values over the boundary, and global variable probe, which defines the values using the global variables as shown in Fig. 6.21.

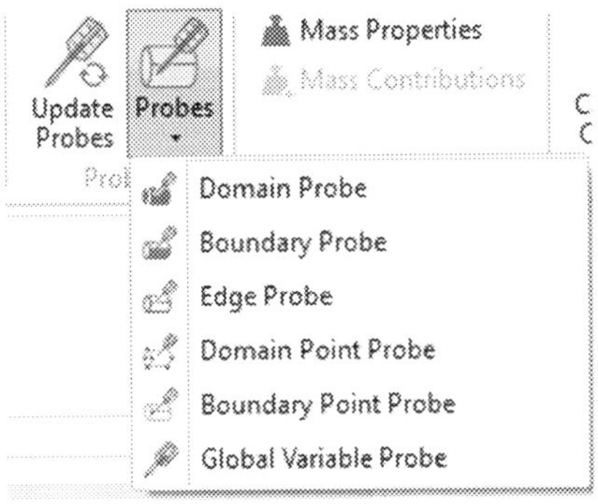

Figure 6.20 Probes section for measuring values.

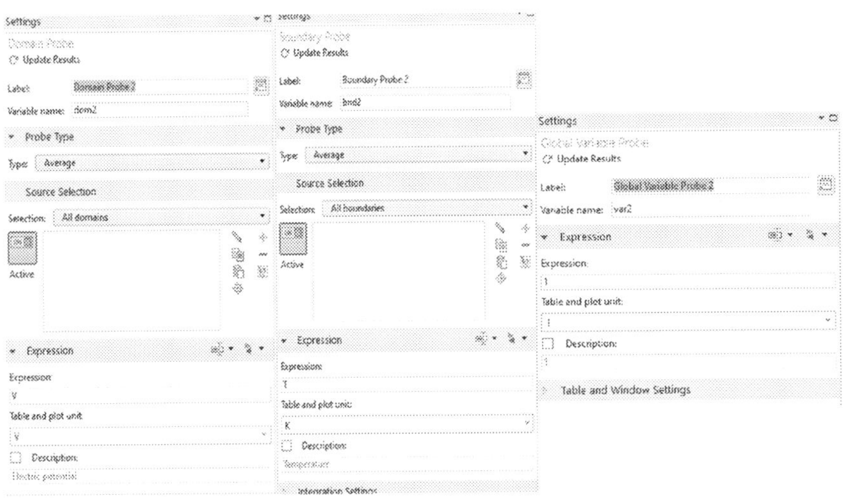

Figure 6.21 Setting windows for some probes.

Also, it is possible to add mass property or contribution nodes to your present model as shown in Fig. 6.22.

Fig. 6.22 explains that the mass density can be obtained from the physics interface as a standard value or inserted by the user. Moreover, integration between the different parts inside the model can be implemented using the coupling operator as shown in Fig. 6.23.

Fig. 6.23 clarifies all the available couplers inside the COMSOL program, such as boundary similarity, which maps an expression defined on a boundary part to another boundary part with the same shape, identify mapping, which maps between the overlapped geometric entities during viewing in different frames, and linear projection, which defines a mapping between the source and destination in a linear map form defined by vertices. Also, the coordinates can be modified according to the target work using the coordinate system as shown in Fig. 6.24.

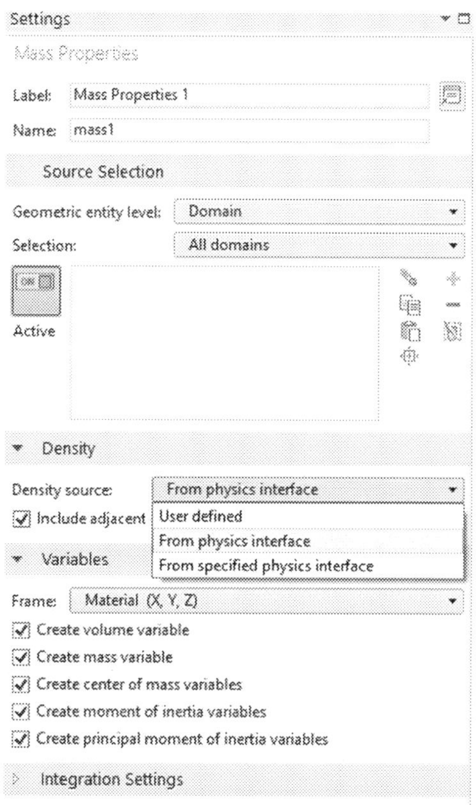

Figure 6.22 Setting the windows mass properties node.

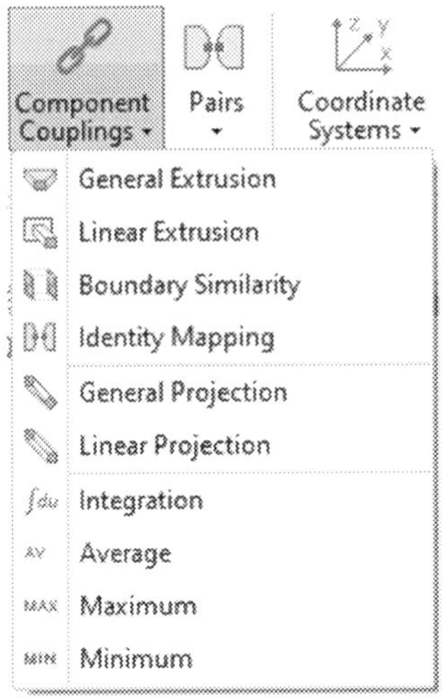

Figure 6.23 Available coupling operators in COMSOL Multiphysics.

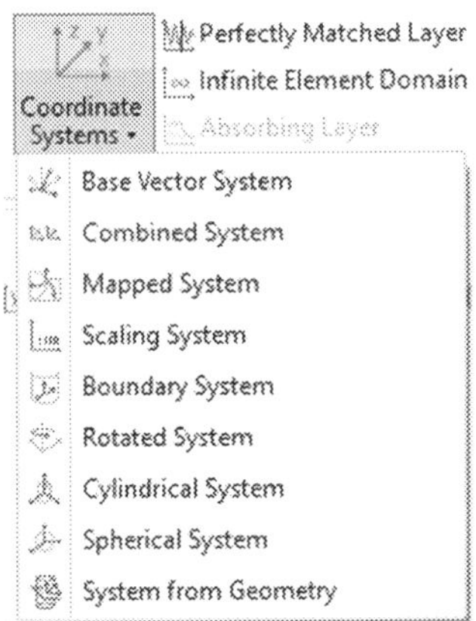

Figure 6.24 Different shapes of the coordinate system in COMSOL Multiphysics.

Fig. 6.24 displays the available coordinate system, which serves the programmer, and facilitates the operations, such as:

- The combined system, which can combine several coordinate systems in different domains;
- The scaling system, which creates a system that can map the geometry into a virtual geometry defined by a virtual coordinate of scaling system;
- The rotated system, which can make a rotation relative to the global system of Cartesian coordinate; and

- The system from geometry, which can be aligned with a work plane in part geometry or 3D geometry.

These systems can be adjusted through the setting window for each coordinate type, as explained in Fig. 6.25.

Moreover, it is possible to expand the working domain to a wider level to facilitate some models and their operations using the perfectly matched layer, which absorbs the outgoing waves by stretching another virtual domain to a complex domain. Also, the infinite element domain, which is considered a virtual domain that is stretched out to infinity acts as an unbounded domain. They are in the definition window and are adjusted from their setting windows, which contain the extra dimensions icon, which allows them to modify the space dimension, in addition, the view icon can appear or hide the lighting, view, transparency, grid, geometry label, units, wireframe rendering, mesh rendering, and axis orientation in the graphic window, as shown in Fig. 6.26.

Fig. 6.26 displays the available parameters which help to adjust the graphic window to be suitable for the model. Accordingly, the geometry window helps in building the model with any shape or size using its icons, as shown in Fig. 6.27.

Fig. 6.27 displays all the available options, which build the geometry according to the target model. Accordingly, the geometry can be made by two methods, including externally through the import icon as shown in Fig. 6.28.

The second method used for building the geometry is manually using some icons inside the geometry window, such as

Figure 6.25 Setting windows of some different coordinate systems.

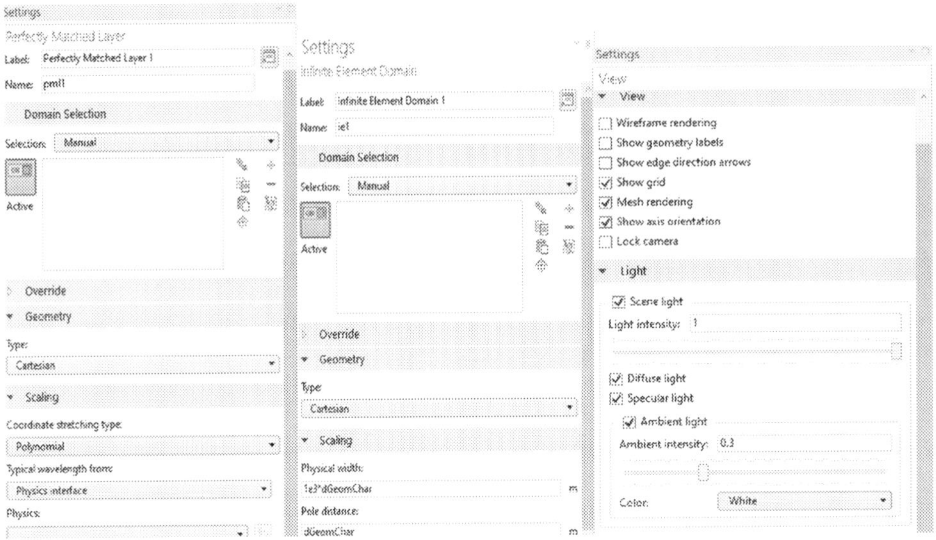

Figure 6.26 View icon in COMSOL Multiphysics.

Figure 6.27 Geometry window in COMSOL Multiphysics.

block, sphere, cone, torus, cylinder, or helix, according to the target model shape. Hence, these shapes are adjusted with their dimensions using the setting window related to each icon, as shown in Fig. 6.29.

There are some other additional shapes with the ability to build any possible geometry, which can build a wide collection of models for many applications, which are found in the more primitives icon as shown in Fig. 6.30.

Fig. 6.30 displays a set of other complex shapes available in the COMSOL Multiphysics, such as:

- Bezier polygon, which creates a solid or curved polygon containing Bezier segments in two or three dimensions;
- Interpolation curve, which creates a curve approximating or interpolating a sequence of points in two or three dimensions; and
- Parametric surface, which creates a parametric surface determined by coordinate expressions in three dimensions, as shown in Fig. 6.31.

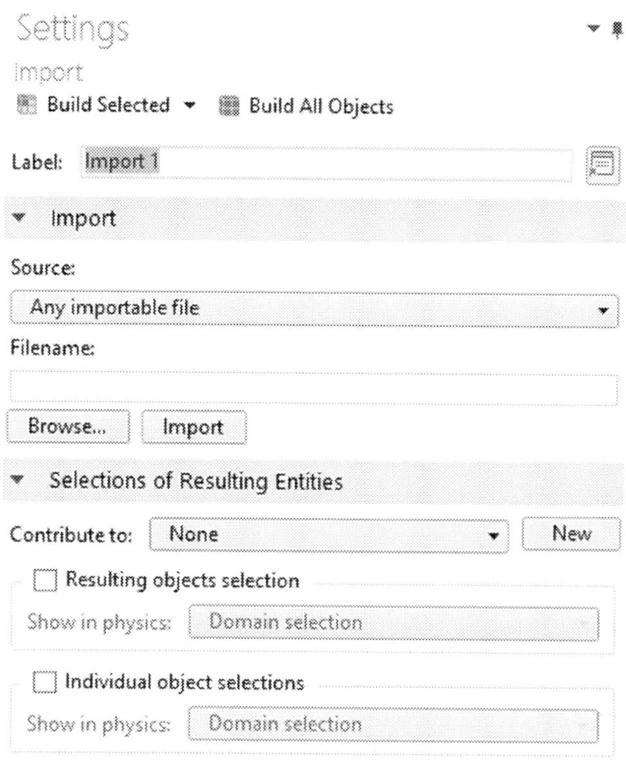

Figure 6.28 External import geometry.

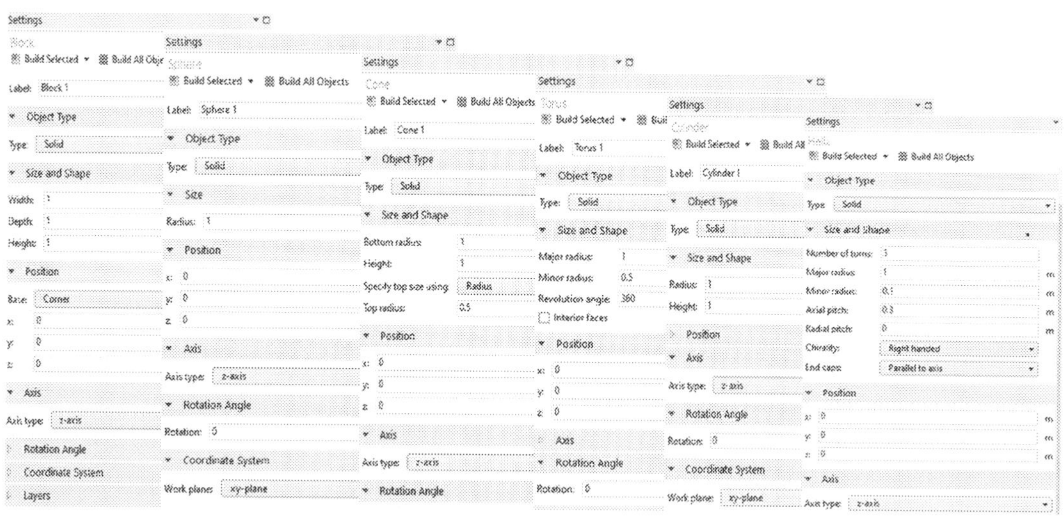

Figure 6.29 Setting windows for different available geometric shapes.

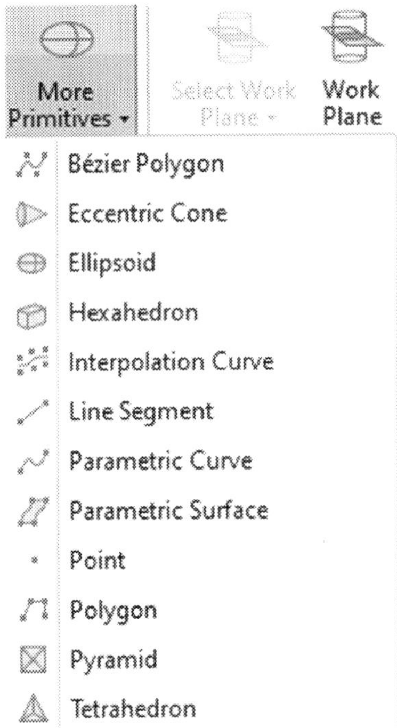

Figure 6.30 Other additional geometric shapes.

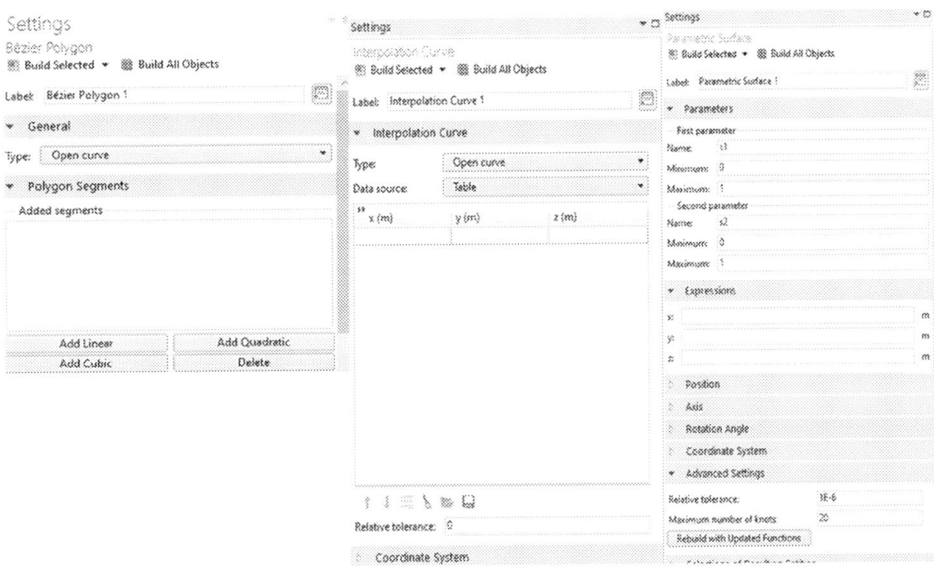

Figure 6.31 Setting windows for other additional geometric shapes.

Furthermore, COMSOL Multiphysics provides a work plane, which can extrude, revolve, and loft the planar 2D faces to 3D objects, and sweep one or multiple faces in a spline curve to a solid in three dimensions, as shown in Fig. 6.32.

Moreover, several operations are available, such as chamfer, which adds chamfer shear corners in the 3D geometries, and the fillet icon, which adds fillets to make rounded corners in the 3D geometries as shown in Fig. 6.33.

Additionally, there are numerous ready operations that can be simply performed, such as Booleans/partitions, transform, and conversion, as shown in Fig. 6.34.

Fig. 6.34 displays several available powerful options for performing many tasks, such as:

- Compose, which makes a geometry using a Boolean formula, array, which creates a linear array of geometry objects, rectangular (2D, 3D) or 3D block shaped;
- Scale icon, which scales the objects around a point;
- Convert to COMSOL, which converts the CAD import module kernel of the geometry objects to the COMSOL kernel;
- Midsurface, which removes the solid object thickness through a midsurface operation, producing a surface object that is useful for a shell interface; and
- Thicken icon, which converts the surface to a solid object by a determined thickness, usually in a normal direction as shown in Fig. 6.35.

Figure 6.32 Setting windows for work plane.

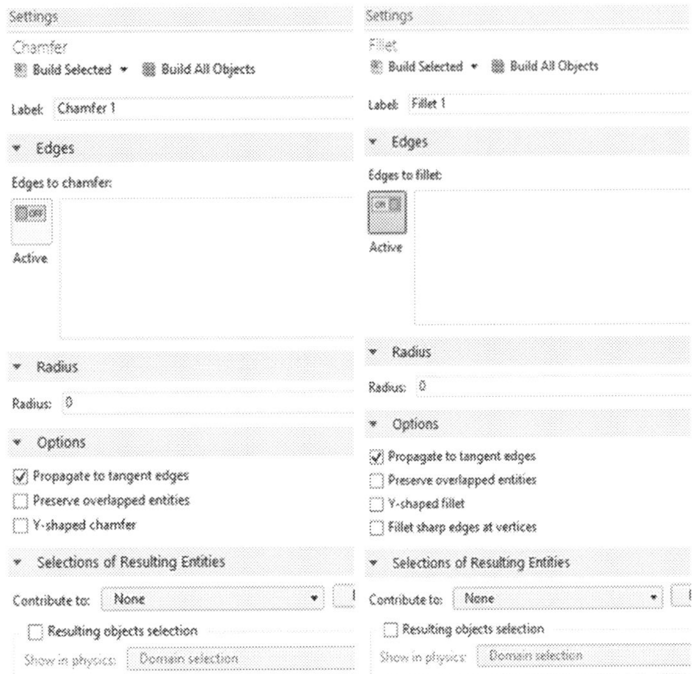

Figure 6.33 Setting windows for chamfer and fillet.

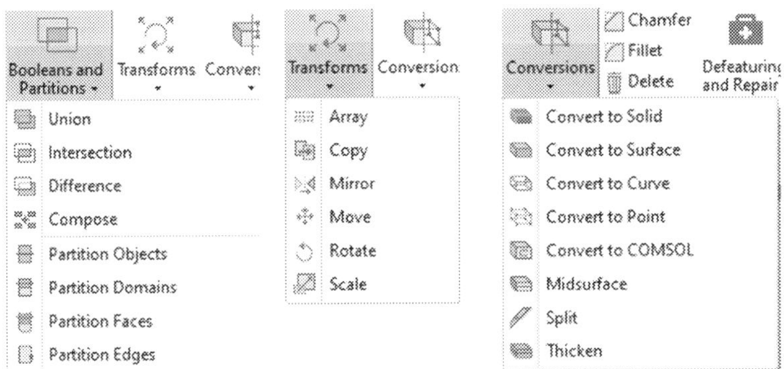

Figure 6.34 Some available operations in COMSOL Multiphysics.

Furthermore, multiple operations can be used and facilitate the model simulation, such as defeaturing and repair and several virtual operations, as shown in Fig. 6.36.

Fig. 6.36 clarifies more additional important operations, including:

- Cap faces, which fills holes in a partition or geometry through capping faces;
- Detach faces, which detaches faces from one geometry object to form another new geometry object;

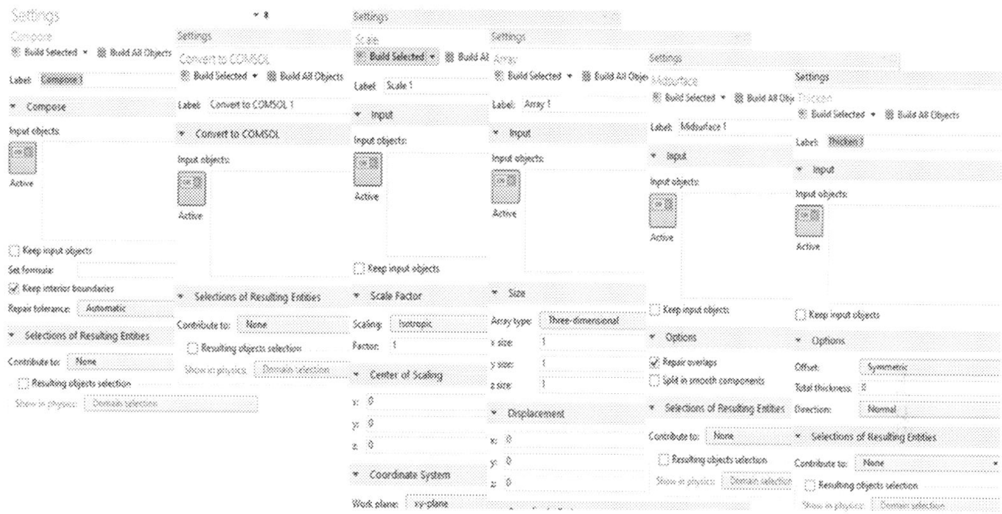

Figure 6.35 Setting windows of additional operations in COMSOL Multiphysics.

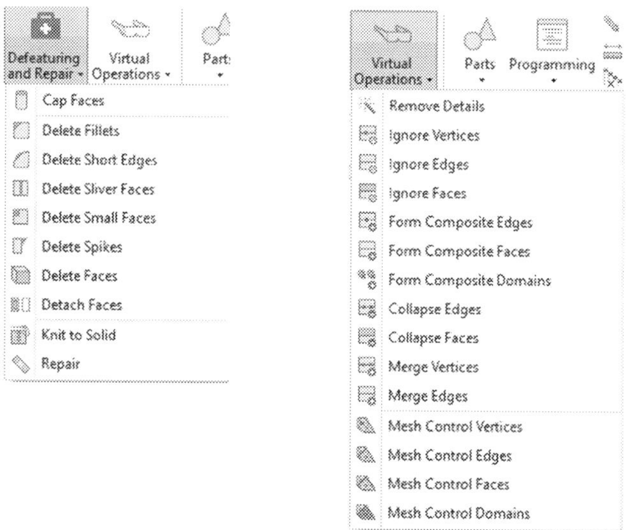

Figure 6.36 Additional operations in COMSOL Multiphysics.

- Repair icon, which repairs the defects in the geometry object by automatic movement of the small details;
- Ignore edges icon, which can remove the selected isolated or adjacent edges to two domains or faces;
- Form composite faces, which can ignore the edges between faces to organize a composite face for the connected face component; and

- Collapse faces, which can remove them or merge their opposite adjacent edges into only one vertex and reconnect the adjacent faces with the merged vertex or edges.

These additional options are adjusted from the setting windows, as shown in Fig. 6.37.

Moreover, it is possible to measure the perimeter, area, volume, and other properties for any selected geometric objects or entities, as shown in Fig. 6.38.

COMSOL Multiphysics provides a selection option, which can adjust the overall geometric system for more accurate results using the selections icon in the geometry window, as shown in Fig. 6.39.

Fig. 6.38 displays several selections options, which are always used inside the model, such as ball selection, which

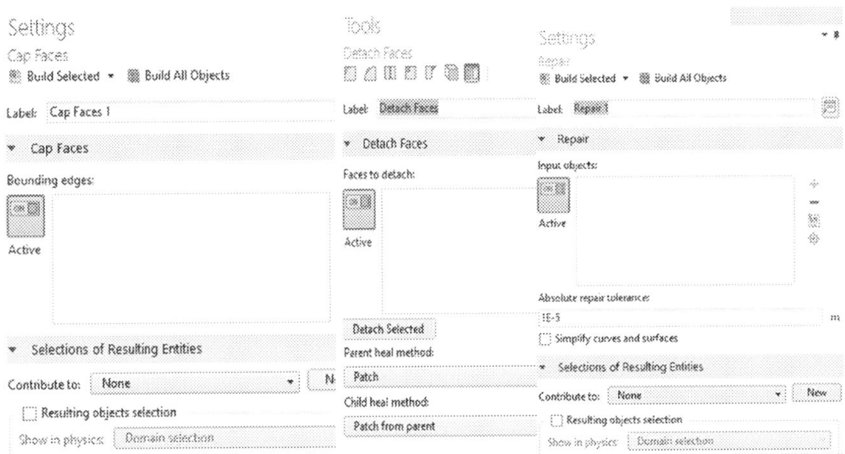

Figure 6.37 Setting windows of some additional operations in COMSOL Multiphysics.

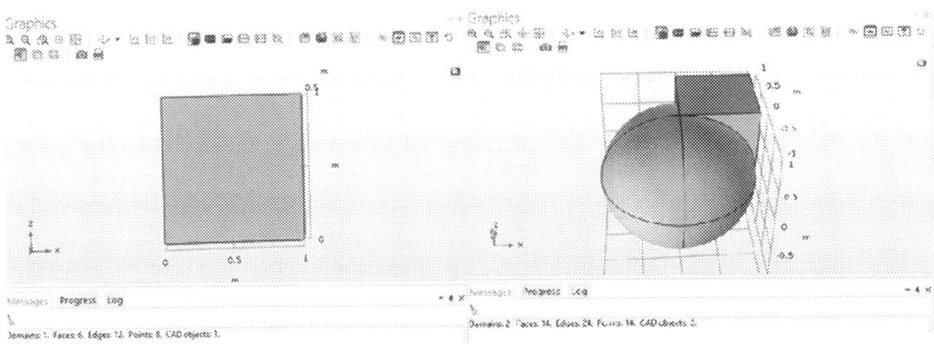

Figure 6.38 Measure geometric properties.

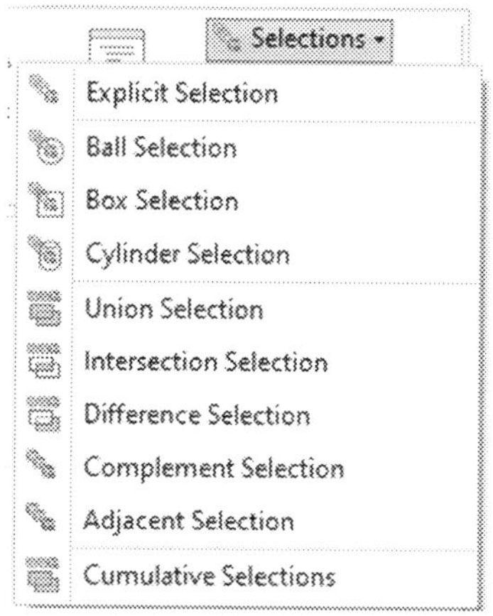

Figure 6.39 Selections option for organizing the geometric model.

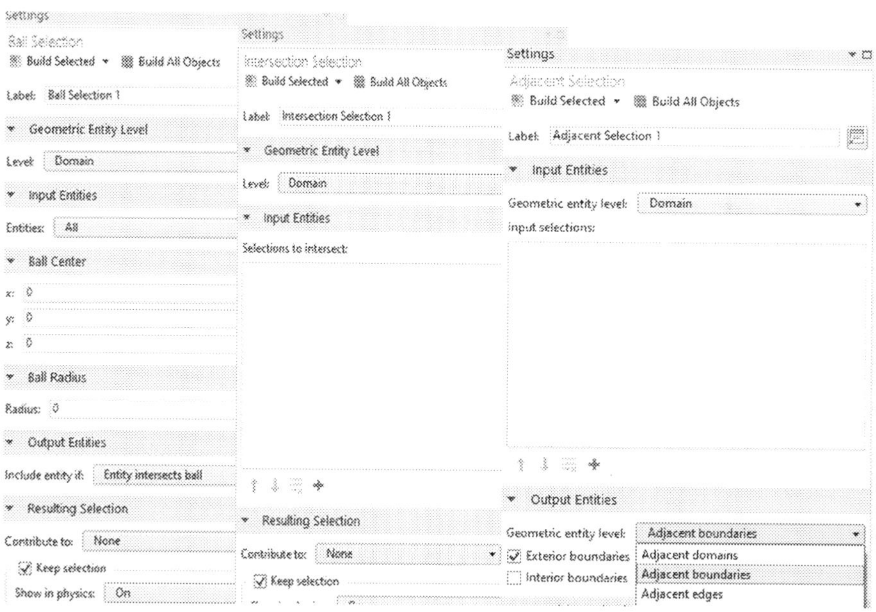

Figure 6.40 Setting windows for some selection tools.

gathers a group of geometric objects or entities inside a ball. Also, the intersection selection, which combines groups of entities using intersection operations. The adjacent selection gathers a group of objects adjacent to determined selections. In addition, there is the cumulative selection, which generates a container node to compose a contributions union from the other selections by adjusting their parameters, as shown in Fig. 6.40.

Accordingly, the geometric model contains different parts with different shapes and dimensions, according to the real model, to obtain accurate results, or with minimum acceptable error. For example, a model containing a human tumorous tissue with different sections and other neighboring objects, such as vessels and bile ducts is built as shown in Fig. 6.41.

Fig. 6.41 displays an internal human tissue system with tumor to be ablated using any ablation treatment techniques, such as RF ablation and microwave (MW) ablation. Therefore each part inside the geometry is made of different materials using the materials window, as shown in Fig. 6.42.

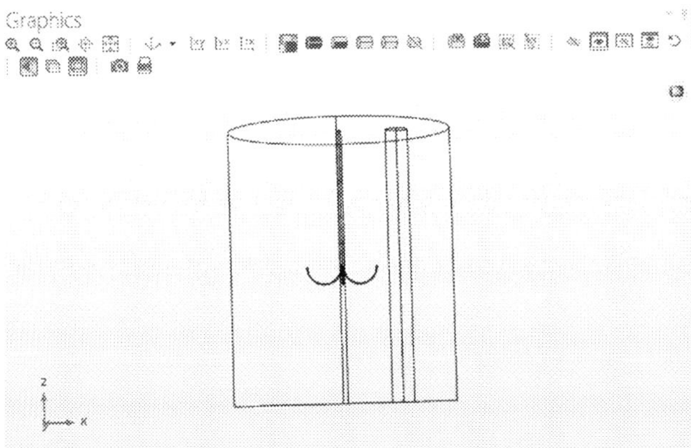

(A)

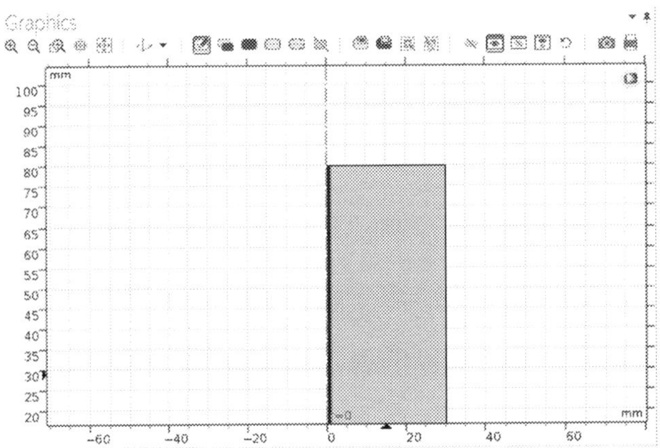

(B)

Figure 6.41 A geometric model for a human organ, blood vessel, and an ablated probe, including (A) radiofrequency probe and (B) microwave probe.

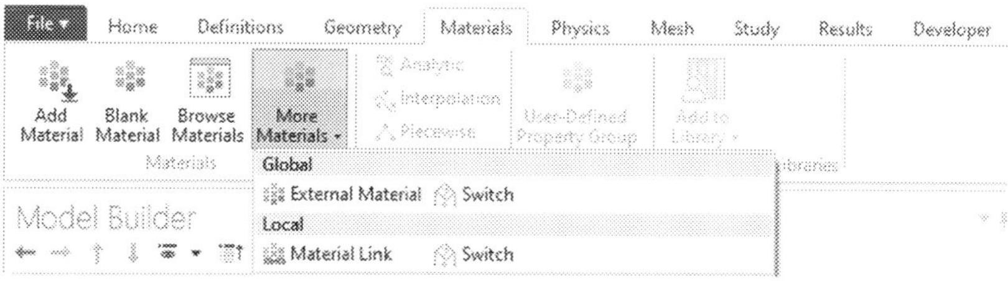

Figure 6.42 Materials window in COMSOL Multiphysics.

Fig. 6.42 displays different methods to add a specific material to a specific part inside the model, including the add material icon, which provides several materials inside the program, such as air, aluminum, and glass in the built-in section, copper and iron in the AC/DC section, bone, lung, and liver in the bioheat section, and stainless-steel material that is used for RF probe fabrication, as shown in Fig. 6.43.

Fig. 6.43 displays a great list of materials which are used in several applications, whether thermal, medical, mechanics, or optics, etc. The material is added to a specific selected part of the model through the setting window of the selected material, as shown in Fig. 6.44.

Fig. 6.44 displays standard properties for air material, such as electrical, thermal conductivities, heat capacity, and density. Also, it is able to change them according to the model. Additionally, the material can be manually added using a blank material icon as shown in Fig. 6.45.

Fig. 6.45 displays an empty material with no properties, where the desired properties and their values can be added to the blank material according to the model. Furthermore, other materials can be added externally and adjusted using the external material setting window as shown in Fig. 6.46.

To add the property group to the current material and add the material to the user-defined library for being available all the time, the user-defined property group and add to library options are used. The user can adjust the material through the setting window as shown in Fig. 6.47.

Additionally, the physics window is considered one of the most important windows to add or reject several physics to the model in multiple fields as shown in Fig. 6.48.

Fig. 6.48 displays the most used physics at the top left of the window. In addition, the Multiphysics window can be used to add Multiphysics coupling to the present geometric model component, as shown in Fig. 6.49.

Figure 6.43 Different materials in COMSOL Multiphysics.

The selected physics can be added as a domain according to the medium type, including solid, fluid, or gas, as a boundary, edge, and points, as shown in the case of heat transfer physics in Fig. 6.50.

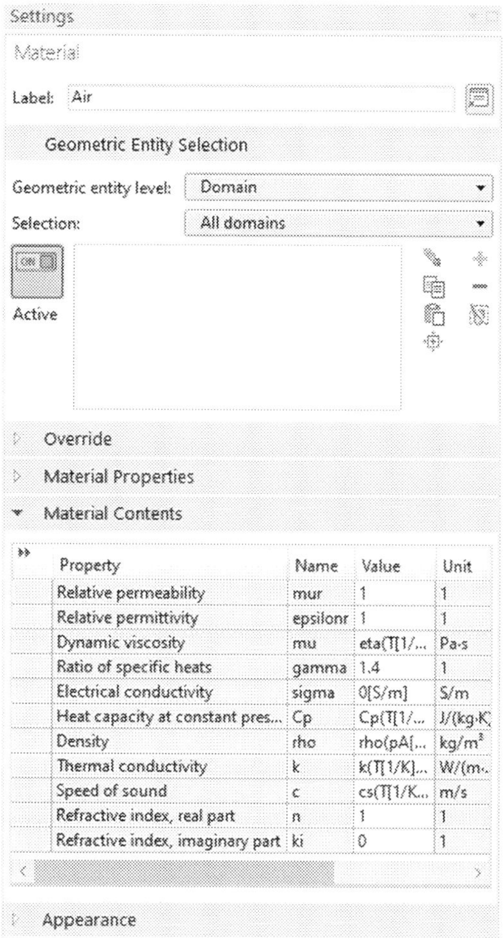

Figure 6.44 The setting window for air with standard properties.

Fig. 6.50 illustrates that the heat transfer mathematical equation changes according to the medium, in which the heat is transferred, where the properties in the equation can be taken from the material, or externally inserted as user defined through the setting window, as shown in Fig. 6.51.

Fig. 6.51 displays the heat transfer equation at constant phase, where the phase change affects the heat transfer equation, or any physics equation generally, according to the apparent formulation of heat capacity through the

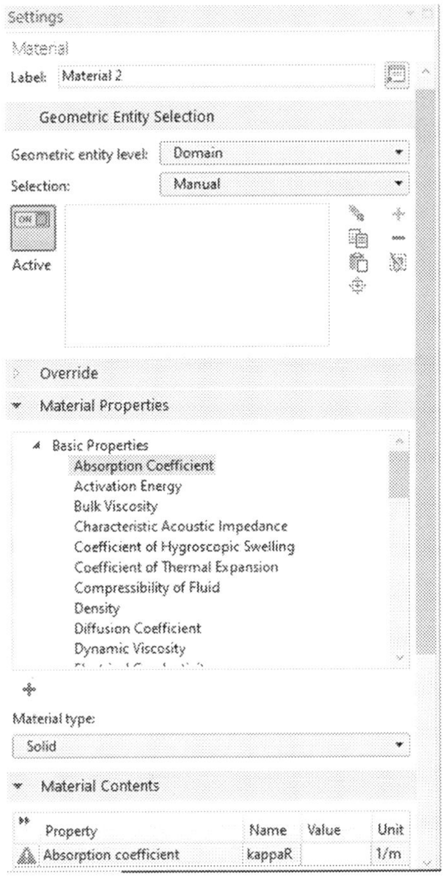

Figure 6.45 The setting window for blank material.

setting window of phase change material as shown in Fig. 6.52.

Moreover, the source and initial values of selected physics, that is, heat transfer physics, are controlled using the setting window of distributed heat source and initial values icon, which determines the initial values for the dependent variables as shown in Fig. 6.53.

The mathematical governed equation at the boundaries of geometry differs from that at the geometric domain, as shown in the heat transfer physics case in Fig. 6.54.

Fig. 6.54 shows several conditions for the selected physics, that is, heat transfer physics, at the boundaries, such as:

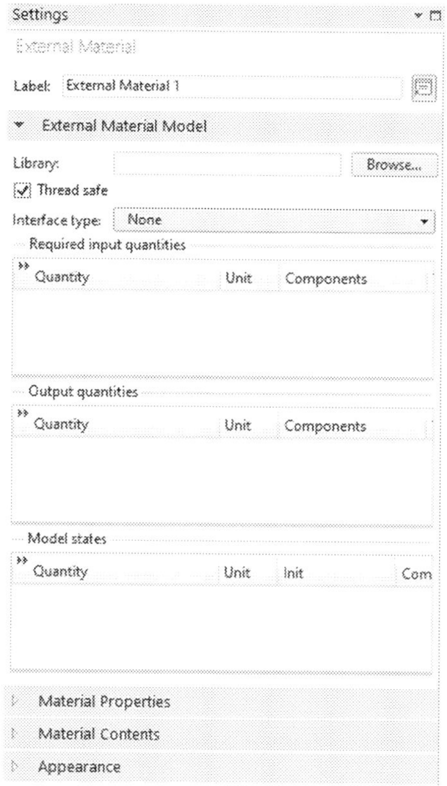

Figure 6.46 The setting window for external material.

- Thermal insulation, which sets a boundary condition of zero conductive heat flux, inflow heat flux due to the incoming external fluid;
- Periodic condition, which periodically repeats the conductive heat flux and temperature at a corresponding boundaries pair, with thermal contact between two domains;
- A thin layer, which can be conductive, resistive, or other using the extra dimensions;
- Deposited beam power, which is generated from narrow beams, such as an electron or laser; and
- Layered shell, in which the heat is transferred.

These conditions are adjusted using their setting windows, as shown in Fig. 6.55.

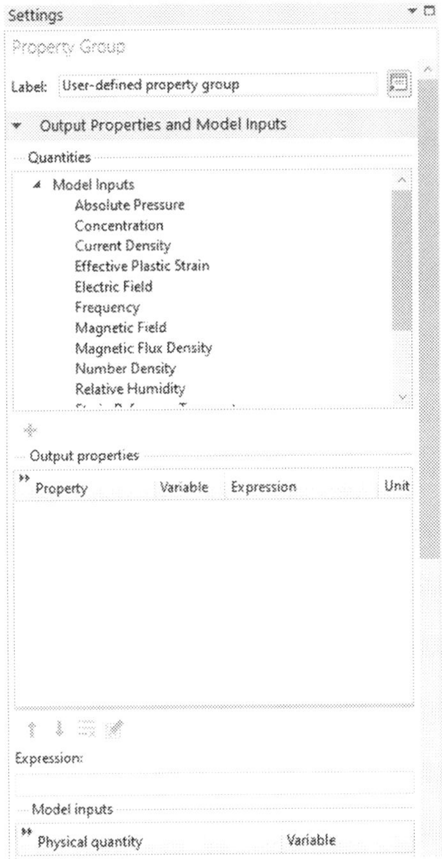

Figure 6.47 The setting window for user defines property group icon.

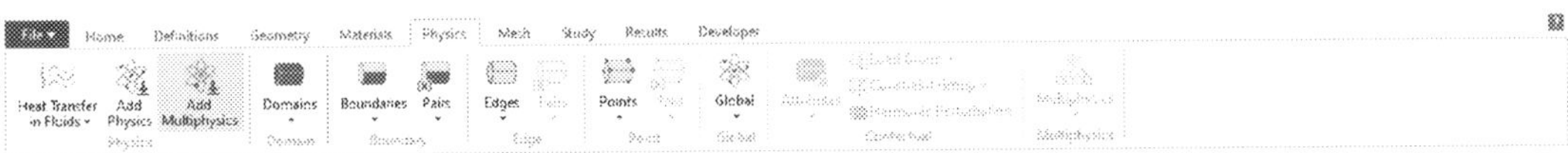

Figure 6.48 Physics window in COMSOL Multiphysics.

Finally, the physical conditions at the geometric edges and points are included, as shown in the heat transfer case in Fig. 6.56.

Fig. 6.56 displays the line heat source, which describes a heat source on a line, including geometrical points in two dimensions or geometrical edges in three dimensions, a heat transfer geometric model for thin rods, and a heat source that is distributed on a geometrical point. These heat transfer

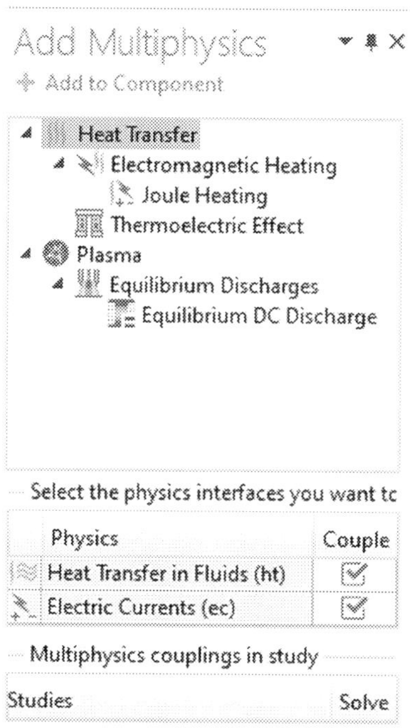

Figure 6.49 Multiphysics window in COMSOL Multiphysics.

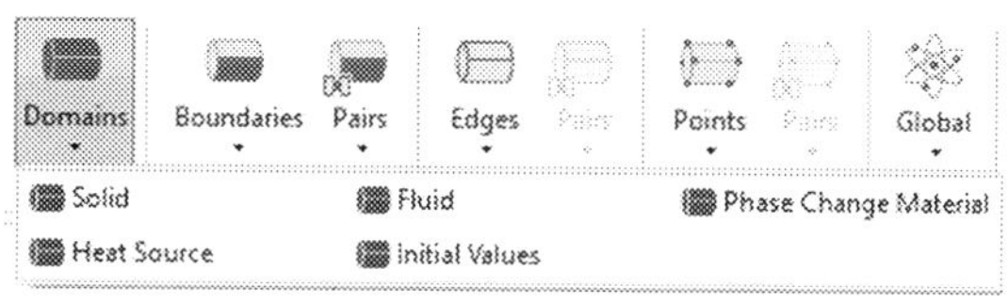

Figure 6.50 Heat transfer physics in different media.

conditions are controlled using their setting windows, as shown in Fig. 6.57.

COMSOL Multiphysics provides the Multiphysics option for couplings between the different physics in the same field, where the available Multiphysics are observed as shown in Fig. 6.58.

Fig. 6.58 displays a list of Multiphysics, which is required during the model simulation to exchange the resulting parameters between the different equations in differently related physics. Hence, the coupling is performed and controlled using the Multiphysics setting windows as shown in Fig. 6.59.

It is imperative to divide the model into small nodes for all parameters to be computed at each node inside the geometry.

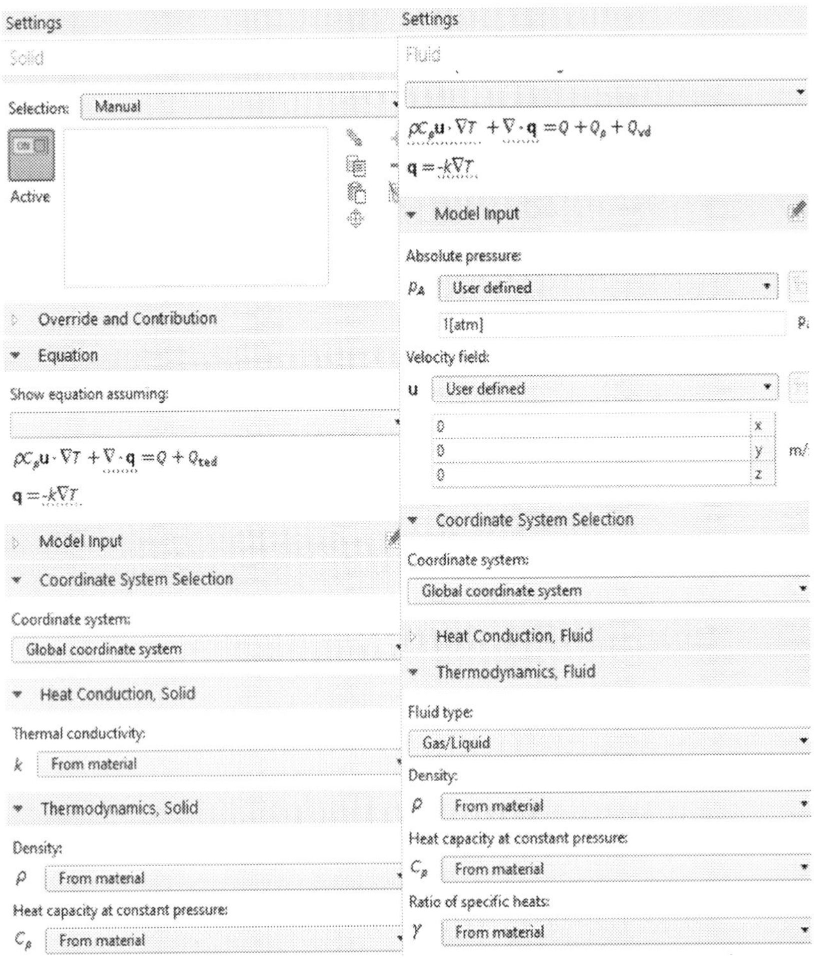

Figure 6.51 Heat transfer equation in different mediums, including (A) solid and (B) fluid.

Thus, the accuracy of the results and the processing time increase with increasing number of nodes. This is performed using the mesh window as shown in Fig. 6.60.

Fig. 6.59 displays the main icons for adding a mesh to the current model, and editing. The mesh type is selected according to the geometry shape in order to avoid empty or/and overlapping regions inside the meshed model. Therefore the most commonly used mesh for the models is considered to be the free

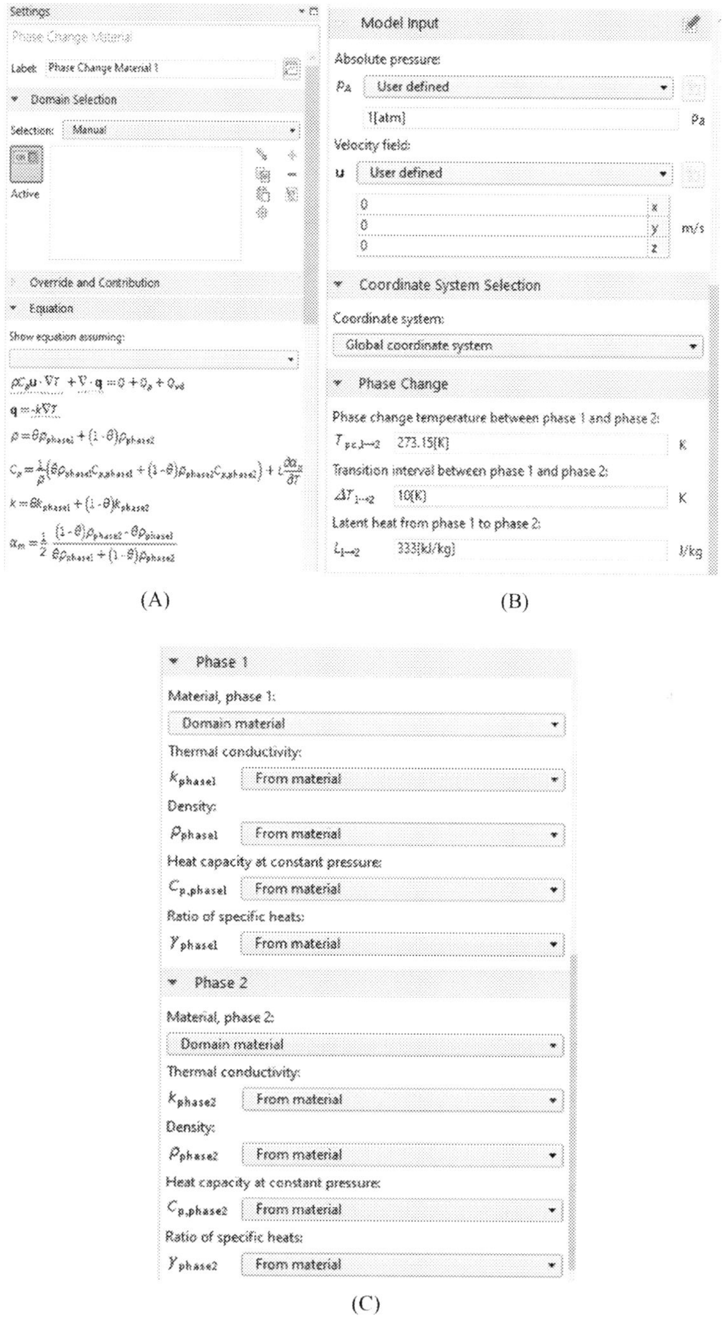

Figure 6.52 The setting window of the phase change material.

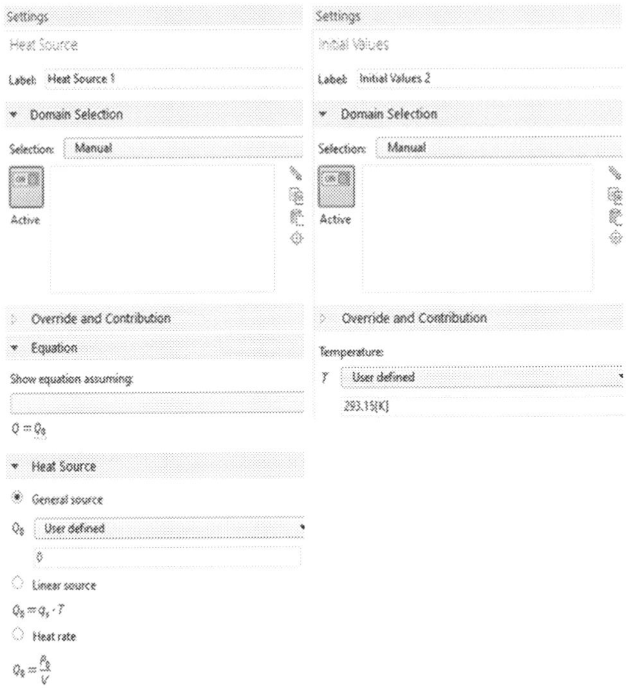

Figure 6.53 Setting windows of the heat source and initial values.

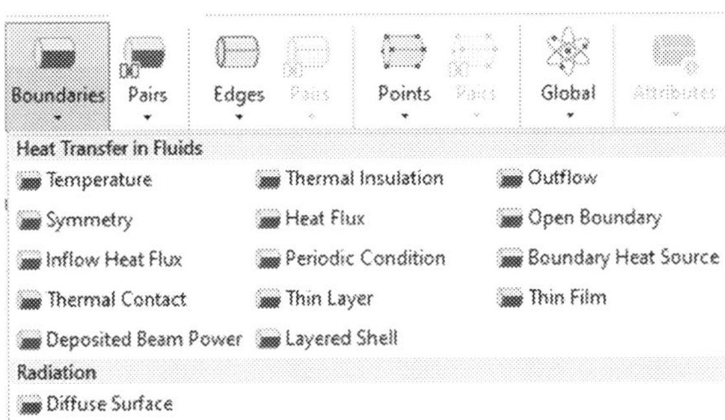

Figure 6.54 Several conditions of the heat transfer at boundaries.

tetrahedral mesh, in which each node contains six edges, four faces, and four vertices, as shown in Fig. 6.61.

Furthermore, there are other types of mesh available inside the program, including swept, which sweeps the 3D mesh from the source face through the domain to the opposite destination face, free triangular, free quad, which produces an unstructured

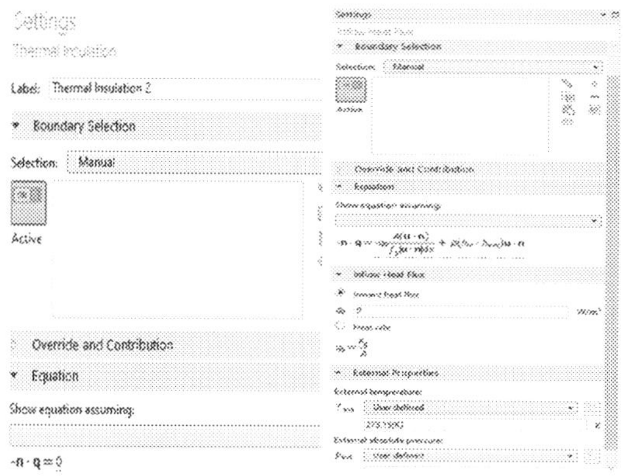

(A)

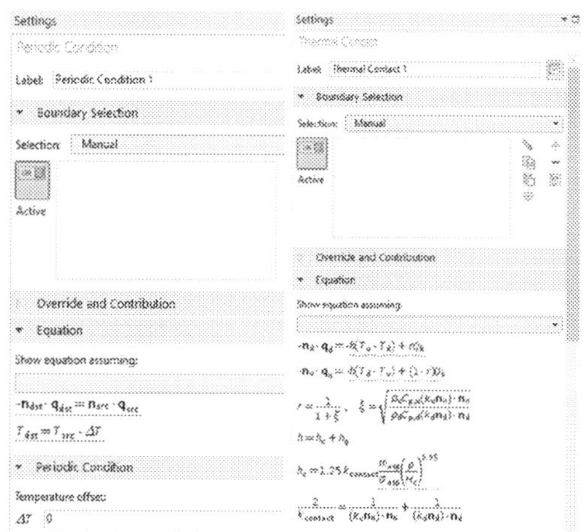

(B)

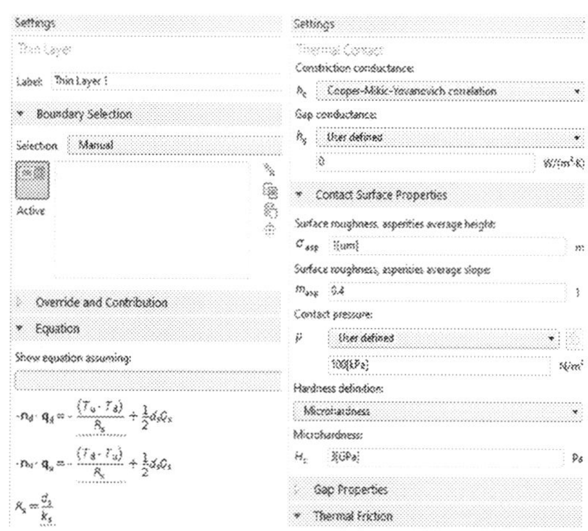

(C)

Figure 6.55 Setting windows of heat transfer at boundaries.

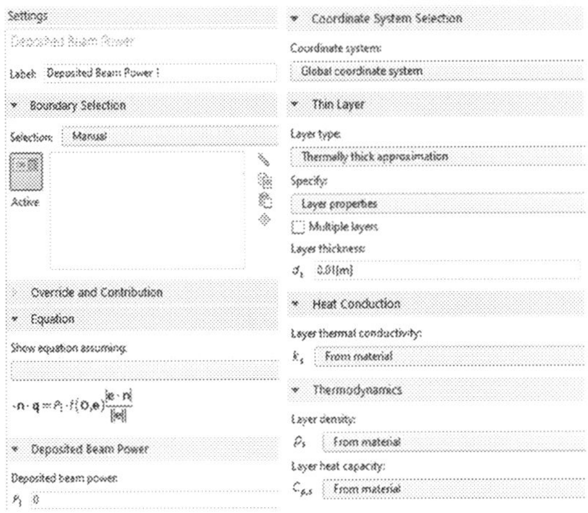

(D)

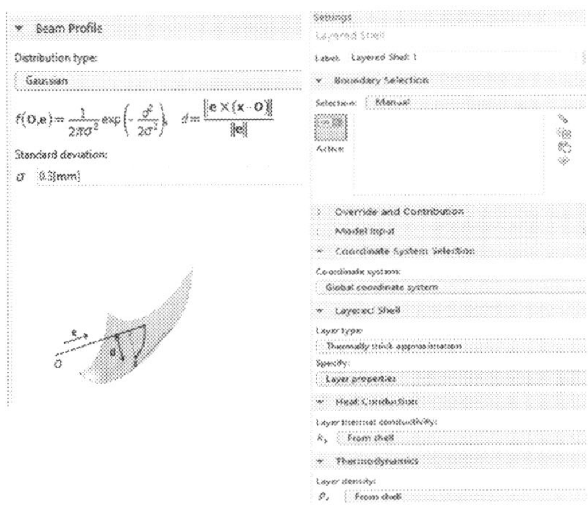

(E)

Figure 6.55 Continued

quad mesh, mapped, which produces a structured quad mesh, edge mesh and boundary layers mesh with their setting windows as shown in Fig. 6.62, and their settings are adjusted as illustrated in Fig. 6.63.

Fig. 6.63 displays the available options to control the mesh structure for all mesh types obtainable after building the final mesh shape, which simulates the geometric models in Fig. 6.62.

Additionally, some parameters related to the mesh, such as the maximum, minimum element size, maximum element growth rate, curvature factor, and the narrow regions resolution are very important to adjust the mesh to obtain more accurate results. These parameters have default values related to the

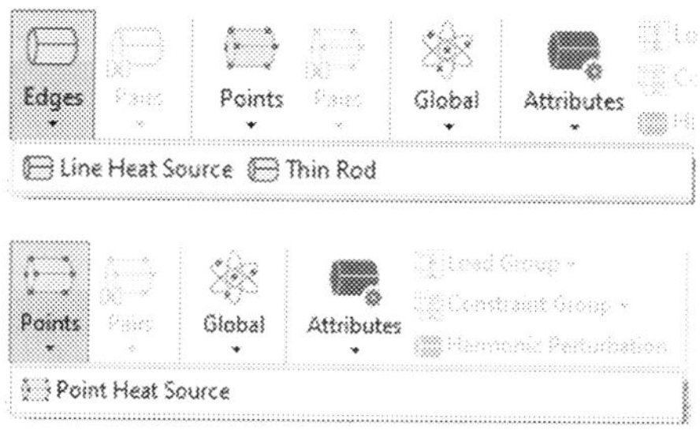

Figure 6.56 Heat transfer condition at edges and points of the geometry.

model type, including general physics, fluid dynamics, plasma, and semiconductor. They can be changed according to the geometry shape to be extremely fine or extremely coarse using the setting window of the mesh size, as shown in Fig. 6.64.

Consequently, other operations can be performed for the mesh structure to be modified in its size, element, and distribution, as shown in Fig. 6.65.

Fig. 6.65 displays several additional options, such as:

- Size expression, which generates a specific expression to determine the local mesh size;
- Distribution, which specifies the distribution properties of the mesh element;
- Corner refinement, which decreases the element size at the sharp corners;
- Convert option, which converts the mesh to become a simple mesh;
- Reference option, which refers to other meshing sequences; and
- Scale, which can scale the mesh size properties, as shown in Fig. 6.66.

Furthermore, copy option, whether a domain mesh, face mesh, or edge mesh, is available in COMSOL Multiphysics rather than building the mesh from the beginning, as shown in Fig. 6.67.

Fig. 6.67 demonstrates the manual method of composing the mesh structure. It is possible to externally import a mesh from a file in your device or other meshing sequences, as shown in Fig. 6.68.

Hence, several operations can be done to modify the imported mesh, including:

- Ball, box, cylinder partitions, and logical expressions, which split their geometric entities by a ball, box, cylinder, and a specified logical expression; and

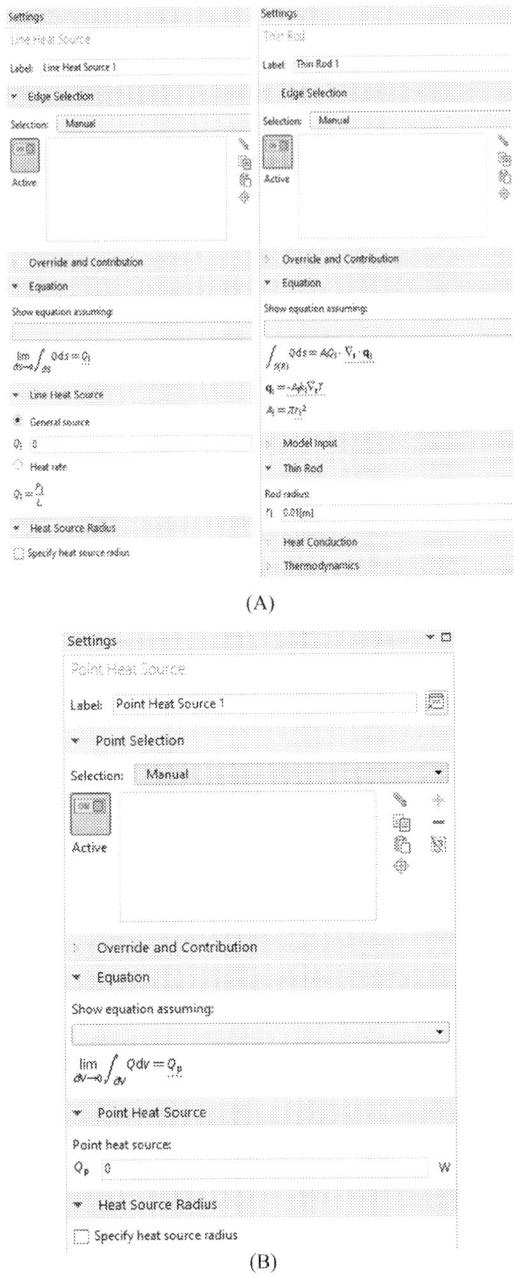

Figure 6.57 Setting windows of the heat transfer condition at geometric edges and points.

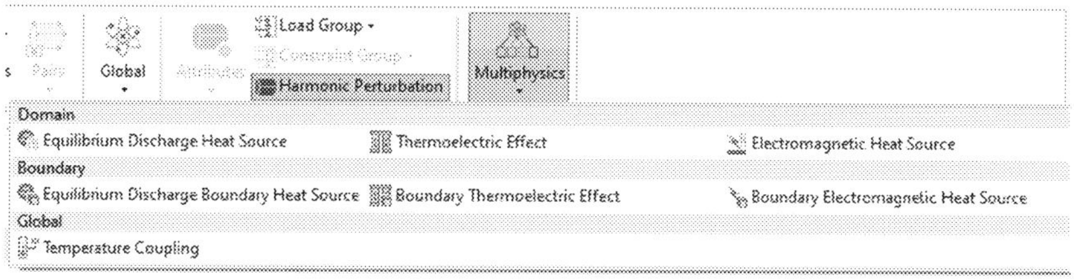

Figure 6.58 Multiphysics in different three conditions, including domain, boundary, and global.

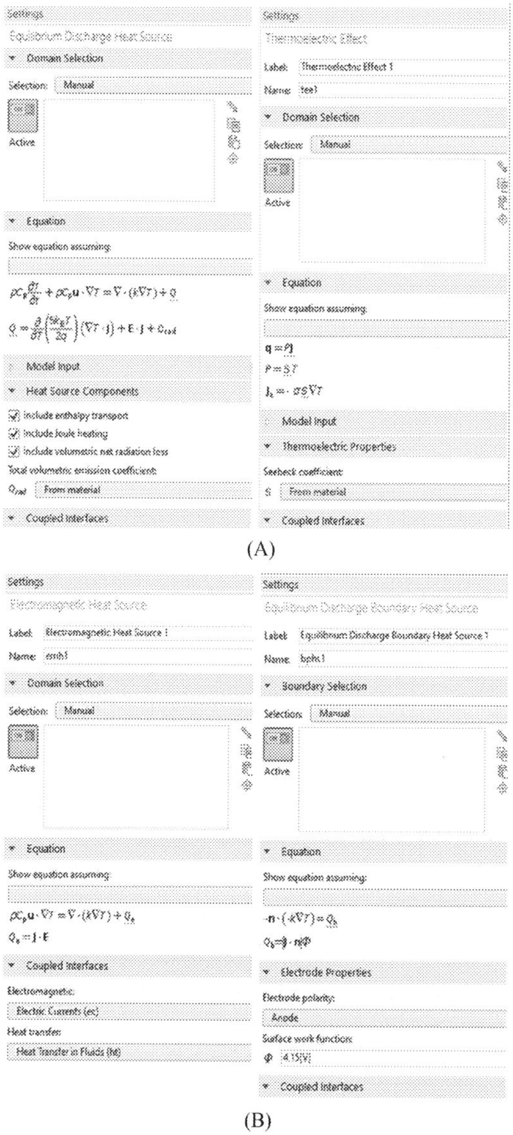

Figure 6.59 Setting windows of the available Multiphysics in COMSOL Multiphysics.

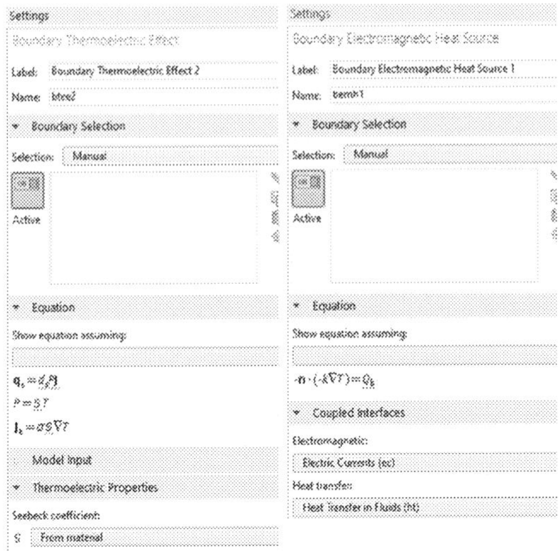

(C)

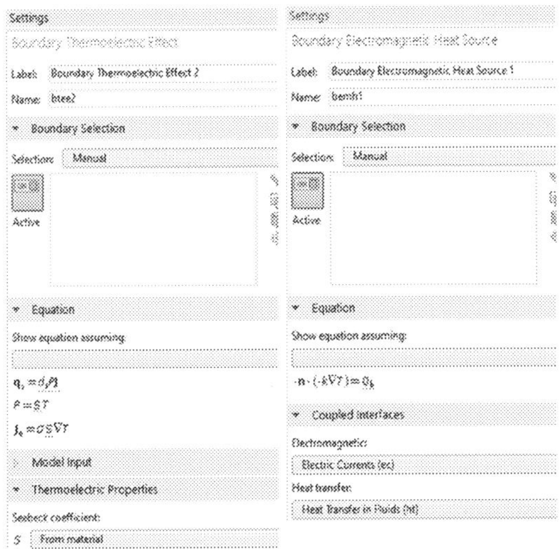

(D)

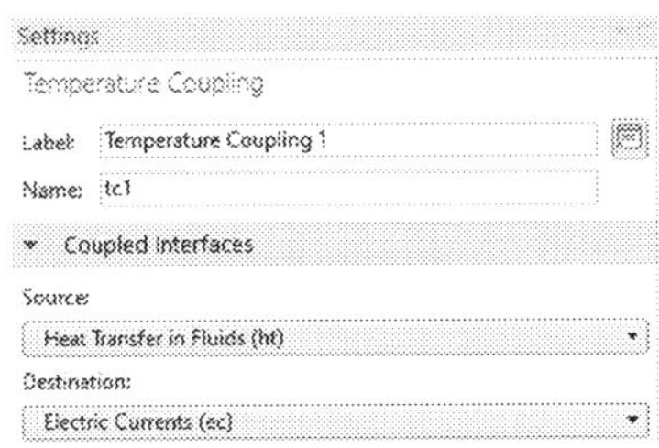

(E)

Figure 6.59 continued

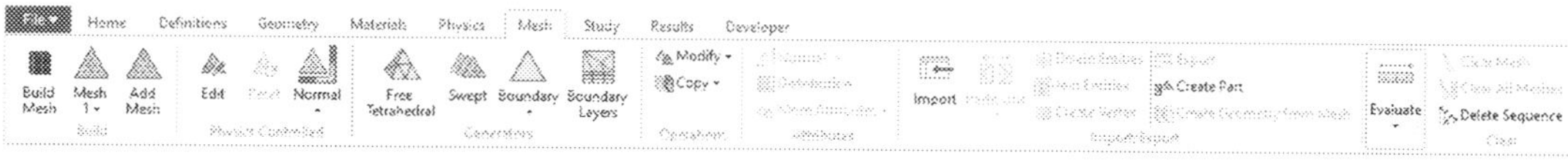

Figure 6.60 Mesh window in COMSOL Multiphysics.

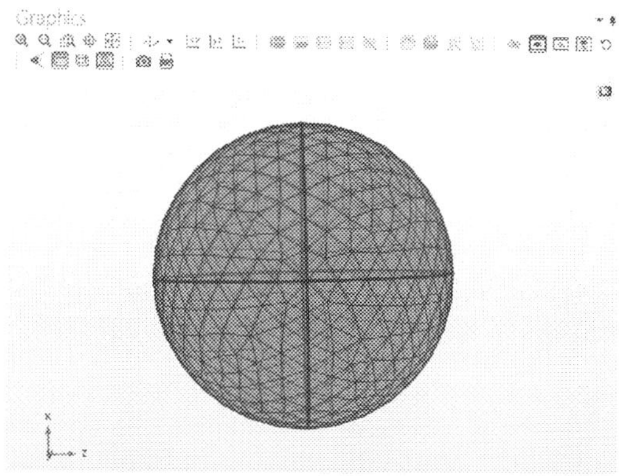

Figure 6.61 Tetrahedral mesh in COMSOL Multiphysics.

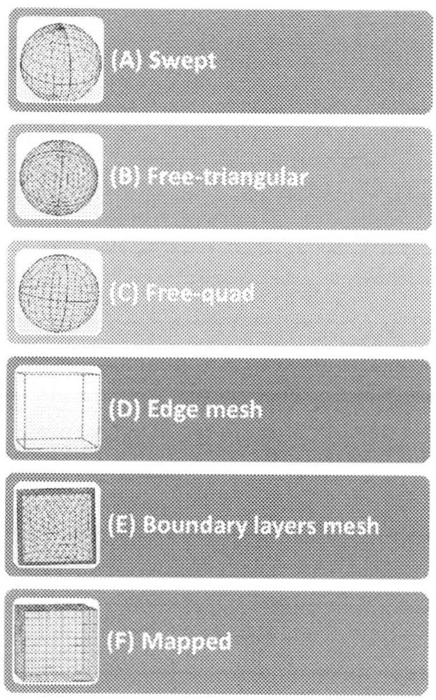

Figure 6.62 Mesh types of FE model in COMSOL.

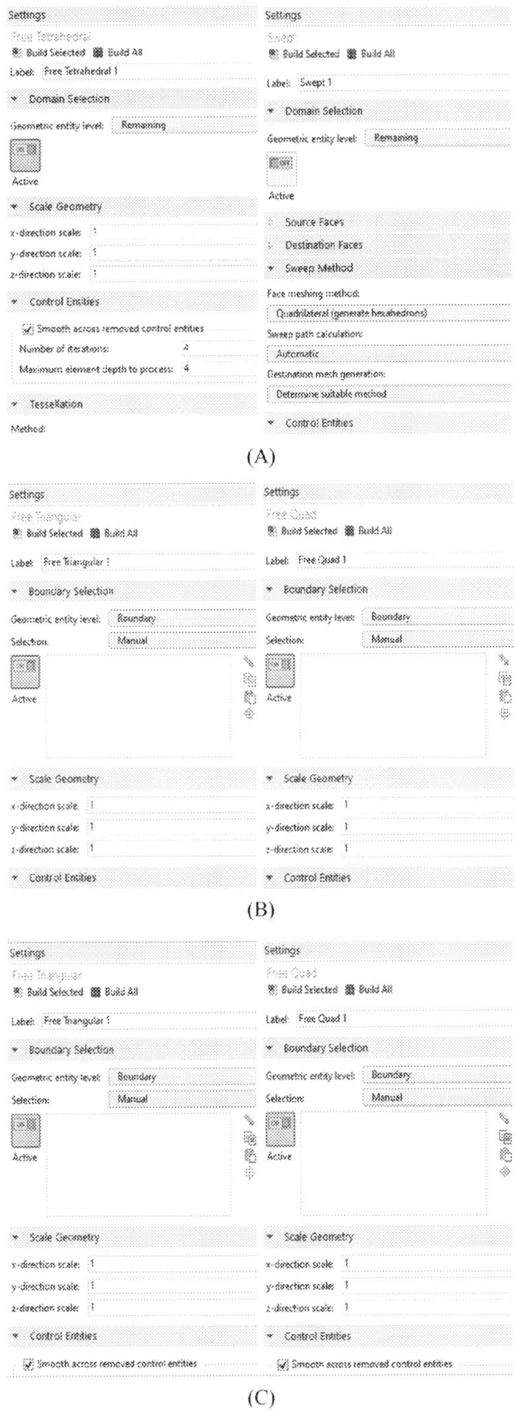

Figure 6.63 Setting windows of different mesh types.

Figure 6.63 continued

- Detect faces, which split the geometric boundary entities of the imported mesh by searching for the shapes inside the mesh that possibly constitute faces.

Furthermore, it is possible to delete or/and join geometric entities or/and create additional vertices in the imported mesh. Also, it is possible to create a new geometric model component from the mesh, as shown in Fig. 6.69.

Moreover, in the geometric model, an evaluation of the measured distance and statistic, can be performed. A sample of the messages window that may appear is highlighted in Fig. 6.70.

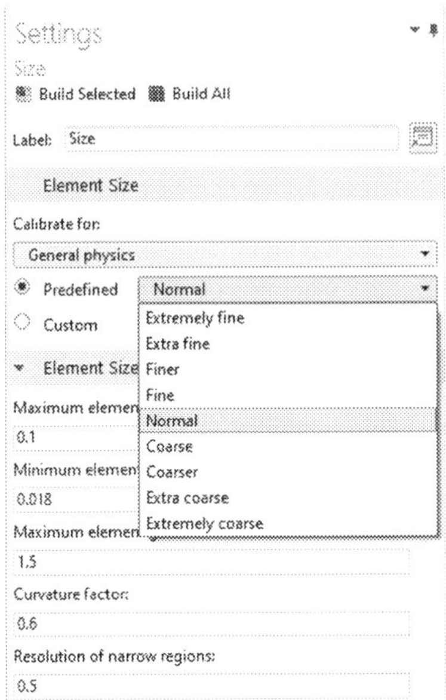

Figure 6.64 The setting window of the mesh size.

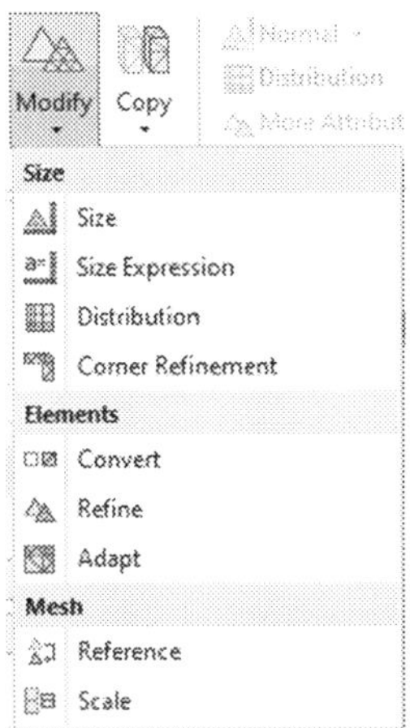

Figure 6.65 Mesh modify operations.

Thus, the model can be studied in any domain, including time and frequency, using the study window shown in Fig. 6.71.

Fig. 6.71 exemplifies the study operations, which are performed on the geometric model to obtain the final results, such as add study, which provides different studies with or without

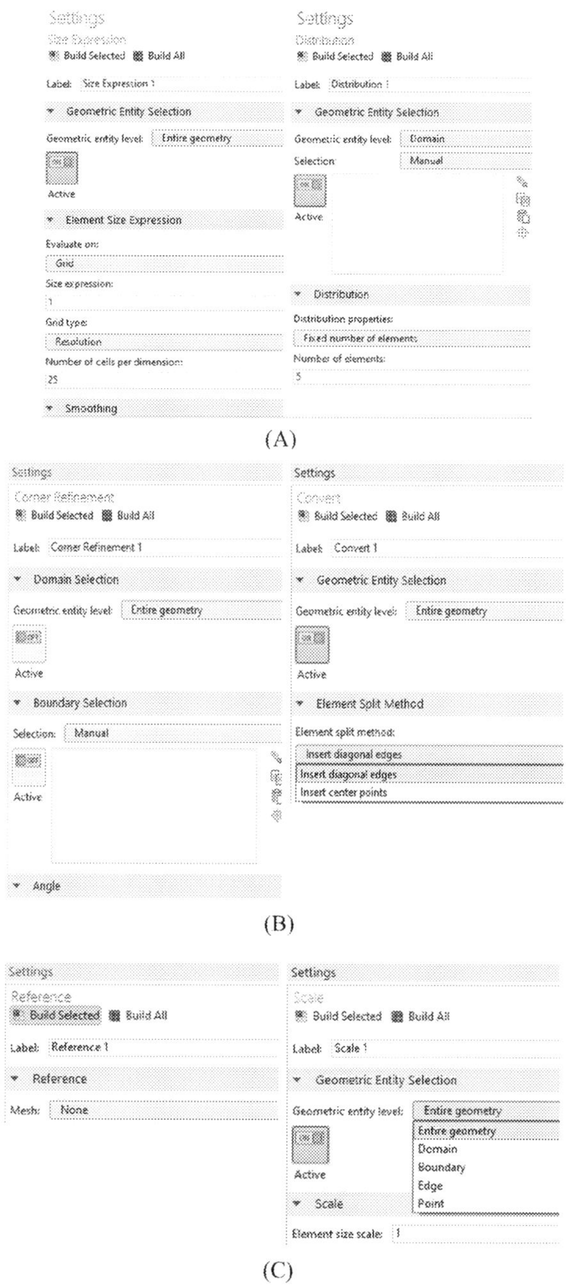

Figure 6.66 Setting windows of mesh modify operations.

the time and frequency, gets the initial value, which generates the solution using only the initial values of dependent variables without solving these dependent values, and shows the default

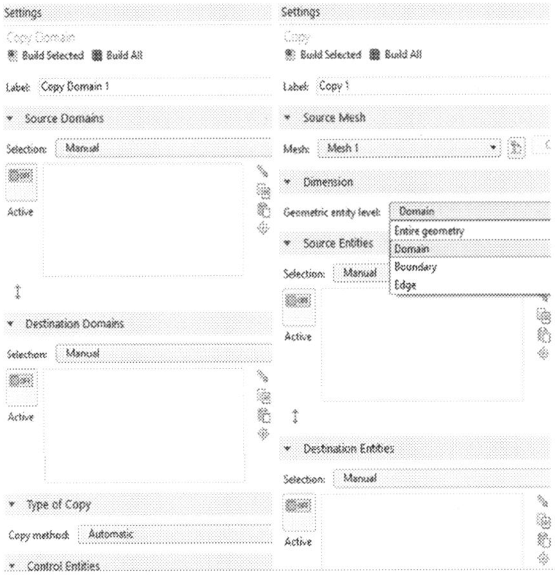

Figure 6.67 Setting windows of copy options.

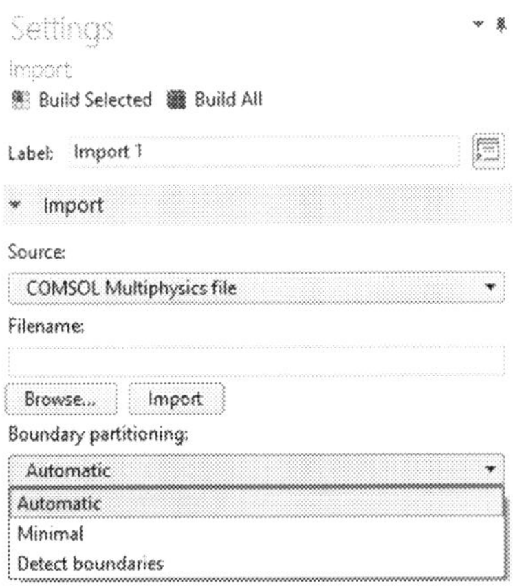

Figure 6.68 Setting windows of mesh import.

solver, which displays the created default nodes in the solver sequences as shown in Fig. 6.72.

In addition, it is possible to change the default solver or add different studies using the study steps, including stationary, time dependent, eigen frequency, frequency domain, and other studies, such as fatigue and frozen rotor, as revealed in Fig. 6.73.

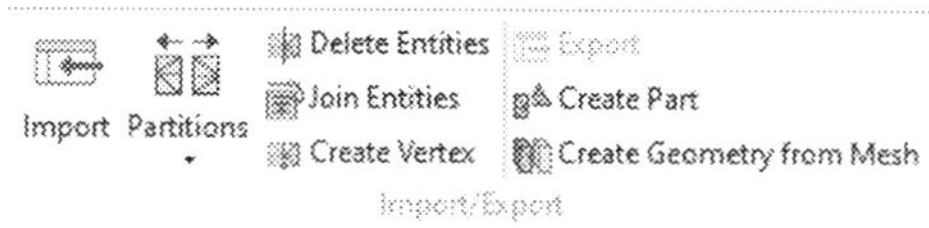

Figure 6.69 Operations on the imported mesh.

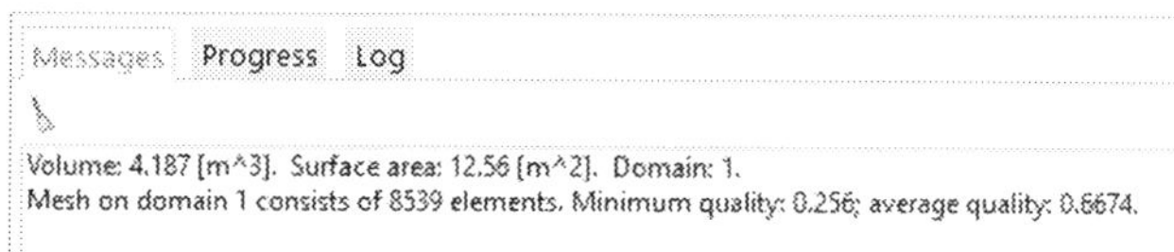

Figure 6.70 Measure and statistics of the mesh in the message window.

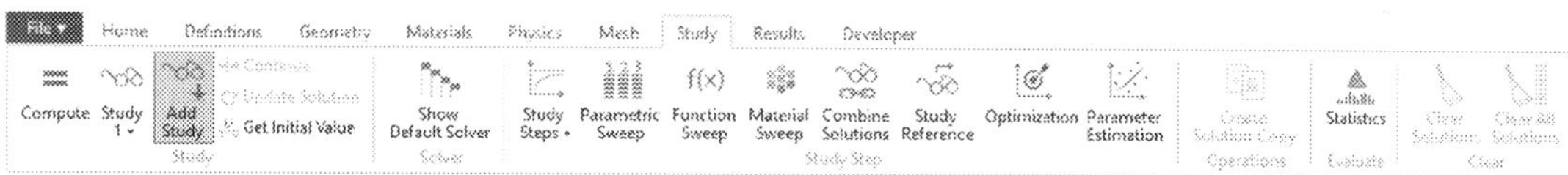

Figure 6.71 Study window in COMSOL Multiphysics.

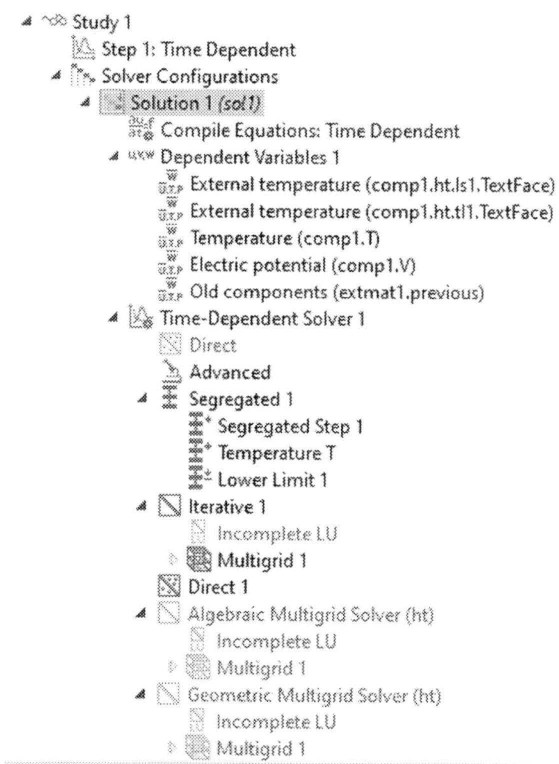

Figure 6.72 Default study solver in COMSOL Multiphysics.

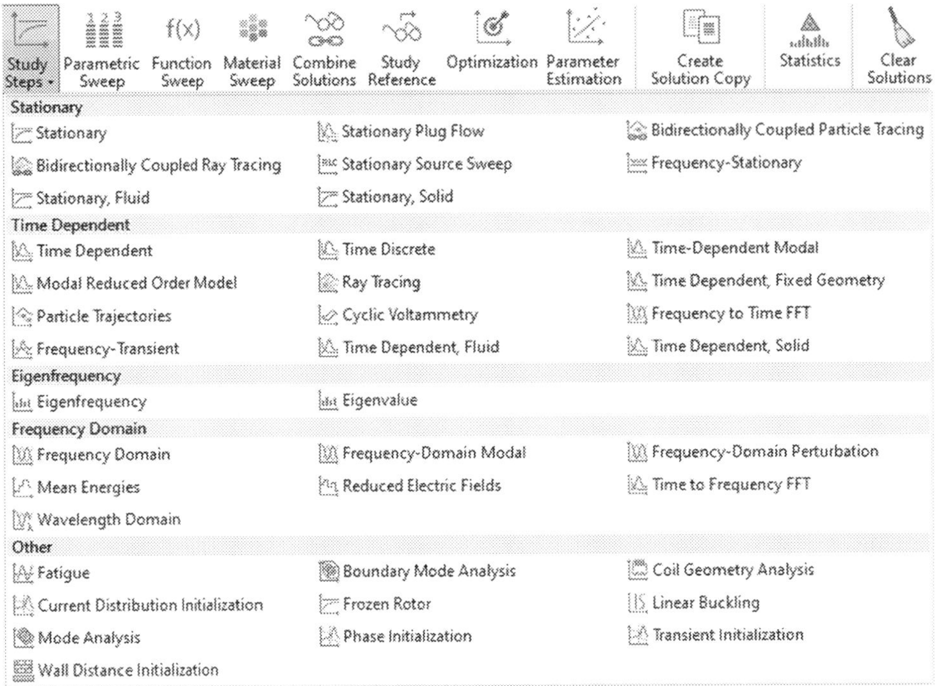

Figure 6.73 Study steps in COMSOL Multiphysics.

Fig. 6.73 displays the most common studies, which are used to solve the current geometric model, including:

- Stationary plug flow, which is used to solve the molar flow rate in the plug flow reactor models produced from the reaction engineering interface;
- Eigen frequency, which is used to compute the eigen frequencies and eigenmodes of a linearized or linear model;
- Frequency domain perturbation study, which contains two study steps, including stationary and frequency domain studies, to study the small oscillations for a biased solution;
- Mean energies, which is used to enter the values array of the mean energy of the electron;
- Time to frequency fast Fourier transform study, which is used to calculate the fast Fourier transformation of the time domain signal to frequency domain;
- Boundary mode analysis, which is used to calculate the propagation constants, wave numbers, and propagation mode shapes at a given frequency; and
- Linear buckling, which estimates the critical load at the structure instability, as shown in Fig. 6.74.

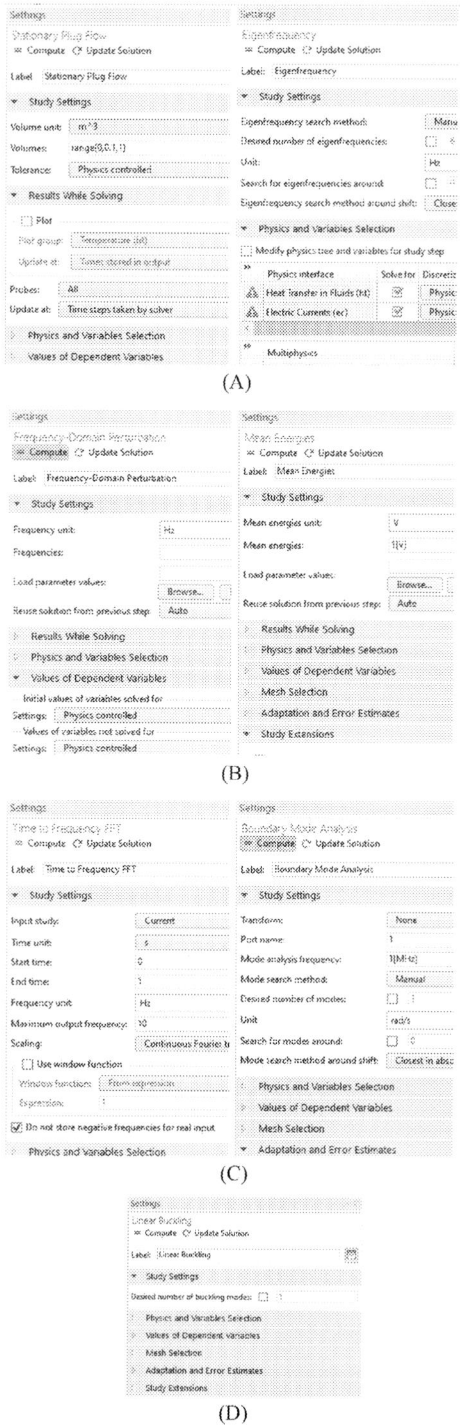

Figure 6.74 Setting windows for some solver studies.

Furthermore, it is possible to provide a change in the parameter's values over a determined range rather than changing them manually using the parametric sweep, as shown in Fig. 6.75.

The function sweep solver is considered a special case of the parametric sweep solver, with sweeps over defined functions in the switch node, as shown in Fig. 6.76.

Materials are selected and added to the model to give the best performance. Hence, instead of manually changing the materials through every simulation, and resolving it multiple times, the process is automatically performed using the material sweep, as shown in Fig. 6.77.

Fig. 6.77 illustrates that a comparative study can be added between different materials to observe their effect on the model results. COMSOL provides combining solutions using summation or concatenation, for example, joining two time-dependent studies, where one covers the range from t0 to t1, and the other

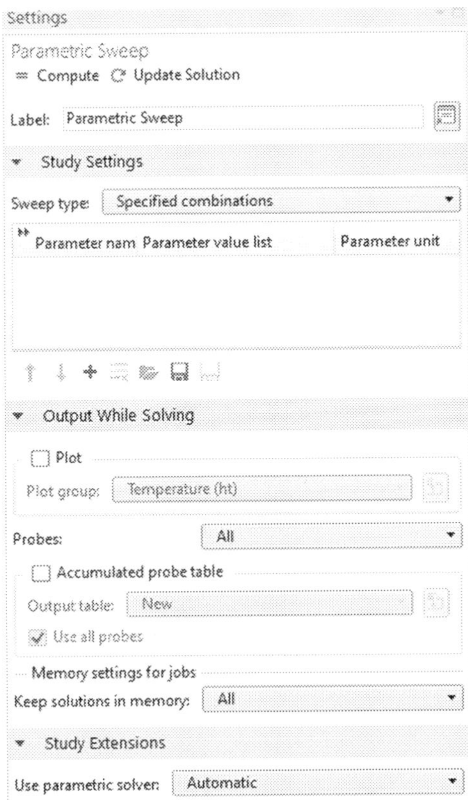

Figure 6.75 Setting windows for the parametric sweep.

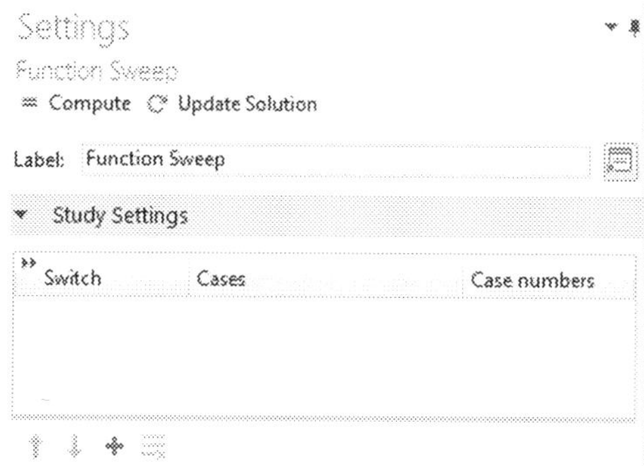

Figure 6.76 Setting windows for the function sweep.

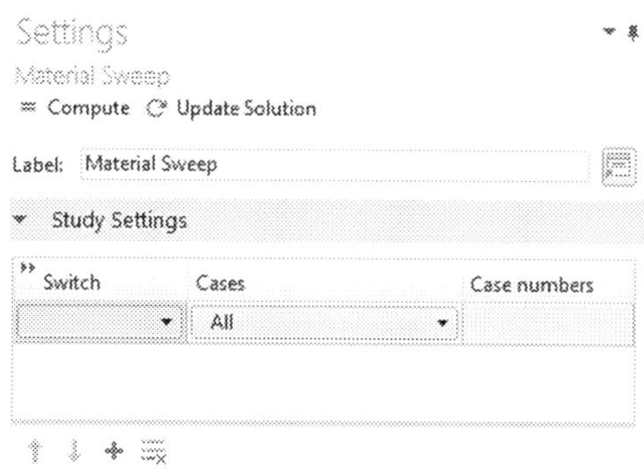

Figure 6.77 Setting windows for the material sweep.

covers from t1 to t2, to obtain only one time-dependent study solution covering from t0 to t2. This process is performed using the combined solutions options in the study window, as shown in Fig. 6.78.

Additionally, several studies can be correlated by referring to one study from the other using the study reference as shown in Fig. 6.79.

Fig. 6.79 demonstrates that one solver study can be considered as a precomputing stage for the subsequent study. The design can be improved by changing its variables to satisfy the constraints using the optimization module with different methods, including coordinate search, Monte Carlo, Nelder-Mead, BOBYQA, COBYLA, SNOPT, MMA, and Levenberg-Marquardt, as shown in Fig. 6.80.

Fig. 6.80 clarifies the common methods for optimization of the design parameters, including part shapes, material distributions, material properties, and geometric dimensions for design

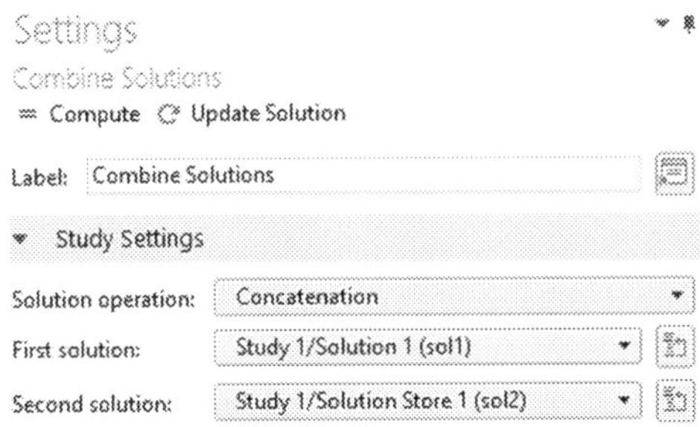

Figure 6.78 Setting windows for the combine solutions.

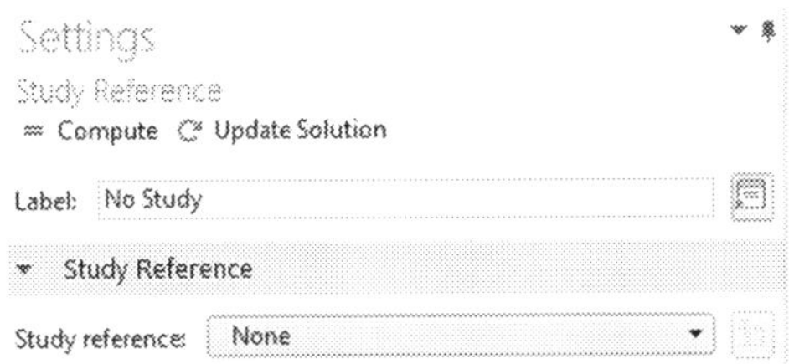

Figure 6.79 Setting windows for the study reference.

improvement. The input values to the geometric model can be estimated to match the computational results with the reference data using the parameter estimation module, as shown in Fig. 6.81.

Finally, the results are obtained, adjusted, formatted, and controlled using the results window shown in Fig. 6.82.

Fig. 6.82 displays the most common tools which allow the user to obtain the final two- and three-dimensional figures, tables, and other data in an organized shape, such as a plot icon of the parameters related to the selected physics, that is, for electric current and heat transfer in fluids physics, as shown in Fig. 6.83.

Moreover, several additional common plots are available to users to explain the results, such as one-, two-, and three-dimensional plot groups with several types, such as line, table, point, graphs, histogram, and annotation for the 1D plot group, surface, line, contour, arrow line, arrow surface, streamline, and particle trajectories for the 2D plot group, and volume, slice, arrow volume, iso-surface, and streamline for the 3D plot group, as shown in Figs. 6.84 and 6.85.

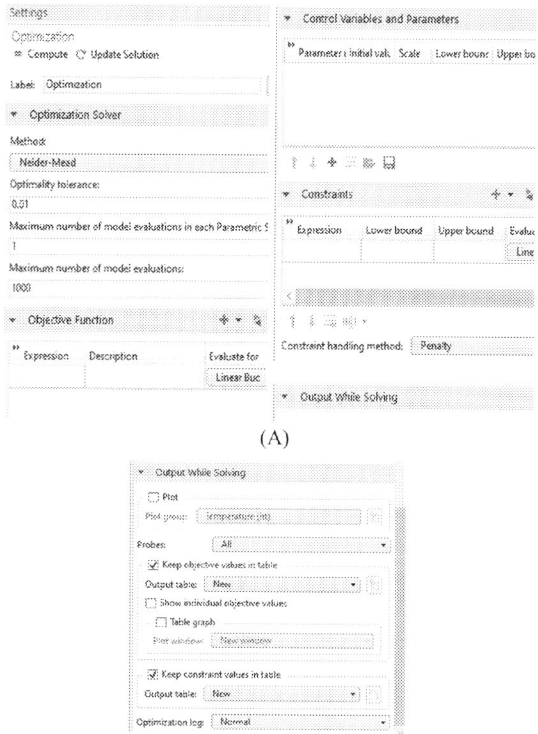

(A)

(B)

Figure 6.80 Setting windows for the optimization module.

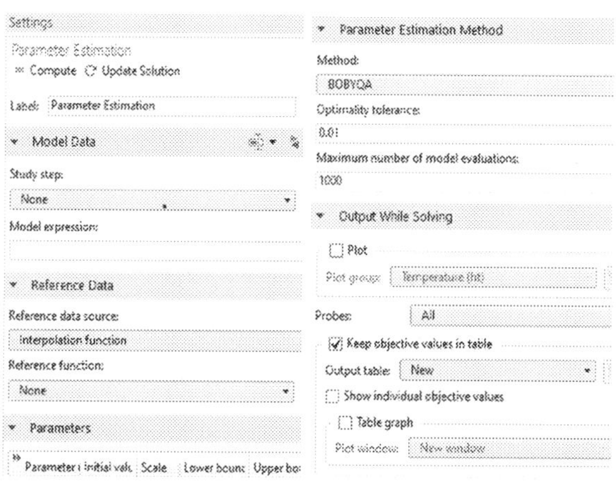

Figure 6.81 Setting windows for the parameter estimation.

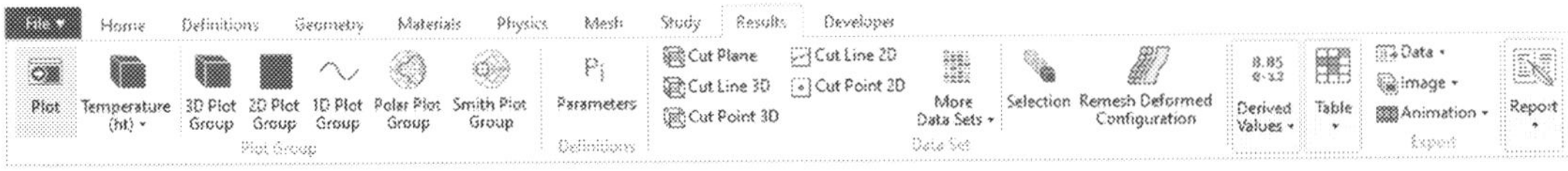

Figure 6.82 Results window in COMSOL Multiphysics.

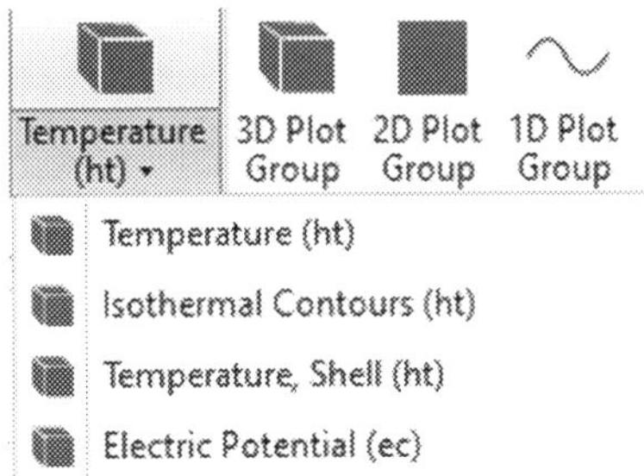

Figure 6.83 Parameters for heat transfer and electric current physics.

There are other one-, two-, and three-dimensional plots that can be added to clarify the results according to the current model, as shown in Fig. 6.86.

Additionally, the image, data, and animation file for any parameter that is changed with a specific variable is obtained and exported, as shown in Fig. 6.87.

The results' windows provide adding polar and smith plot groups, which are used for the smith charts of the impedance, admittance, and reflection, as shown in Figs. 6.88 and 6.89.

Moreover, COMSOL can locally add available parameters to the results section for parameterized postprocessing using the parameters modules in the results window, as shown in Fig. 6.90.

Additionally, a cutting option is also available, including planes, lines, and points, whether in the 2D or 3D plane using the setting windows as shown in Fig. 6.91.

Different types of data sets can be obtained, such as:

- Join data sets, which gathers two data sets to obtain several quantities, such as difference;
- Shell, which creates a shell data to visualize the bottom, and top surfaces of the shell;
- Intersection point 3D, which creates intersection points 3D data set to evaluate the expressions at the intersection points of ray trajectories or particle, and the surface; and
- Parametric extrusion 2D, which extends a 2D solution data set using parameters, such as dimension and time, as shown in Figs. 6.92 and 6.93.

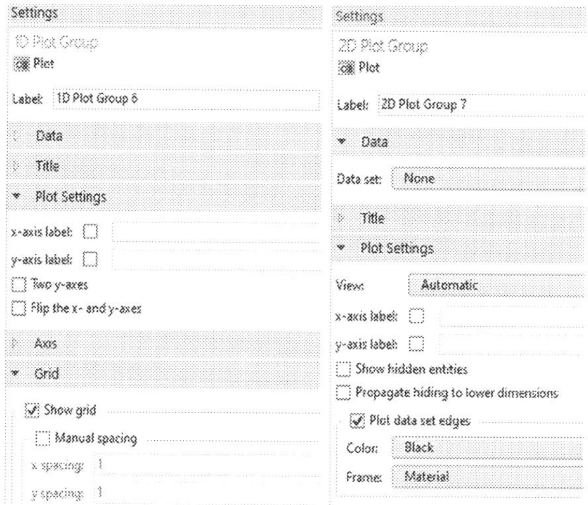

(A)

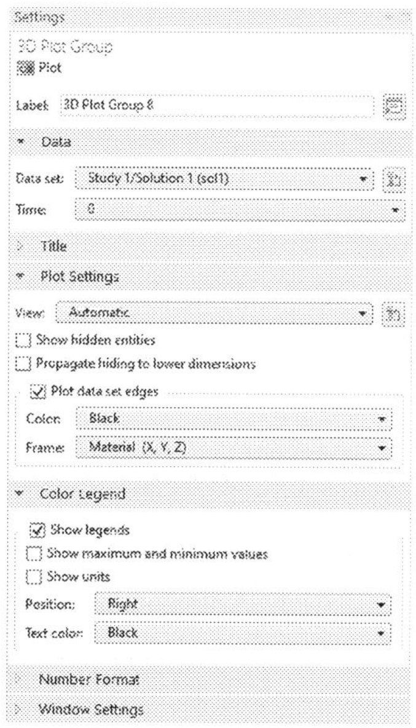

(B)

Figure 6.84 Setting windows for the one-, two-, and three-dimensional plot groups.

In addition, the most common functions, such as volume, integration, average, maximum, minimum, and other for the optics, particle trajectories, and Zernike coefficients are as shown in Figs. 6.94 and 6.95.

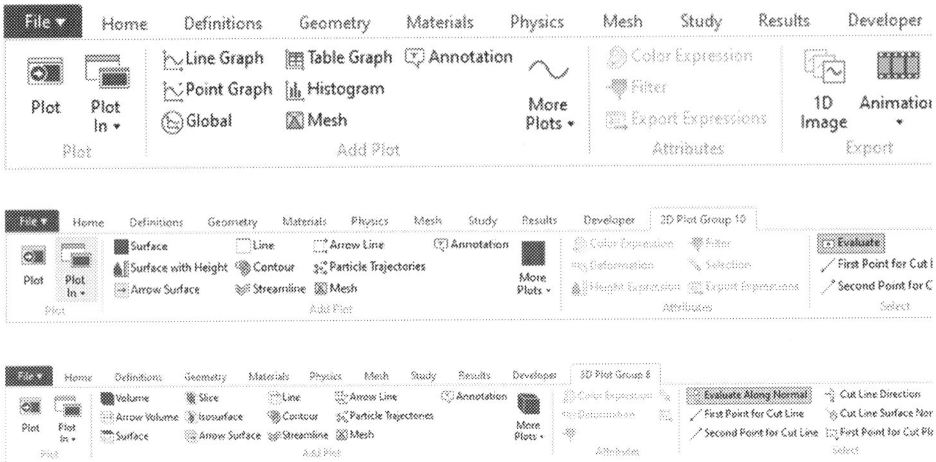

Figure 6.85 The one-, two-, and three-dimensional plot windows.

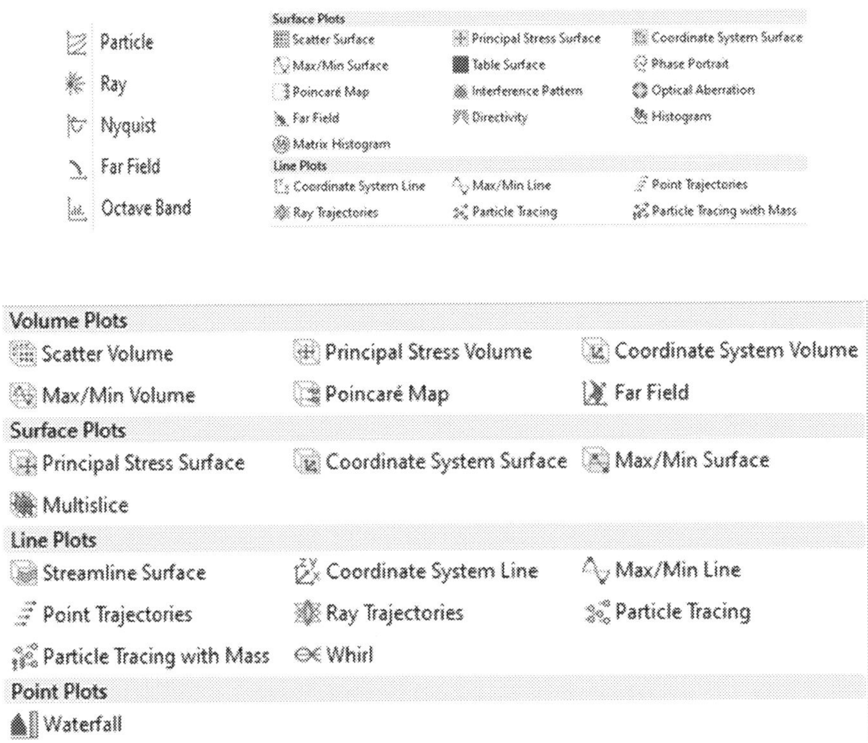

Figure 6.86 Additional plots in one, two, and three dimensions.

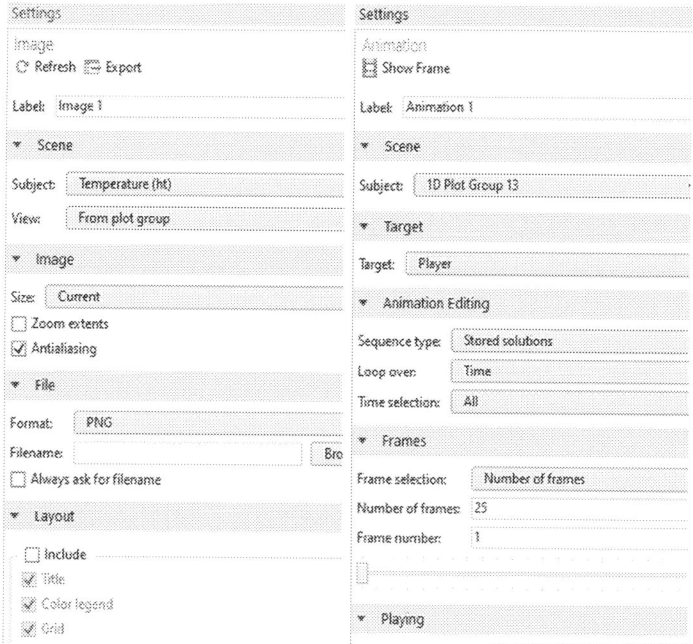

Figure 6.87 The setting window of the image and animation export tool.

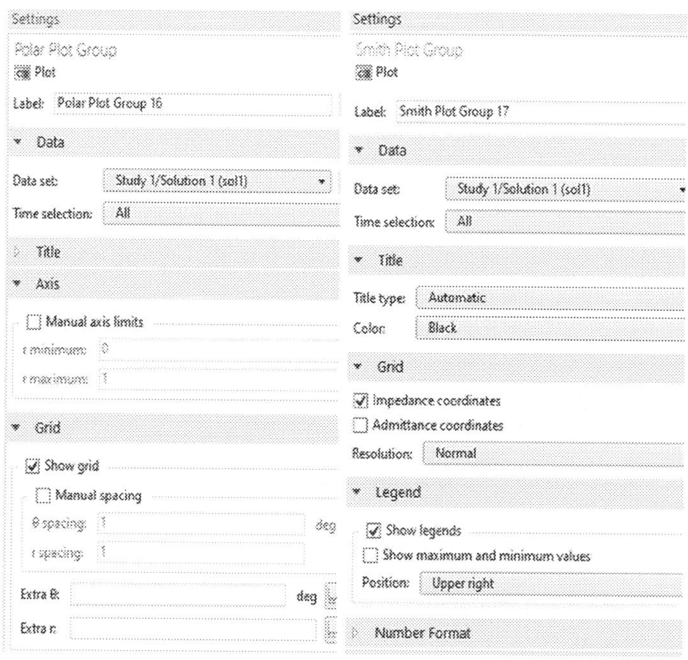

Figure 6.88 Setting windows of the polar and smith plots.

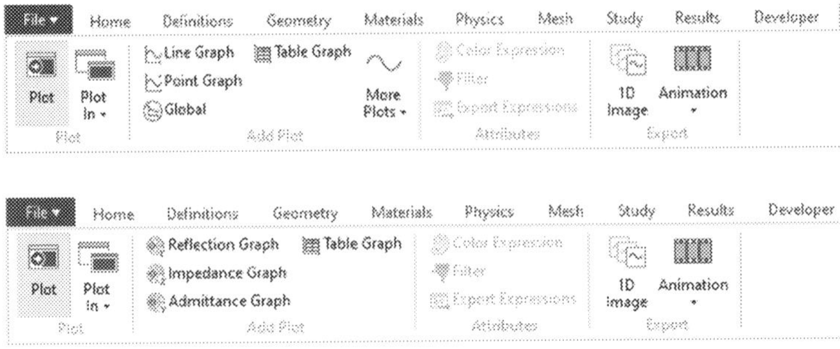

Figure 6.89 Windows of the polar and smith plots.

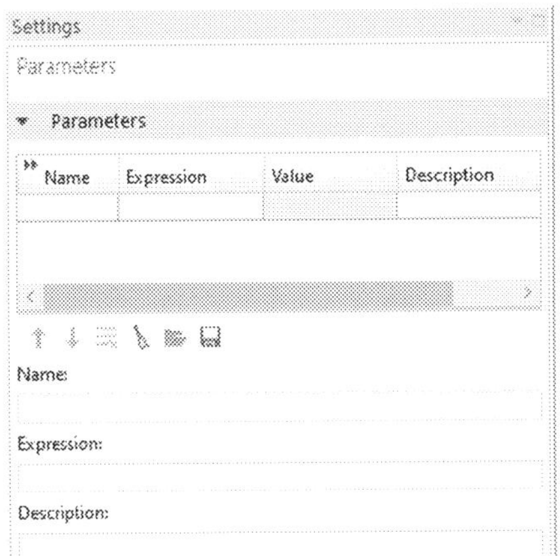

Figure 6.90 The setting window of the parameters module.

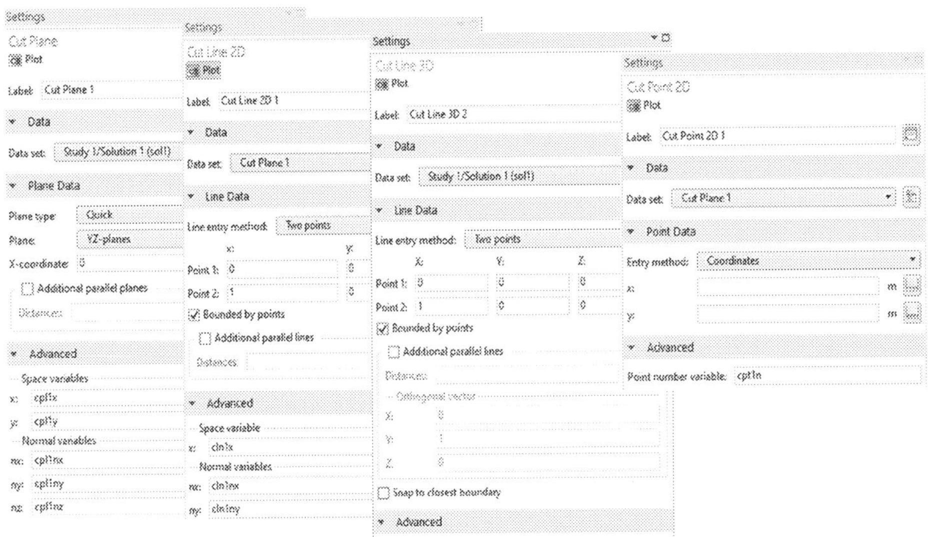

Figure 6.91 Setting windows of the cut modules.

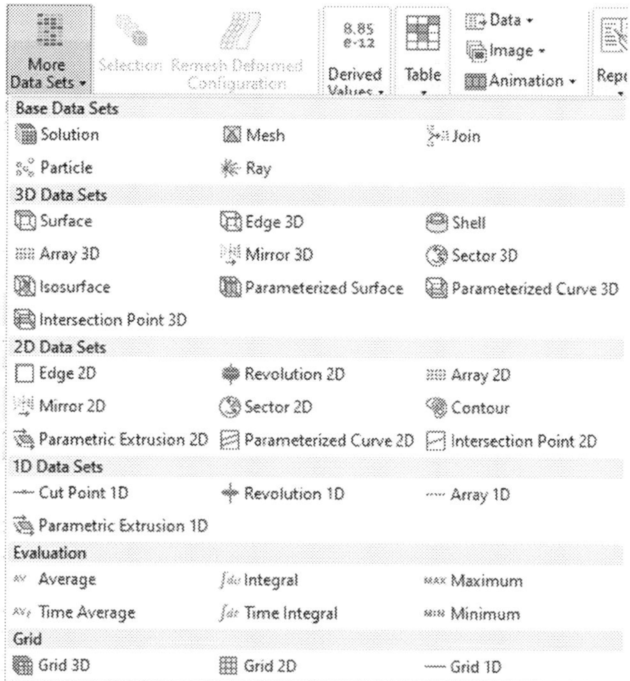

Figure 6.92 Data sets in COMSOL Multiphysics.

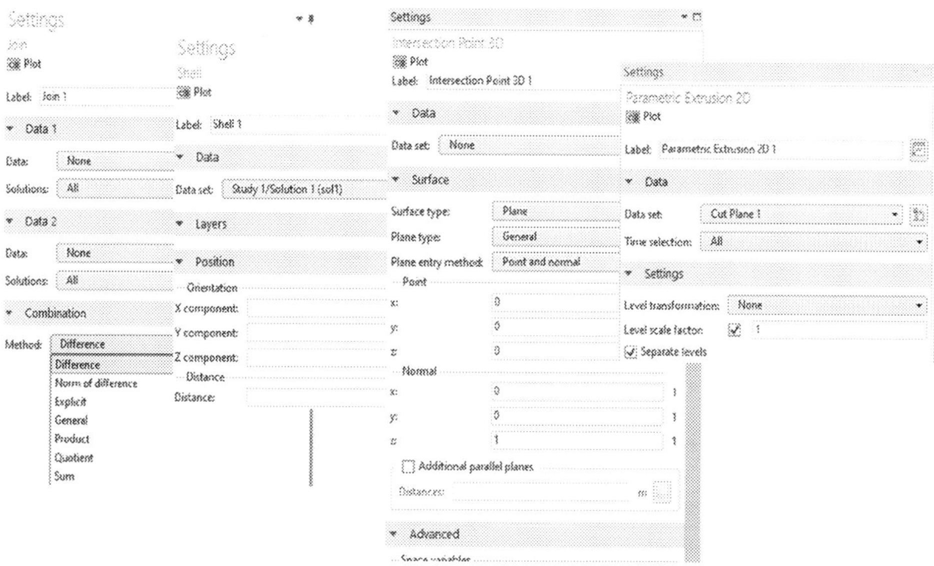

Figure 6.93 Setting windows of data sets in COMSOL Multiphysics.

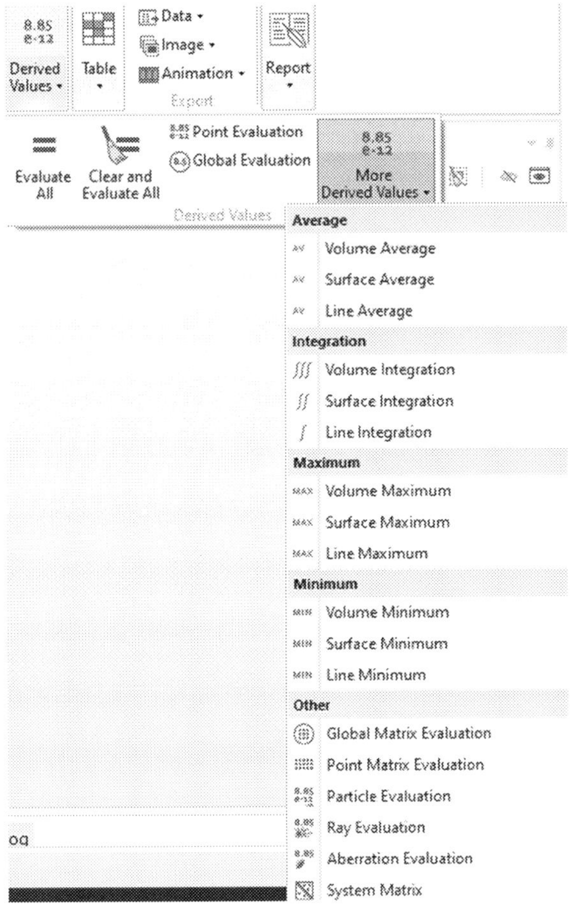

Figure 6.94 Most common functions in COMSOL Multiphysics.

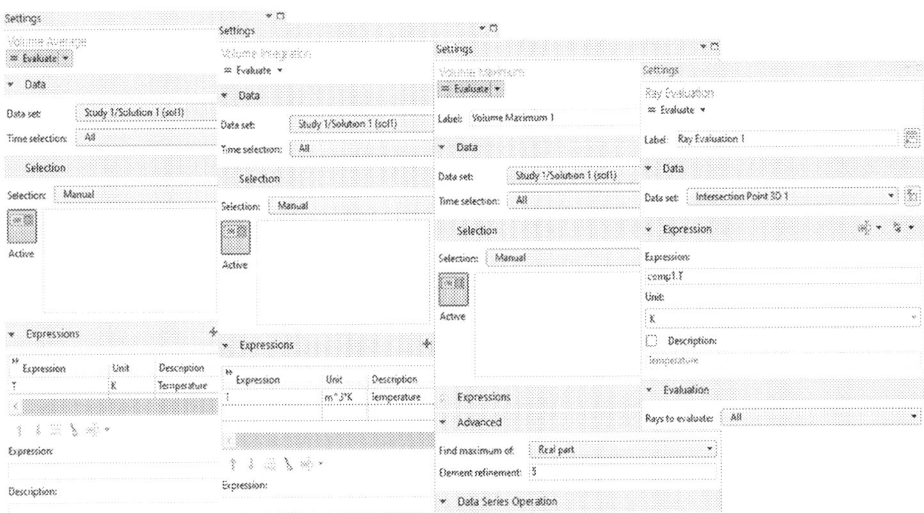

Figure 6.95 Setting windows of the most common functions in COMSOL Multiphysics.

Based on the previously detailed explanation of the different tools and options available in COMSOL Multiphysics, any simulation model related to ablation therapy can be designed. Several reports can be obtained, including a brief report, an intermediate report, which includes an intermediate level of the data, a complete report, and a custom report, which is empty and can be manually filled with the contents, as shown in Fig. 6.96.

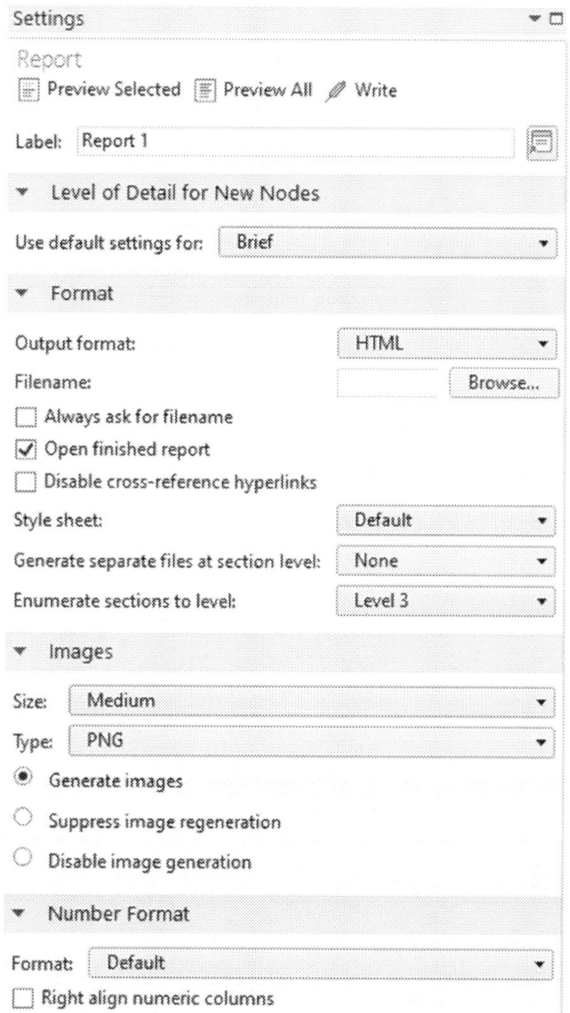

Figure 6.96 Setting windows of the report types.

6.4 Conclusion

For cancer treatment, hyperthermia oncology has a heating effect on the biological tissue properties. In hyperthermia therapy, the spatial position and power of the EM device of the heating source must be adapted based on the tumor location, electrical/thermal properties of the tumor/healthy tissues, and blood perfusion, and so simulation modeling is essential to study and design the planned process to achieve complete removal of tumor tissues without damaging the nearby tissues.

To find the best design of a device-based noninvasive therapy, such as RF probes or thin MW antennas injected directly into the tumor zone via the patient's skin, numerical simulation is essential. Measurement of the thermal damage due to the heating source of the biological tissues assists and guides optimization of the clinical setup. An FEM model for steered ablation of a tumor is the key solution to controlling the temperature diffusion direction during the thermal ablation process. Several simulation software tools can be used to model the ablation system, aiming at finding the optimal design of the ablation applicator and the settings for the ablation process parameters. COMSOL Multiphysics software is considered the most efficient numerical simulation software for the ablation process. It is a simulation platform incorporating all the modeling workflow phases, starting by defining the material properties, geometries, and the descriptive physics of a specific phenomenon to solve and postprocess the model with trustworthy results.

COMSOL has different features, tools, calculations, and reports for the accurate evaluation of a modeled ablation system. Also, the transient analysis involving EM heating multiphysics coupling, the electric current interface, and the bioheat transfer interface can be employed to optimize the applicator position and input power for tissue damage conditions. The model solves the EM waves and frequency domain interface to determine the most suitable model. A time-dependent study can be conducted including the bioheat transfer interface to calculate the temperature change over time and the resultant thermal damage. Generally, in biological tissue modeling, several analyses are included during the simulation process using COMSOL, such as heat transfer, thermal damage, the Arrhenius kinetics model, and prediction of the level of tissue damage.

References

Ahmed, M., Liu, Z., Humphries, S., Nahum Goldberg, S., 2008. Computer modeling of the combined effects of perfusion, electrical conductivity, and thermal conductivity on tissue heating patterns in radiofrequency tumor ablation. Int. J. Hyperth. 24, 577–588.

Ashour, A.S., Asran, M.I., Mohamed, W.S., Fotiadis, D.I., 2021. Optimal localization of a novel shifted 1T-ring based microwave ablation probe in hepatocellular carcinoma. IEEE Trans. Biomed. Eng. 68 (2), 505–514.

Bazyar, M.H., Talebi, A., 2015. Scaled boundary finite-element method for solving non-homogeneous anisotropic heat conduction problems. Appl. Math. Model. 39, 7583–7599.

Berjano, E.J., 2006. Theoretical modeling for radiofrequency ablation: state-of-the-art and challenges for the future. Biomed. Eng. Online 5, 24.

Bonomo, A.L., Isakson, M.J., 2016. Modeling the acoustic scattering from axially symmetric fluid, elastic, and poroelastic objects due to nonsymmetric forcing using COMSOL Multiphysics. In: Proc. COMSOL Conf. Boston, MA.

Chang, I.A., Nguyen, U.D., 2004. Thermal modeling of lesion growth with radiofrequency ablation devices. Biomed. Eng. Online 3, 27.

Chen, M.-H., Yang, W., Yan, K., Zou, M.-W., Solbiati, L., Liu, J.-B., et al., 2004. Large liver tumors: protocol for radiofrequency ablation and its clinical application in 110 patients—mathematic model, overlapping mode, and electrode placement process. Radiology 232, 260–271.

de Oliveira, M.M., Song, B., Ahfock, T., Li, Y., Wen, P., 2018. Human head modelling simulation applied to electroconvulsive therapy. Multi-Modality Imaging. Springer.

Duan, B., Wen, R., Fu, Y., Chua, K.-J., Chui, C.-K., 2016. Probabilistic finite element method for large tumor radiofrequency ablation simulation and planning. Med. Eng. Phys. 38, 1360–1368.

Ghose, B., Balasubramaniam, K., Krishnamurthy, C., Rao, A.S., 2011. COMSOL based 2D FEM model for ultrasonic guided wave propagation in symmetrically delaminated unidirectional multilayered composite structures. In: Proceedings of the National Seminar and Exhibition on Nondestructive Evaluation.

Habash, R.W., Bansal, R., Krewski, D., Alhafid, H.T., 2007. Thermal therapy, Part IV: electromagnetic and thermal dosimetry. Crit. Rev. Biomed. Eng. 35, 123–182.

Haemmerich, D., Staelin, T., Tungjitkusolmun, S., Lee, F.T., Mahvi, D.M., Webster, J.G., 2001. Hepatic bipolar radio-frequency ablation between separated multiprong electrodes. IEEE Trans. Biomed. Eng. 48, 1145–1152.

Hall, S.K., Ooi, E.H., Payne, S.J., 2015. Cell death, perfusion and electrical parameters are critical in models of hepatic radiofrequency ablation. Int. J. Hyperth. 31, 538–550.

Horton, P., Peet, D.J., Sutton, D.G., Martin, C.J., 2014. Radiotherapy: brachytherapy and unsealed radionuclide therapy. In: Practical Radiation Protection in Healthcare, OUP, Oxford, p. 399.

Huang, H.-W., Liauh, C.-T., 2011. Therapeutical applications of heat in cancer therapy. J. Med. Biol. Eng. 32, 1–11.

Jiang, Y., Zhao, J., Li, W., Yang, Y., Liu, J., Qian, Z., 2017. A coaxial slot antenna with frequency of 433 MHz for microwave ablation therapies: design, simulation, and experimental research. Med. Biol. Eng. Comput. 55, 2027–2036.

Kengne, E., Hamouda, F.B., Lakhssassi, A., 2013. Extended generalized Riccati equation mapping for thermal traveling-wave distribution in biological tissues through a bio-heat transfer model with linear/quadratic temperature-dependent blood perfusion. Appl. Math. 4, 1471–1484.

Kuzhagaliyev, T., Clancy, N.T., Janatka, M., Tchaka, K., Vasconcelos, F., Clarkson, M.J., et al., 2018. Augmented reality needle ablation guidance tool for irreversible electroporation in the pancreas. Medical Imaging 2018: Image-Guided Procedures, Robotic Interventions, and Modeling. International Society for Optics and Photonics, p. 1057613.

Liu, K.-C., Chen, H.-T., 2009. Analysis for the dual-phase-lag bio-heat transfer during magnetic hyperthermia treatment. Int. J. Heat. Mass. Transf. 52, 1185–1192.

Liu, P., Liu, J., Duan, H., 2012. Thermal modelling for endocardiac radiofrequency ablation: comparison of hyperbolic bioheat equation and Pennes bioheat equation with finite element method. arXiv preprint arXiv 1204.0098.

Mi, Y., Rui, S., Li, C., Yao, C., Xu, J., Bian, C., et al., 2017. Multi-parametric study of temperature and thermal damage of tumor exposed to high-frequency nanosecond-pulsed electric fields based on finite element simulation. Med. Biol. Eng. Comput. 55, 1109–1122.

Mohammed, Y., Verhey, J.F., 2005. A finite element method model to simulate laser interstitial thermo therapy in anatomical inhomogeneous regions. Biomed. Eng. Online 4, 2.

O'Hara, P., Duarte, C., Eason, T., 2009. Generalized finite element analysis of three-dimensional heat transfer problems exhibiting sharp thermal gradients. Comput. Methods Appl. Mech. Eng. 198, 1857–1871.

Perera-Bel, E., Yague, C., Mercadal, B., Ceresa, M., Beitel-White, N., Davalos, R.V., et al., 2020. EView: an electric field visualization web platform for electroporation-based therapies. Comput. Methods Prog. Biomed. 197, 105682.

Prakash, P., Diederich, C.J., 2012. Considerations for theoretical modelling of thermal ablation with catheter-based ultrasonic sources: implications for treatment planning, monitoring and control. Int. J. Hyperth. 28, 69–86.

Primavesi, F., Swierczynski, S., Klieser, E., Kiesslich, T., Jäger, T., Urbas, R., et al., 2018. Thermographic real-time-monitoring of surgical radiofrequency and microwave ablation in a perfused porcine liver model. Oncol. Lett. 15, 2913–2920.

Pryor, R.W., 2009. Multiphysics Modeling Using COMSOL®: A First Principles Approach. Jones & Bartlett Publishers.

Salimpour, M.R., Shirani, E., 2017. Heat transfer analysis of skin during thermal therapy using thermal wave equation. J. Therm. Biol. 64, 7–18.

Tungjitkusolmun, S., Staelin, S.T., Haemmerich, D., Tsai, J.-Z., Cao, H., Webster, J.G., et al., 2002. Three-dimensional finite-element analyses for radio-frequency hepatic tumor ablation. IEEE Trans. Biomed. Eng. 49, 3–9.

Villard, C., Soler, L., Gangi, A., 2005. Radiofrequency ablation of hepatic tumors: simulation, planning, and contribution of virtual reality and haptics. Comput. Methods Biomech. Biomed. Eng. 8, 215–227.

Wang, X., Gao, H., Wu, S., Bai, Y., Zhou, Z., 2018. RF ablation thermal simulation model: parameter sensitivity analysis. Technol. Health Care 26, 179–192.

Woo, E.J., Tungjitkusolmun, S., Cao, H., Tsai, J.-Z., Webster, J.G., Vorperian, V.R., et al., 2000. A new catheter design using needle electrode for subendocardial RF ablation of ventricular muscles: finite element analysis and in vitro experiments. IEEE Trans. Biomed. Eng. 47, 23–31.

Zampolli, M., Tesei, A., Jensen, F.B., Malm, N., Blottman III, J.B., 2007. A computationally efficient finite element model with perfectly matched layers

applied to scattering from axially symmetric objects. J. Acoust. Soc. Am. 122, 1472–1485.
Zhang, C., Johnson, D.T., Brazel, C.S., 2008. Numerical study on the multi-region bio-heat equation to model magnetic fluid hyperthermia (MFH) using low Curie temperature nanoparticles. IEEE Trans. Nanobiosci. 7, 267–275.
Ziegle, J., Audigier, C., Krug, J., Ali, G., Kim, Y., Boctor, E.M., et al., 2018. RF-ablation pattern shaping employing switching channels of dual bipolar needle electrodes: ex vivo results. Int. J. Comput. Assist. Radiol. Surg. 13, 905–916.

7

Simulation-driven modeling of radiofrequency ablation systems

Chapter outline

7.1 Introduction

In radiofrequency ablation (RFA), an alternating current is applied to lead to high frequency and so avoid exciting nerve cells. Such a process occurs by using pair of radiofrequency (RF) electrodes, namely a large dispersive electrode for closing the electrical circuit and an active electrode with a small surface area, which is positioned on the target tumor zone (Bayer et al., 2016; Chen et al., 2016). To improve the currently used RFA techniques and develop new RFA methods targeting more efficient ablation, researchers use design simulation models for theoretical and phantoms models before ex vivo and/or in vivo experimental and clinical studies. The simulation models can provide energetic information on the thermal and electrical behaviors of the ablation process in a cost-effective, rapid, and safe way while measuring the effect of different inherent and extrinsic factors on the temperature distribution and electrical current (Singh and Melnik, 2020). They have a crucial role in optimizing the design of the electrode shape to avoid unwanted hot spots. In addition, transient simulations can envisage the temperature distribution evolution in tissues as well as the vessels and blood based on the location of the tumor.

Thermal Ablation Therapy. DOI: https://doi.org/10.1016/B978-0-12-819544-4.00001-0
© 2021 Elsevier Inc. All rights reserved.

The simulations are effective in supporting the therapy quality by studying the effects of different parameters, including the size of the damaged zone, electrode positioning, duration of ablation cycle, and blood flow rates. Accordingly, simulation models enable the possibility to assess new RF electrode geometries, and/or new protocols for electrical power delivering the volume of the ablated tumor, and the duration of the ablation process. A number of research groups are developing computer models to analyze the RFA procedure and assisting RFA device designers to predict the electric fields and thermal distribution in the affected tumor tissues (Liu et al., 2005). Nevertheless, during simulation, there are some challenges in appropriate integration between the Multiphysics analysis and defining realistic inputs to the model.

Typically, during RFA, an RF current (300 kHz–1 MHz) is delivered through the RF needle, which is embedded in the tumorous tissues that are to be treated. Volumetric heating is induced close to the electrode tip due to the generated high current density and its Joule effect. Hence, the temperature of tissues surrounding the needle increases uniformly above 50°C, where a temperature limit of 100°C to avoid tissue microexplosions and charring must be applied (Chou, 2018). The simulation model simulates the clinical procedure, where the physician uses a bipolar or monopolar electrode to inject current from a generator around and into the tumor. This current varies sinusoidally with time and has frequencies of 300–500 kHz.

Theoretical simulations modeling has been carried out to study the RF heating procedures including the temperature distributions and current density in the tissues under the RF electrode focusing on new ablation therapies, including pancreatic cancer ablation (Song et al., 2016), pancreas ablation (Rustagi and Chhoda, 2017), liver tumor ablation (Voglreiter et al., 2018; Chen et al., 2018; Ooi et al., 2019), and ablation of other tissues (Singh and Repaka, 2017; Shao et al., 2017; Zhang et al., 2017; Kath et al., 2019). Different configurations of the RF electrode can be designed, ranging from flat/patch plates and a single needle to a cluster of needles. The selected design depends on the required shape of the heated zone in the tumor and around it. The complicated problem facing the engineer designing an RFA device is to achieve the required shape of the heated tissue based on the number of electrodes, and the desired thermal energy absorption distribution of the neighboring tissues over time. The pointed, sharp ends of the electrode needle complicate the design process, and it is a required characteristic that guarantees high current densities. Accordingly, to model the RFA procedure appropriately, a model of the electric current flow, temperature rising, and heat generation via the tissue is essential. Therefore COMSOL Multiphysics modeling of an RFA problem is carried out using an electrode targeting tumor tissues to predict the temperature variation and the damaging evolution of the tumor tissues.

7.2 Theoretical simulation modeling stages of RFA

The main methodology to build and use a theoretical, simulation model to study RF heating and electrodes design in the RFA process includes the following stages, where generally finite element method (FEM) models are used.

- Stage 1: Observe and simplify the physical situation of the problem under consideration.
- Stage 2: Formulate the mathematical governing equations of the electrical and thermal phenomena.
- Stage 3: Determine the thermal and electrical boundary conditions.
- Stage 4: Specify the thermal and electrical physical characteristics of the biological tissues of the targeted tumor and its surrounding healthy tissues.
- Stage 5: Specify the material characteristics of the used applicator included in the model.
- Stage 6: Choose the numerical method for simulation to analytically/computationally realize a solution.
- Stage 7: Design the structure for inserting the applicator inside the tumor and/or healthy tissues, then set the mesh to suit this structure.
- Stage 8: Conduct the post-processing of the computed results.
- Stage 9: Assess the post-processing output variables and the damaged tumor.

Fig. 7.1 illustrates a block diagram of the overall general workflow of the RFA system modeling using simulation. The model can simulate healthy tissues only, tumor tissues only, or both by embedding the tumor in the healthy tissues.

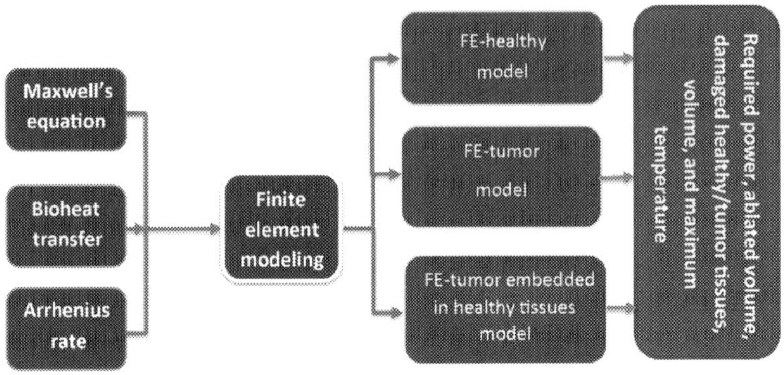

Figure 7.1 Workflow of simulation of an RFA system.

The RFA process includes the following mathematical equations: (1) *Maxwell's equations*, which control the alternative waves generated from the RF source, (2) the *bioheat transfer equation*, which controls the heat transfer across the tissues, and (3) the *first-order Arrhenius rate equations*, which are used to calculate the injured degree at each point inside the tumor tissue.

7.3 COMSOL to simulate RFA: a case study

This chapter displays a case study, which is simulated with COMSOL Multiphysics, including liver tissues. Several designs of the radiofrequency probes can be used, such as plain, wet, cooled, coiled expandable, multitines expandable, and bipolar. The proper design is selected according to the individual patient, tumor position with respect to the adjacent blood vessels or bile ducts, tumor size, and tumor shape. Therefore a three-dimensional (3D) finite element (FE) model is simulated in this chapter, which contains liver tumor tissue with its properties, and a radiofrequency probe. The used RF applicator is made of stainless steel for the active part, and polyurethane for the insulated part.

In this chapter, different probe types are designed to be is inserted inside tumorous liver tissues for ablation. To implement the proposed model in COMSOL, the model starts by identifying the parameters inside the used equations, for example, the electric current, and bioheat transfer equations in the RFA in the present case study, and their values according to the tissue type, for example, liver, as shown in Fig. 7.2.

Fig. 7.2 displays the essential parameters, such as the external applied alternative voltage, the required ablation time adequate for complete ablation, and the initial conditions. These parameters exist in any model and can be easily modified and changed using the parameter setting window. Hence, the model builder of the COMSOL file contains geometry, material, physics, mesh, study, and results in sections, as shown in Fig. 7.3.

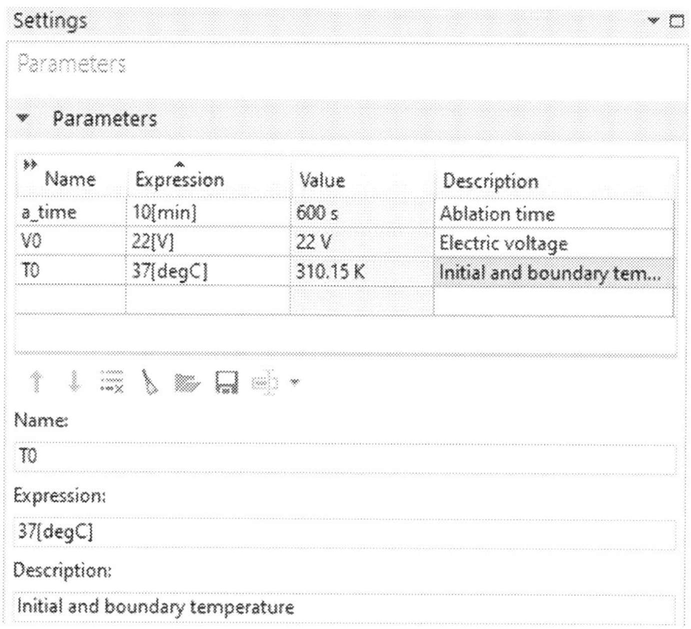

Figure 7.2 Parameters identified in the settings window.

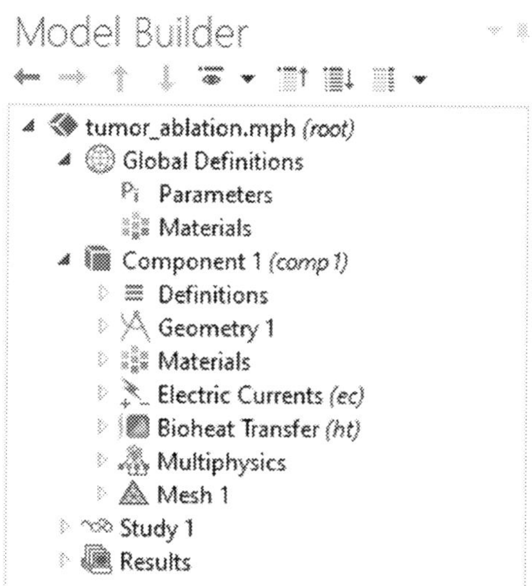

Figure 7.3 Model builder in COMSOL Multiphysics.

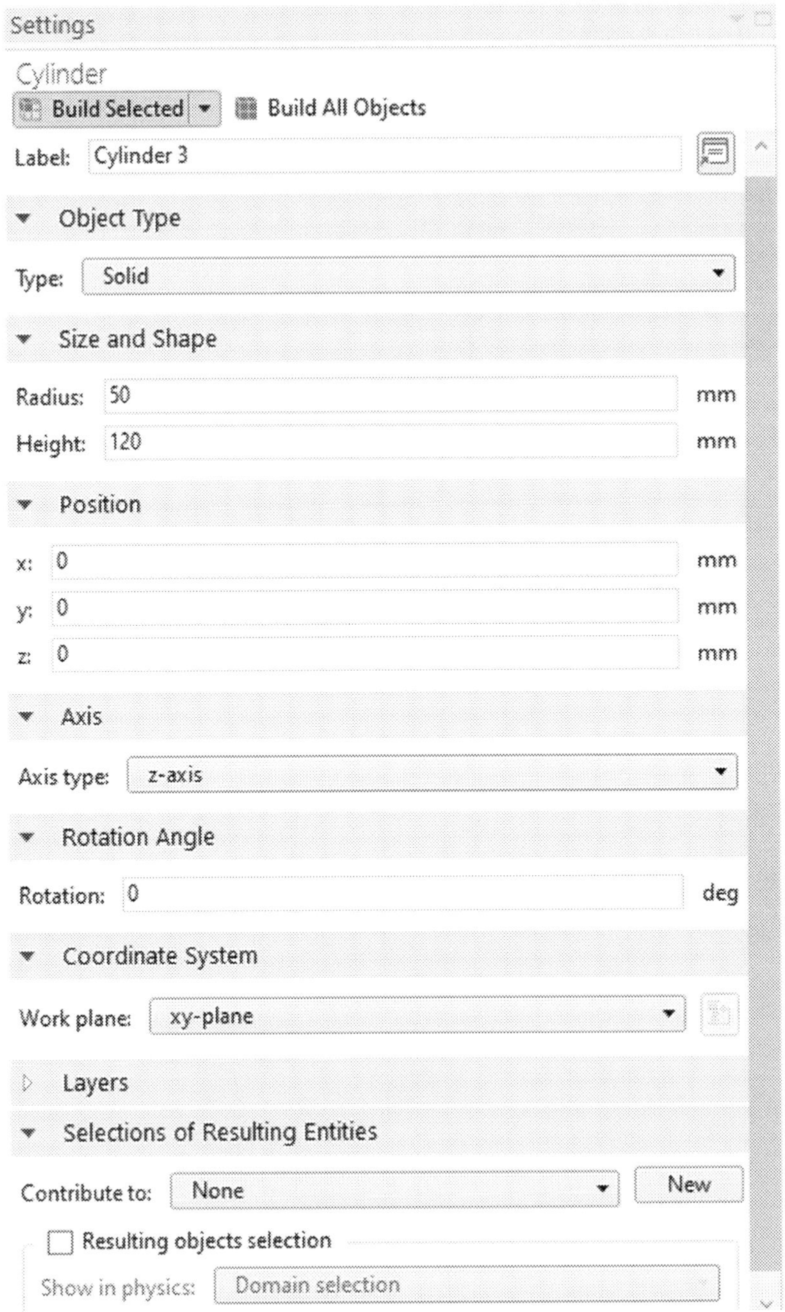

Figure 7.4 Setting window of the cylinder representing the liver tumorous tissue.

Fig. 7.3 demonstrates the overall view of the simulated COMSOL file, where the geometry can be built according to the real model, including a cylinder representing liver tumor tissue with dimensions, position, and work plane using the setting window of a cylinder as shown in Fig. 7.4.

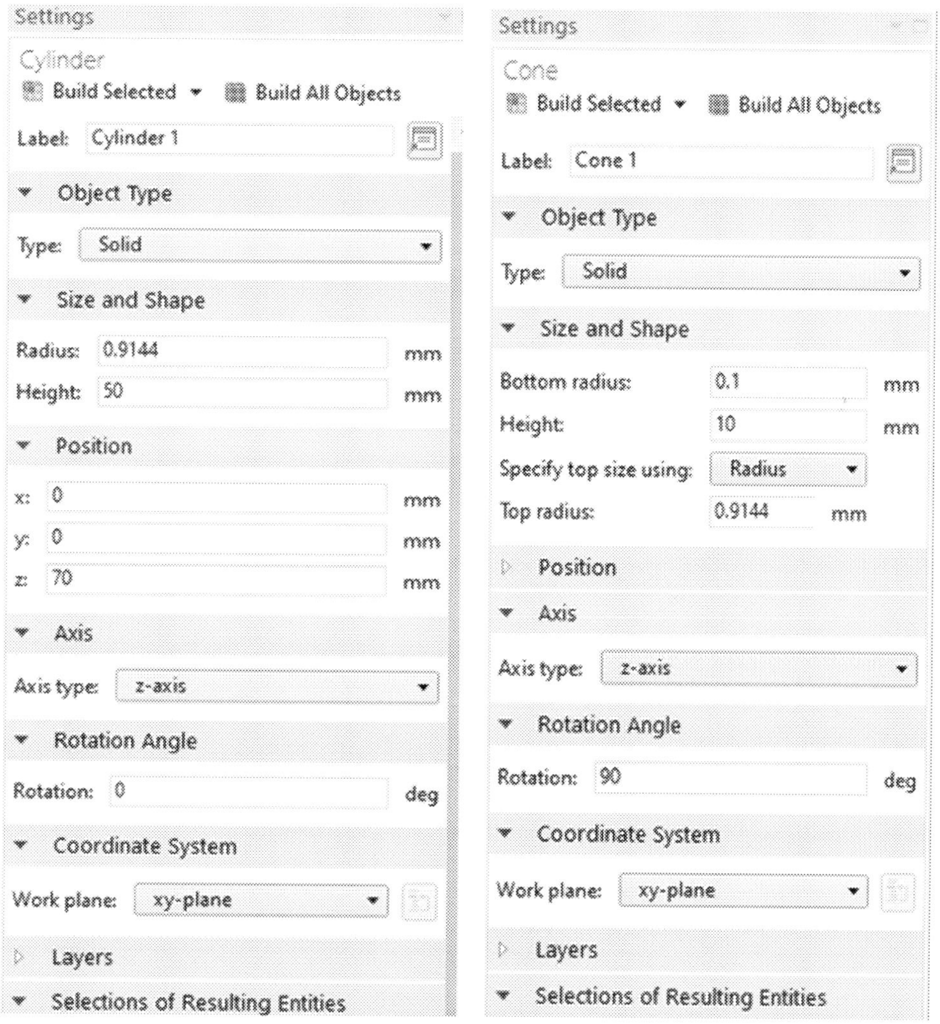

Figure 7.5 Setting windows of the geometry representing the plain probe.

Moreover, the geometry section includes another cylinder and cone, which represent the insulated and active parts of the plain probe, respectively, and their dimensions, position, and rotation are adjusted using their setting windows, as shown in Fig. 7.5.

Figs. 7.4 and 7.5 illustrate that all geometry components are built to produce the geometric model as shown in the graphic window in Fig. 7.6.

Since the plain probe ablates a small tumor size of less than 3 mm diameter, the wet probe is used to increase the necrotic region. A saline solution injection can be applied to increase the tumor tissue electrical conductivity and delays the occurrence of the roll-off point. Hence, the wet probe is designed using a cylinder, which represents the insulated part and has a layer of 10 mm in length as well as a cone

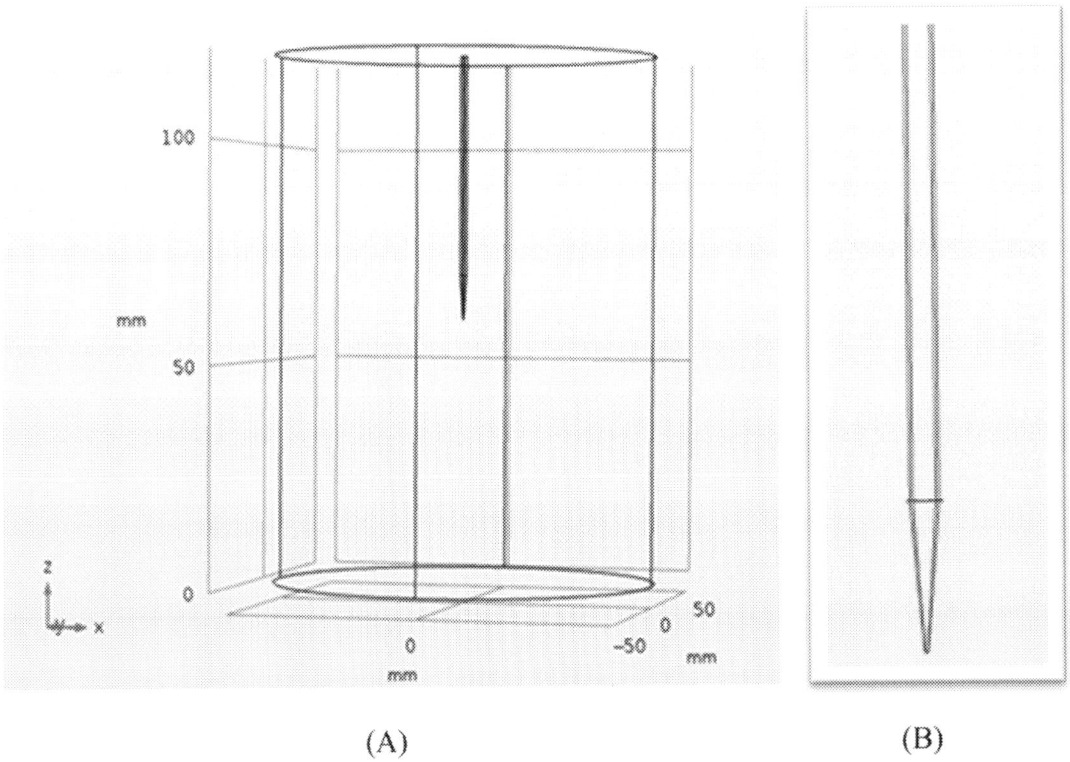

Figure 7.6 Graphic window of the geometric model, including (A) liver tumorous tissue and plain probe, and (B) side view of the plain probe.

representing the active part with dimensions as shown in Fig. 7.7.

Additionally, a cylindrical cavity inside the probe is designed for saline solution injection by making a cylinder inside the probe and cutting it from the main cylinder using the difference module as shown in Fig. 7.8.

Some holes are opened through the wet probe to allow the saline solution to get out of the internal cavity and enter the tumorous tissues. Hence, four holes are designed by opening

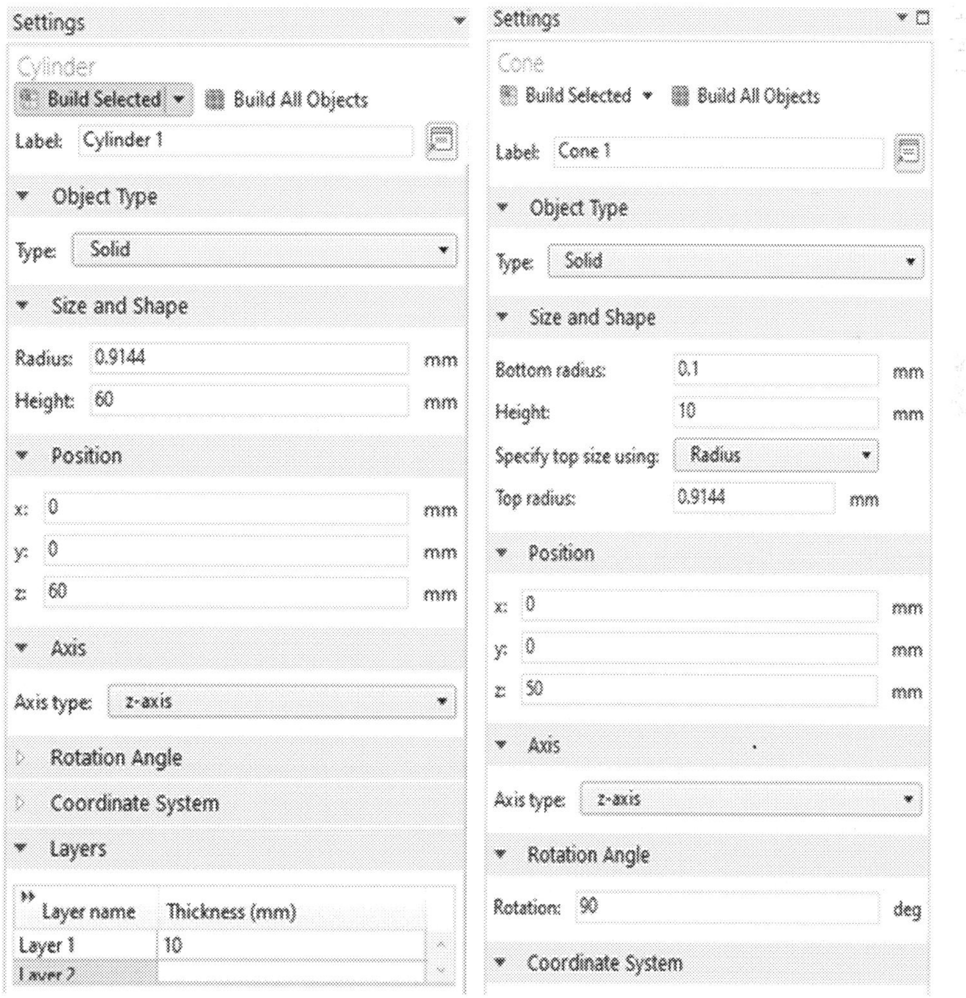

Figure 7.7 Setting windows of the geometry representing the wet probe.

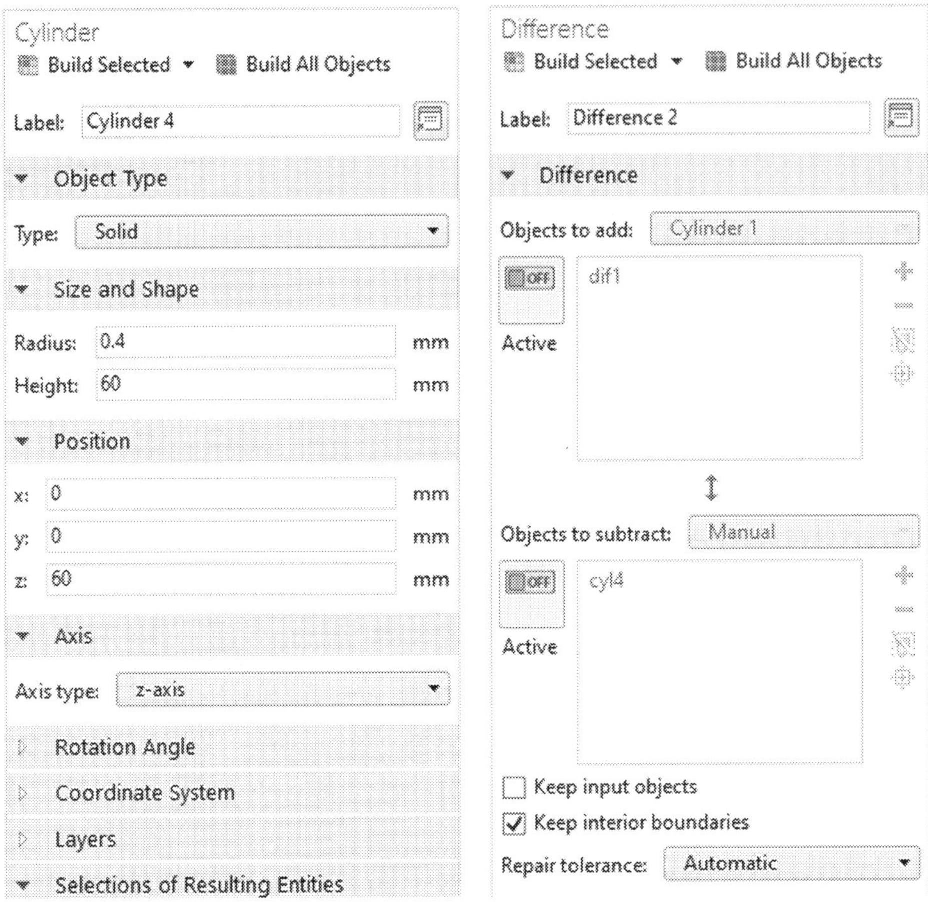

Figure 7.8 Setting windows of the internal cavity.

four small spheres at the surface of the wet probe and cutting them using the difference module as shown in Fig. 7.9.

Figs. 7.7–7.9 demonstrate that all geometry components are built to obtain the wet probe for ablating tumors with a volume larger than those ablated using the traditional plain probe as shown in Fig. 7.10.

Likewise, the cooled probe consists of:

- Inner lumen to pump the chilled saline solution into the probe tip; and
- Outer lumen to return the solution to an embedded external collection unit, which can be designed with a cylinder representing the probe's insulated part, with a layer of 10 mm,

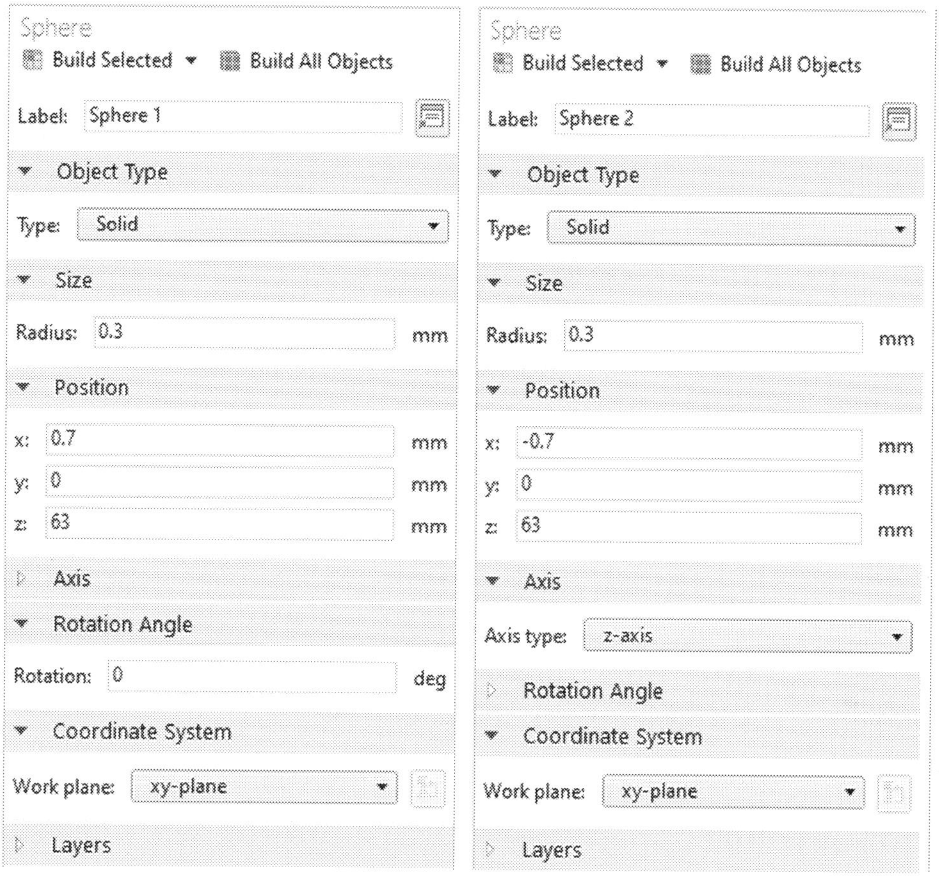

Figure 7.9 Setting windows of the four holes on the wet probe surface.

as well as a cone representing the probe's active part as previously mentioned.

The inner, and outer cavities are designed using three cylinders and two difference modules with dimensions as shown in Fig. 7.11.

Figs. 7.7 and 7.11 illustrate that all geometry components are built to get the cooled probe which prevents the tissue carbonization problem, as shown in Fig. 7.12.

Furthermore, the energy deployment and its direction can be controlled using the coiled expandable probe, which has a spring allowing a hollow probe tip to move out. Hence, the coiled expandable probe is designed using a cylindrical insulated base as previously mentioned, and a helix, which

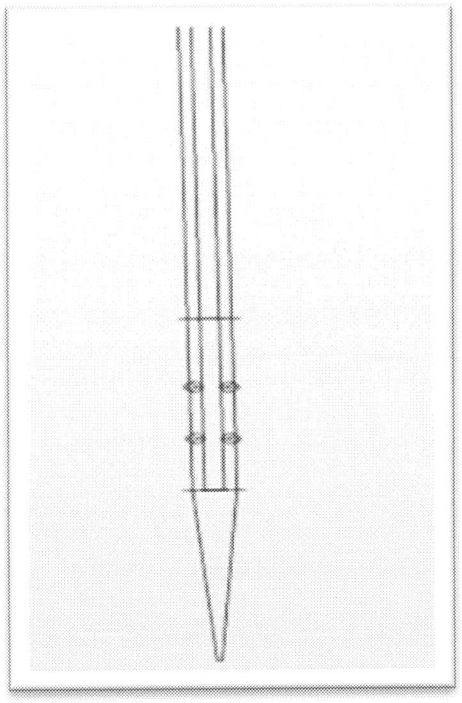

Figure 7.10 Side view of the wet probe.

represents the vertical spring active part with dimensions adjusted using its setting window, as shown in Fig. 7.13.

Figs. 7.7 and 7.13 show that all geometric components for the insulated part and the spring active part are built to obtain the coiled expandable probe as shown in Fig. 7.14.

To prevent the skin burns resulting from the ground pads position, a bipolar probe is used with two positive and negative electrodes both inside the tumorous tissues without using grounding pads on the patient's skin. Hence, the bipolar probe is designed using a main cylindrical trocar base and an active positive electrode, as previously described, as well as a negative electrode, which is designed using a cylinder with the same radius as the main cylinder, a short length of 10 mm, and a separation distance of 15 mm from the positive electrode, as shown in Fig. 7.15.

To separate the negative electrode from the insulated trocar base, a difference module which subtracts it from the main

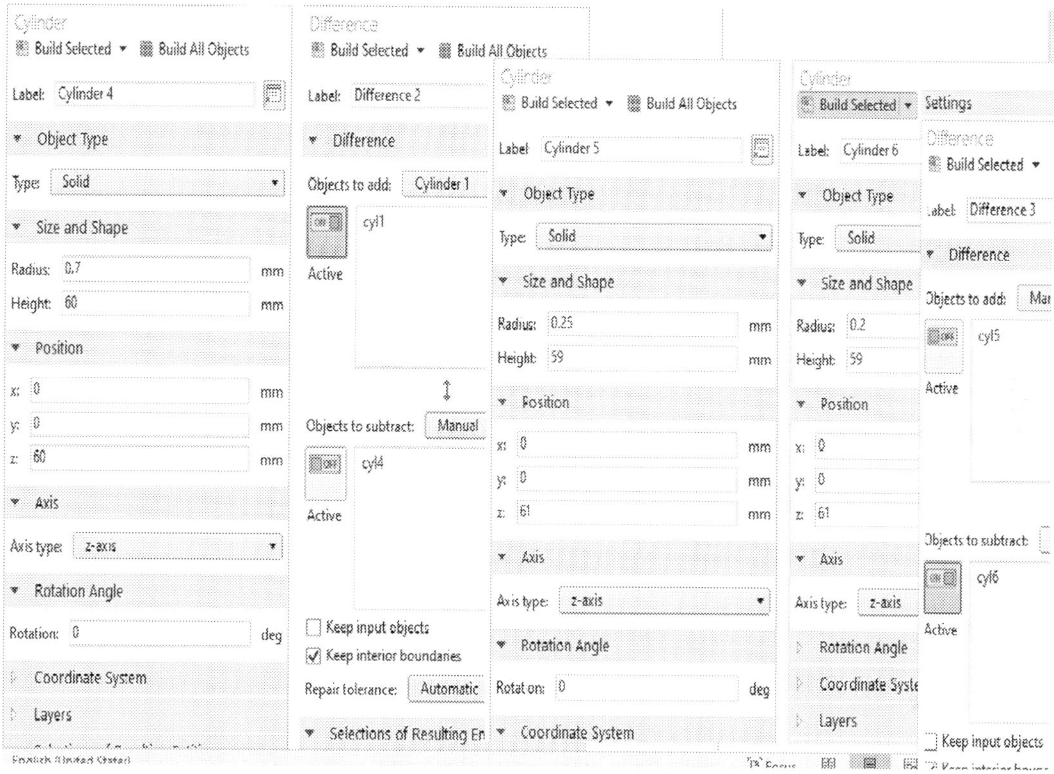

Figure 7.11 Setting windows of the inner and outer cavities inside the cooled probe.

cylindrical trocar base is used, then, the same cylinder representing the negative electrode is added, as shown in the setting windows in Fig. 7.16.

Figs. 7.7, 7.15, and 7.16 show the setting of the geometric components of a bipolar probe using COMSOL to build the required final probe design as illustrated in Fig. 7.17.

Finally, a multitines expandable probe is the most commonly used type for many patients due to its larger necrotic volume and spherical parameter. Consequently, it is designed using a cylindrical trocar base representing the probe's insulated part, with a layer 10 mm in length as previously described. Additionally, a specific number of electrodes representing the probe's active part is designed using a torus with dimensions, position, and angle as shown in Fig. 7.18.

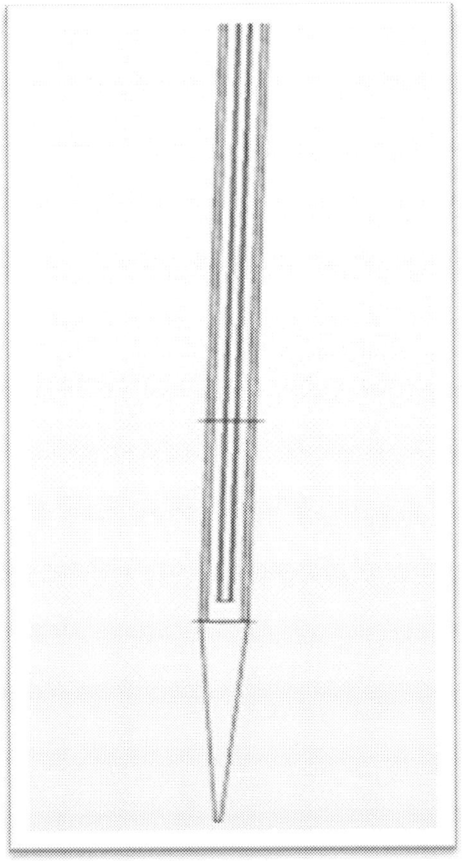

Figure 7.12 Side view of the cooled probe.

Fig. 7.18 illustrates one curved electrode with a revolution angle of 180 degrees. Hence, it is rotated to obtain a determined number of curved electrodes according to the target model using a rotate module as shown in its setting window in Fig. 7.19.

Figs. 7.7, 7.18, and 7.19 illustrate the settings selection of the different geometric components in COMSOL, which are related to the multitines expandable probe, inside liver tumorous tissue, to build the required design shown in Fig. 7.20.

Additionally, each part inside the geometric model must be defined with a special name to easily deal with it during the simulation by selecting the target part and adjusting the model's initial condition, as shown in Fig. 7.21.

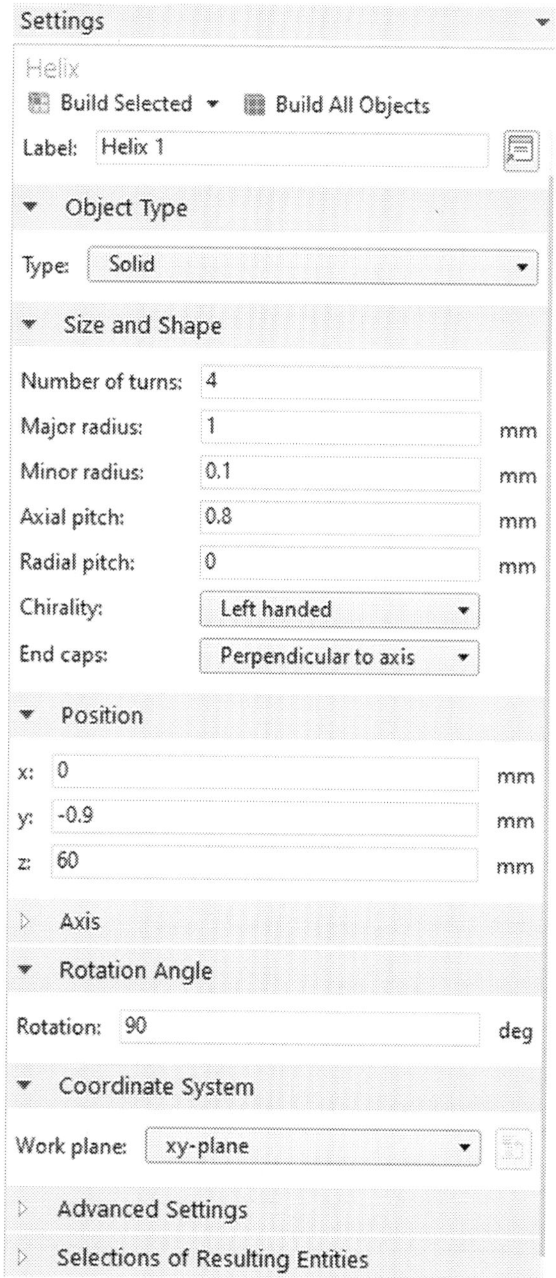

Figure 7.13 Setting window of the helix representing the spring of a coiled expandable probe.

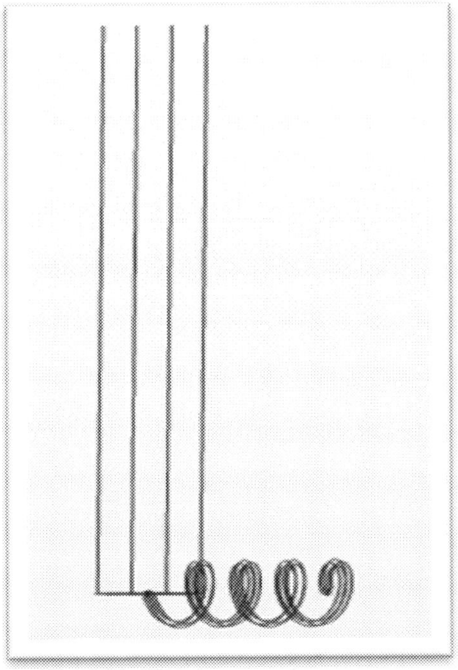

Figure 7.14 Side view of the coiled expandable probe.

Furthermore, more than one explicit part can be gathered to compose a single section, which is called a union, as shown in Fig. 7.22.

Fig. 7.22 determines the overall sections of a multitines expandable probe to be easily selected during the alternative current generation, and the overall geometric model to be easily selected during the heat transfer. Therefore the external boundaries of explicit or unions can be determined through the adjacent modules, as shown in Fig. 7.23.

Fig. 7.23 illustrates setting of the external boundaries of specific determined sections inside the geometric model for easily dealing with them during the simulation, such as the exterior boundaries of the trocar tip and four electrodes active part, to generate the radiofrequency voltage. Additionally, different quantities can be determined through point probe expression and measured at any point inside the model using the domain point probe as shown in Fig. 7.24.

In addition, all parts inside the geometric model are specified with selected materials and their properties, such as the

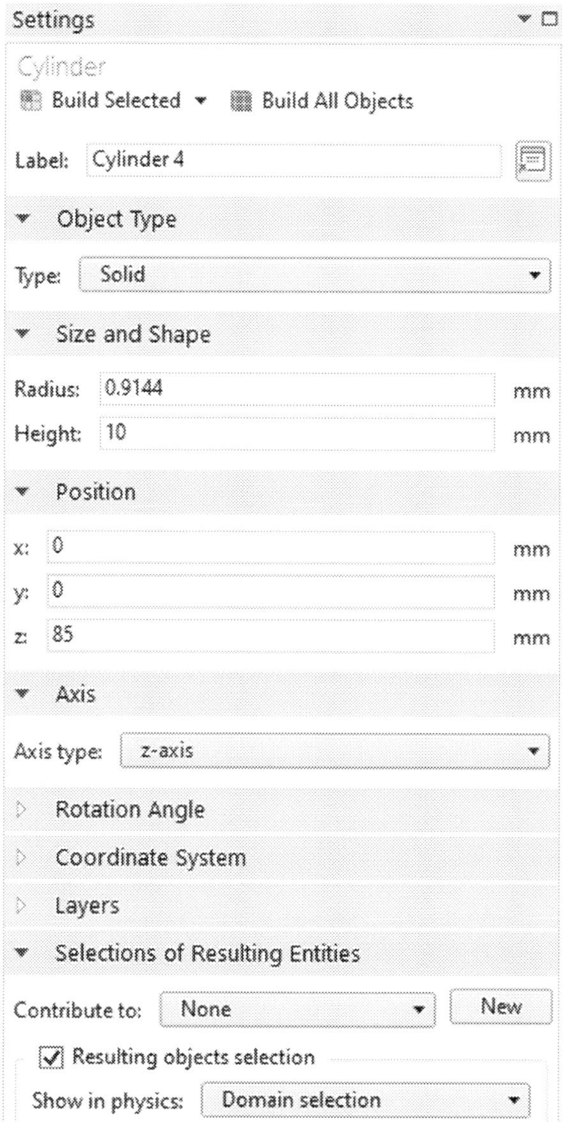

Figure 7.15 Setting window of the cylinder representing the negative electrode of the bipolar probe.

liver tumorous tissue section with properties at 500 kHz as shown in Fig. 7.25.

Fig. 7.25 displays the density, heat capacity, thermal, and electrical conductivities related to liver tumor tissue as well as its activation energy, which expresses the energy amount for molecule activation in one mole of the material, and the

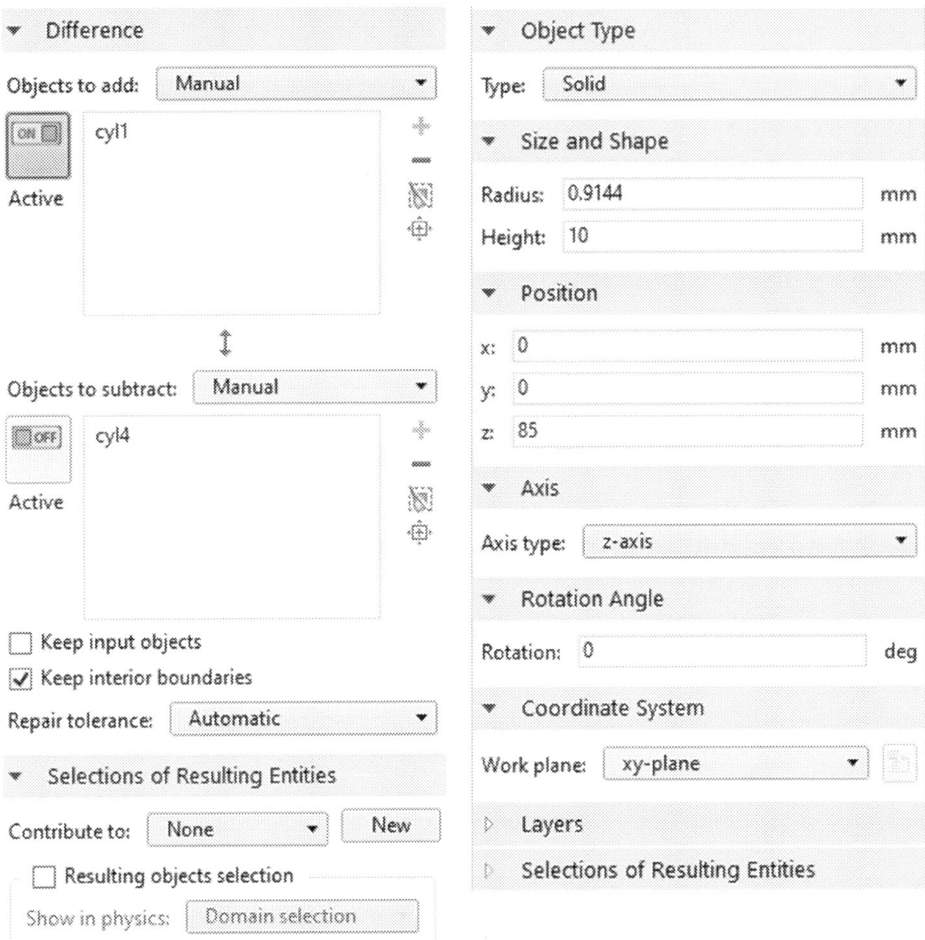

Figure 7.16 Setting windows of the bipolar probe.

frequency factor, which describes the collision times of the molecules in 1 second. Moreover, the trocar tip and four electrodes are specified with stainless-steel material and the properties as shown in Fig. 7.26.

Fig. 7.26 describes the material determined for the probe's active part, which must be conducted for the current generation. Finally, the trocar base, which represents the probe's insulated part, is specified with polyurethane material and its properties as shown in Fig. 7.27.

The physics are added according to the simulation study type, for example, electric current and bioheat transfer for the

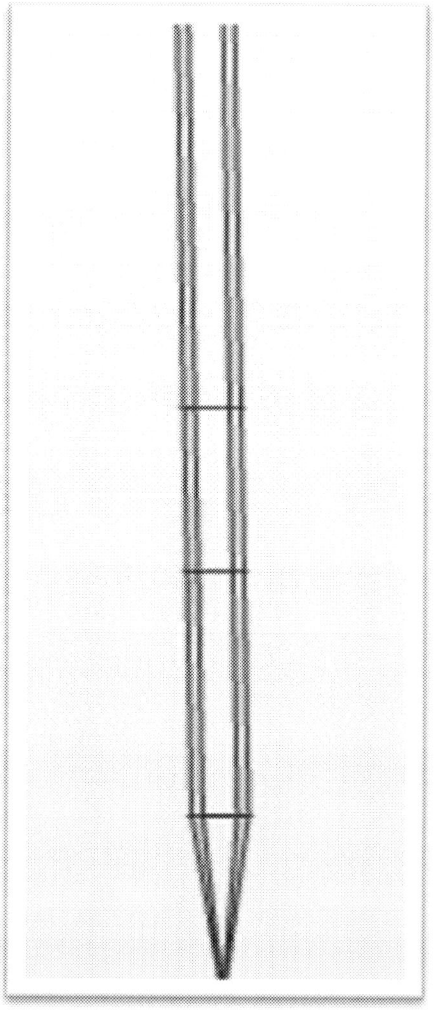

Figure 7.17 Side view of the bipolar probe.

RFA. Hence, electric current physics is responsible for producing an alternative voltage in the radiofrequency range as shown in Fig. 7.28.

Fig. 7.28 displays Maxwell's equations in the frequency domain case, which is applied to all domains, and controls the current generation and the conductive path between the probe's active part, and the model external boundaries, which depends on the tissue electrical conductivity obtained from the defined material, as shown in Fig. 7.29.

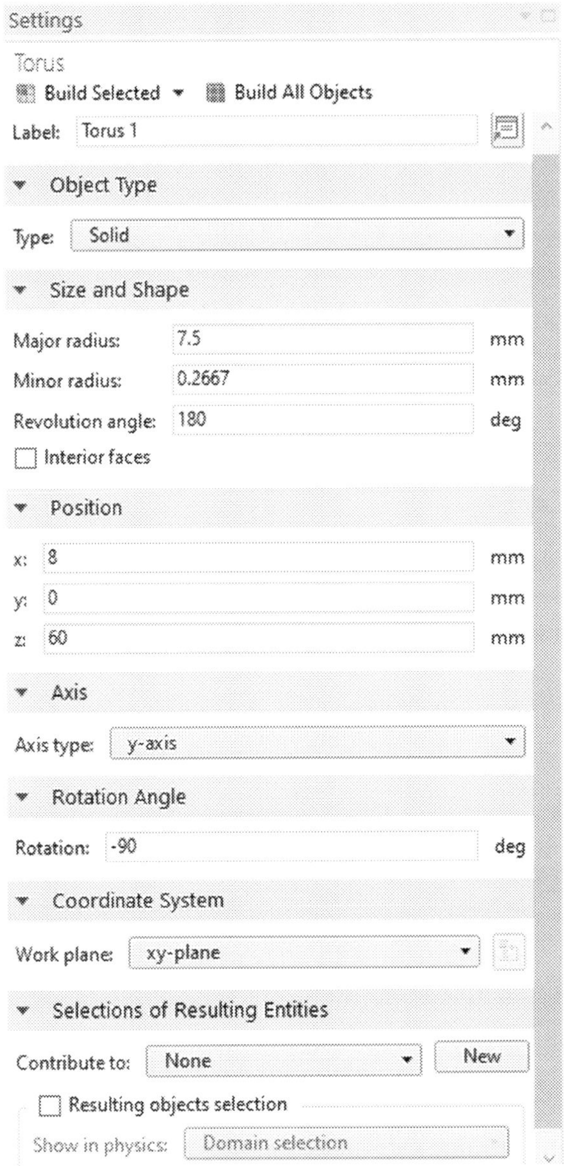

Figure 7.18 Setting window of geometric torus representing one curved electrode.

At the beginning of the ablation procedure, the voltage initial value is set at all domains as shown in Fig. 7.30.

Furthermore, the external boundaries of the geometric model are set to the ground to allow a conductive path between the needle and ground padding as shown in Fig. 7.31.

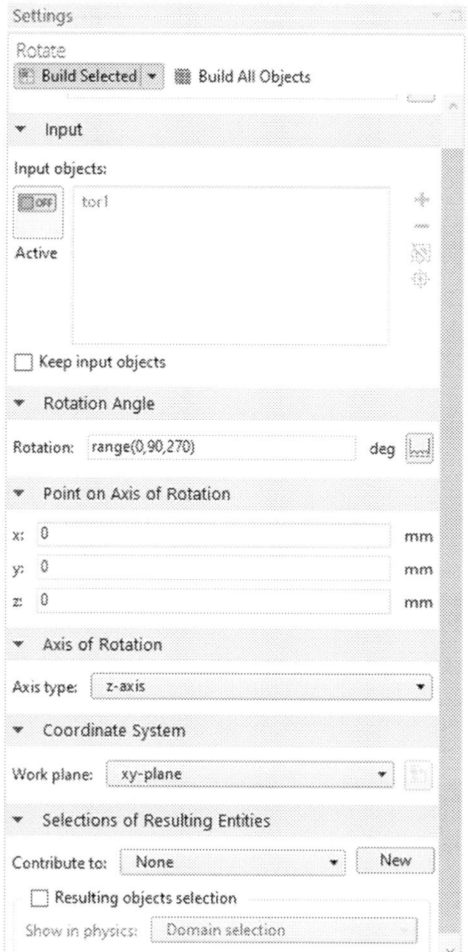

Figure 7.19 Setting window of the rotate module.

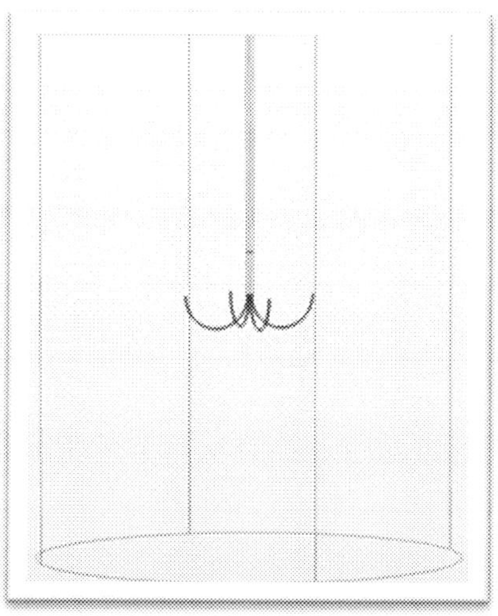

Figure 7.20 Geometric model of the multitines expandable probe and liver tumorous tissue.

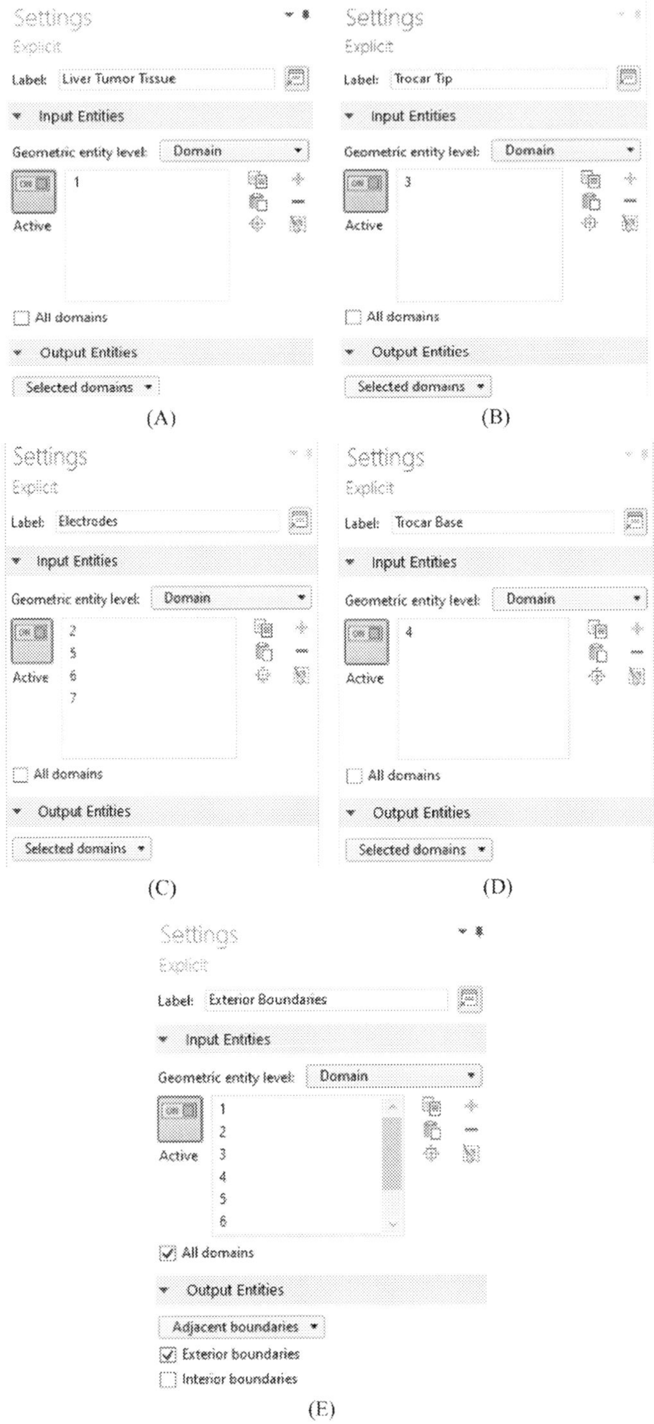

Figure 7.21 Explicit windows for different five parts in the geometric model, including (A) liver tumor tissue, (B) trocar tip, (C) four electrodes, (D) trocar base, and (E) external boundaries.

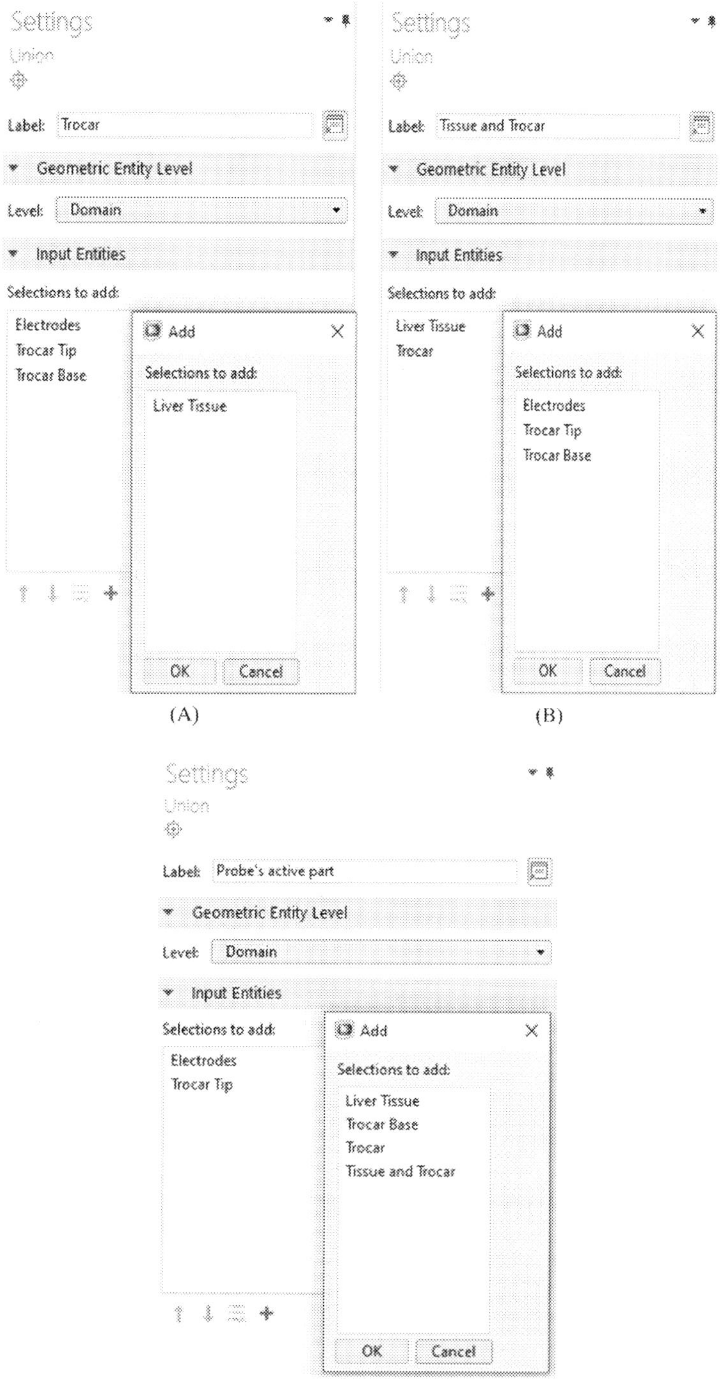

(C)

Figure 7.22 Union windows for two different sections in the geometric model, including (A) trocar, (B) tissue and trocar, and (C) probe's active part.

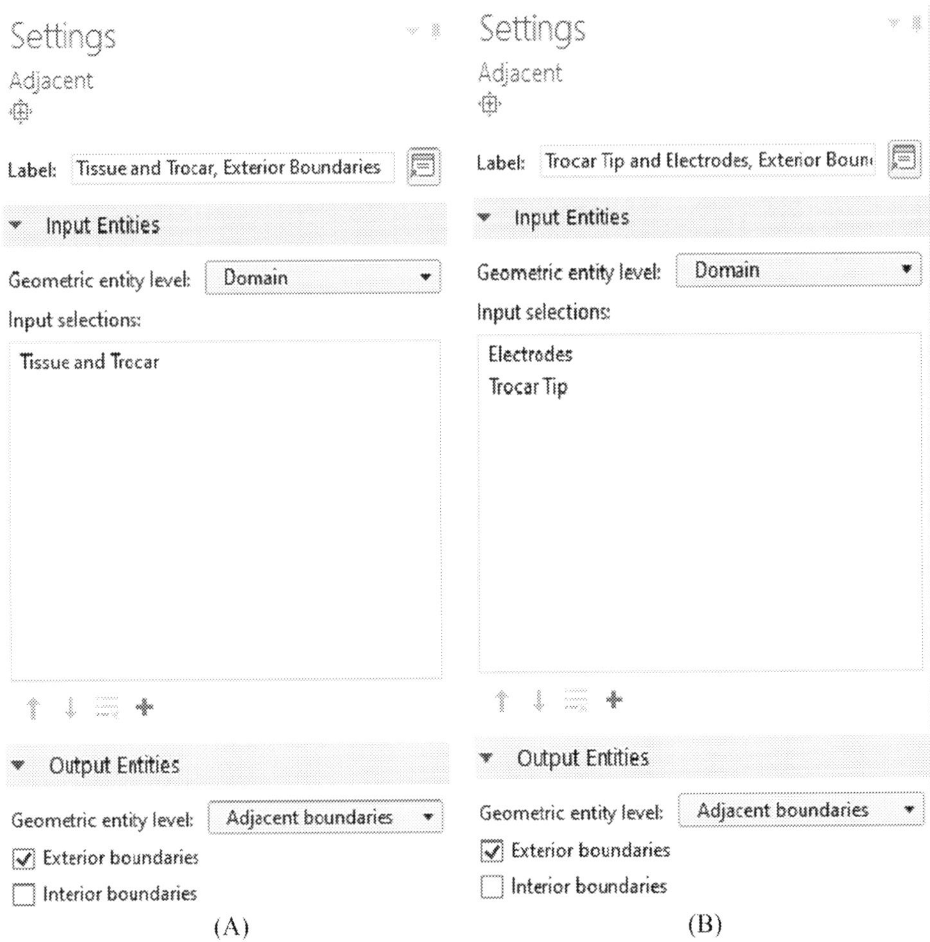

Figure 7.23 Adjacent setting windows for two different sections in the geometric model, including (A) tissue and trocar and (B) active part external boundaries.

The ablated alternative voltage is applied to the trocar tip and the four electrodes active part of the probe with a radiofrequency of 500 kHz and value suitable for the tumor size, as shown in Fig. 7.32.

The change in current direction causes ion friction, which produces heat to be transferred through the tissue due to its thermal conductivity. Hence, the temperature at each point inside the model and at each time is calculated using Penn's bioheat transfer equation, as shown in Fig. 7.33.

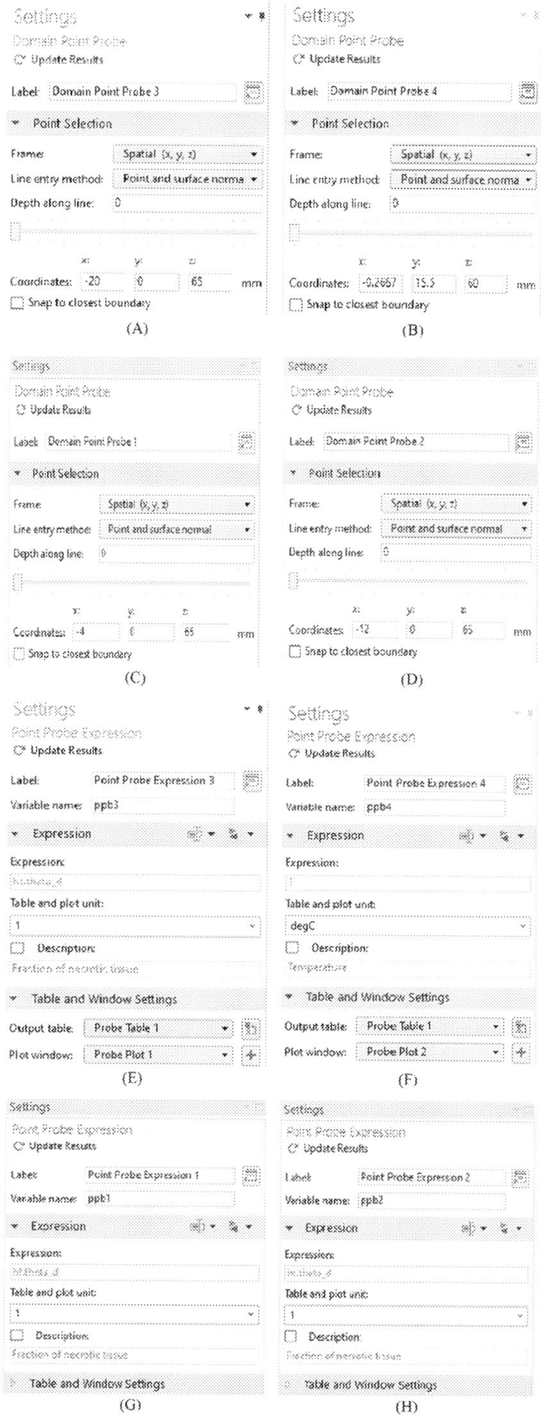

Figure 7.24 Setting windows for two different domain point probes at two different points, including (A) (−20,0,65), (B) (−0.2667,15.5,60), (C) (−4,0,65), and (D) (−12,0,65) in mm with two different quantities, including (E) level of damage at (−20,0,65), (F) temperature at (−0.2667,15.5,60), (G) level of damage at (−4,0,65), and (H) level of damage at (−12,0,65).

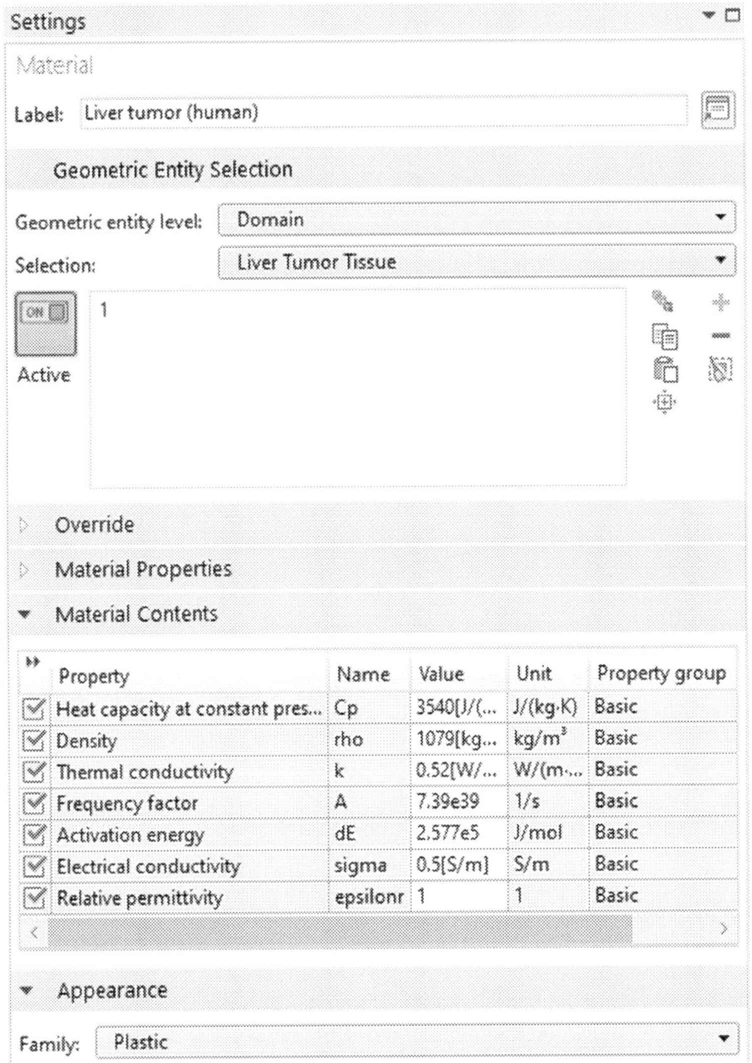

Figure 7.25 Material setting window of the liver tumor tissue with its properties.

Fig. 7.33 displays the heat transfer equation, which is a function of time and tissue properties, such as density, specific heat, and thermal conductivity, accordingly, the values obtained are from the defined material, as shown in Fig. 7.34.

Furthermore, the biological energy is obtained due to the metabolism, which has a small value for the external applied

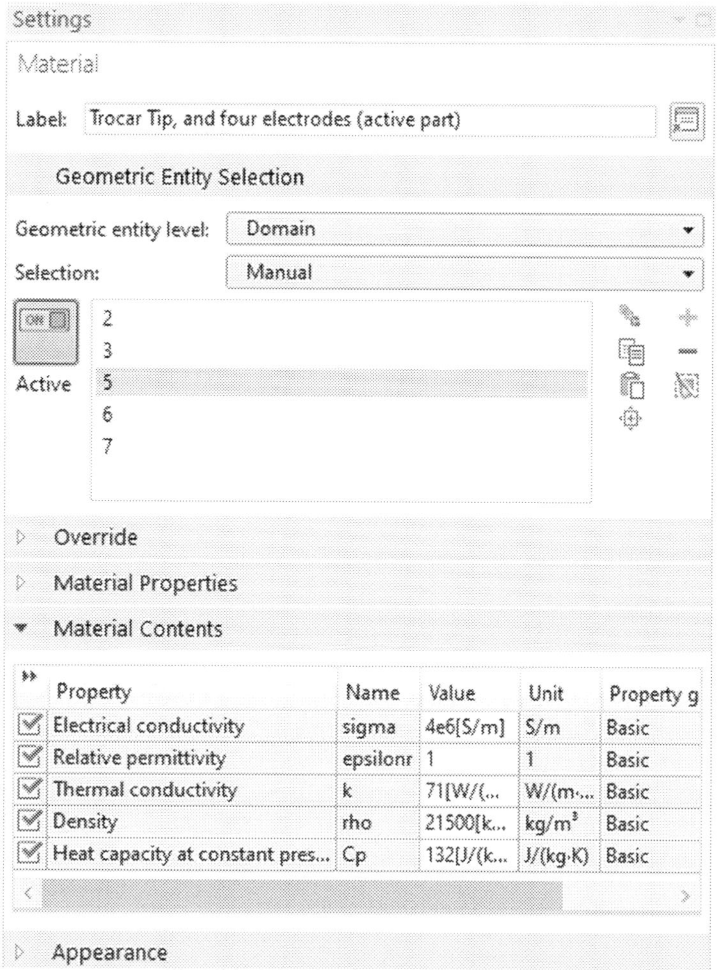

Figure 7.26 Material setting window of the trocar tip and four electrodes representing the probe's active part with its properties.

energy that can be neglected, and the energy absorbed through the existed neighboring blood vessels is controlled according to the blood properties and blood perfusion, as shown in Fig. 7.35.

Fig. 7.35 displays the main parameters that greatly affect the temperature distribution through the model, which are adjusted according to the target system. Moreover, the initial temperature value and boundary condition must be set as shown in Fig. 7.36.

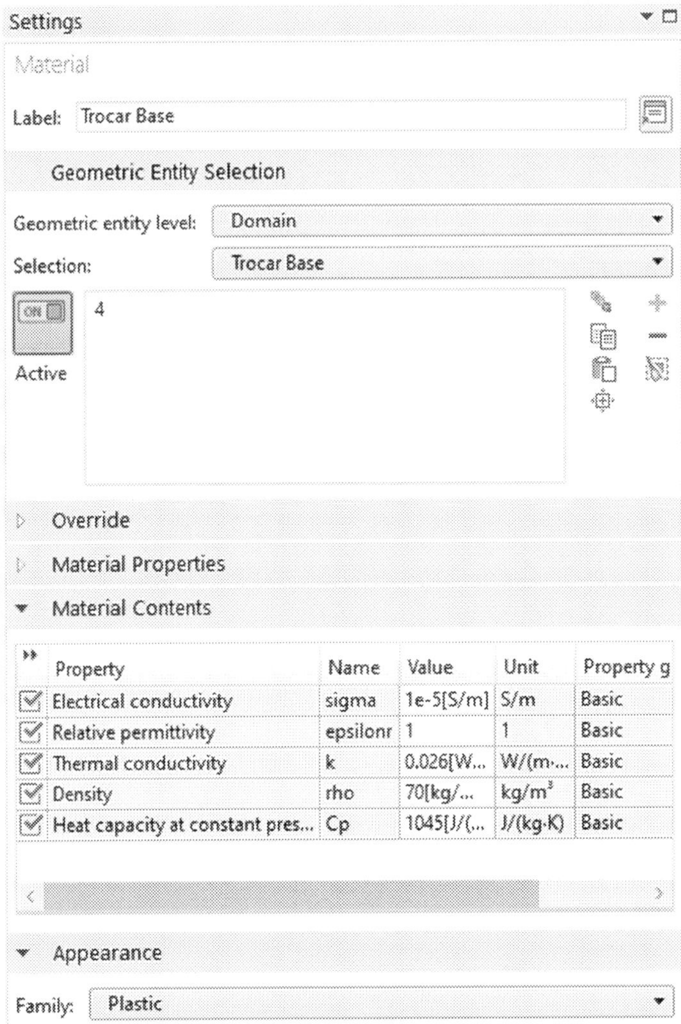

Figure 7.27 Material setting window of the trocar base with its properties.

Furthermore, the bioheat transfer equation is applied to the probe's active part, which is considered to be the heat source for the neighboring tissue. Thus, the properties of the probe manufactured material represent one of the main parameters that greatly affect the ablation process and are specified from the defined material, as shown in Fig. 7.37.

Accordingly, the electric current and bioheat physics must be correlated using the electromagnetic heat source to send and

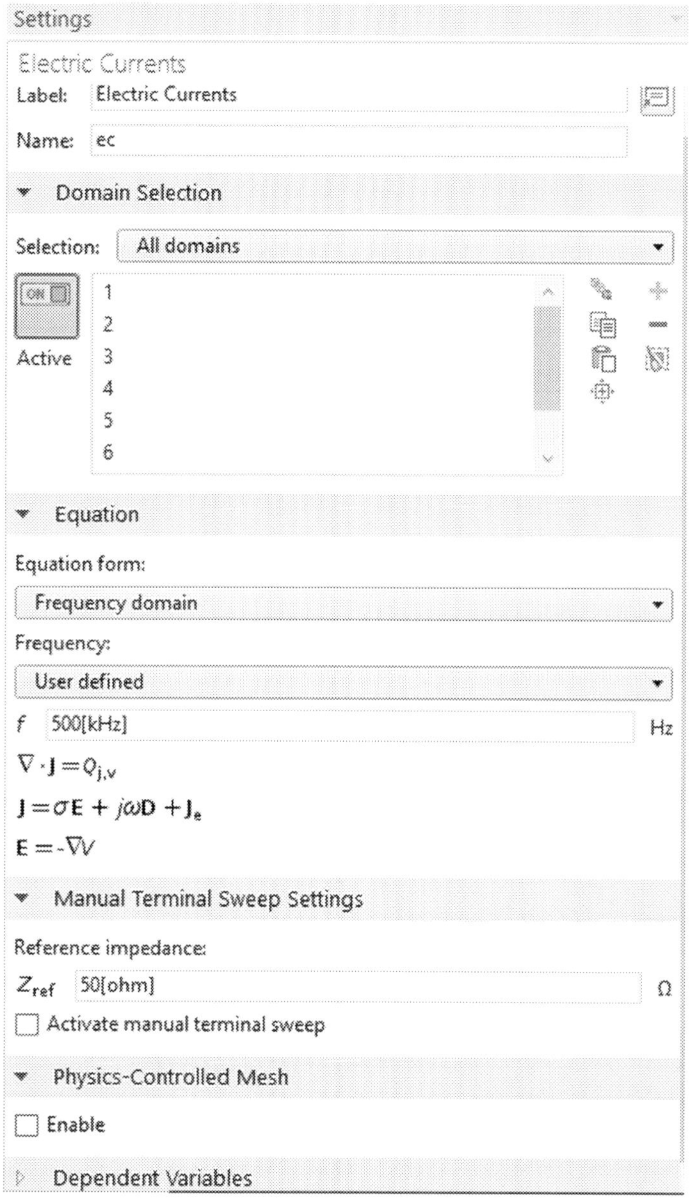

Figure 7.28 Setting window of the electric current physics.

receive the values between the interface for the two equations, as shown in Fig. 7.38.

Fig. 7.38 displays the correlation technique between the two selected physics to obtain accurate results. Accordingly, to

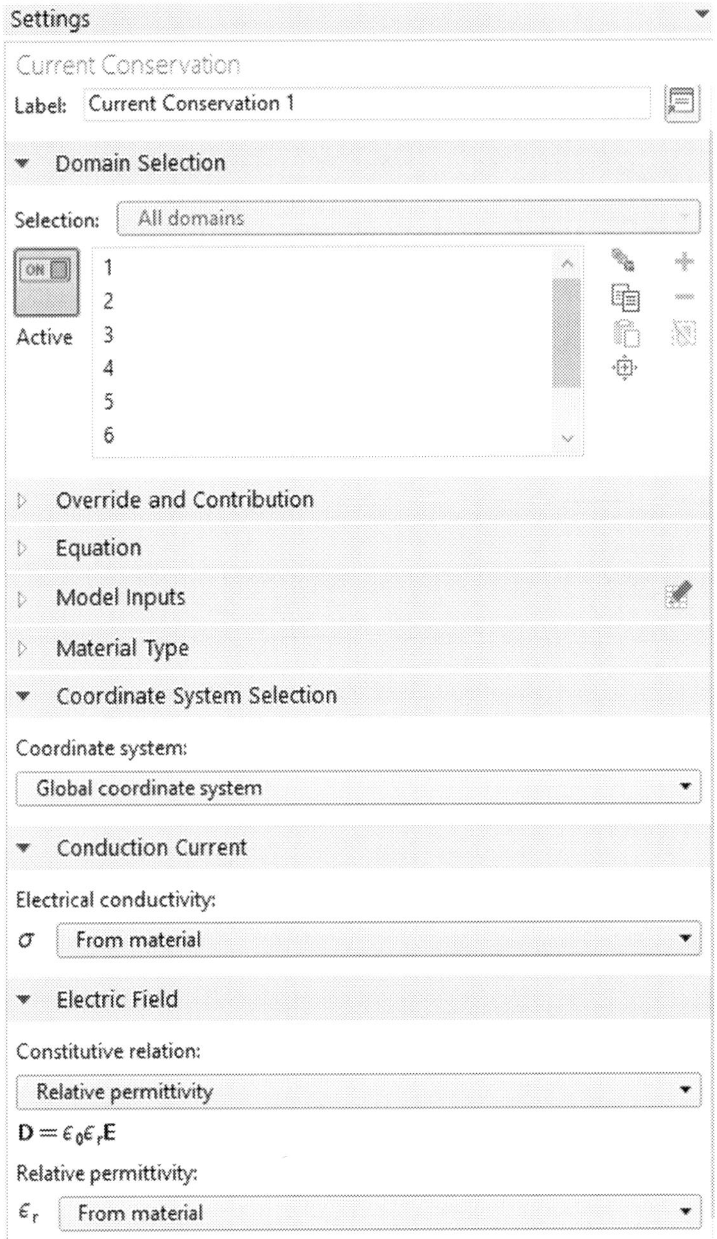

Figure 7.29 Setting window of the current conservation.

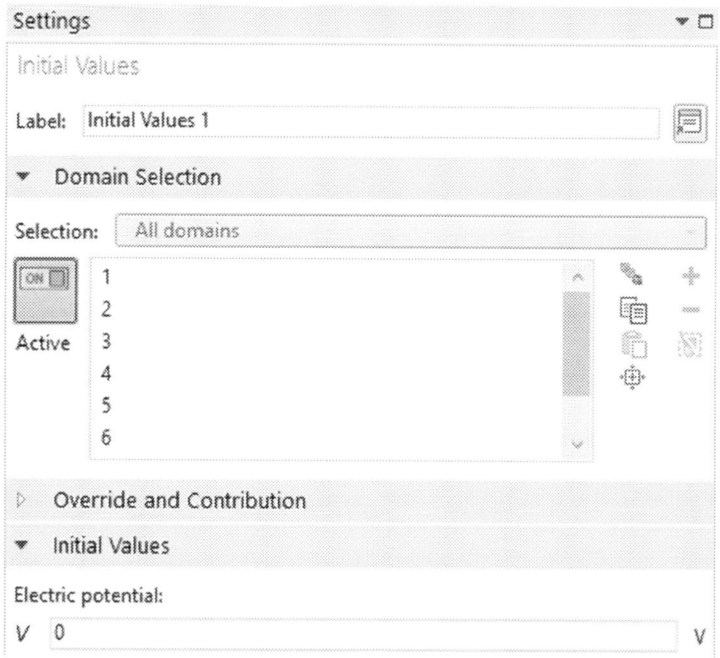

Figure 7.30 Setting window of the voltage initial value.

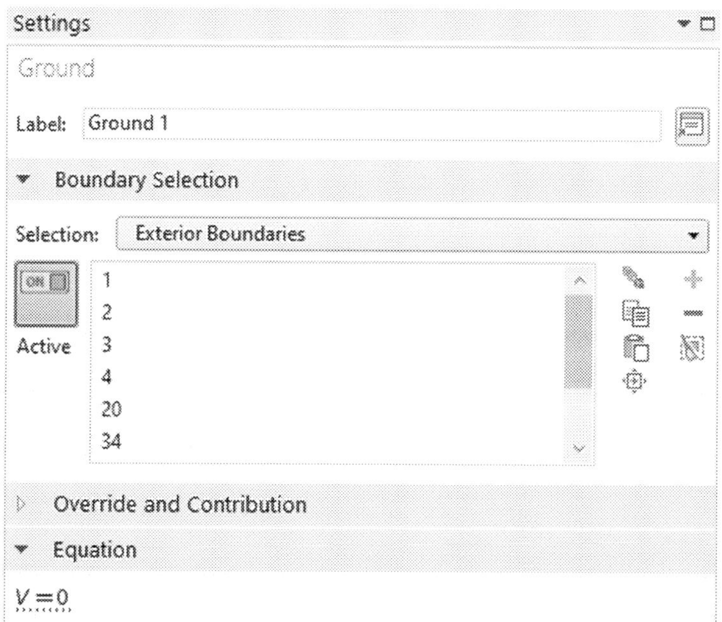

Figure 7.31 Setting window of the ground.

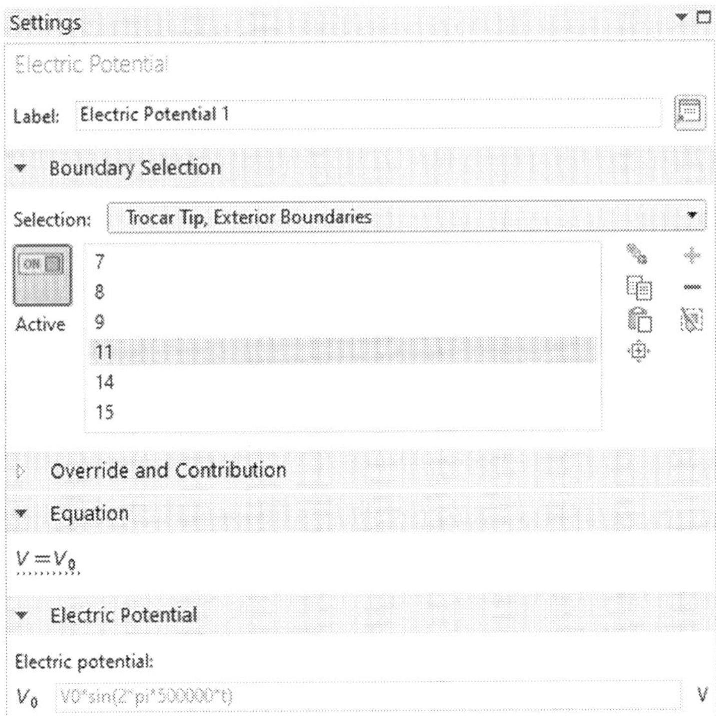

Figure 7.32 Setting window of the electric alternative voltage.

calculate the temperature of the damaged level at all points inside the model, the total geometric model is divided into very small elements using the mesh as shown in Fig. 7.39.

Fig. 7.39 displays the general standard element size for the geometric model. Therefore it is possible to change the size to be suitable for the target model, where the element size is selected to be very small in the critical operating region, however it is large in the farther region to enable a tradeoff between the processing time and obtaining more accurate results. Hence, the model is divided into free tetrahedral elements as shown in Fig. 7.40.

Fig. 7.40 clarifies the most commonly used mesh type, which does not produce empty or/and overlapping regions, consequently, three sizes are made, including the custom option for the trocar base, extra fine for the probe's active part, and finer for the liver region, as shown in Fig. 7.41.

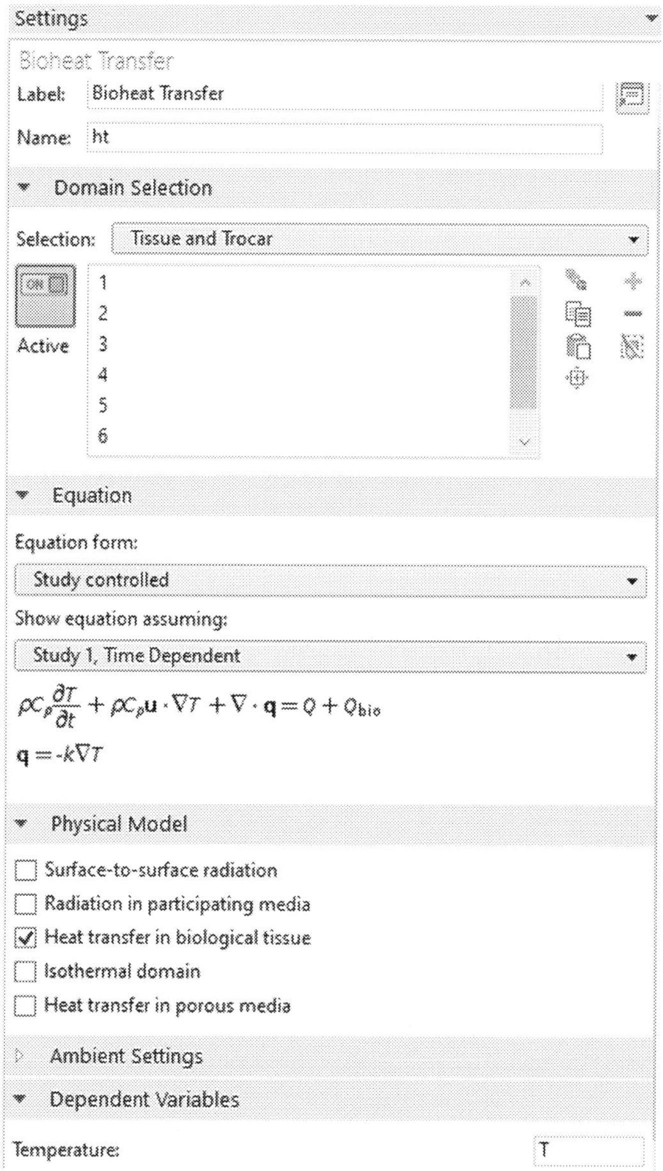

Figure 7.33 Setting window of the bioheat transfer equation.

Fig. 7.41 demonstrates that all subsections inside the geometric model are regularly divided into free tetrahedral elements, as shown in Fig. 7.42.

Therefore the model is studied through a time-dependent study to produce the results and compute the

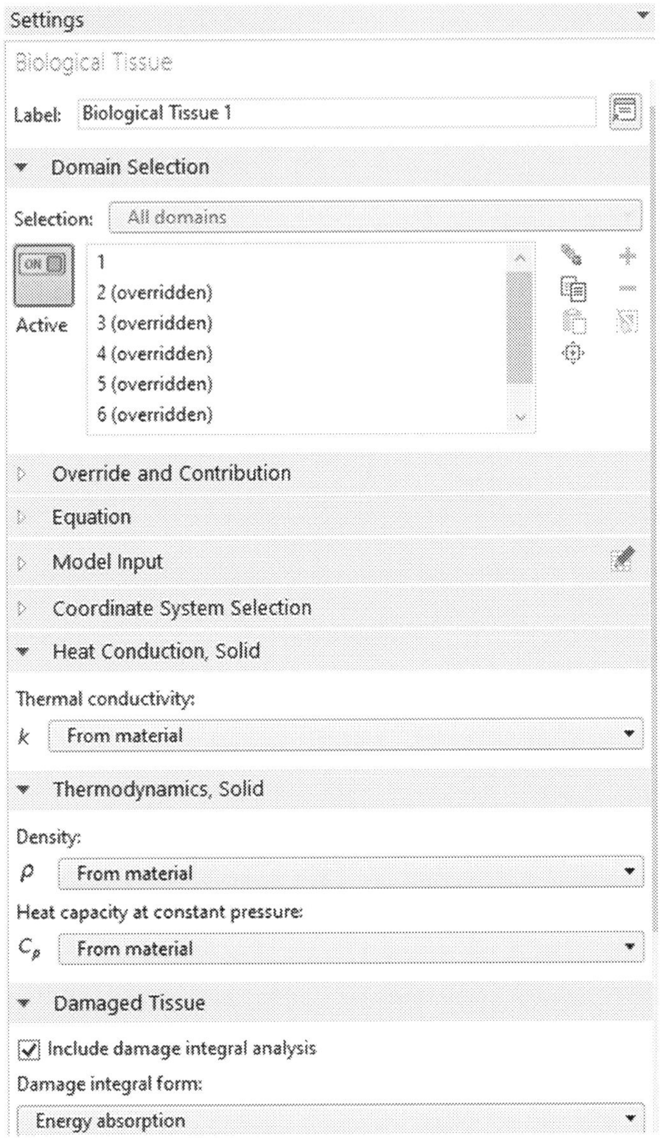

Figure 7.34 Setting window of the biological tissue.

temperature and level of damage for every minute, as shown in Fig. 7.43.

Finally, the geometric model is computed to display all possible obtained values for the electric current and bioheat

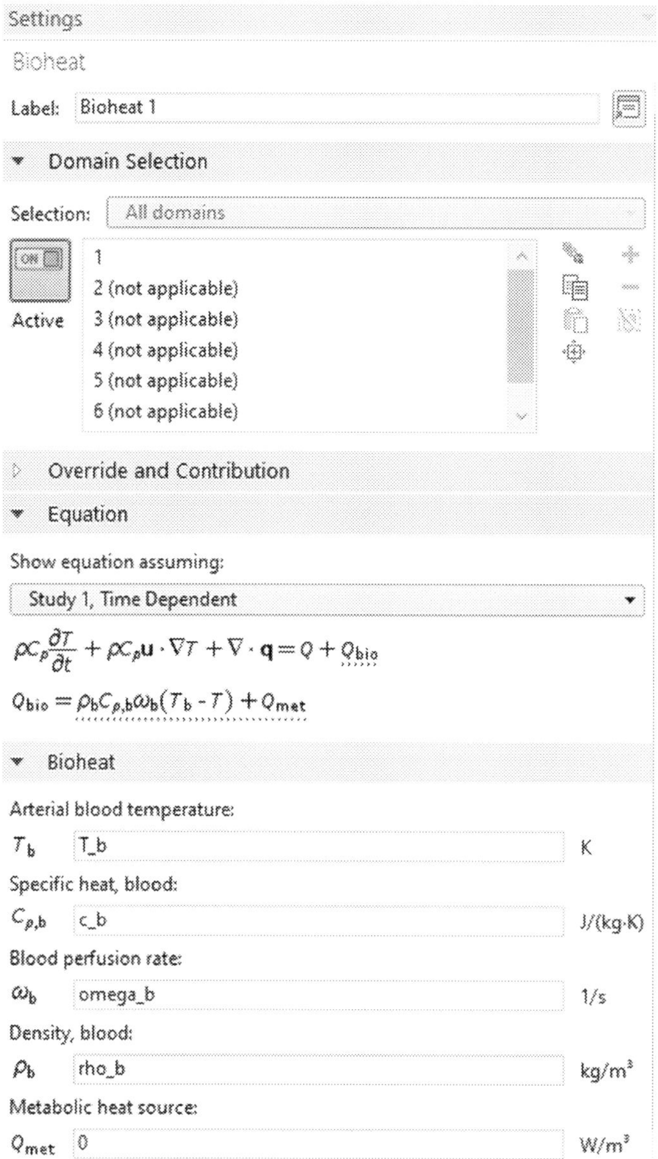

Figure 7.35 Setting window of the bioheat module.

physics, such as the 2D/3D figure, tables, and thermal contours. Hence, the probe table is produced to obtain the temperature and level of damage at four points around the multitines probe, as shown in Fig. 7.44.

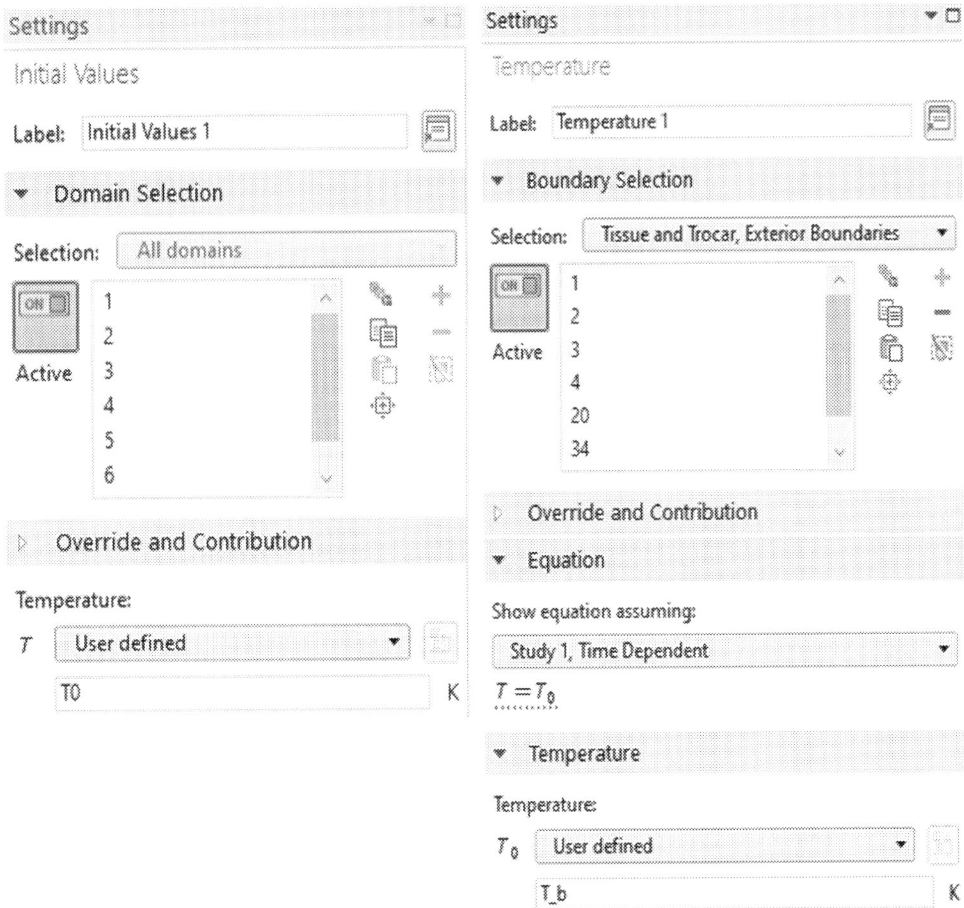

Figure 7.36 Setting window of the initial and boundary condition values for the bioheat transfer equation.

Fig. 7.44 displays a list of points at a random location inside the geometric model to describe the heat transfer as shown in Fig. 7.45.

Fig. 7.45 illustrates that the level of damage increases with increasing time and/or decreasing distance from the probe. Fig. 7.46. illustrates the setting window of the electric potential 3D plot.

Fig. 7.46 illustrates that the electric voltage 3D plot can be formatted in the figure itself, and its background. Accordingly, the colorful trace is displayed through a multislice module as shown in Fig. 7.47.

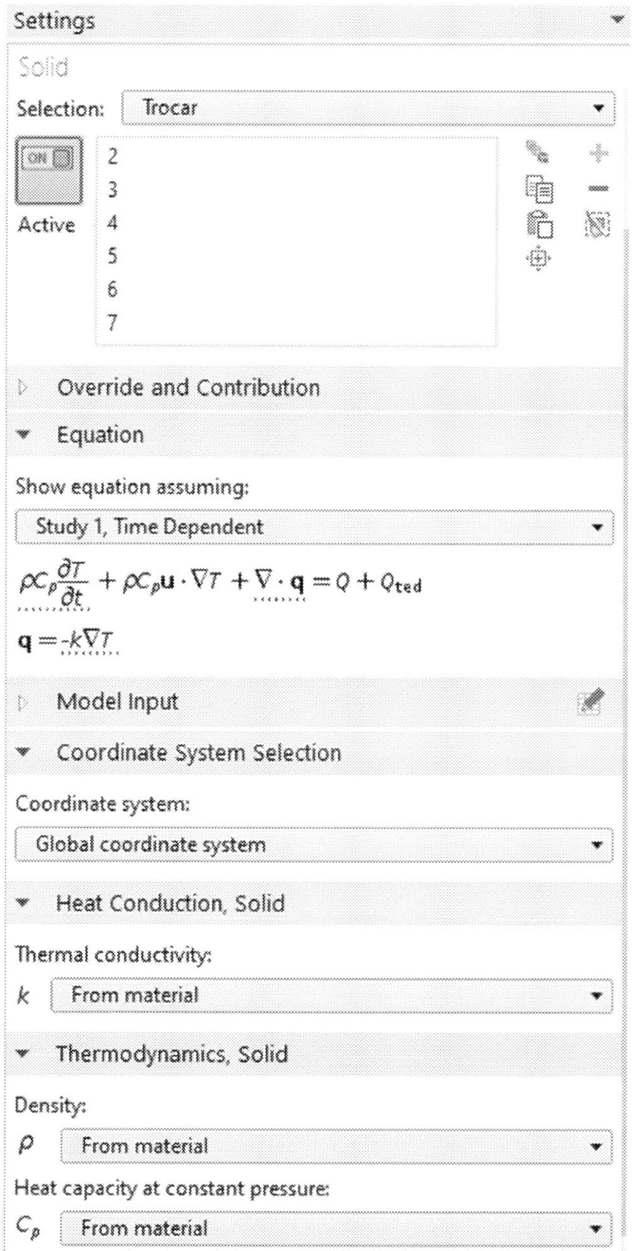

Figure 7.37 Setting window of the solid bioheat transfer equation.

Fig. 7.47 demonstrates that the target expression can be selected to draw the electric voltage in any number of planes in the three coordinates, that is, one plane in each coordinate X, Y, and Z planes. In addition, the rainbow color table,

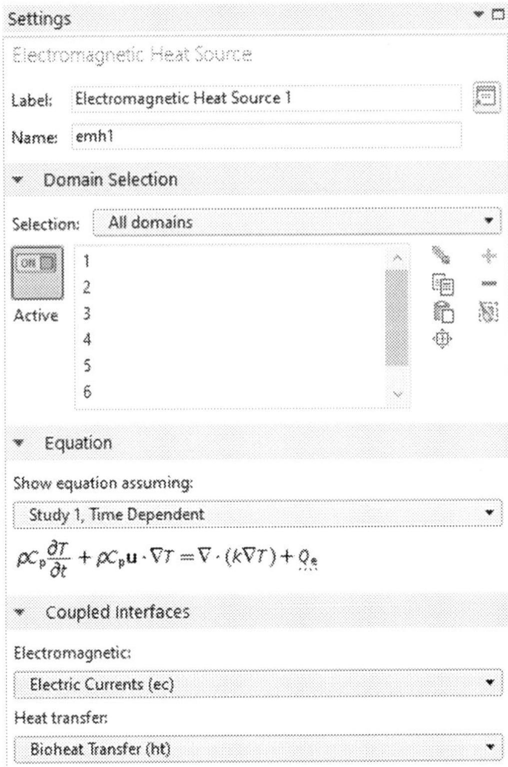

Figure 7.38 Setting window of the electromagnetic heat source.

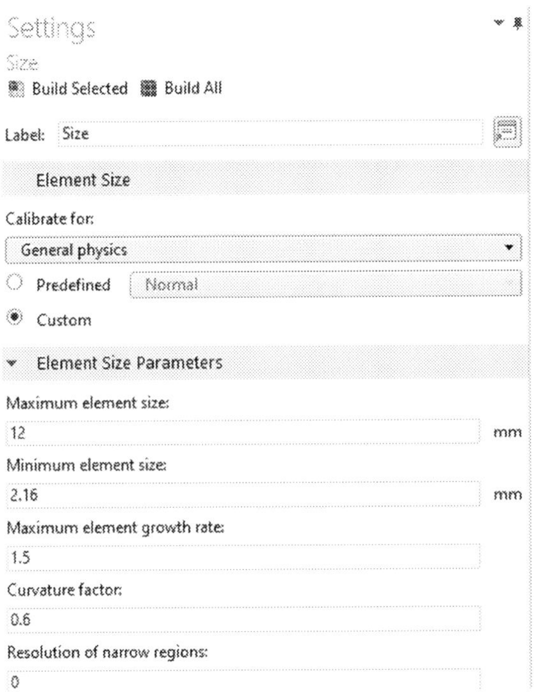

Figure 7.39 Setting window of the mesh.

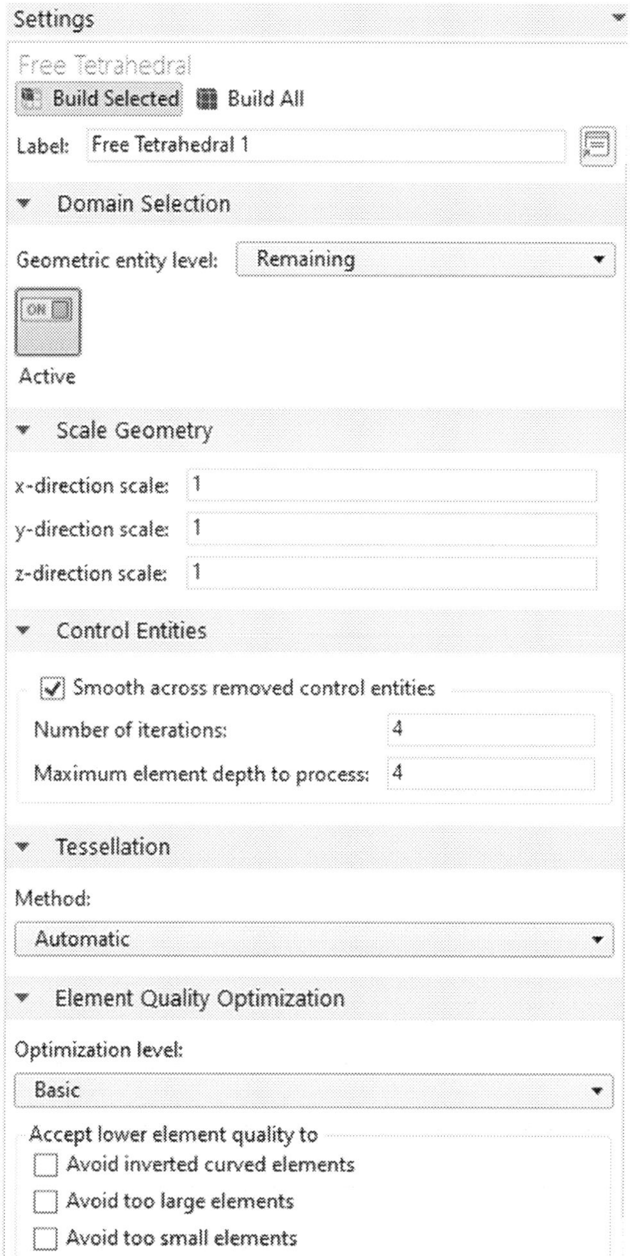

Figure 7.40 Setting window of the free tetrahedral element.

resolution, quality, and smoothing are adjusted as shown in Fig. 7.48.

Figs. 7.47 and 7.48 display the setting window of the electric voltage for plotting the 3D figure as shown in Fig. 7.49.

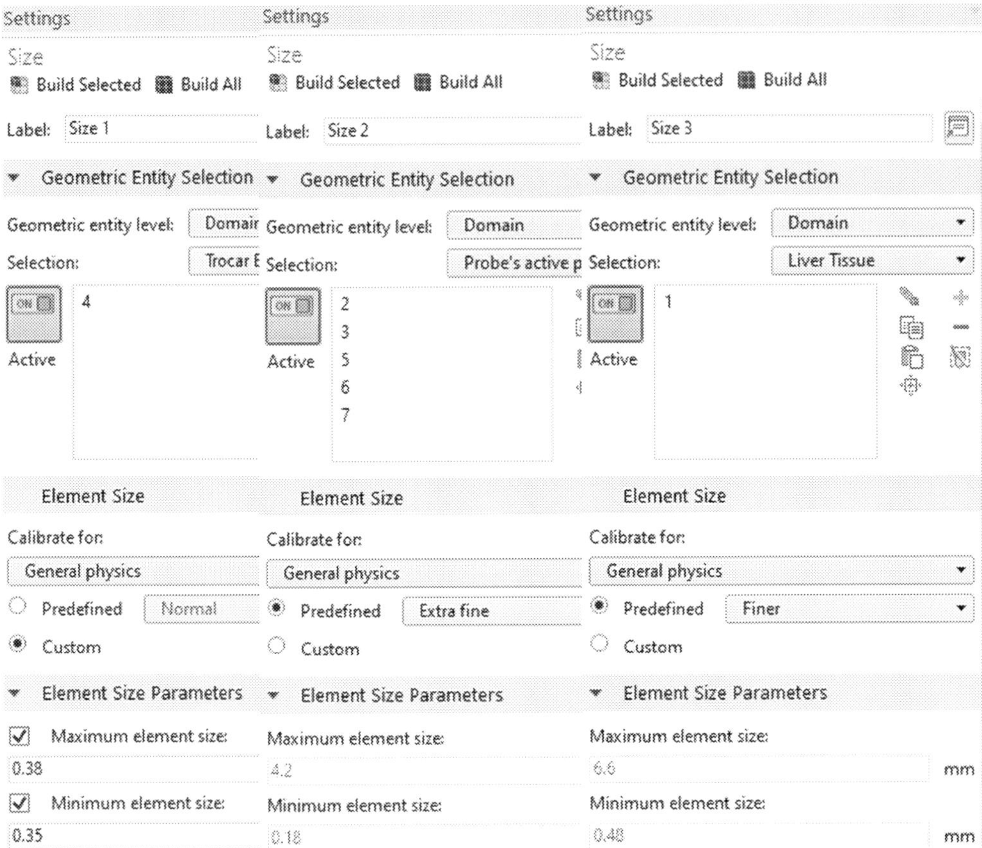

Figure 7.41 Setting window of three mesh sizes, including trocar base, probe's active part, and the liver tumorous tissue.

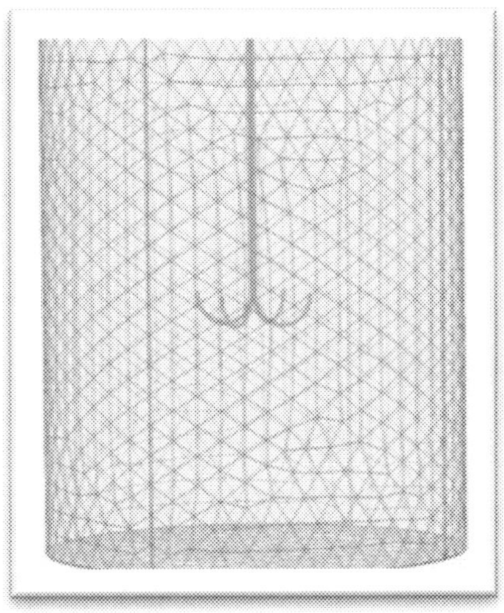

Figure 7.42 Mesh for the geometric model, including liver tumor tissue and multitines expandable probe.

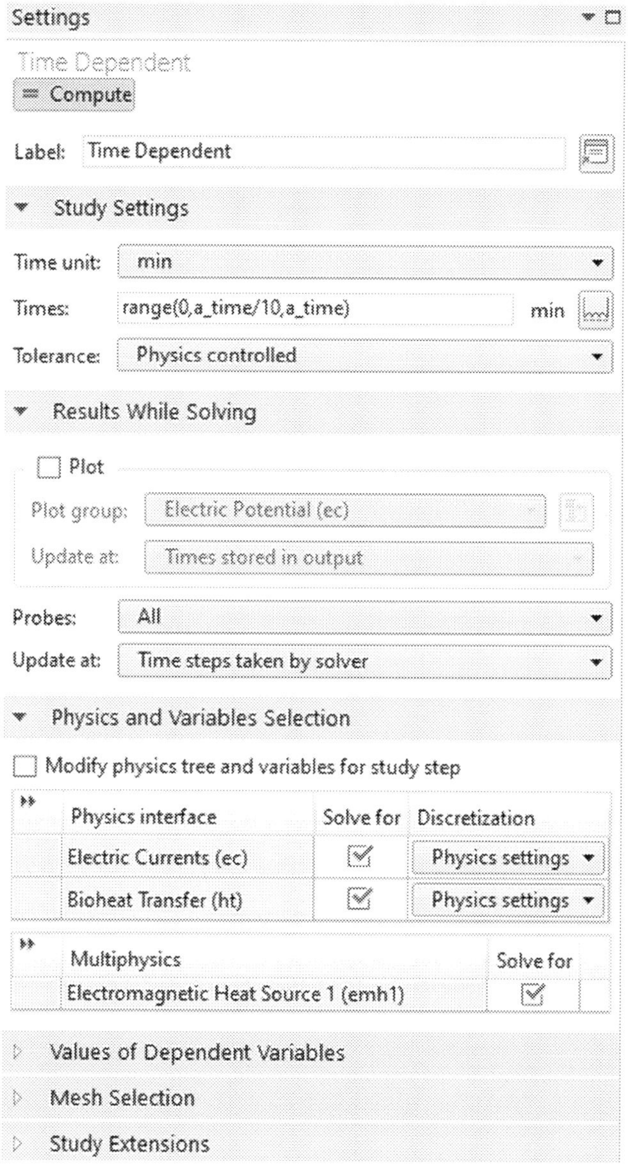

Figure 7.43 Setting window of the time-dependent study.

Fig. 7.49 illustrates that the electric voltage has a maximum value at the probe's active part and decreases with increasing distance from the probe until it reaches the minimum value at the grounding pads of the model's external boundaries.

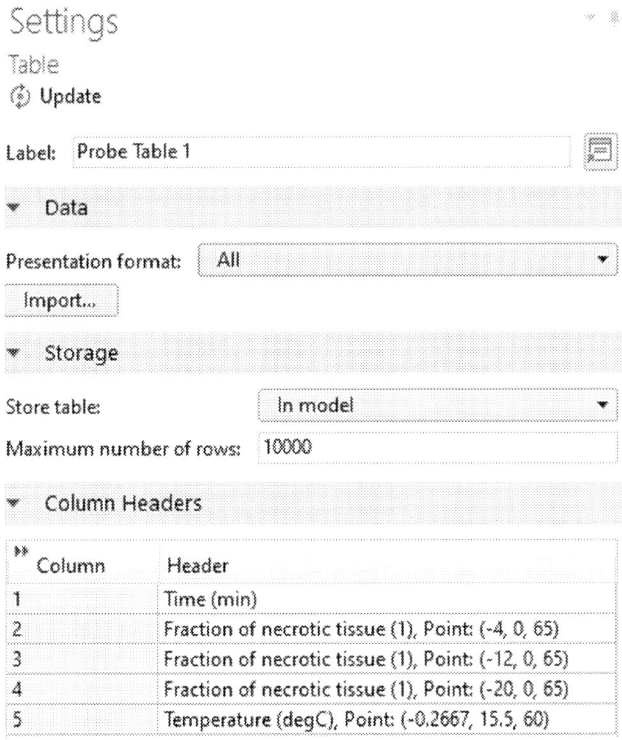

Figure 7.44 Setting window of the probe table.

Time (min)	Fraction of necrotic tissue (1), Point: (-4, 0, 65)	Fraction of necrotic tissue (1), Point: (-12, 0, 65)	Fraction of necrotic tissue (1), Point: (-20, 0, 65)	Temperature (degC), Point: (-0.2667, 15.5, 60)
0.27961	0.0097671	0.0054389	0.0047865	70.916
0.30138	0.011209	0.0059150	0.0052141	71.761
0.34492	0.014577	0.0068975	0.0056457	72.543
0.38847	0.018658	0.0079228	0.0065222	73.952
0.43201	0.023558	0.0089943	0.0074170	75.186
0.47555	0.029385	0.010115	0.0083312	76.279
0.51909	0.036252	0.011289	0.0092659	77.256
0.56263	0.044271	0.012519	0.010222	78.139
0.64972	0.064397	0.015161	0.011201	78.941
0.72809	0.087516	0.017758	0.013231	80.360
0.80647	0.11588	0.020578	0.015143	81.456
0.88484	0.14980	0.023634	0.017139	82.415
0.96322	0.18944	0.026941	0.019223	83.265
1.1200	0.28603	0.034367	0.021397	84.023
1.2767	0.40034	0.042925	0.026022	85.336
1.4335	0.52361	0.052679	0.031016	86.420
1.5902	0.64482	0.063675	0.036380	87.331
1.9037	0.84338	0.089579	0.042108	88.106
2.2172	0.95085	0.12053	0.054604	89.377
2.8442	0.99858	0.19626	0.068394	90.353
3.4712	0.99999	0.28504	0.099216	91.783
4.0982	1.0000	0.37955	0.13342	92.707
4.7252	1.0000	0.47291	0.16984	93.322
5.3522	1.0000	0.55999	0.20742	93.744
6.1856	1.0000	0.66086	0.24533	94.040
7.0189	1.0000	0.74331	0.29509	94.310
7.8522	1.0000	0.80810	0.34337	94.479
8.6856	1.0000	0.85767	0.38950	94.585
9.5189	1.0000	0.89497	0.43306	94.653
10.352	1.0000	0.92276	0.47399	94.697

Figure 7.45 Probe table of the damaged volumes and temperature at four points with the time change.

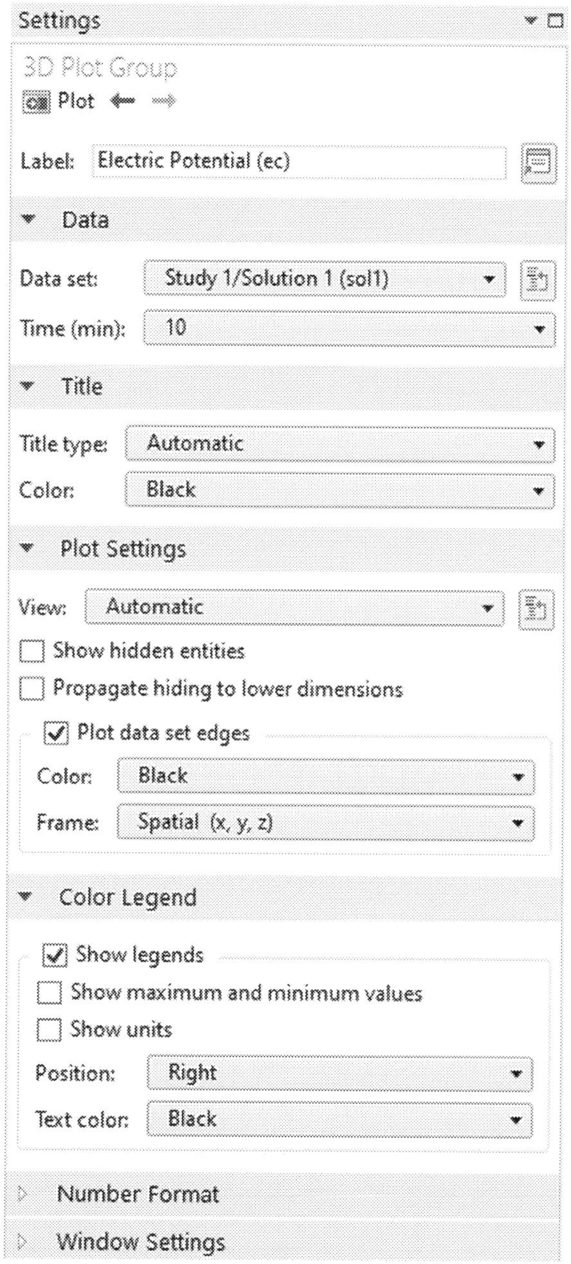

Figure 7.46 Setting window of the electric potential 3D plot.

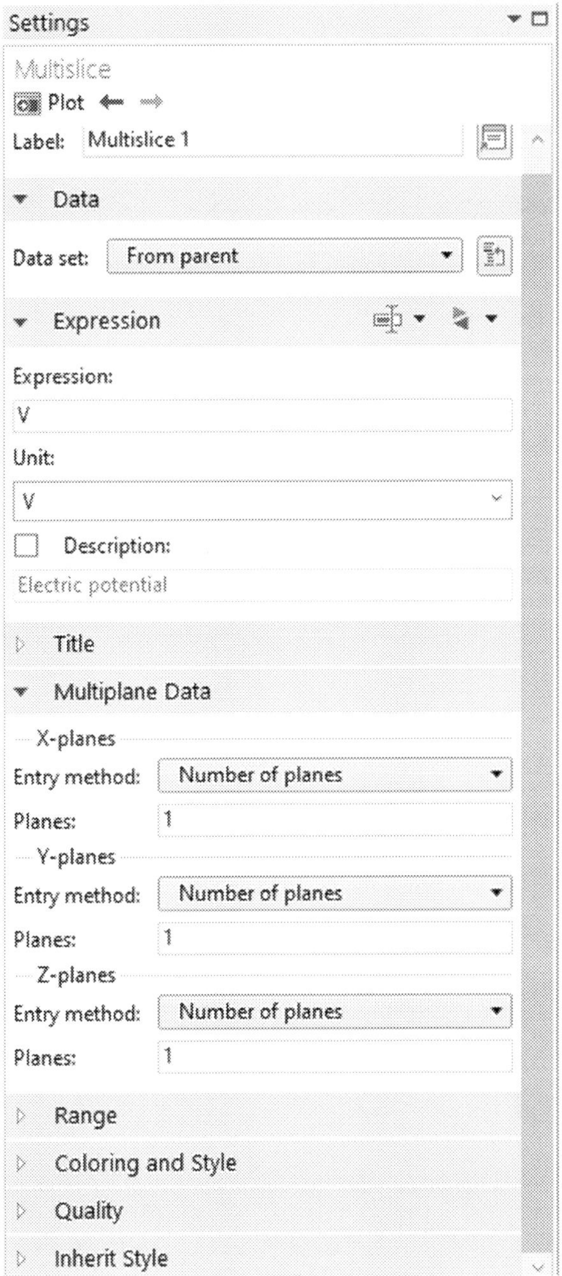

Figure 7.47 Setting window of a multislice for the electric potential.

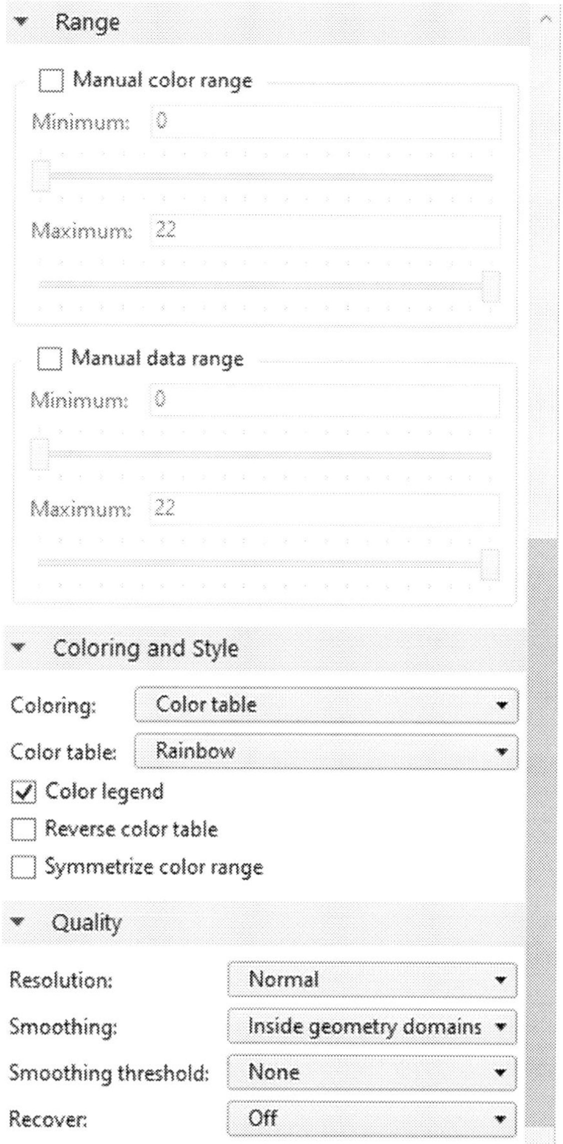

Figure 7.48 Adjustment of the quality, smoothing, and resolution of the electric potential multislice.

Furthermore, the temperature distribution at any time is obtained using the setting window related to the temperature parameter, as shown in Fig. 7.50.

Fig. 7.50 illustrates that the temperature values, which are used for plotting, are obtained from the resulting study solution.

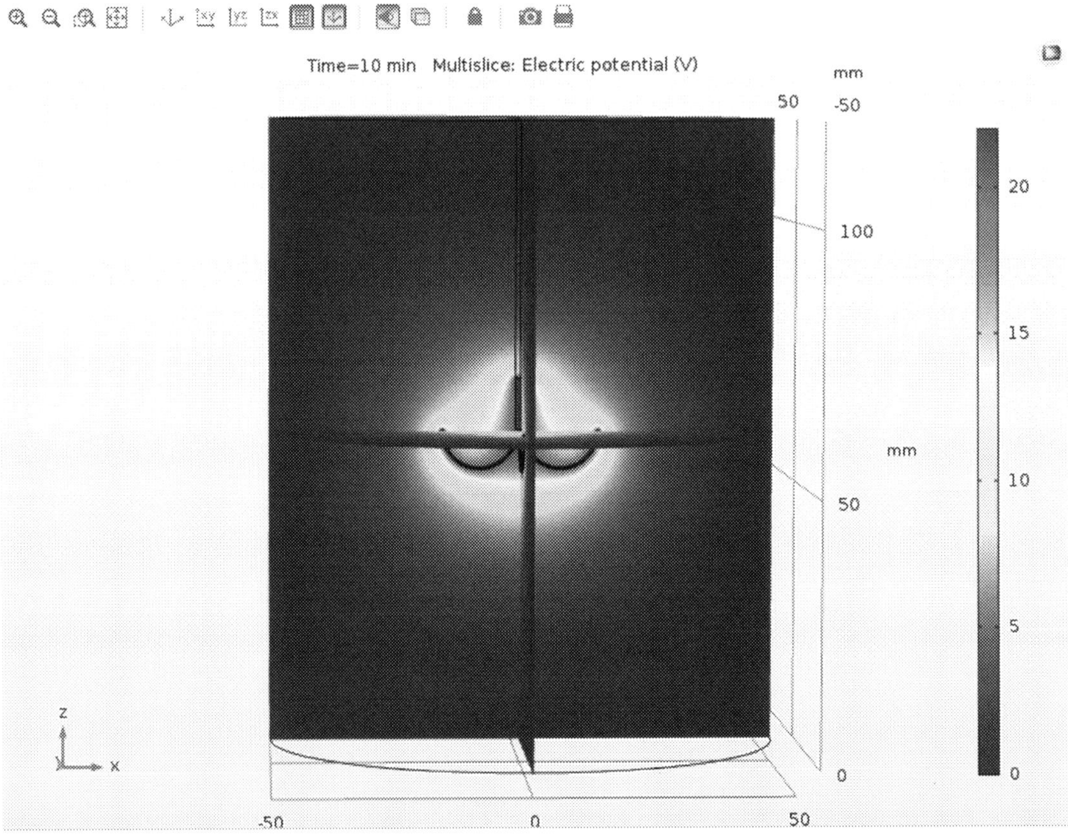

Figure 7.49 Graphic window for 3D electric potential at time 10 min.

Accordingly, two slices at the YZ and ZX planes are used to clarify the temperature profile around the probe, as shown in Fig. 7.51.

Fig. 7.51 displays the adjusted options for the temperature parameter, which is selected from the available parameters, as shown in Fig. 7.52.

Fig. 7.52 displays all parameters available for the electric current and bioheat transfer physics, in which the temperature is selected to be plotted as a 3D figure, as shown in Fig. 7.53.

Fig. 7.53 displays the temperature distribution inside the geometric model at time 10 minutes, where the maximum

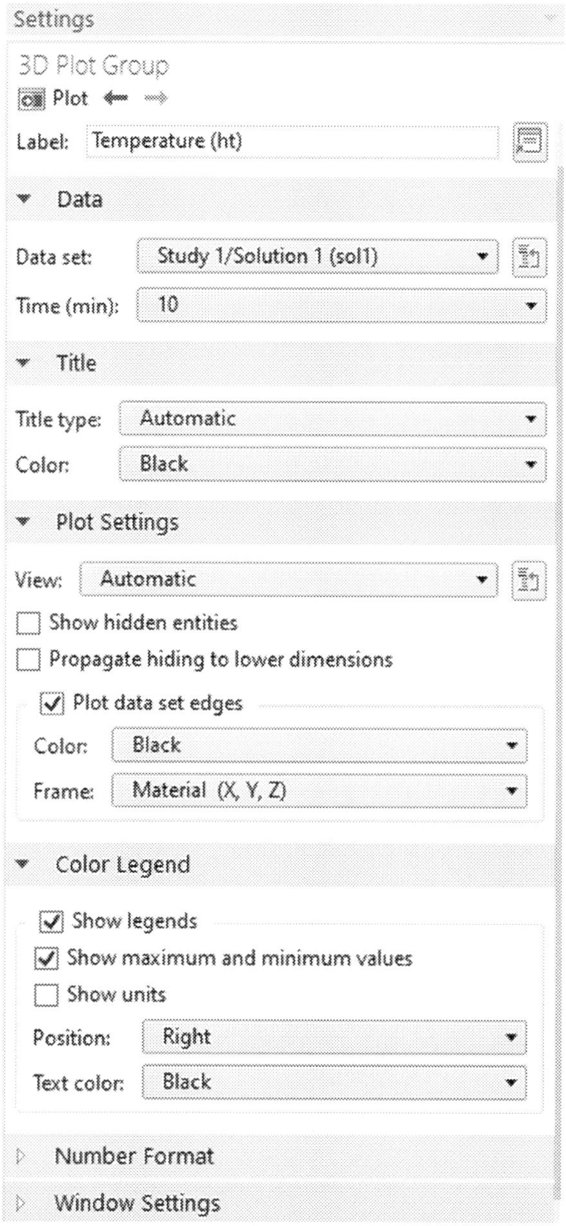

Figure 7.50 Setting window of the temperature parameter.

temperature in this case reaches 95°C, which is less than the forbidden value (100°C). For lower values, to be on the safe side and achieve complete tumor ablation, the ablation time is

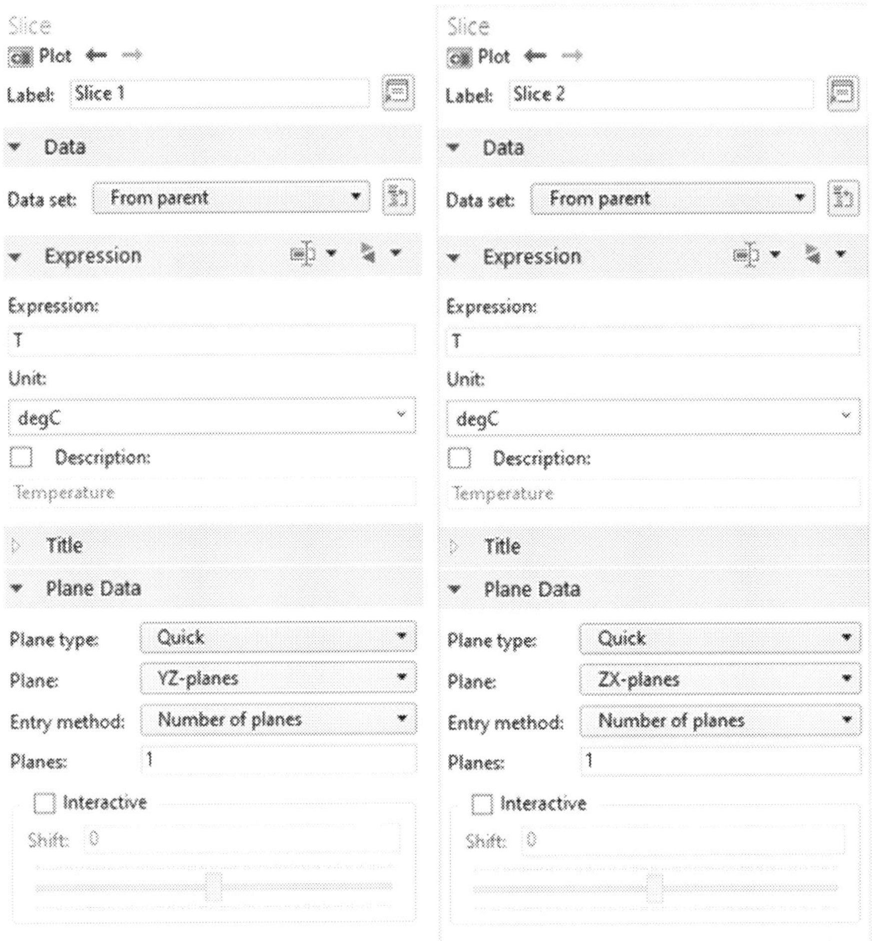

Figure 7.51 Setting windows of two slices at the YZ and ZX planes related to temperature.

reduced as shown in the temperature profiles at several periods in Fig. 7.54.

Fig. 7.54 demonstrates that the temperature is greatly affected by the time, as shown in the GIF file in Fig. 7.55.

Additionally, some specific points can be determined and several options related to the temperature parameter, as shown in Fig. 7.56.

Fig. 7.56 shows the most important parameters, which can be calculated for the geometric model, such as the maximum and minimum temperature values, that are determined as a

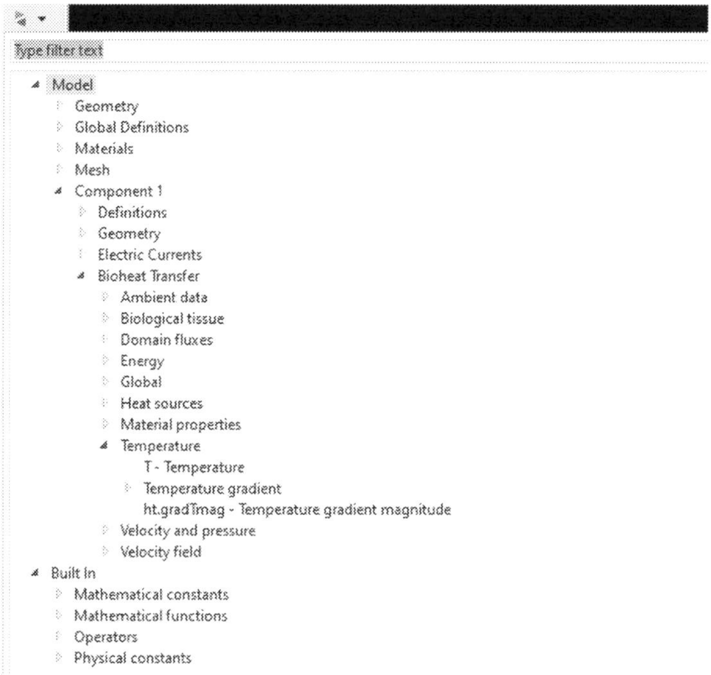

Figure 7.52 Available parameters related to the electric current and bioheat transfer physics.

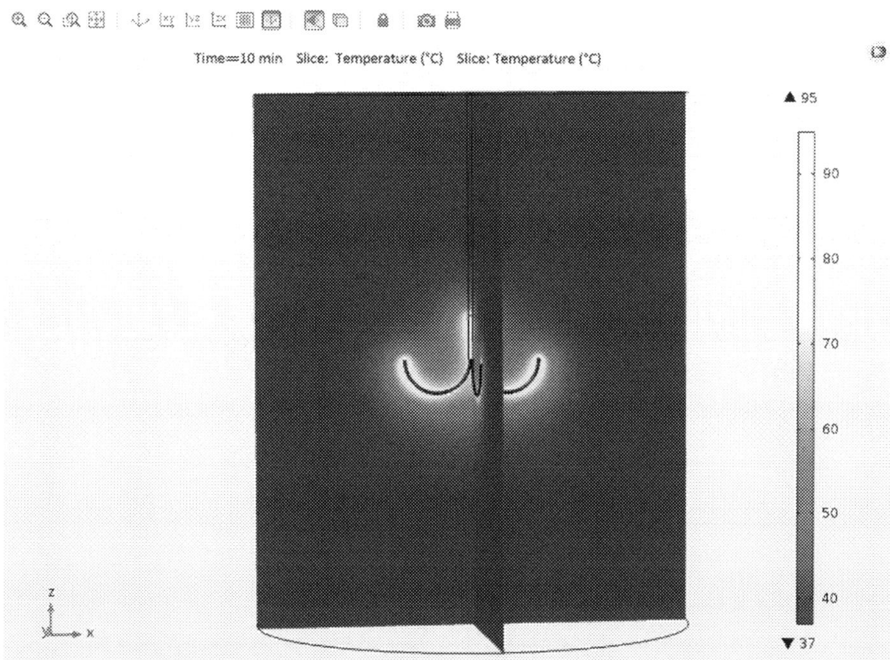

Figure 7.53 Graphic window for the temperature profile at a time of 10 min.

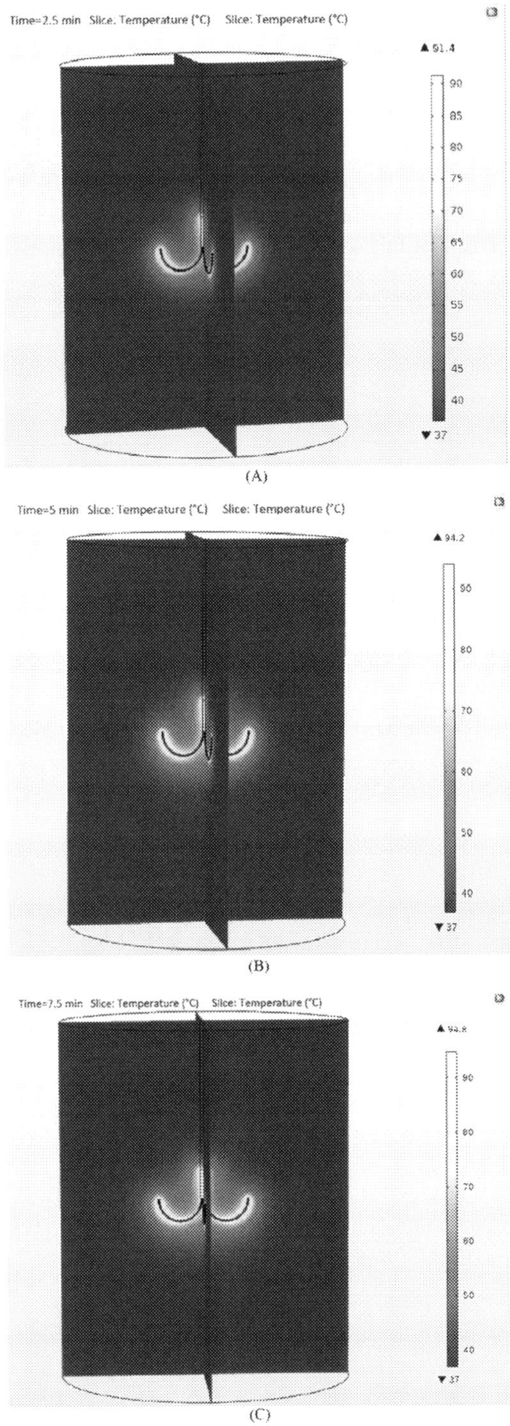

Figure 7.54 Graphic windows for the temperature profile at several periods, including (A) 2.5 min, (B) 5 min, and (C) 7.5 min.

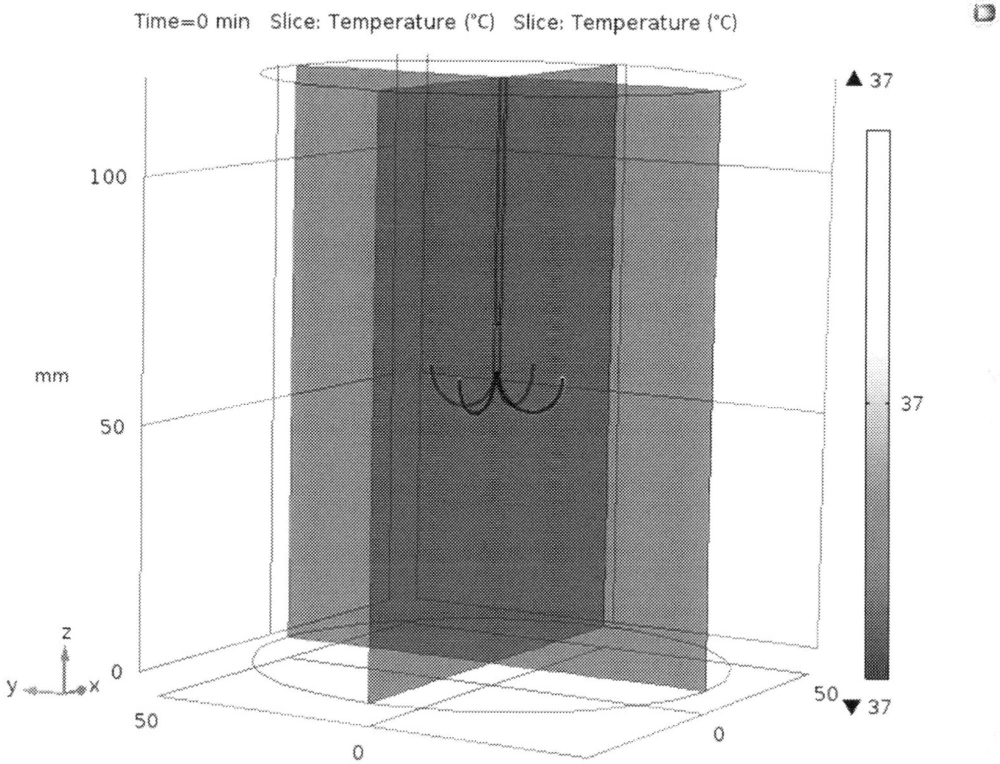

Figure 7.55 Temperature with time.

value and location using the setting windows, as shown in Fig. 7.57.

Fig. 7.57 illustrates that the maximum and/or minimum temperature points are specified inside the model, as shown in Fig. 7.58.

Fig. 7.58 illustrates that the maximum value is located at the probe's active part, which has a coordinate of (0.13,15.7,60), and the minimum value is at the external boundaries of the model, which has a coordinate of (−50,0,0). Thereafter, the temperature values at different levels are determined using the temperature contour as shown in its setting window in Fig. 7.59.

Fig. 7.59 illustrates that the temperature profile through the model can be divided into levels from the initial temperature to the maximum temperature, as shown in Fig. 7.60.

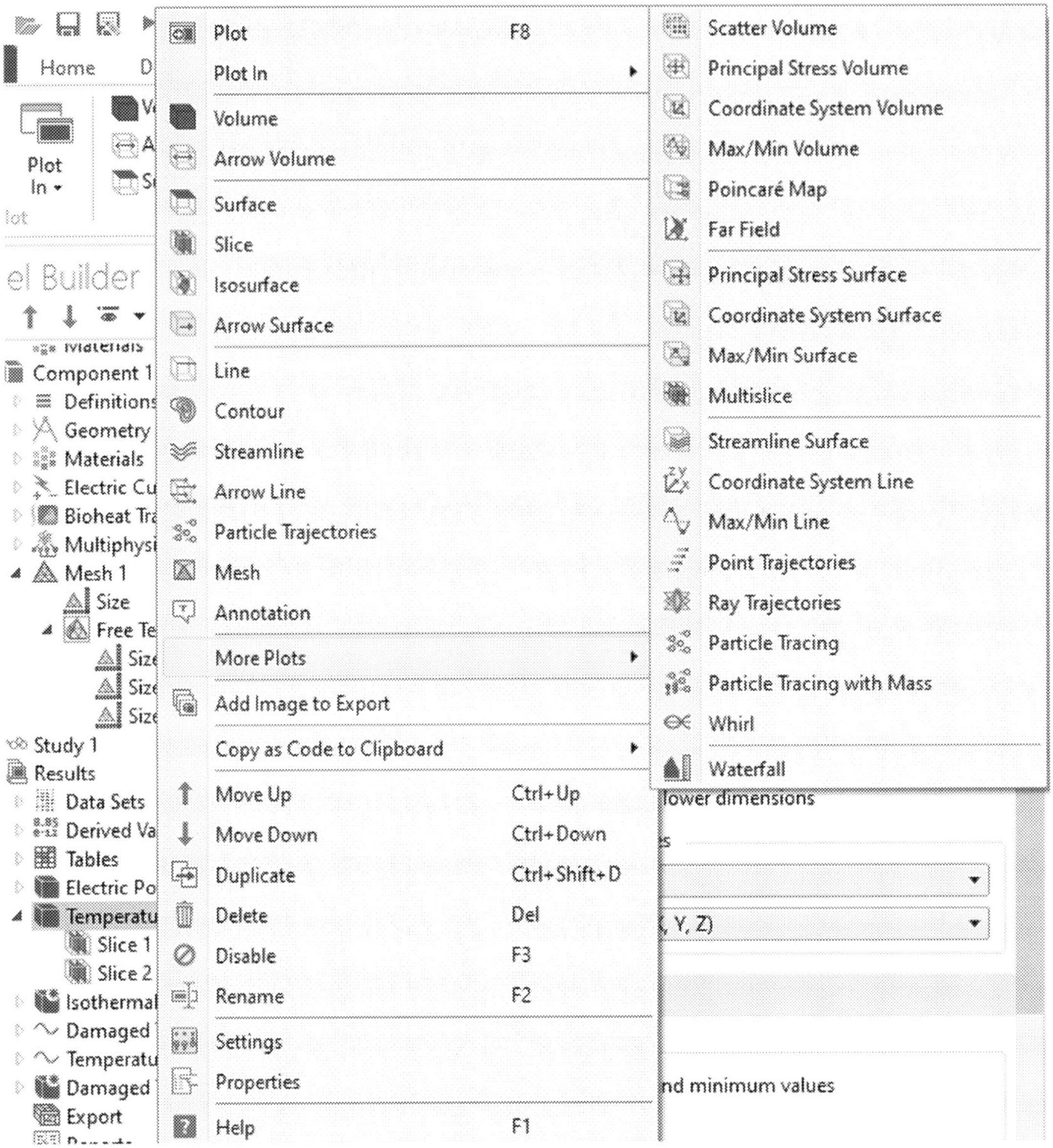

Figure 7.56 Available plots for the temperature parameter.

Fig. 7.60 confirms that the temperature has a maximum value at the spot point, and gradually decreases with increasing distance from the spot point. In addition, to determine the level at which the ablation procedure is

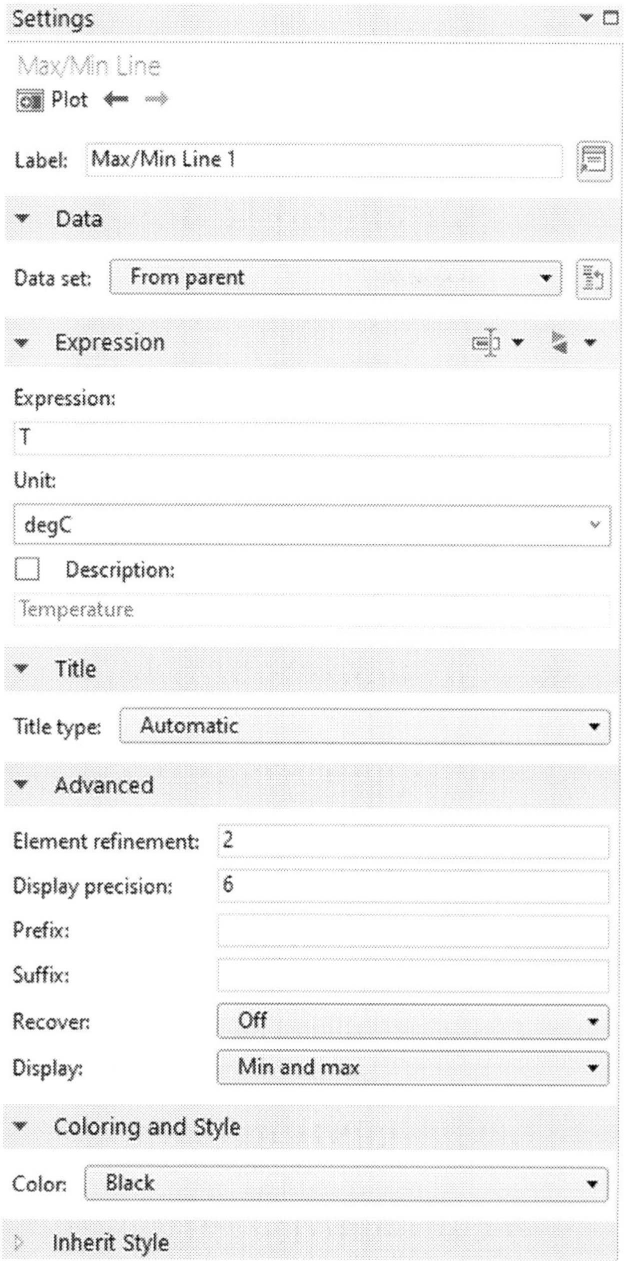

Figure 7.57 Setting window of the maximum and/or minimum points.

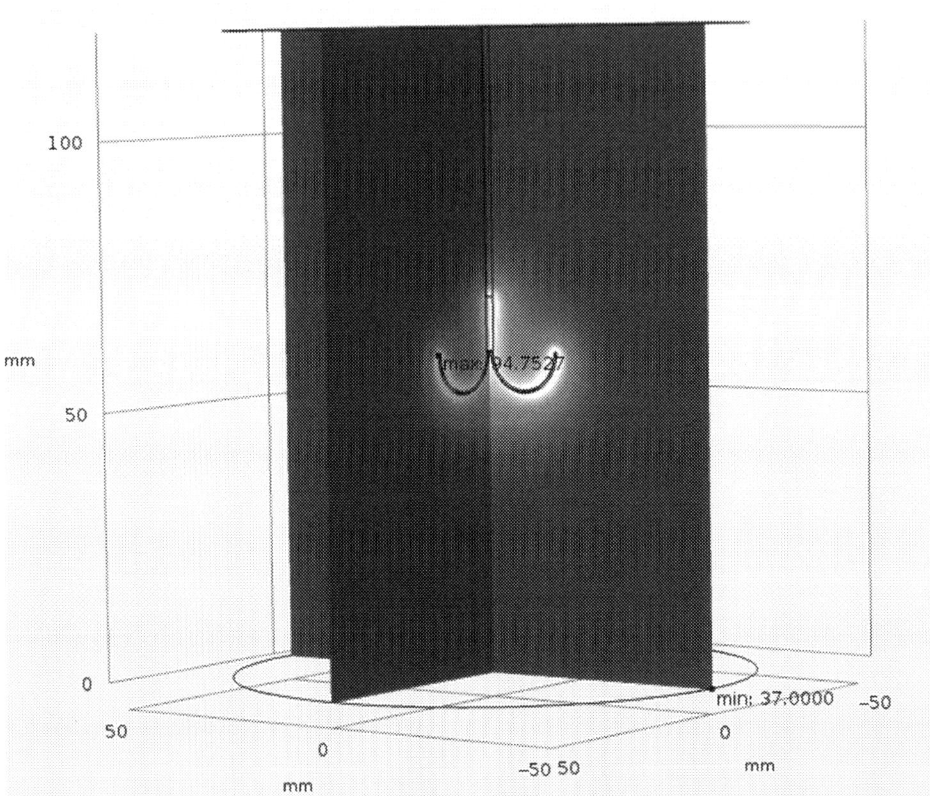

Figure 7.58 Geometric temperature window, including the maximum and minimum values and their locations.

performed, the isothermal contours are used as shown in Fig. 7.61.

Fig. 7.61 displays the formatting parameters for the isothermal contours graphic window, including the title, the time at which the isothermal contour is displayed, and the color legend. Accordingly, the borderline, which divides the domain into two sections, is determined using the isosurface as shown in its setting window in Fig. 7.62.

Fig. 7.62 illustrates that the threshold temperature value, which is sufficient for tumor ablation, is set at 45°C, as shown in Fig. 7.63.

Fig. 7.63 displays two regions with the temperature above and below 45°C at time 10 minutes, accordingly it responds to the time using animation, as shown in Fig. 7.64.

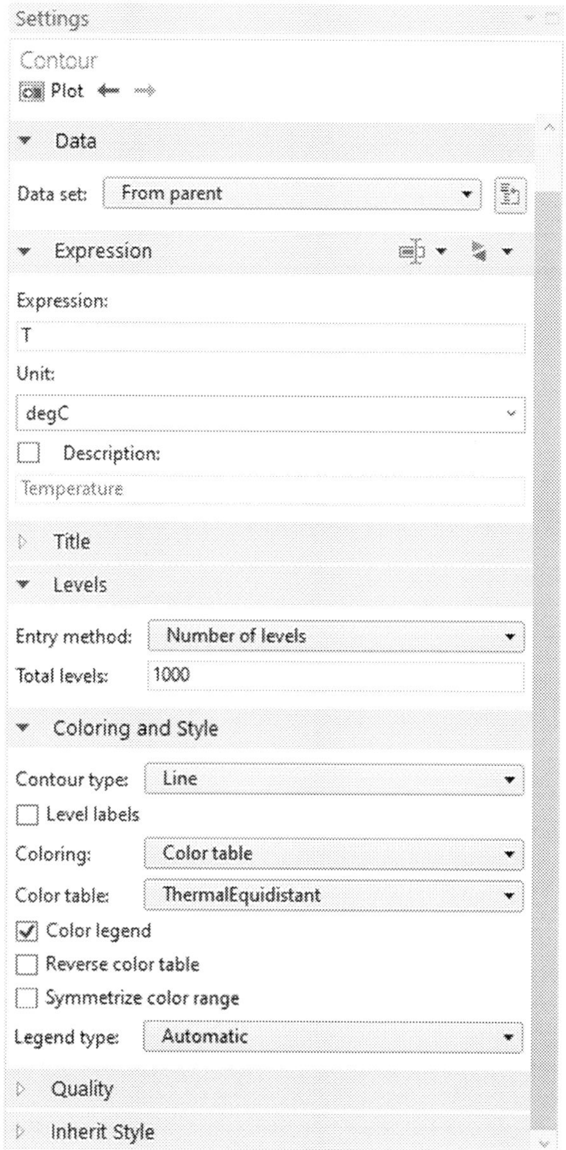

Figure 7.59 Setting window of the temperature contour.

The temperature profile can be divided into more than two regions by changing the entry method from the isosurface setting window, as shown in Fig. 7.65.

Fig. 7.65 illustrates that the temperature distribution is in five sections inside the geometric domain, where each section

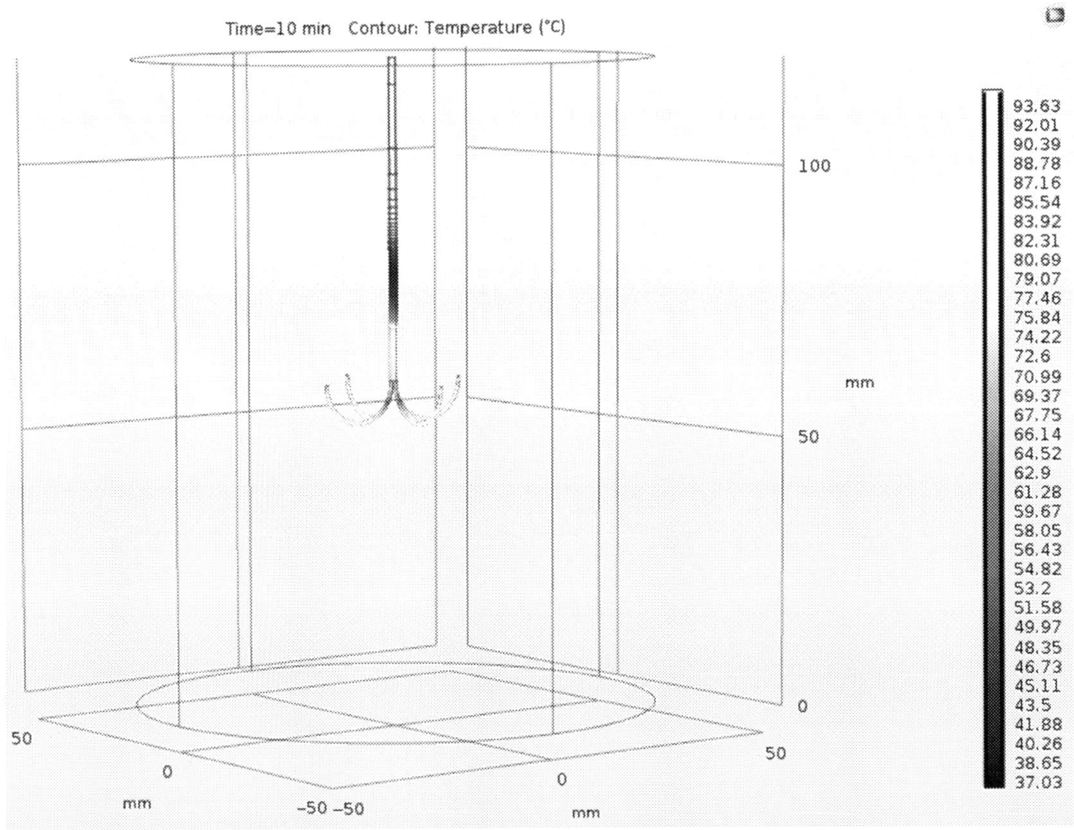

Figure 7.60 Geometric window of the temperature contour.

differs from the other in the start and end temperature values, as shown in Fig. 7.66.

The borderline is controlled by several parameters, such as the probe type, externally applied voltage, ablated time, adding other chemical materials, and cooling with the saline solution. Thus, the ablated region, which has a temperature greater than 45°C, is controlled. Therefore the level of damage at several points inside the geometric model is explained using the one-dimensional (1D) plot group and table graph, as shown in their setting window in Fig. 7.67.

Fig. 7.67 demonstrates that the level of damage with time is explained at three specified selected points inside the model, as shown in Fig. 7.68.

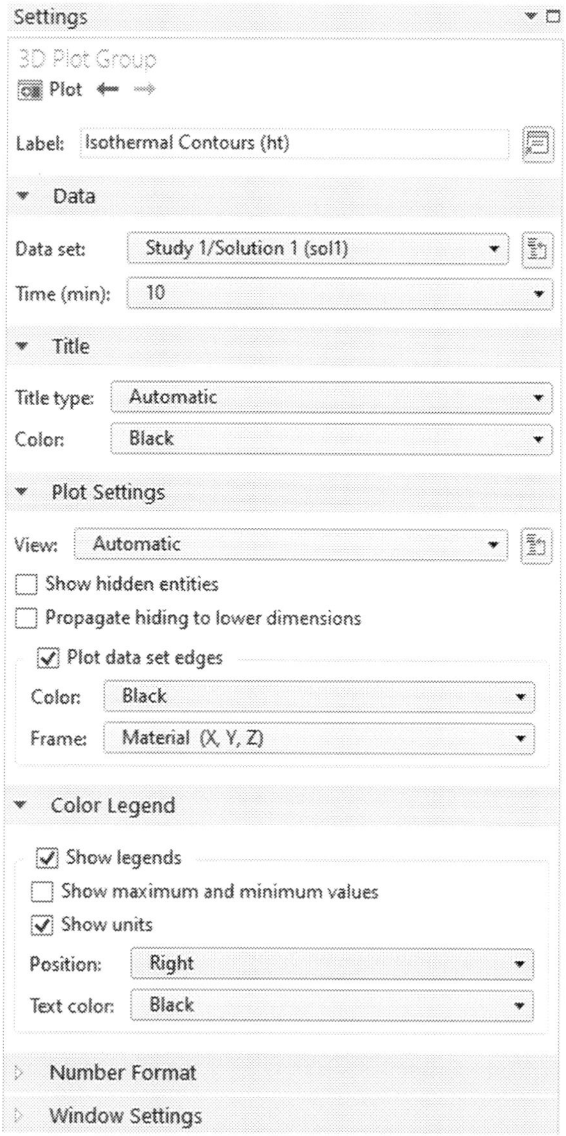

Figure 7.61 Setting window of the isothermal contours 3D plot for the temperature.

Furthermore, for example, at point (−0.2667,15.5,60), the temperature is plotted using the 1D plot group and table graph, as shown in their setting window in Fig. 7.69.

Fig. 7.69 illustrates the relationship between temperature and time at a determined point inside the model, as shown in Fig. 7.70.

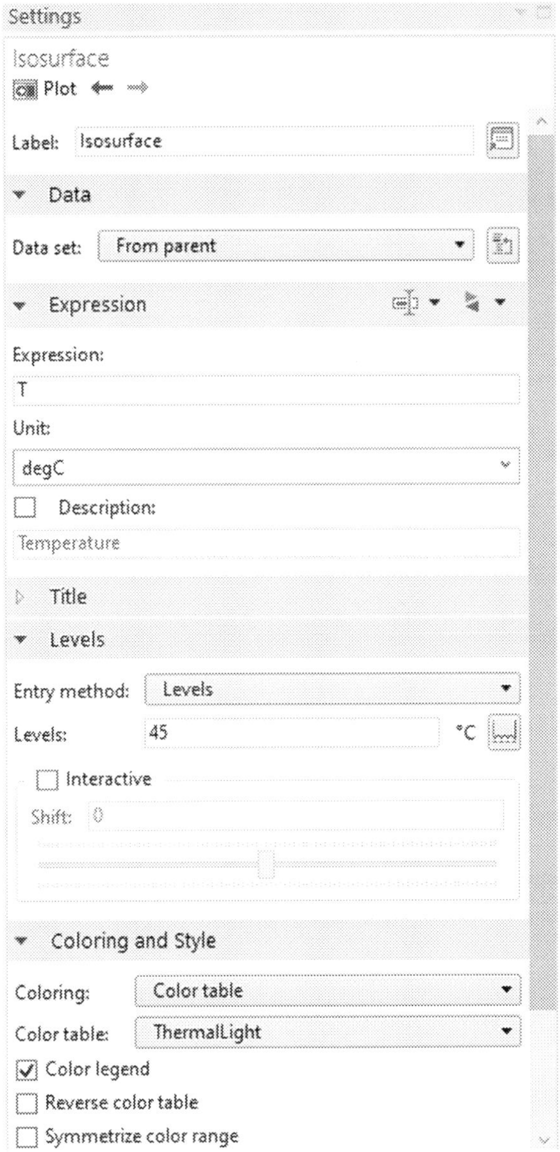

Figure 7.62 Setting window of the isosurface.

Fig. 7.70 displays the temperature at the spot point, which is considered the maximum value inside the geometric model. The damage profile is obtained using the setting window of a 3D plot group related to the damaged tissues, as shown in Fig. 7.71.

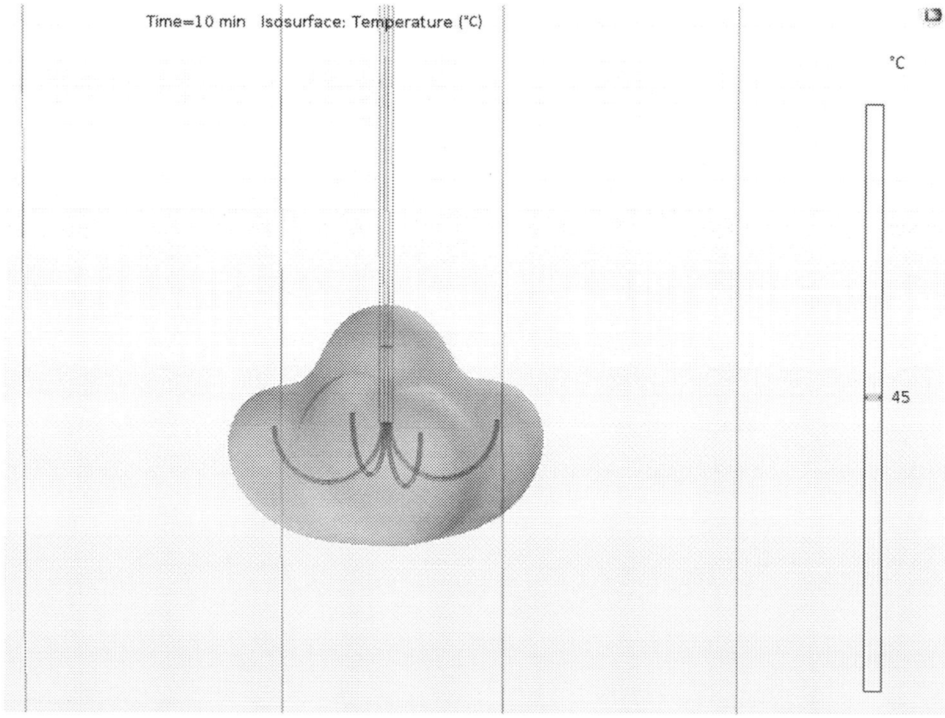

Figure 7.63 Geometric window of the isosurface at time 10 min.

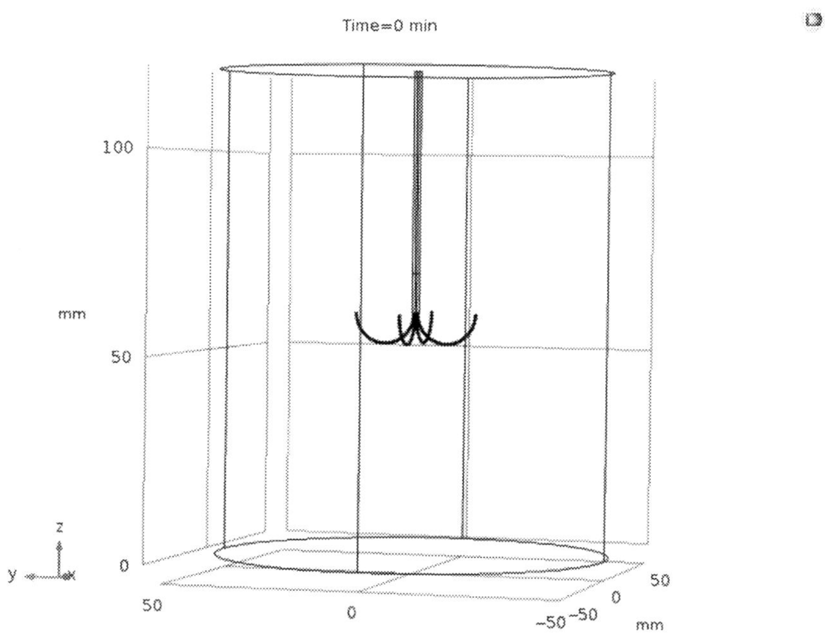

Figure 7.64 Graphic animation window for isothermal contours.

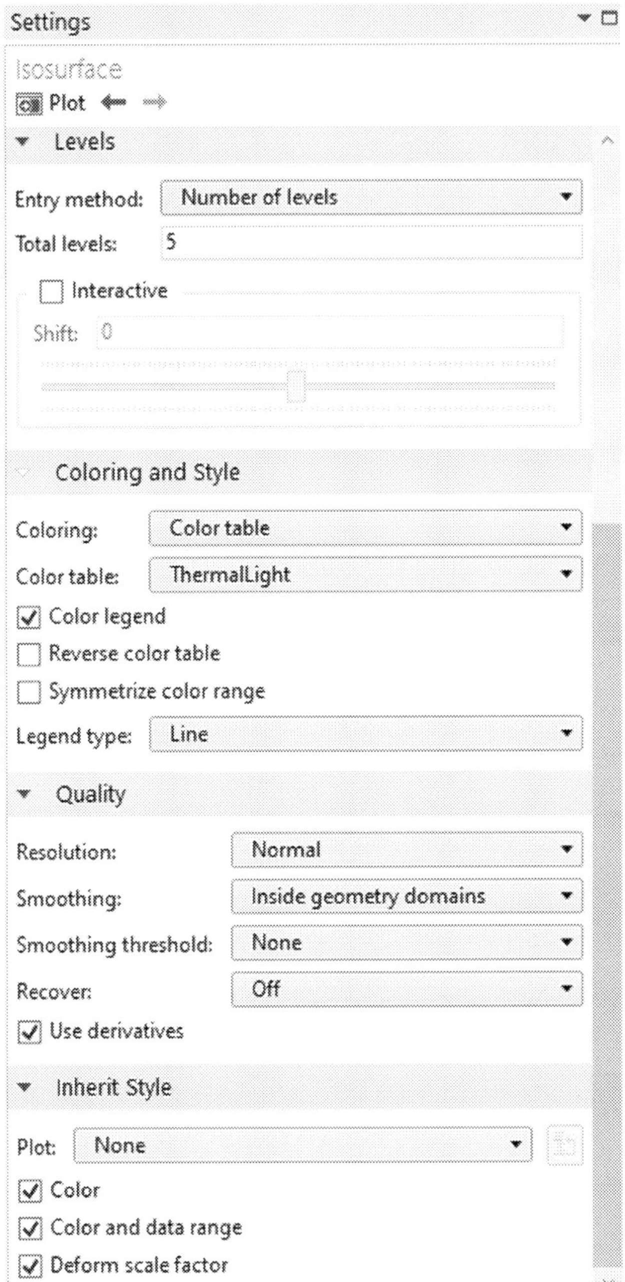

Figure 7.65 Setting window for a number of levels method for the isosurface.

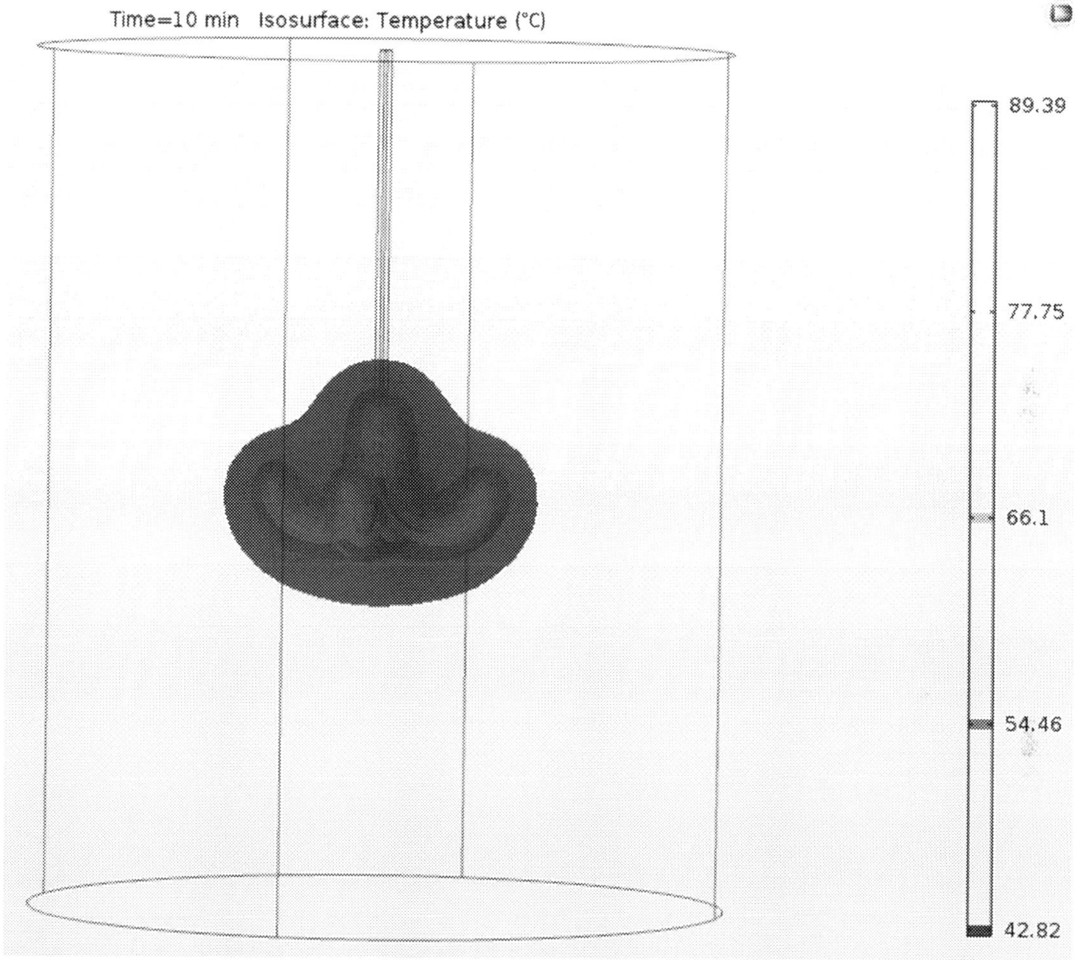

Figure 7.66 Graphic window of five levels for the isothermal profile.

Hence, the distribution of the level of damage is displayed using two slices at two different perpendicular planes through their setting windows, as shown in Fig. 7.72.

Fig. 7.72 displays the formatting for planes representing the damaged tissue profile to obtain a 3D shape, as shown in Fig. 7.73.

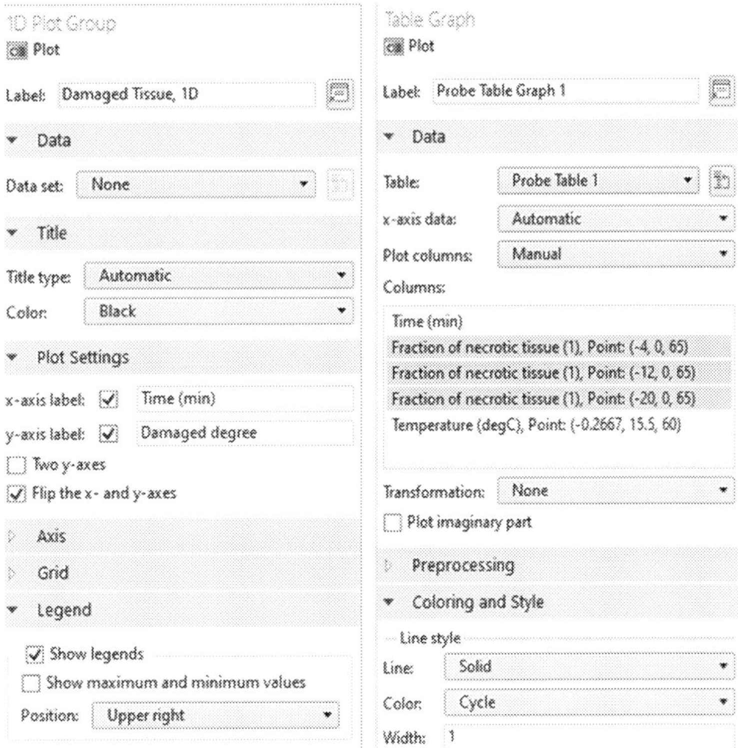

Figure 7.67 Setting windows for 1D plot group and table graph for the relationship between the level of damage and time at some determined points around the probe.

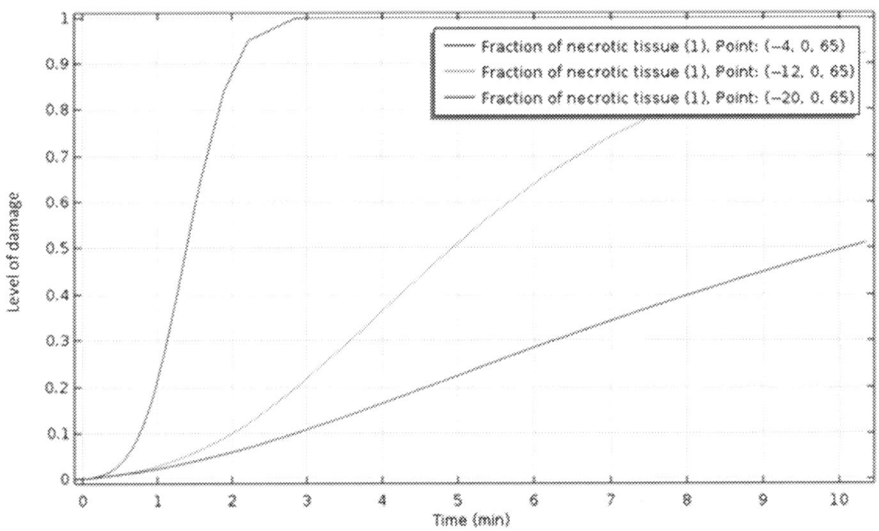

Figure 7.68 Relationship between the level of damage and time at three points around the probe.

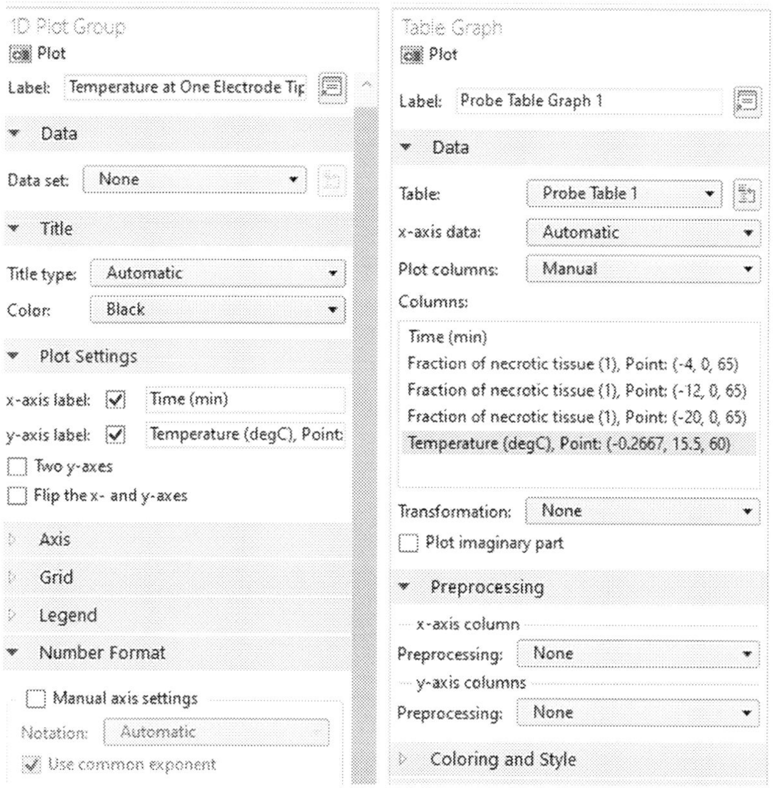

Figure 7.69 Setting windows of a 1D plot group and table graph for the relationship between the temperature and time at (−0.2667,15.5,60).

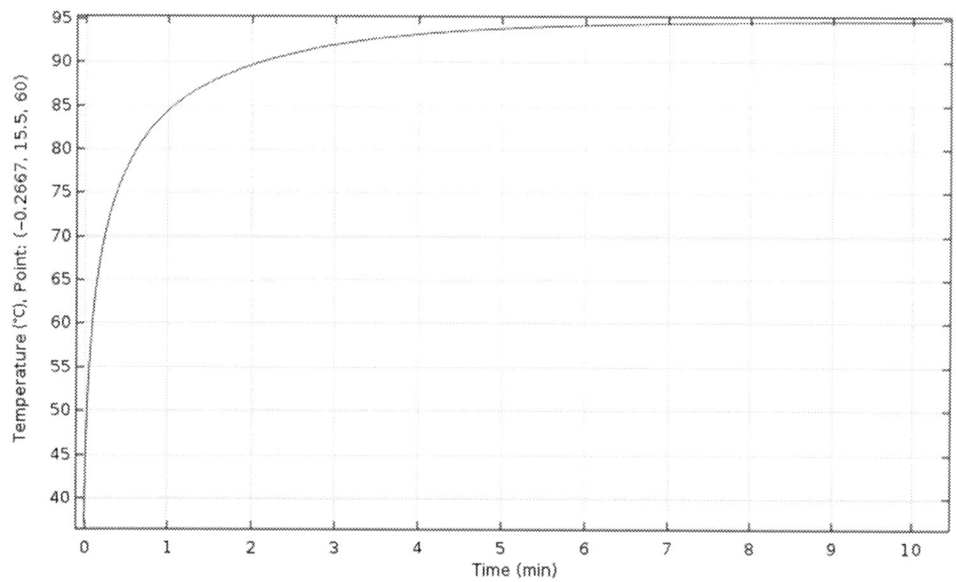

Figure 7.70 Relationship between the temperature and time at (−0.2667,15.5,60).

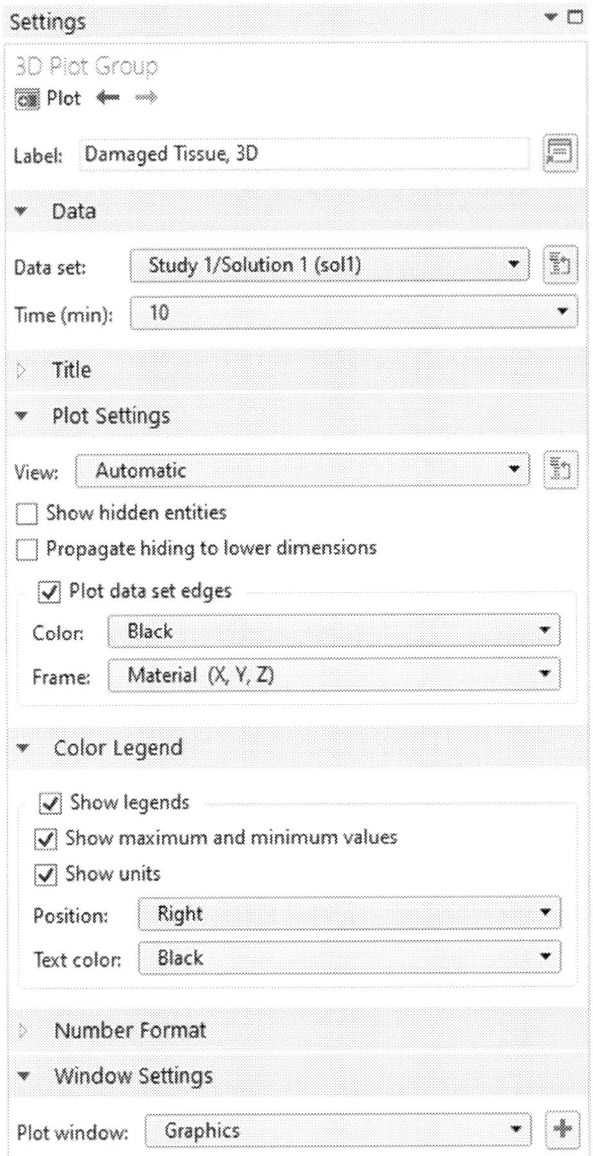

Figure 7.71 Setting window of the damaged tissue 3D plot group at time 10 min.

Fig. 7.73 demonstrates that the level of damage is at its maximum near the probe's active part and decreases with increasing distance from the probe. Moreover, its shape mainly follows the probe shape, that is, an umbrella shape using this multi-tine expandable probe, and it is greatly affected by time. Therefore the damaged tissue at the different ablating times is observed as shown in Fig. 7.74.

Fig. 7.74 illustrates the damaged necrotic region with the time change, which can be explained using the animation module, as shown in Fig. 7.75.

Fig. 7.75 shows the necrotic region increase with increasing ablation times, accordingly, the damaged volume can be calculated using the volume integration module, as shown in Fig. 7.76.

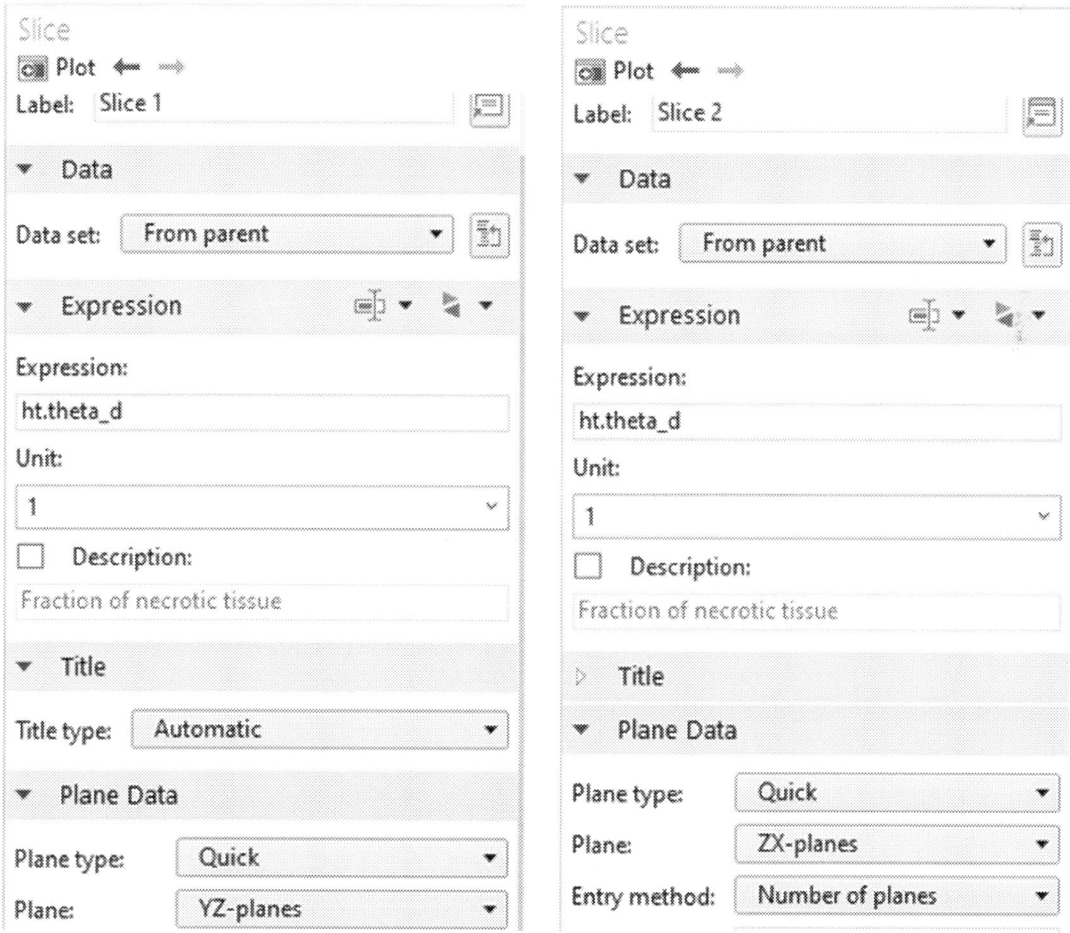

Figure 7.72 Setting windows of two slices related to the level of damage.

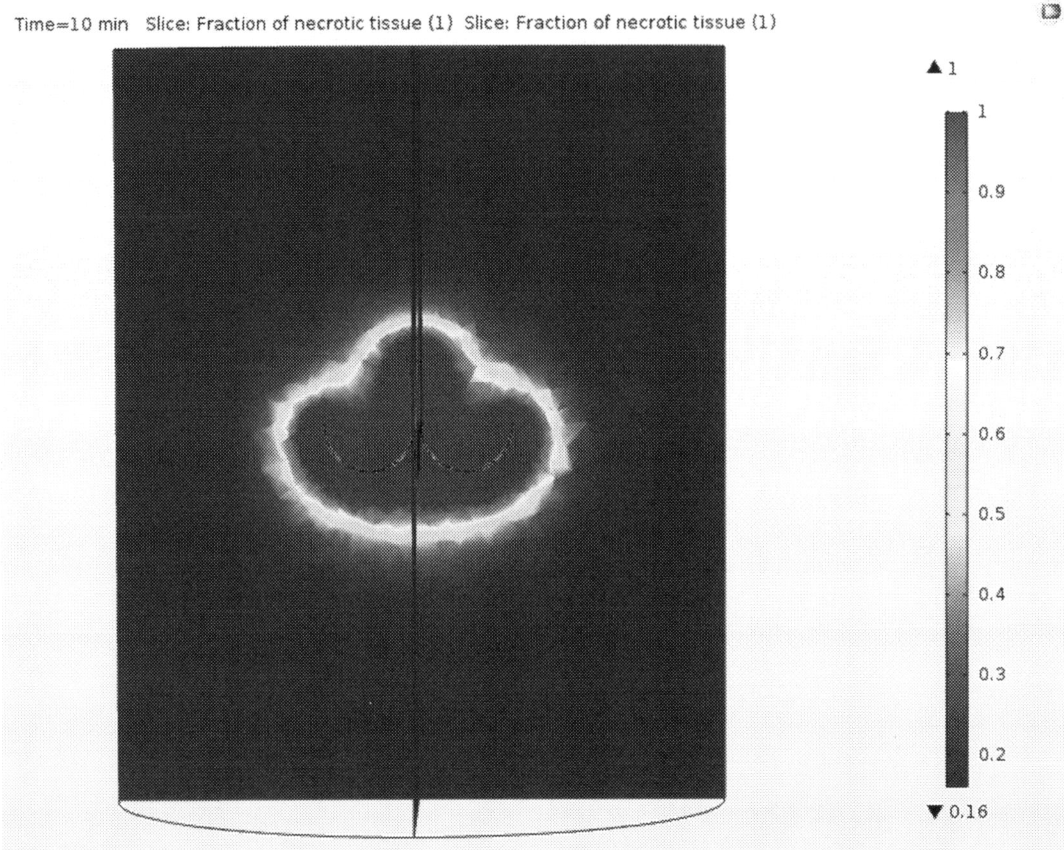

Figure 7.73 Graphic window for the damaged tissues at time 10 min.

Fig. 7.76 explains that the tumor is ablated when the level of degree at all points is greater than or equal to 1. Thus, the damaged volume is computed at several periods as shown in Fig. 7.77.

Fig. 7.77 reports the necrotic area inside the liver tissue region. Moreover, the volume can be calculated with other techniques using the threshold temperature as shown in Fig. 7.78.

Fig. 7.78 clarifies that the tumor is ablated when the temperature at all points inside the tumor is greater than or equal 43.7°C. Therefore the damaged area is calculated at several periods, as shown in Fig. 7.79.

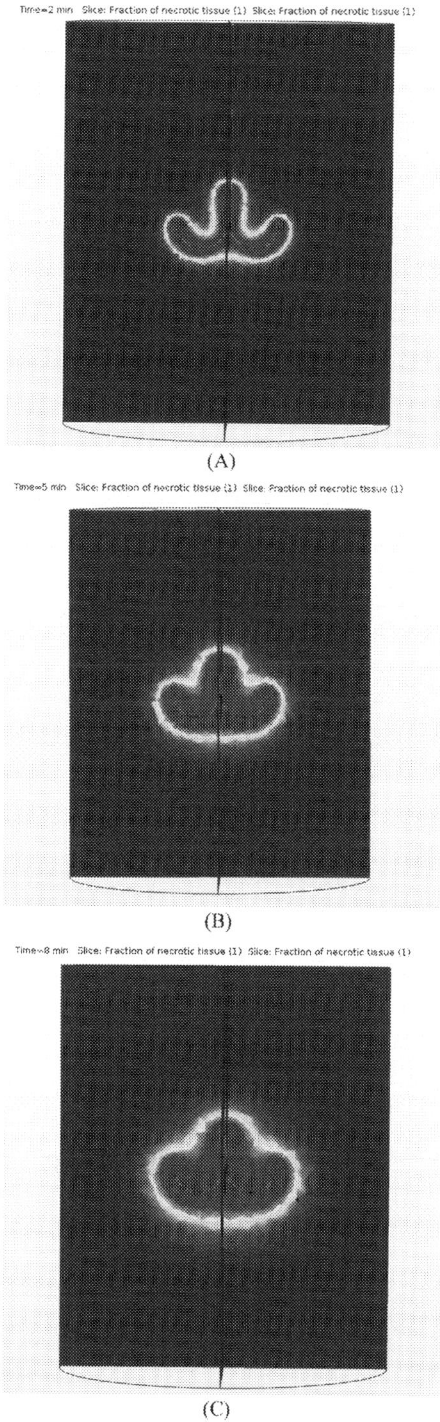

Figure 7.74 Graphic windows for the damaged tissues at different ablating times, including (A) 2 min, (B) 5 min, and (C) 8 min.

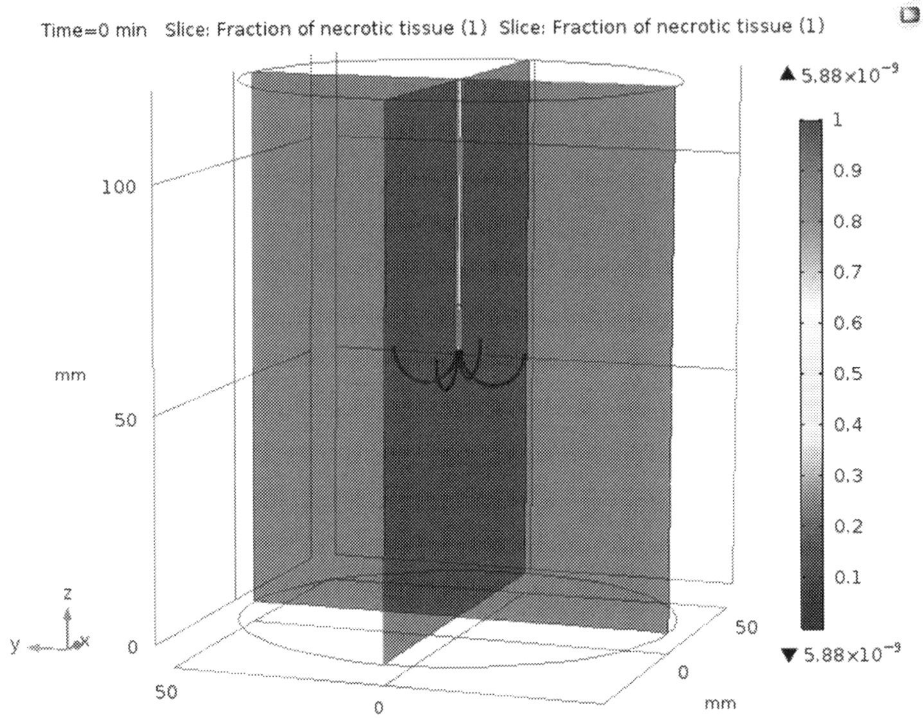

Figure 7.75 Damaged tissues with the time change.

Computer theoretical modeling of RFA offers numerous unbeatable advantages compared to experimental approaches. Consequently, it has become an indispensable tool to counterpart experimental studies on RFA practices. It is not only more cost effective and quicker than in vivo and ex vivo experiments, it also tolerates time evolution and spatial distribution for the analysis of the physical variables, which cannot be monitored during clinical trials. These advantages provide great support to designers and researchers in understanding RFA procedures, guiding the ablation procedure and providing guidance during the manufacturing of RFA systems. RFA modeling allows the evaluation of different issues, including designs of the electrodes, applicator settings, and the required power.

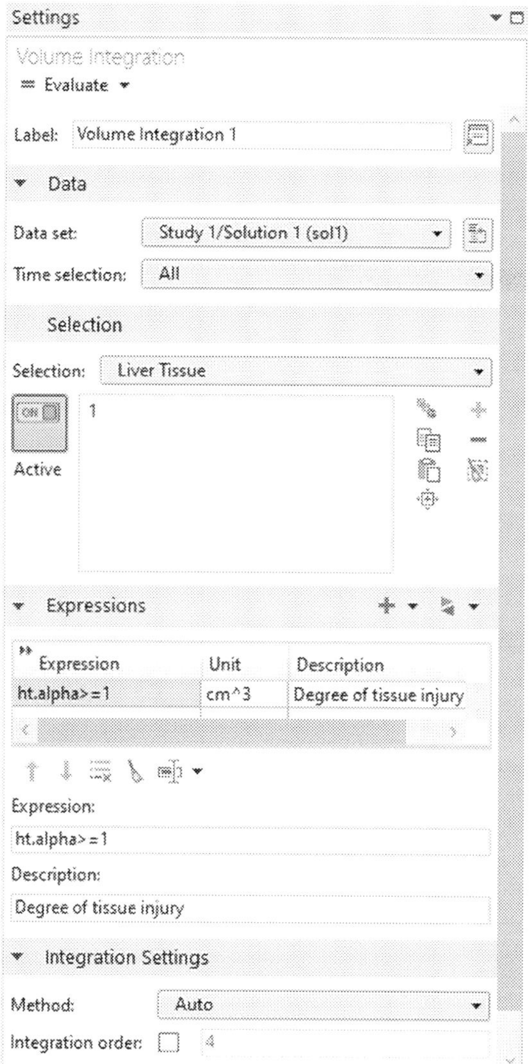

Figure 7.76 Setting window of the volume integration using the threshold level of damage.

Time (min)	Degree of tissue injury (cm^3)
0.0000	0.0000
1.0000	1.2836
2.0000	4.1807
3.0000	7.3886
4.0000	10.431
5.0000	13.024
6.0000	15.357
7.0000	17.445
8.0000	19.382
9.0000	21.218
10.000	22.953

Figure 7.77 Ablated volumes at different ablation times.

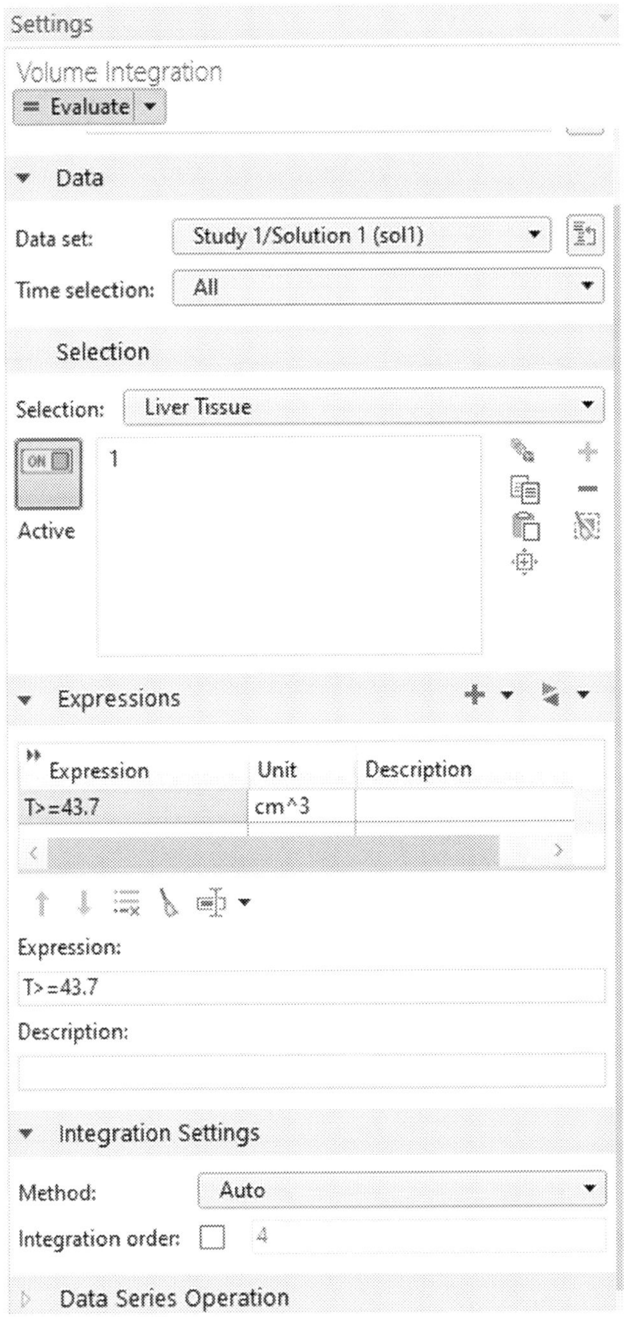

Figure 7.78 Setting window of the volume integration using the threshold temperature.

Time (min)	T≥316.7 (cm^3)
0.0000	0.0000
1.0000	6.8681
2.0000	13.022
3.0000	16.941
4.0000	19.343
5.0000	20.820
6.0000	21.749
7.0000	22.324
8.0000	22.654
9.0000	22.832
10.000	22.956

Figure 7.79 Ablated volumes at different ablation times at a temperature greater than or equal to 43.7°C.

7.4 Conclusions

Several models have been developed to predict the ablated tumor size during RF ablation, which is affected by a number of factors, such as the electrode insertion depth in the tumor, the organ to be ablated, location of the tumor, and temperature. To evaluate the effect of such parameters, and to design the most appropriate RF applicator for complete ablation of the tumor with minimal damage to the surrounding healthy tissues, simulation and theoretical models have become essential. Accordingly, 3D FE models are used to study the effect of RF ablation on tumorous and healthy tissues. The FE model typically includes the RF applicator and the specifications of the tumor characteristics, which differ from one organ to another, along with the required settings. This chapter highlights the main phases of simulation using COMSOL of the RFA procedure by reporting the steps one by one to be followed during the design of an RFA system. It also demonstrates the outputs and available results to be displayed showing the performance of the ablation process.

References

Bayer, J.D., Roney, C.H., Pashaei, A., Jais, P., Vigmond, E.J., 2016. Novel radiofrequency ablation strategies for terminating atrial fibrillation in the left atrium: a simulation study. Front. Physiol. 7, 108.

Chen, L., Sun, J., Yang, X., 2016. Radiofrequency ablation-combined multimodel therapies for hepatocellular carcinoma: current status. Cancer Lett. 370, 78–84.

Chen, R., Lu, F., Wu, F., Jiang, T.A., Xie, L., Kong, D., 2018. An analytical solution for temperature distributions in hepatic radiofrequency ablation incorporating the heat-sink effect of large vessels. Phys. Med. Biol. 63, 235026.

Chou, C., 2018. Therapeutic heating applications of radio frequency energy. Biol. Med. Asp. Electromagn. Fields, 413, 413–425.

Kath, N., Handels, H., Mastmeyer, A., 2019. Robust GPU-based virtual reality simulation of radio-frequency ablations for various needle geometries and locations. Int. J. Comput. Assist. Radiol. Surg. 14, 1825–1835.

Liu, Z., Lobo, S.M., Humphries, S., Horkan, C., Solazzo, S.A., Hines-Peralta, A.U., et al., 2005. Radiofrequency tumor ablation: insight into improved efficacy using computer modeling. Am. J. Roentgenol. 184, 1347–1352.

Ooi, E.H., Lee, K.W., Yap, S., Khattab, M.A., Liao, I.Y., Ooi, E.T., et al., 2019. The effects of electrical and thermal boundary condition on the simulation of radiofrequency ablation of liver cancer for tumours located near to the liver boundary. Comput. Biol. Med. 106, 12–23.

Rustagi, T., Chhoda, A., 2017. Endoscopic radiofrequency ablation of the pancreas. Dig. Dis. Sci. 62, 843–850.

Shao, Y., Leo, H., Chua, K., 2017. Studying the thermal performance of a bipolar radiofrequency ablation with an improved electrode matrix system: in vitro experiments and modelling. Appl. Therm. Eng. 116, 623–635.

Singh, S., Melnik, R., 2020. Thermal ablation of biological tissues in disease treatment: a review of computational models and future directions. Electromagn. Biol. Med. 1–40.

Singh, S., Repaka, R., 2017. Temperature-controlled radiofrequency ablation of different tissues using two-compartment models. Int. J. Hyperth. 33, 122–134.

Song, T.J., Seo, D.W., Lakhtakia, S., Reddy, N., Oh, D.W., Park, D.H., et al., 2016. Initial experience of EUS-guided radiofrequency ablation of unresectable pancreatic cancer. Gastrointest. Endosc. 83, 440–443.

Voglreiter, P., Mariappan, P., Pollari, M., Flanagan, R., Sequeiros, R.B., Portugaller, R.H., et al., 2018. RFA guardian: comprehensive simulation of radiofrequency ablation treatment of liver tumors. Sci. Rep. 8, 1–13.

Zhang, B., Moser, M.A., Zhang, E.M., Luo, Y., Zhang, W., 2017. A new approach to feedback control of radiofrequency ablation systems for large coagulation zones. Int. J. Hyperth. 33, 367–377.

8

Simulation-driven modeling of a microwave ablation system

Chapter outline

8.1 Introduction

Microwave ablation (MWA) is recommended in several cases, such as ablating tissues of high resistance, including bone and lung tissue, and eradicating large tumors in different organs with a high success rate without surgical incision (Lorentzen et al., 2011; Clark et al., 2007). In contrast to radiofrequency ablation (RFA), MWA is independent of the electrical current, leading to no ground-pad being required in the MWA system and no effect from tissue resistance. During the ablation process, electromagnetic (EM) waves are used to elevate the temperature in the tissues close to the microwave (MW) probe based on fluctuating the polar molecules and causing tissue necrosis (Simon et al., 2005; Carrafiello et al., 2008; Jones et al., 2011). Although the MWA has several advantages compared to other methods, some disadvantages limit its performance, including backward heating, the irregular shape of radiation, and the high temperature which leads to damaging large zones in the healthy tissues nearby the tumor. Accordingly, simulation studies are recommended to determine the different parameters' values for designing efficient MWA system. In addition, MWA has several procedures and MW probe designs, leading to the necessity for instigating finite element (FE) models to simulate the different ablation approaches for studying and analyzing their effects, essential settings of the used probe's design/parameters, and temperature distribution during the ablation process. With finite element method (FEM), modeling has a great impact in reducing the error rate before performing any

Thermal Ablation Therapy. DOI: https://doi.org/10.1016/B978-0-12-819544-4.00004-6
© 2021 Elsevier Inc. All rights reserved.

clinical trials. The FE models offer mathematical results and allow accurate comparisons between the different ablation procedures.

An FE method can be applied to study different effects of any proposed design, such as:

- The required maximum temperature;
- The temperature effect in the MWA therapy;
- The required power for complete tumor ablation;
- The tumor size;
- The best MW frequency;
- The required initial temperature and its change over time (temperature distribution);
- The relation between the temperature and the rate of tissue destruction;
- The relation between the probe design and heat transfer; and
- The probe design's effect on the features of the tumor in different biological tissues.

During the simulation, the process using COMSOL, the EM wave equations, and the bioheat transfer equations are applied while varying the ablation duration, and also the frequency for studying the effect of using different MW frequencies on the temperature of tissues. Furthermore, the tissue damage equation is applied to determine the degree of tissue destruction in the tissues surrounding the microwave probe.

Several simulation studies were carried out to show the effect of using MWA, for example, the MWA energy effect on tissue heating was explored by Meloni et al. (2017). Another study was conducted to study the larger diameter to be ablated by the MWA system (Veltri et al., 2015; Thamtorawat et al., 2016; Ibitoye et al., 2018). Other studies have been conducted to compare the use of RFA and MWA for treatment of different organs (Brace, 2009; Poulou et al., 2015; Izzo et al., 2019). The impact of the MW frequency on the MWA process has been studied (Sawicki et al., 2017; Jiang et al., 2017; Curto et al., 2015; Wu et al., 2016). In addition, the effect of MWA on heat transfer and blood flow was studied (Rattanadecho and Keangin, 2013).

8.2 COMSOL-based finite element method to simulate MWA: a case study

To solve engineering problems using modeling, such as simulating physical phenomena, the FEM is considered the most commonly applied method. For MWA, the FEM splits the system into simple and small parts, called FEs to provide simultaneous calculation, an inclusive set of results, and a visual representation of several parameters including damaged tissues, power, and temperature profile. It also allows analysis of the probe design with the ability to adjust its design for comparative studies to reach the optimal design. Using the FEM provides fast calculations, and saves time/cost before clinical trials.

Accordingly, a symmetric 2D model is presented in this chapter to design the whole MWA system structure, including the probe design, tumor, and/or healthy tissues close to the probe tip to simulate the specific human organ. Furthermore, the governing equations of the EM waves and the basic bioheat transfer are used in the time/frequency domain in the FE model based on the EM heat source. This chapter presents a case study to simulate the MWA of hepatic tumor tissues using COMSOL Multiphysics. Typically, several microwave probe designs can be used, such as a monopole, dipole, single slot, and multislot. The appropriate design is chosen according to the patient's condition and the tumor site in relation to the adjacent bile ducts, blood vessels, or tumor shape/size. Consequently, a 2D FE model is simulated to model liver tumor tissues using their properties, where the single-slot microwave probe is used in this case study.

There are several types of microwave probes, including single slot, a single slot with 1T ring, multislot, monopole, dipole, triaxial, floating sleeve, choked, cap choke, double slot cap choke, and helix. These probe designs are different in the design, generated radiation pattern/shape, and components/composition, leading to different performances during the MWA process. Fig. 8.1 illustrates the flow of modeling of any MWA system.

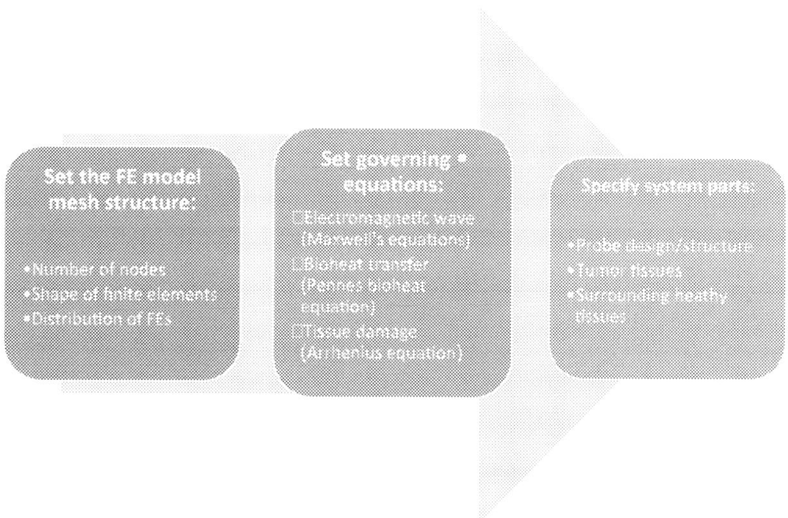

Figure 8.1 The main sequence during the simulation process using COMSOL.

In this chapter, the FEM model is used to simulate the tumor of liver tissues. The number of elements in the FE model of different shapes and dimensions can be used to find the optimal solution. Typically, the element shape and dimensions in the mesh are based on the size of the parts in the simulation model. The more complex the probe shape, the more FEs required. A large number of FEs are used to simulate small parts in the model, such as the probe tip. Thus, the design of the mesh is denser near the probe to guarantee accurate calculations of the necrotic tissues, temperature, electric field, and energy absorption. However, a large number of elements leads to increased computational complexity and calculation time. In addition, the boundary conditions are defined in the FE model, which is the core temperature of 37°C of the tissues.

In this chapter, different types of probes are designed to be inserted into the tumor tissues of the liver for ablation. To implement the proposed model in COMSOL, the model started by defining the parameters within the used equations, including the EM wave and bioheat transfer equations in the MW ablation in the current case study, and their values according to the type of tissue, as shown in Fig. 8.2.

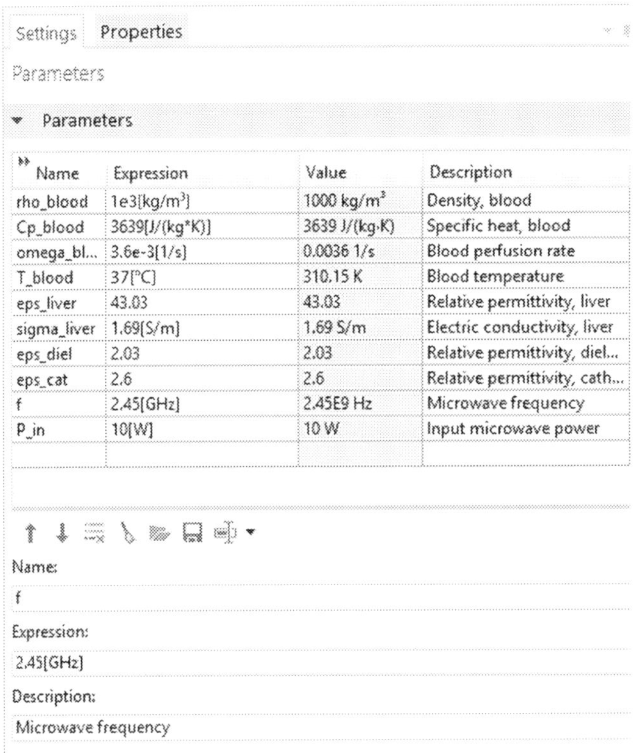

Figure 8.2 Main parameter setting window.

Fig. 8.2 shows the basic parameters, such as the microwave frequency, the required ablation time sufficient for complete ablation, the amount of power used, the initial conditions, and some values of materials/tissue parameters to facilitate later handling. These parameters can be modified using this parameter setting window. Henceforth, the model builder for the COMSOL file includes the material, geometry, mesh, and results as shown in Fig. 8.3.

Fig. 8.3 shows the general view of the simulated COMSOL file, where the geometry can be constructed based on the real model, including a rectangle to represent the liver tumor tissue with dimensions of 30 mm width with 80 mm height, position, and orientation using the setup window of the rectangle as shown in Fig. 8.4.

In addition, the geometry section contains other rectangles, and a triangle representing the different parts of the single-slot probe, and their dimensions, position, and rotation are adjusted using adjustment windows, as shown in Figs. 8.5–8.9.

Figs. 8.4–8.9 show that all the geometry components are built to produce the geometric model as shown in the graphical window in Fig. 8.10.

Fig. 8.10 represents a single-slot probe in tumor tissue. To simulate a monopole probe with a metal tip the modeling steps are illustrated in Figs. 8.11–8.16.

Another modeling example introducing the composition of a choked coaxial probe is illustrated in Figs. 8.17–8.19.

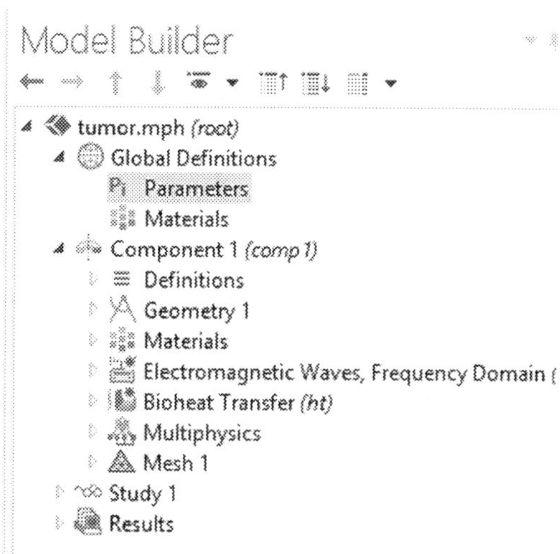

Figure 8.3 Model builder in COMSOL Multiphysics.

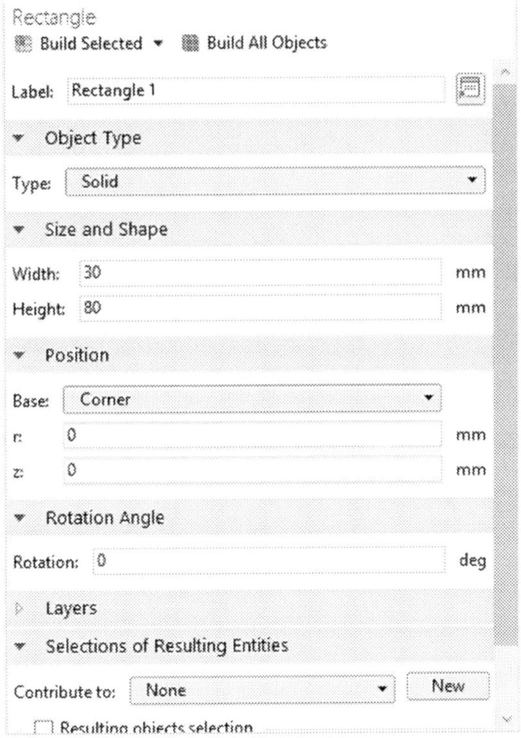

Figure 8.4 The setting window of rectangular liver tumor tissues.

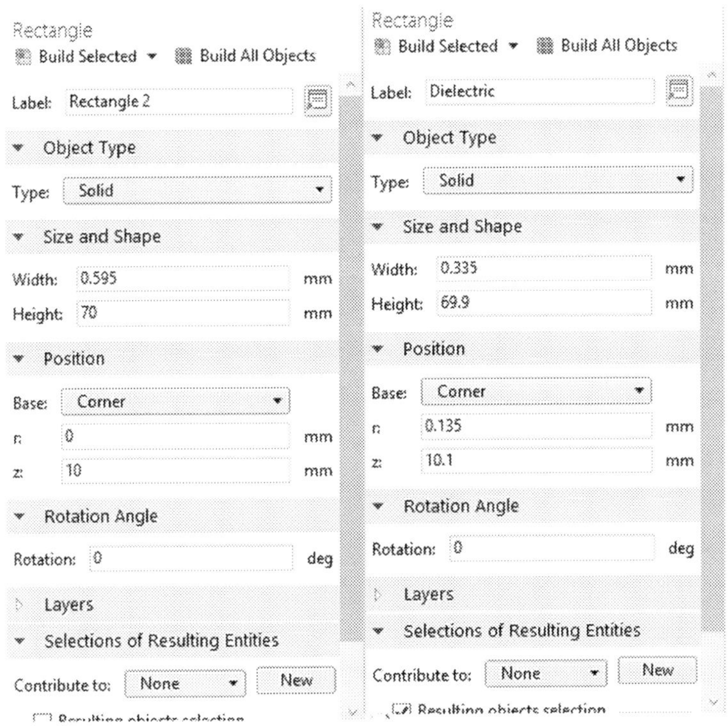

Figure 8.5 Setting windows of the geometry representing the dielectric part of the probe.

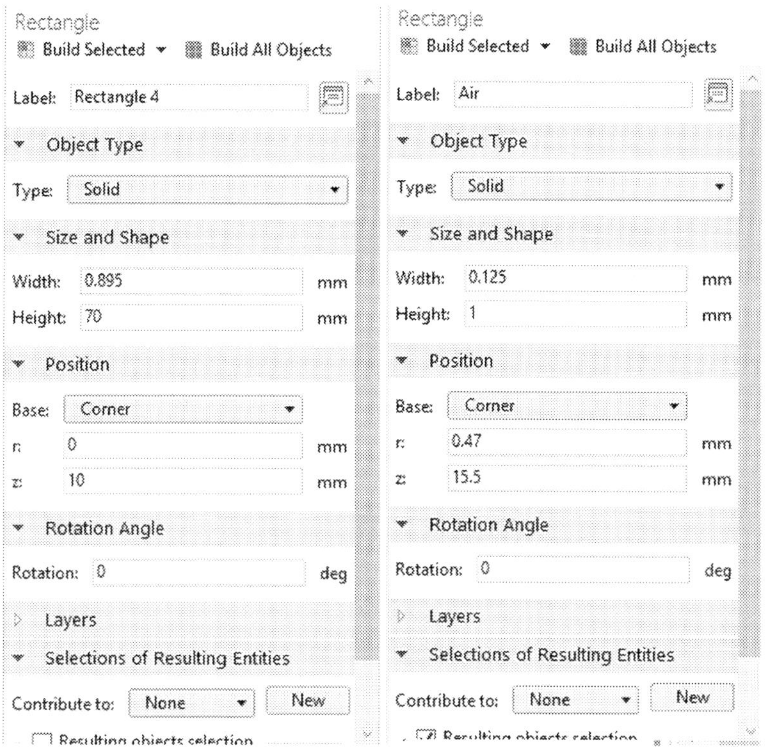

Figure 8.6 Setting windows of the geometry representing the catheter and air parts of the probe.

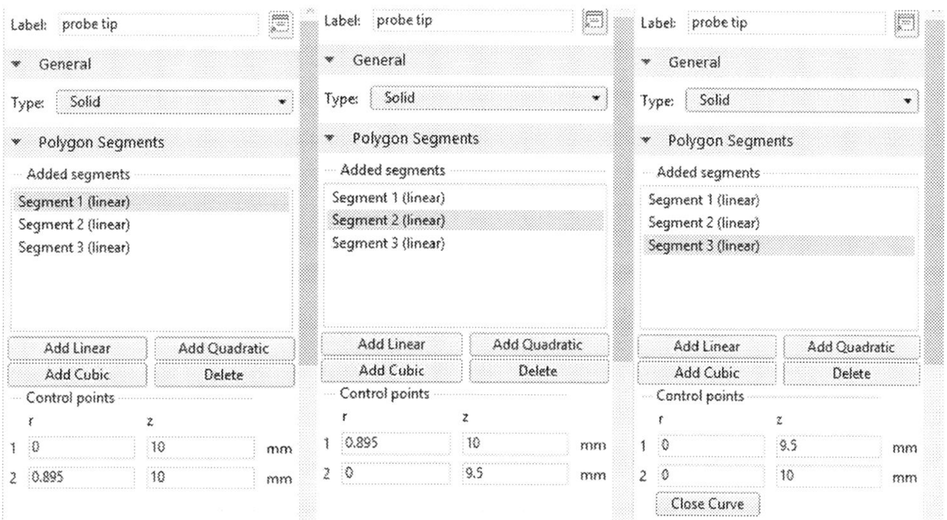

Figure 8.7 Setting windows of the geometry representing the tip of the probe.

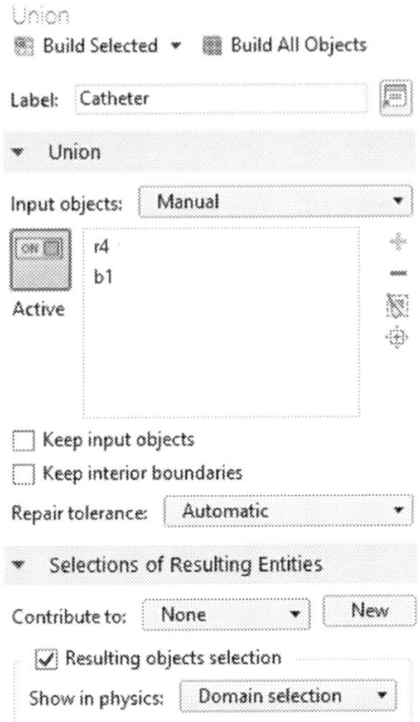

Figure 8.8 Setting windows of the geometry representing the union of two parts to obtain the catheter part of a probe.

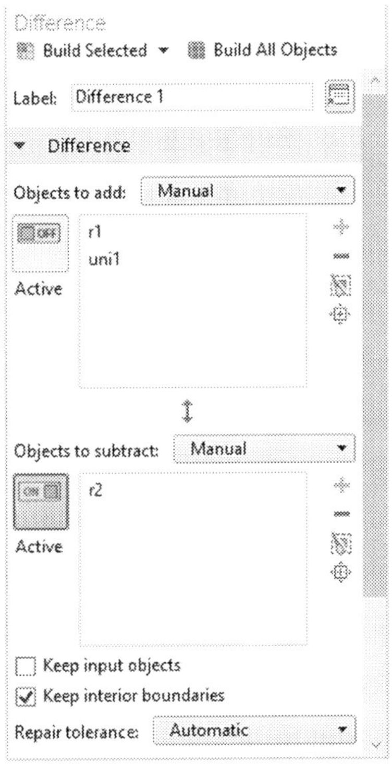

Figure 8.9 Setting windows of the geometry representing the difference of some parts to obtain the probe.

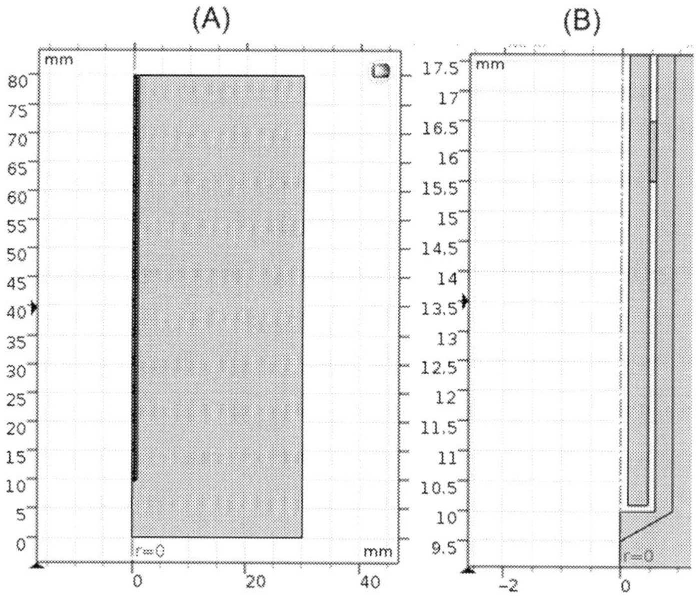

Figure 8.10 A graphic window of the geometric model, including (A) liver tumor tissues and the single-slot probe, and (B) near view of the probe.

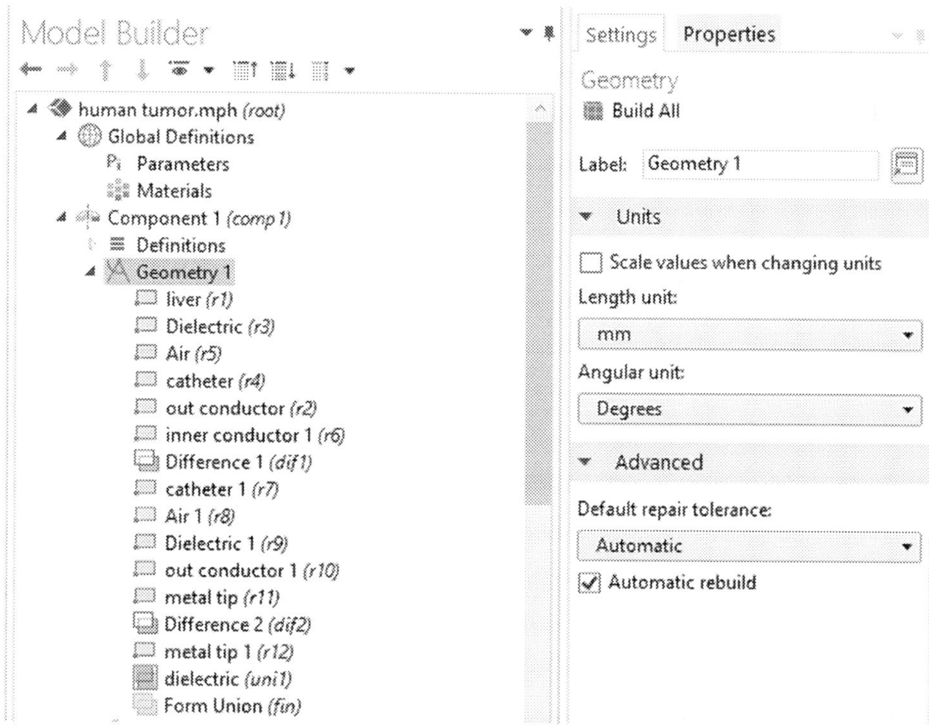

Figure 8.11 The geometry components, dimensions, and angles for the monopole probe.

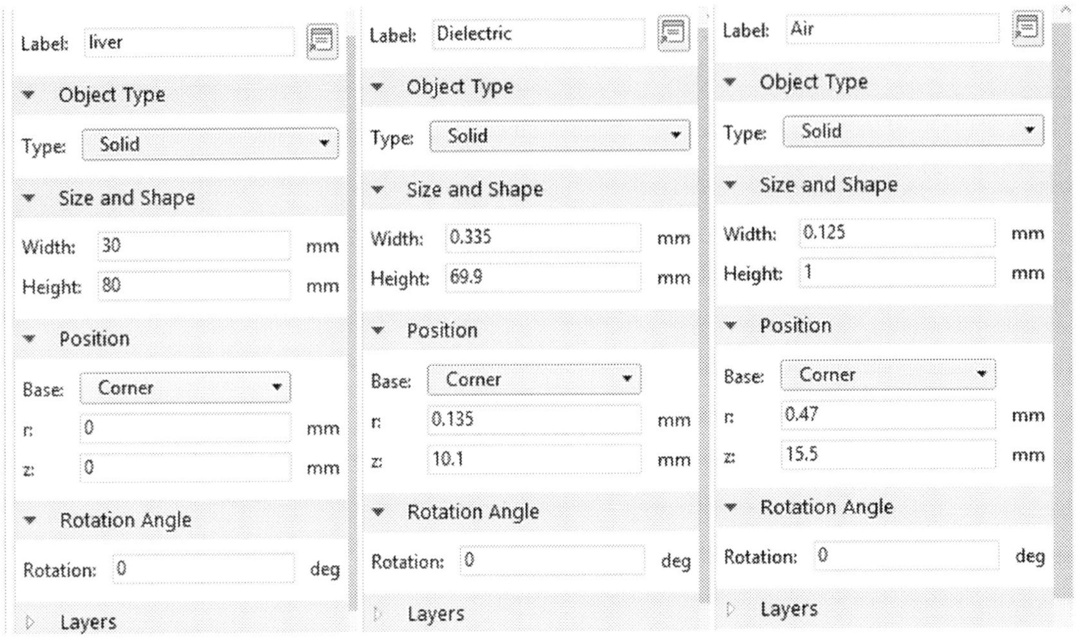

Figure 8.12 Setting windows of the geometry representing the tumor liver, the dielectric, and the air parts of the probe.

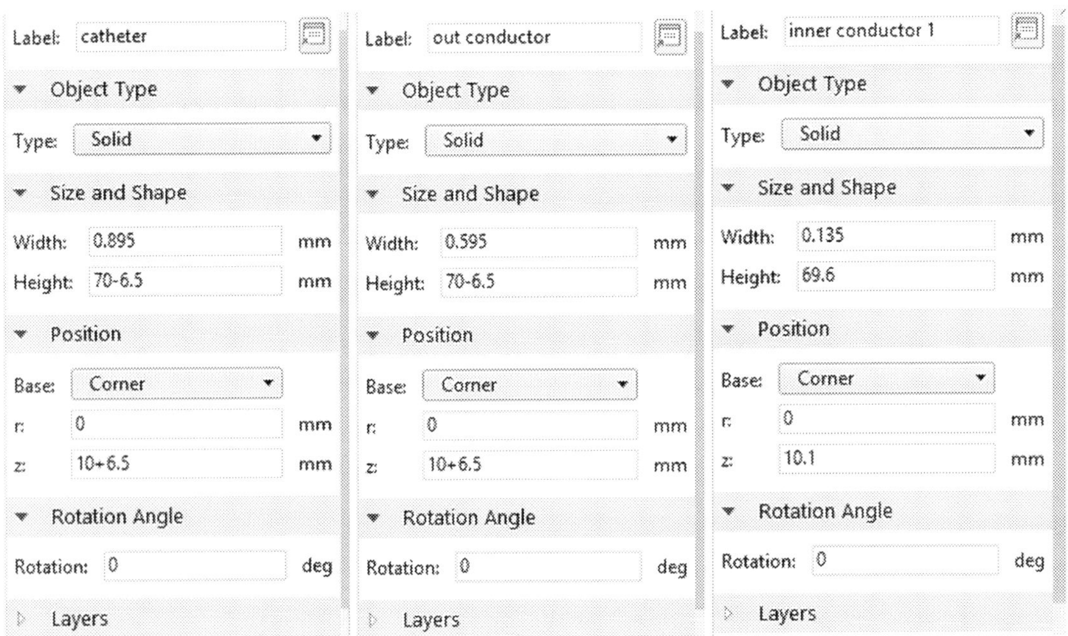

Figure 8.13 Setting windows of the geometry representing the catheter, the outer conductor, and the inner conductor parts of the probe.

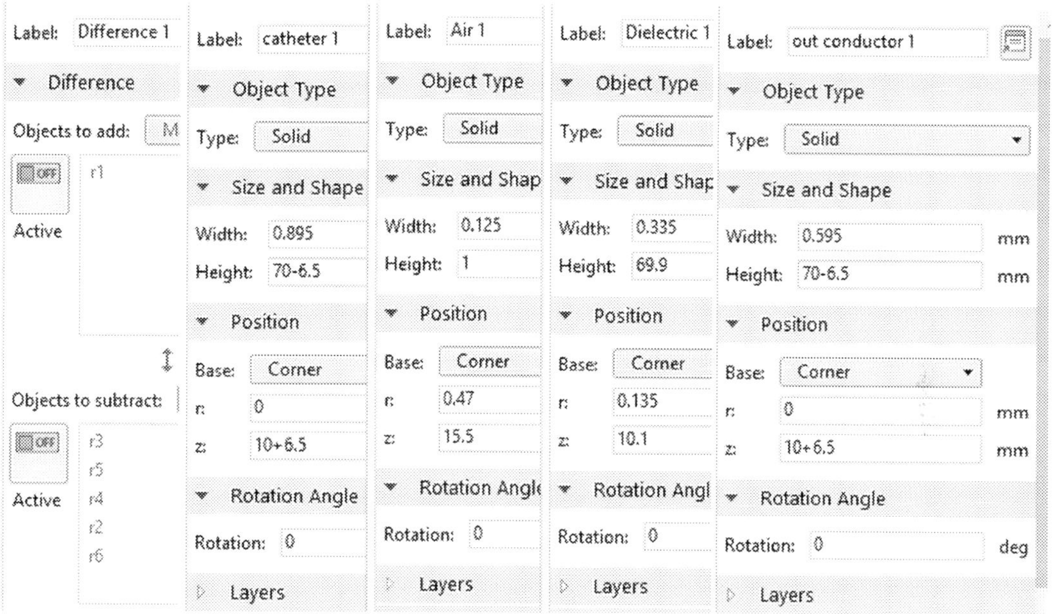

Figure 8.14 Setting windows of the geometry representing some parts of the monopole probe.

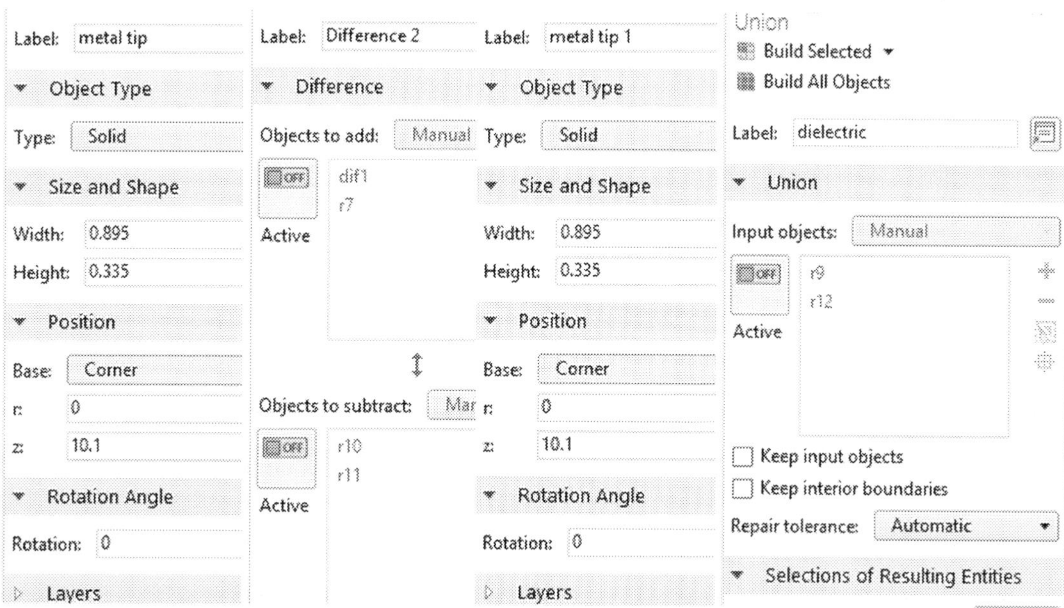

Figure 8.15 Setting windows of the geometry representing some parts of the monopole probe.

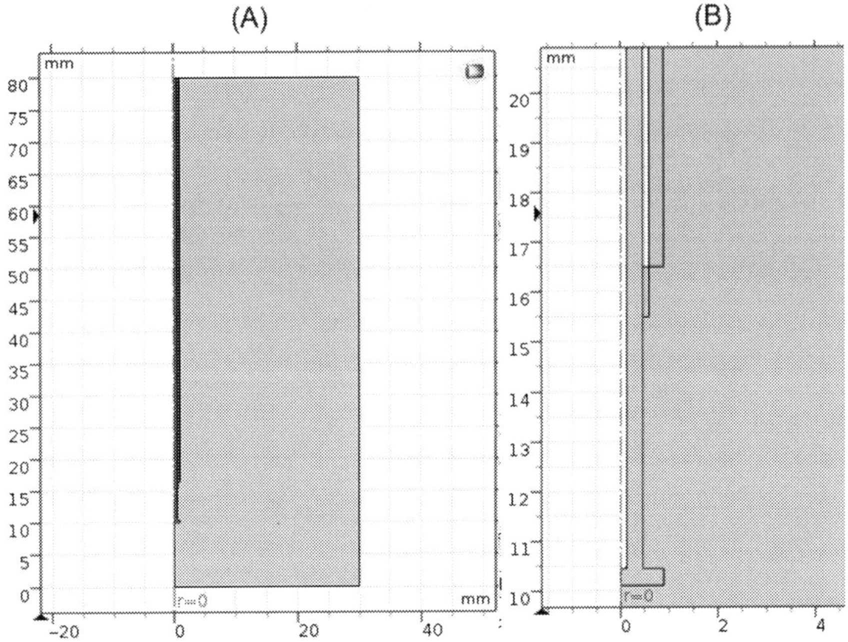

Figure 8.16 A graphic window of the geometric model, including (A) liver tumor tissue and the monopole probe with a metal tip, and (B) near view of the probe.

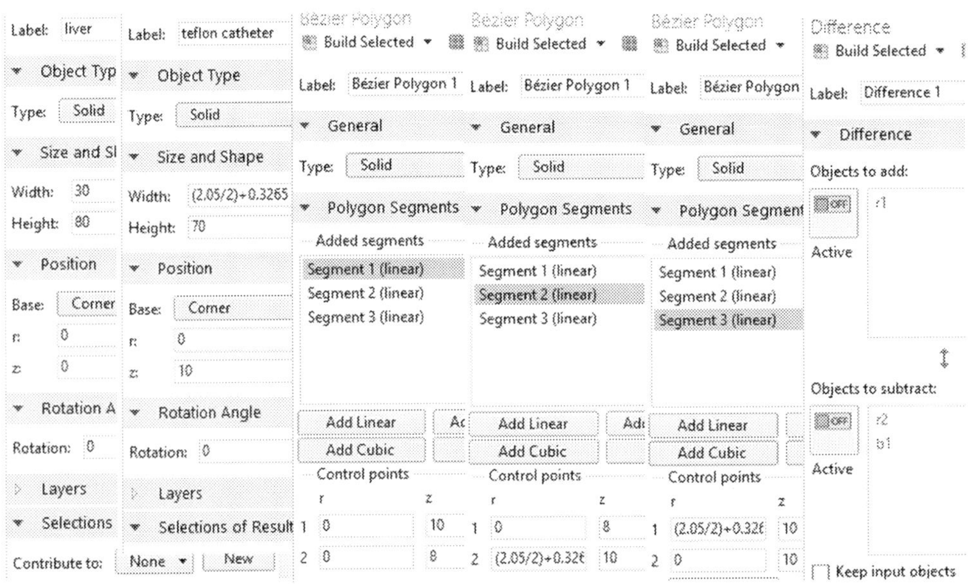

Figure 8.17 Setting windows of the geometry representing some parts of the chocked probe ablation system, including describing the liver, teflon catheter, and the required mesh design.

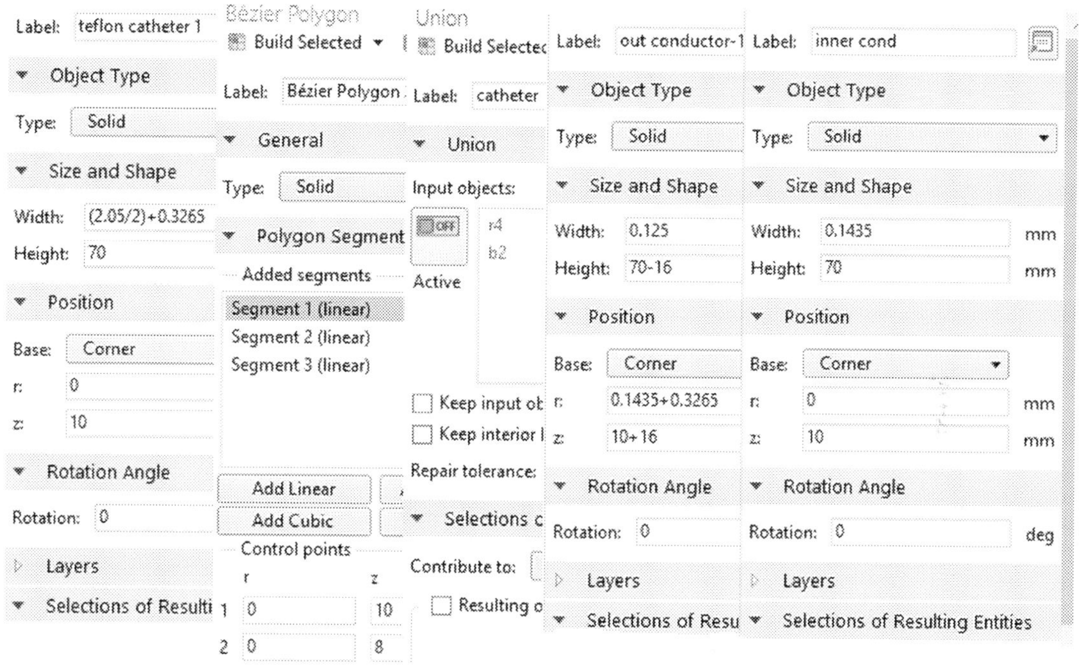

Figure 8.18 Setting windows of the geometry representing some parts of the chocked probe ablation system, including the catheter, the inner conductor, and the outer conductor.

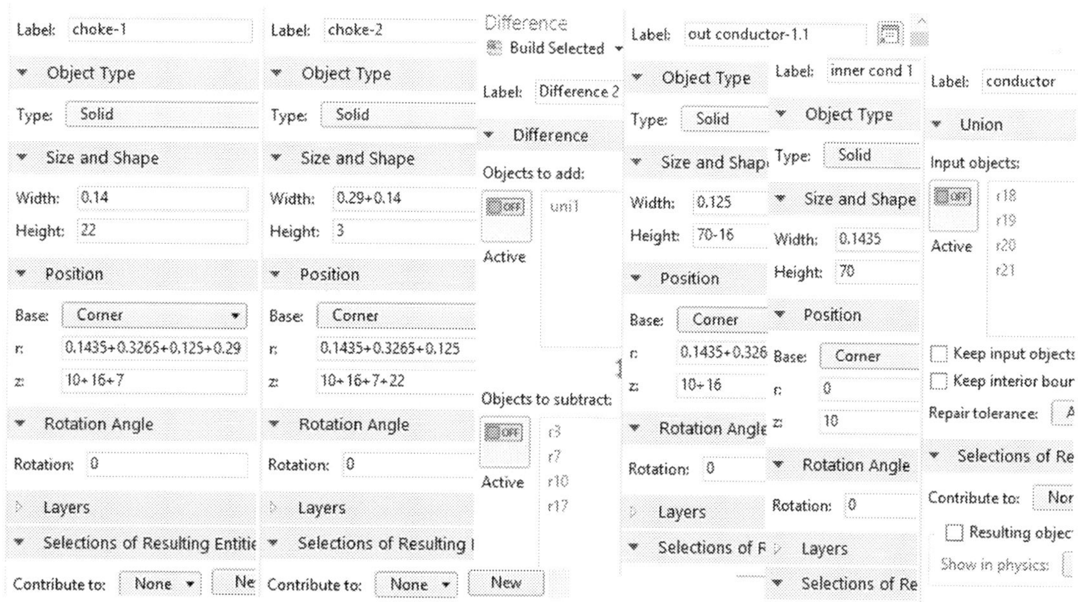

Figure 8.19 Setting windows of the geometry representing some parts of the chocked probe ablation system, including the different parts in the choke probe.

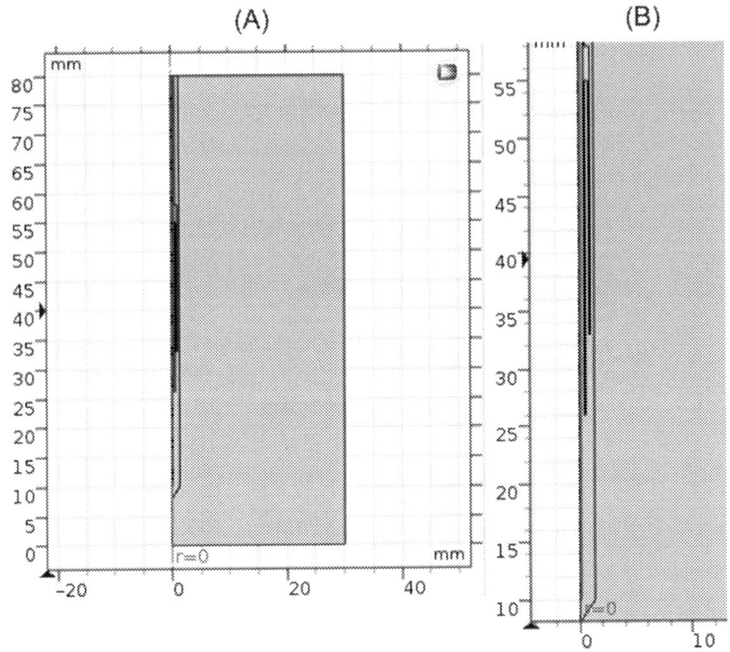

Figure 8.20 A graphic window of the geometric model, including (A) liver tumor tissue and the chocked probe, and (B) near view of the probe.

Fig. 8.20 illustrates the mesh of the modeled chocked probe.

Additionally, all parts within the geometrical model area are specified by adding the materials characteristics of each part, such as the tumor tissue section of the liver with their properties at the used microwave frequency of 2.45 GHz, as illustrated in Fig. 8.21.

Fig. 8.21 shows the density, heat capacity, and thermal and electrical conductivity related to liver tumor tissues as well as the activation energy, which expresses the amount of energy to activate molecules in one mole of material, and the frequency factor describing the collision times of molecules in 1 second. The material of the outer body of the probe (the catheter) is also determined using a good conducting material (teflon) having its properties as shown in Fig. 8.22.

Fig. 8.22 describes the specific material for the outer part of the probe. The dielectric part uses the insulating material of polytetrafluoroethylene material with its properties as shown in Fig. 8.23. Also, the material setting window to specify the air with its properties is shown in Fig. 8.24.

The physics are added based on the simulation study, for example, EM waves and bioheat transfer for the MWA. Hereafter, the EM wave physics is responsible for producing EM waves in the MW frequency range as shown in Fig. 8.25.

At the beginning of the ablation procedure, the electric field initial value is set at all domains as shown in Fig. 8.26.

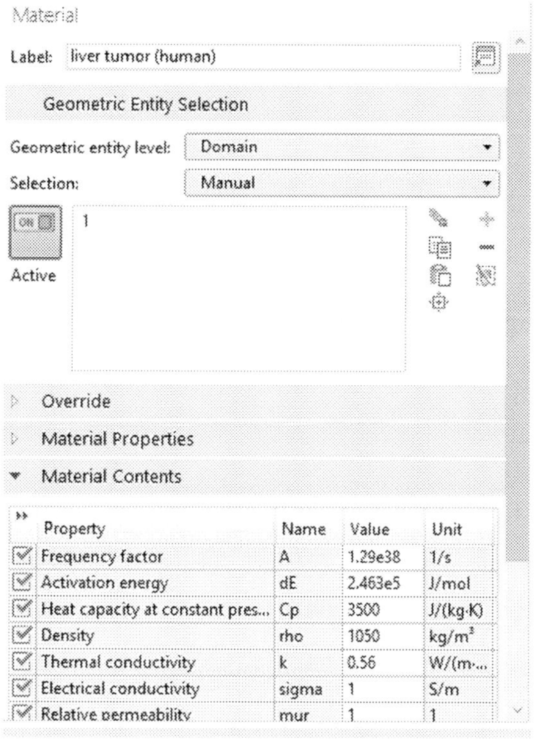

Figure 8.21 A material setting window of the liver tumor tissue with its properties.

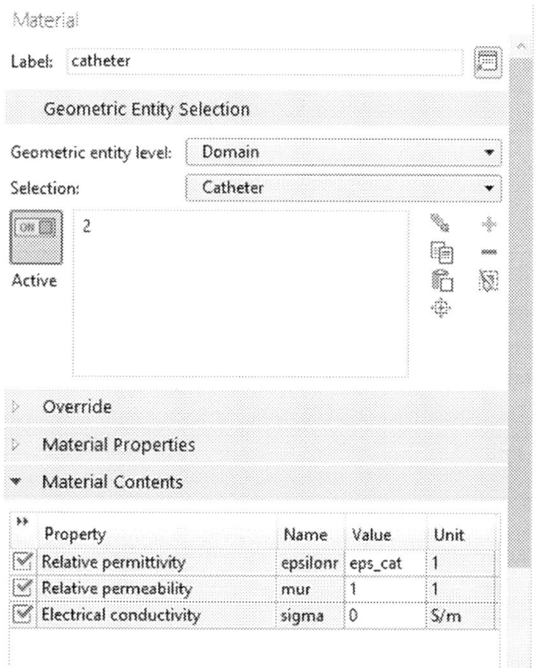

Figure 8.22 A material setting window of the teflon catheter material with its properties.

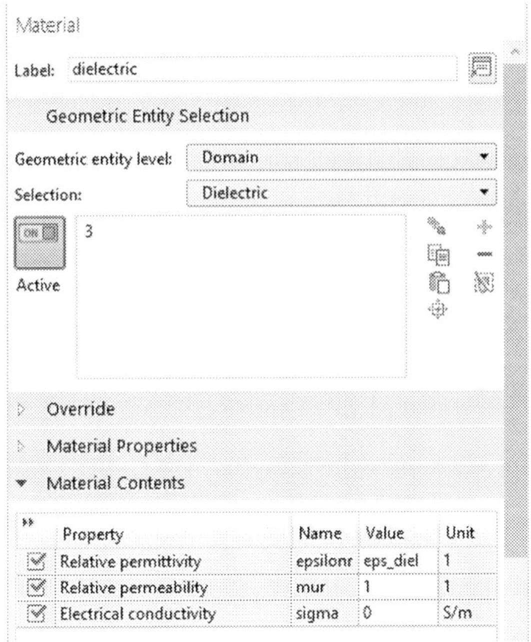

Figure 8.23 Material setting window of the dielectric with its properties.

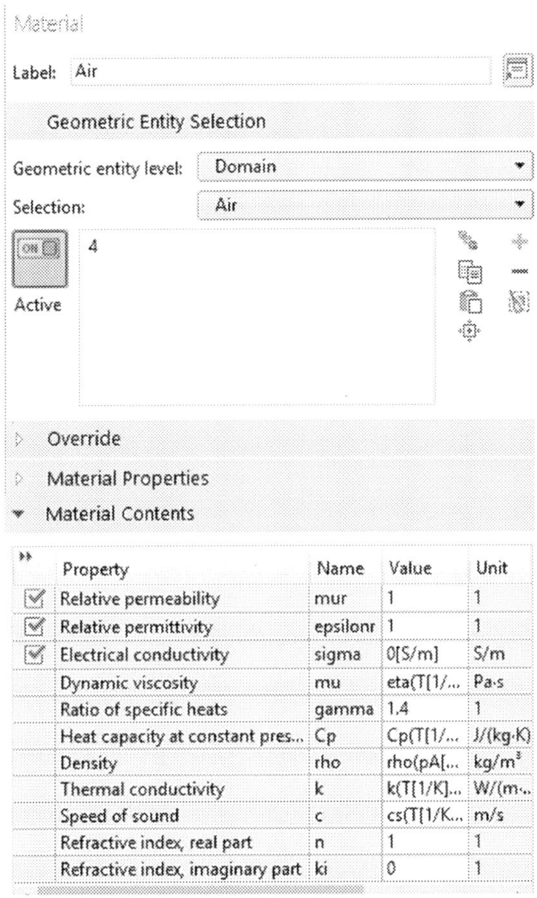

Figure 8.24 A material setting window of the air with its properties.

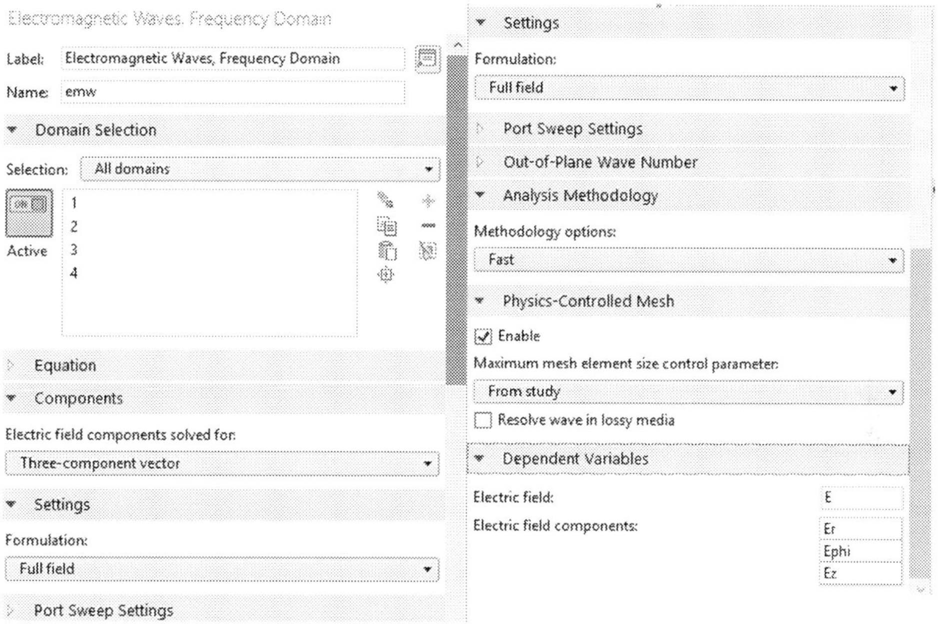

Figure 8.25 A setting window of the physics of the electromagnetic waves.

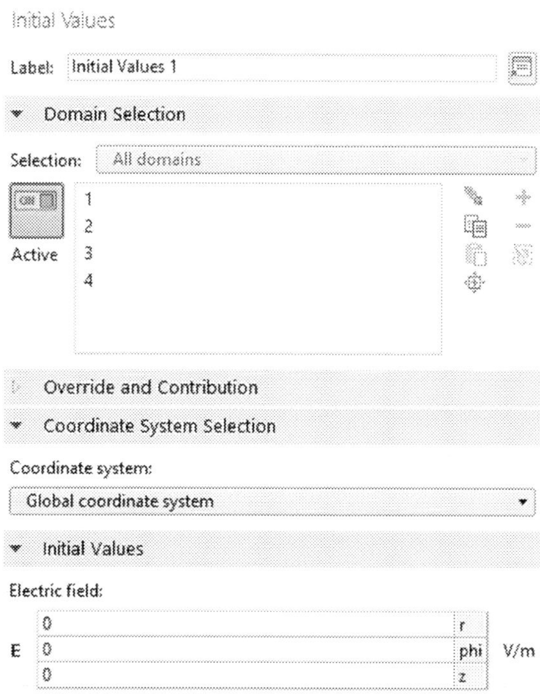

Figure 8.26 A setting window of the electric field initial value.

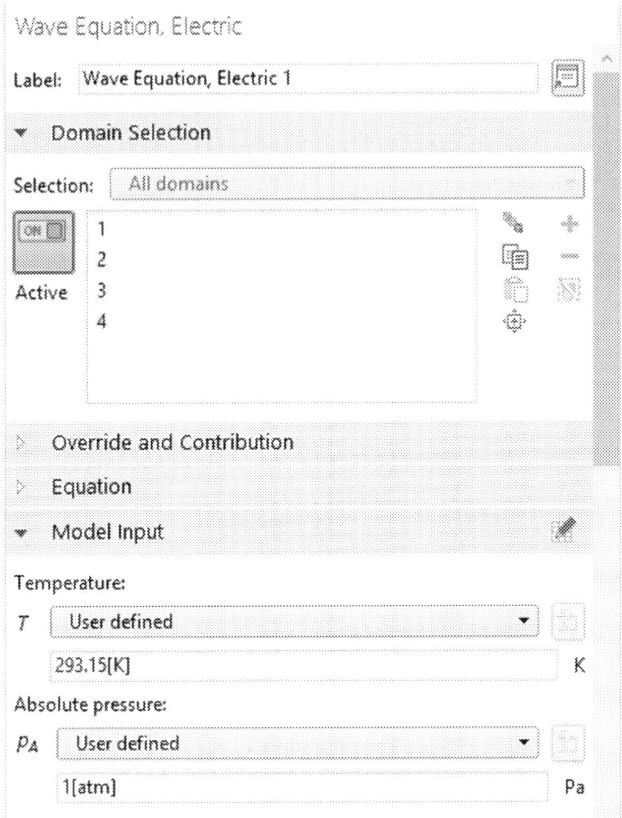

Figure 8.27 A setting window of the model input parameters.

Furthermore, the input of the geometric model, which is the temperature and the absolute pressure, is set to the values as shown in Fig. 8.27.

Then, the port type and location of the port are chosen, as shown in Fig. 8.28.

Also, the scattering boundary conditions are selected as shown in Fig. 8.29.

The EM waves lead to the friction of particles as a result of their vibrations at high speed, resulting in heat transfer through tissues due to thermal conductivity. Hence, the temperature is calculated at each point within the model and each time it is calculated using the Pennes equation for bioheat transfer, as shown in Fig. 8.30.

Fig. 8.30 displays the heat transfer equation, which is a function of time and tissue properties, such as heat, density, and thermal conductivity. Accordingly, their values are obtained from the specific substance, as shown in Fig. 8.31.

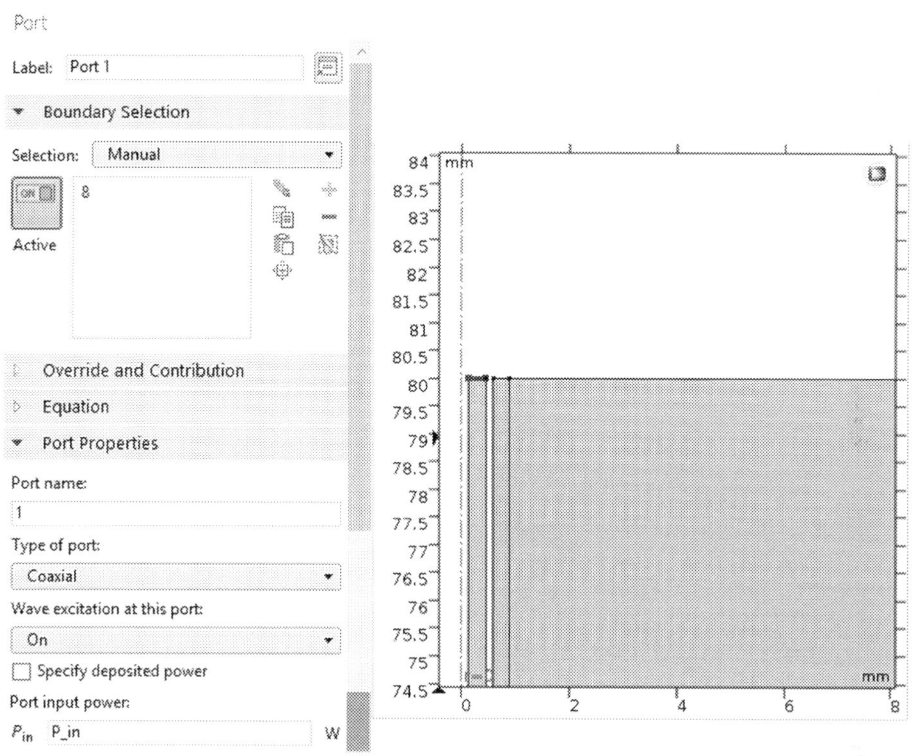

Figure 8.28 A setting window of the port settings and the location of the port.

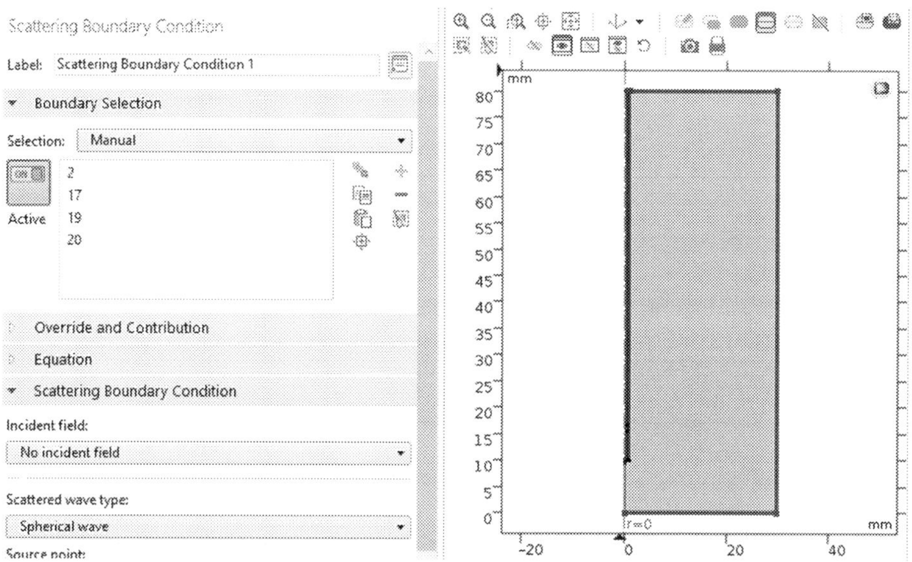

Figure 8.29 A setting window of the scattering boundary conditions settings and the location of the selected boundaries.

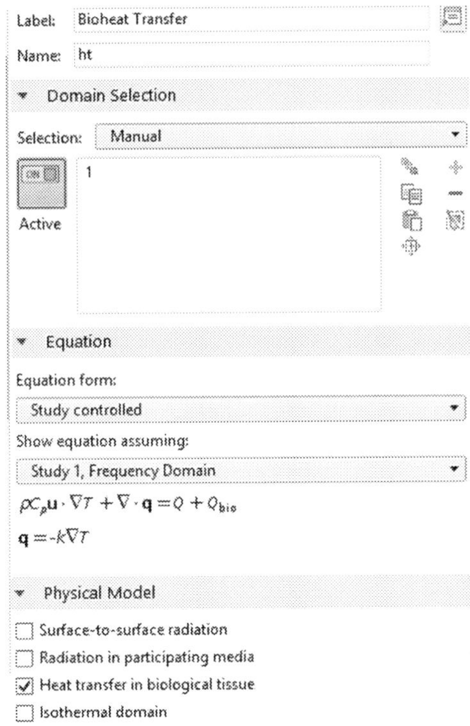

Figure 8.30 A setting window of the bioheat transfer equation.

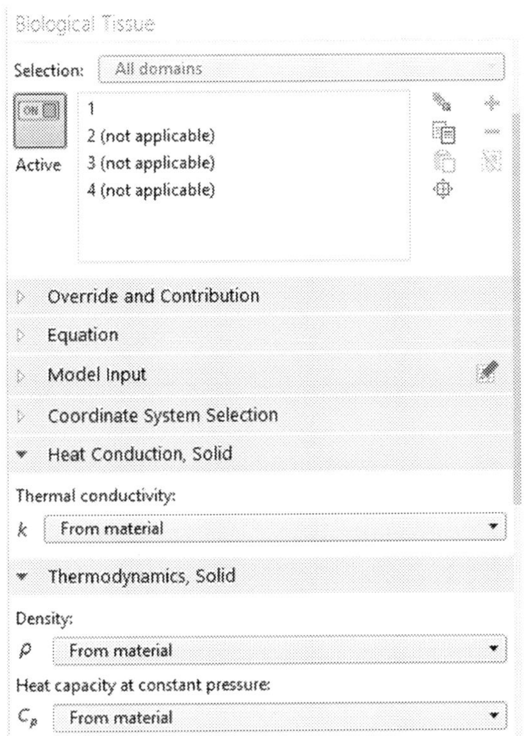

Figure 8.31 A setting window of the biological tissue.

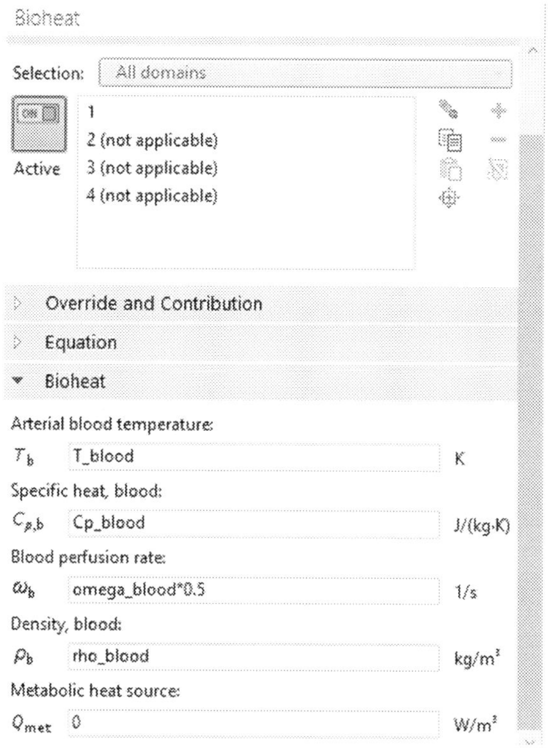

Figure 8.32 A setting window of the bioheat module.

Likewise, the biological energy obtained due to metabolism, which has a small value with respect to the external energy applied, can be neglected, and the energy absorbed through the adjacent blood vessels present is controlled according to blood characteristics and blood perfusion, as shown in Fig. 8.32.

Fig. 8.32 displays the main parameters that significantly affect the temperature distribution through the model, which are modified according to the target system. Additionally, the initial temperature value and thermal insulation have to be set as shown in Fig. 8.33.

Accordingly, the EM waves have to be correlated with the bioheat physics using the EM heat source to send and receive values between the interface of the two equations, as shown in Fig. 8.34.

Fig. 8.34 shows the correlation technique between the physics chosen to obtain accurate results. Accordingly, to calculate the temperature for the level of damage at all points within the model, the total geometric model is divided into very small elements using the mesh as shown in Fig. 8.35.

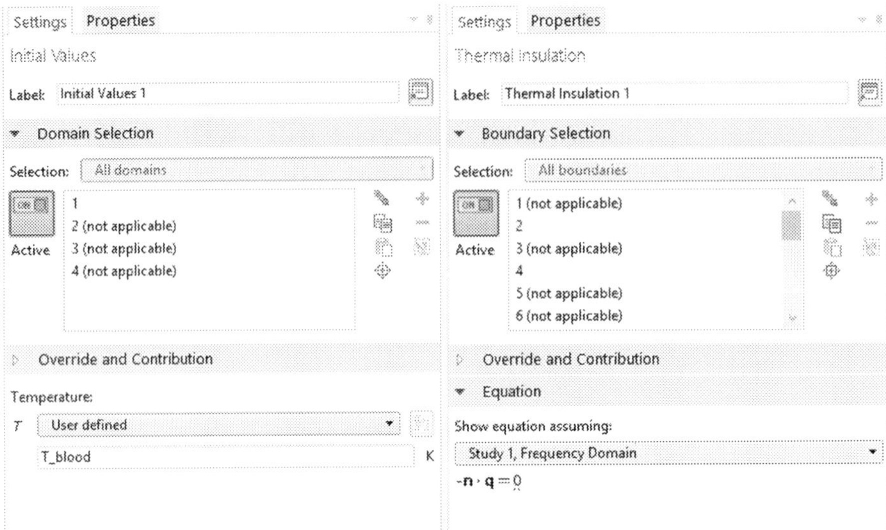

Figure 8.33 Setting window of the initial and thermal insulation values for the bioheat transfer equation.

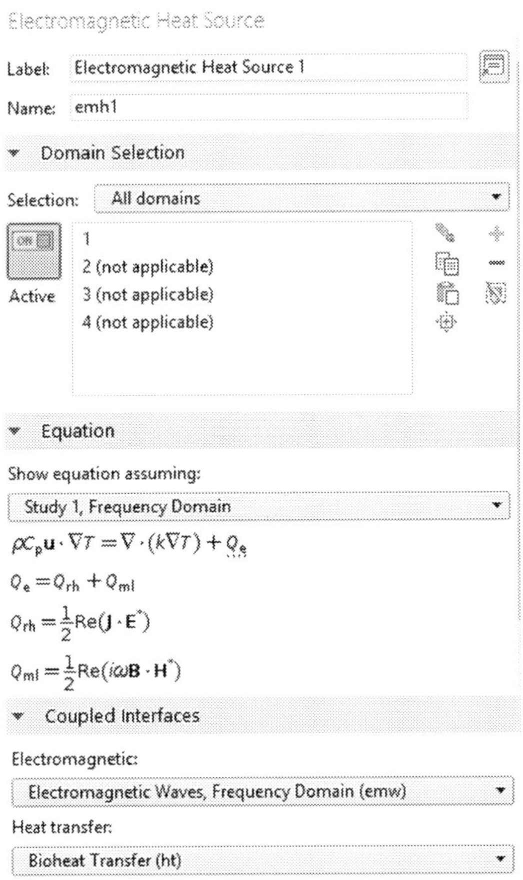

Figure 8.34 Setting window of the electromagnetic heat source.

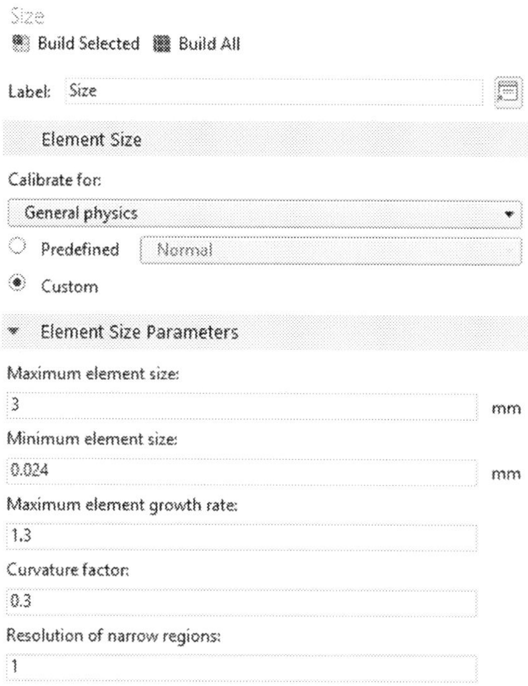

Figure 8.35 Setting window of the mesh.

Fig. 8.35 shows the general standard element size of the geometric model, which can be resized to fit the target model, as the element size is determined to be too small in the critical operating area. However, too large FEs in the outermost region can be used to solve the trade-off between the processing time and obtaining more accurate results. Hence, the model contains other smaller triangular elements, as demonstrated in Fig. 8.36.

All subsections inside the geometric model are regularly divided into free triangular elements, as shown in Fig. 8.37.

Consequently, the model is studied using both time-dependent and frequency-dependent measures to obtain the results and calculate the temperature and level of damage every minute, as shown in Fig. 8.38.

The frequency domain is used when dealing with EM waves, and the time domain is used to calculate the level of damage resulting from the transfer of heat in tissues, which is dependent on time. Finally, the geometric model is computed to display all possible obtained values related to the EM waves and bioheat transfer physics, such as 2D/3D figures, tables, and isothermal contours. Fig. 8.39 illustrates the setting window of the electric field 2D plot.

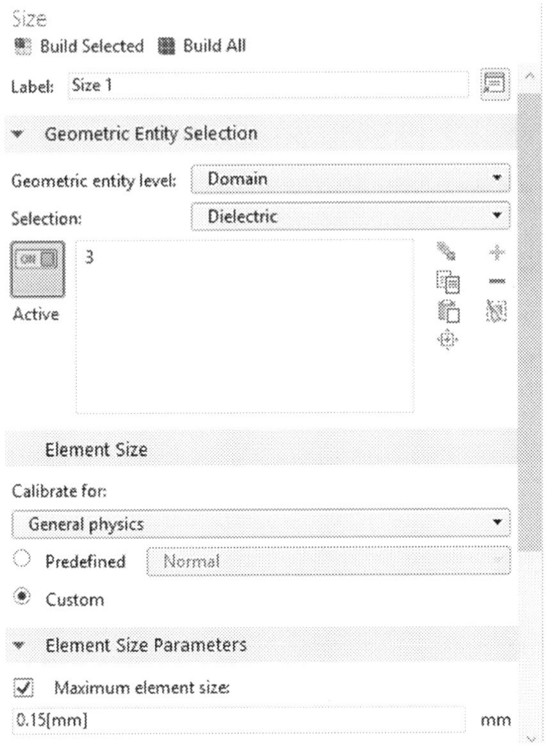

Figure 8.36 Setting window of the free triangular element.

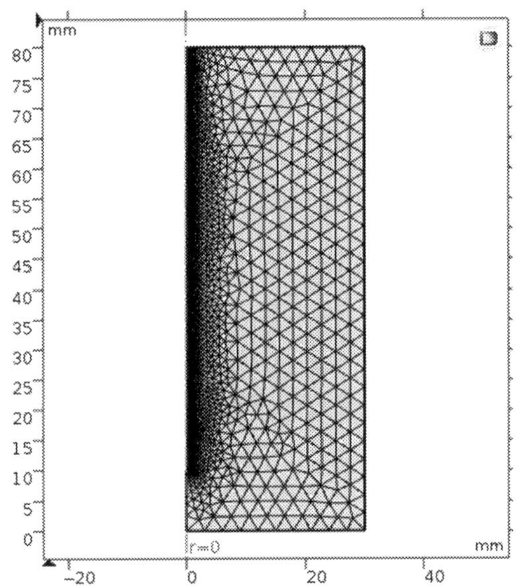

Figure 8.37 Mesh for the geometric model, including liver tumor tissue and single-slot microwave probe.

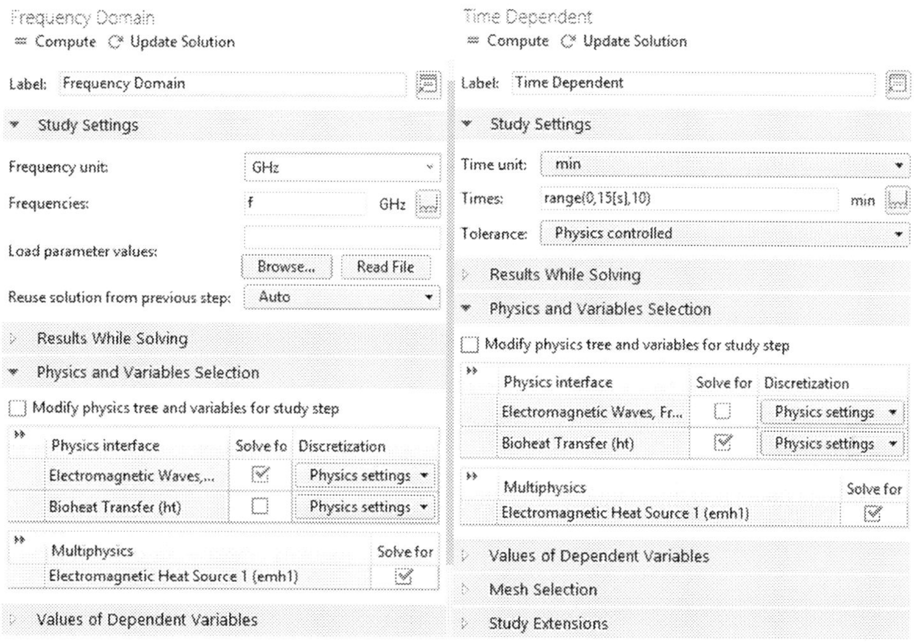

Figure 8.38 Setting window of the frequency- and time-dependent study.

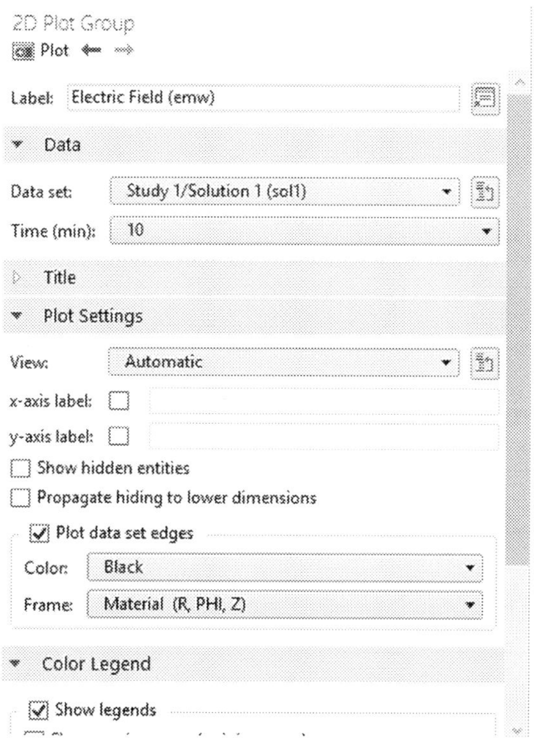

Figure 8.39 Setting window of the electric field 2D plot.

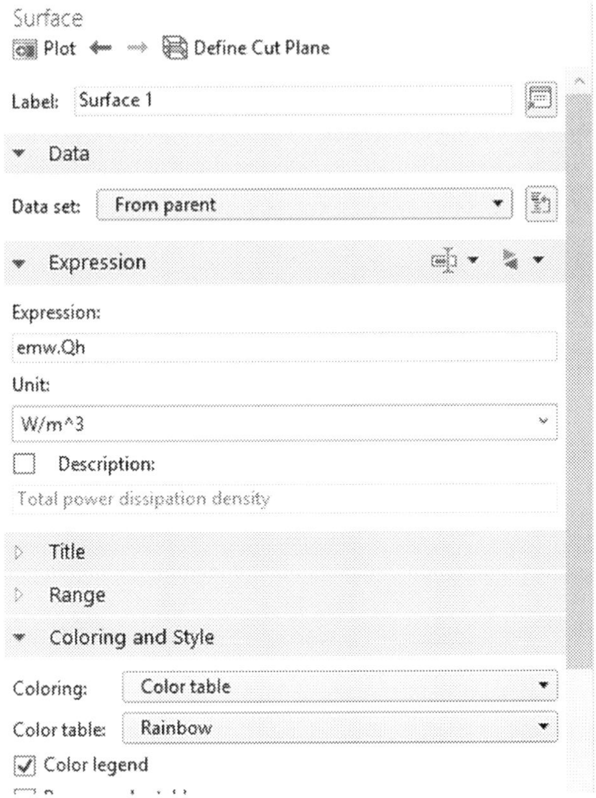

Figure 8.40 Setting window of the electric field surface settings.

The settings and equation of the plot can be managed as shown in Fig. 8.40.

Fig. 8.41 illustrates that the electric field has a maximum value around the probe and decreases with increasing distance from the probe until reaching the minimum value at the model's external boundaries. The shape of the electric field varies according to the type of probe. Thus, Fig. 8.42 shows the shape of the electric field resulting from the monopole probe and the choked probe.

Fig. 8.42 shows that the shape of the electric field is different due to the different composition of the probe, where the electric field of the monopole is more concentrated around the axis of the probe, and the electric field resulting from the chocked probe is more circular, more concentrated, and larger around the axis of the probe. Furthermore, the temperature distribution at any

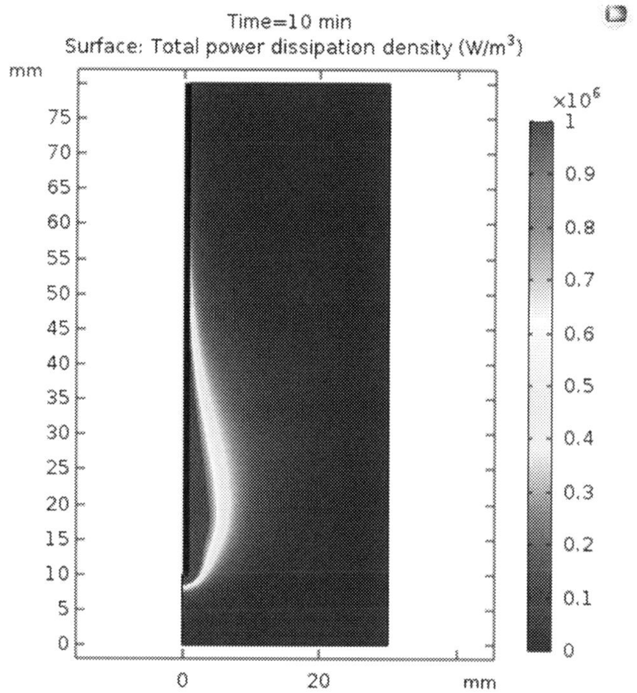

Figure 8.41 Graphic window for the 2D electric field at time 10 min for the single-slot probe.

time is obtained using the setting window related to the temperature parameter (Fig. 8.43), where Fig. 8.44 illustrates the temperature profile of the single slot probe at the time of 10 minutes.

Fig. 8.45 demonstrates the temperature profile after 10 minutes, where (A,B) are the 2D and 3D plots of the monopole probe, and (C,D) are the 2D and 3D plots of the chocked probe.

Fig. 8.45 shows the temperature profile of the two probes, monopole and chocked, showing the resulting thermal radiation shape. The results illustrate that the monopole probe produces a higher temperature, but is more concentrated at the active point of the probe. However, the chocked probe has more radiation around the probe, but its temperature is lower. To show the maximum and/or minimum temperature points in the model, the setting window in Fig. 8.46 is used.

Fig. 8.46 illustrates that the maximum and/or minimum temperature points are specified inside the model as shown in Fig. 8.47 for the single-slot probe.

Fig. 8.47 illustrates that the maximum value is located at the probe's active part, and the minimum value is at the external

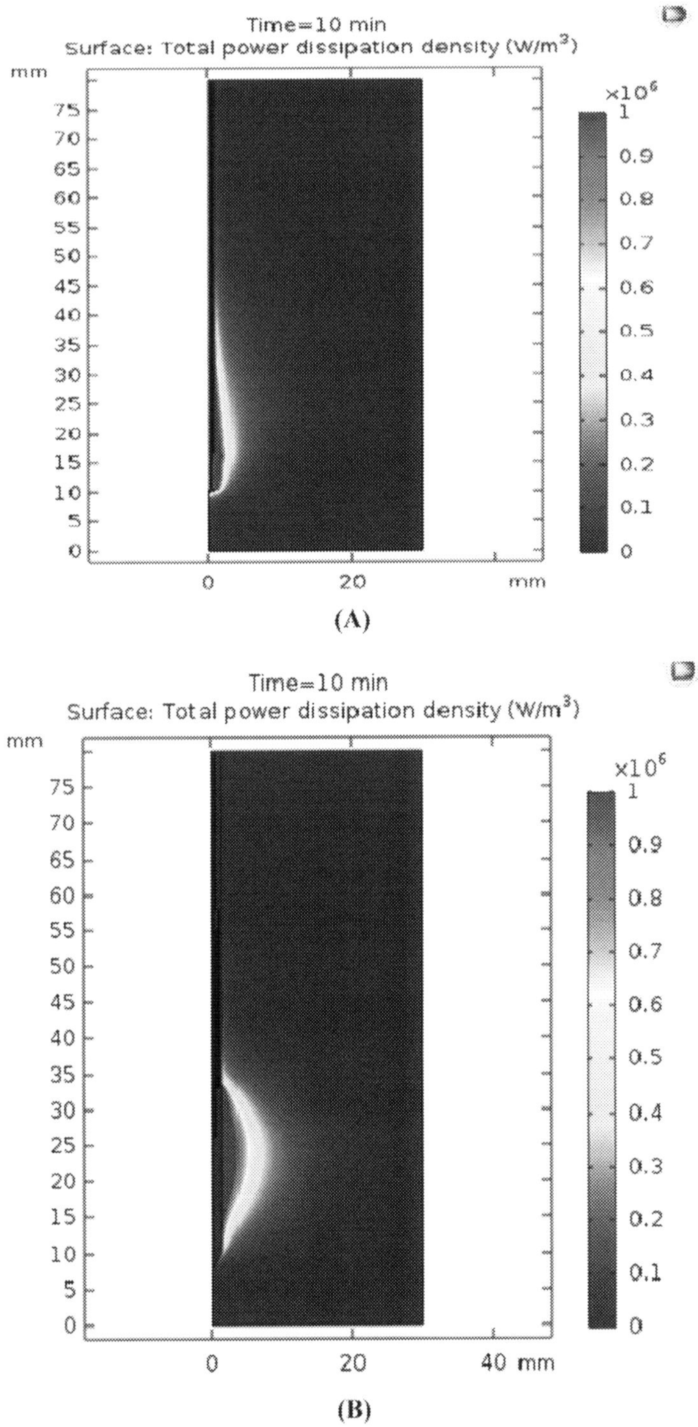

Figure 8.42 A 2D Graphic window for an electric field after 10 min for (A) the monopole probe (B) the chocked probe.

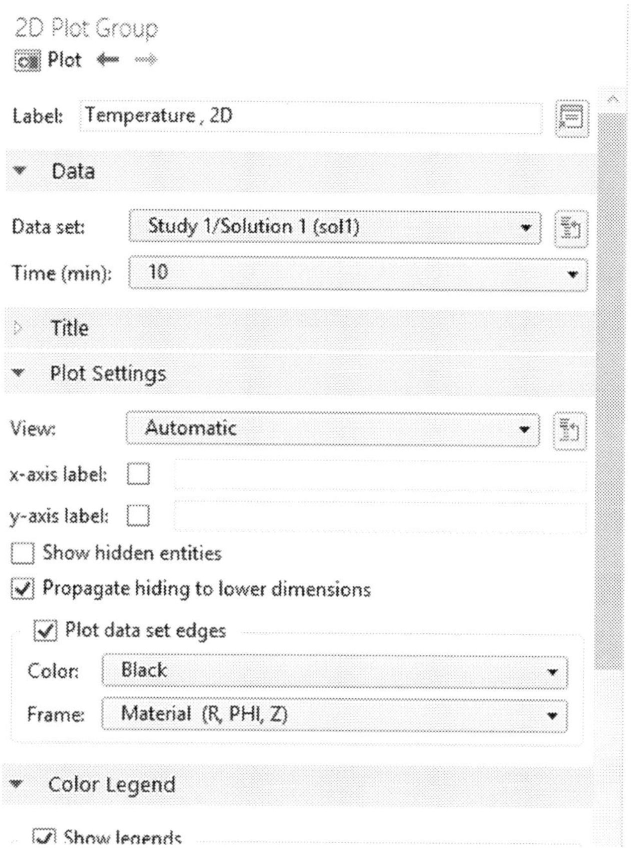

Figure 8.43 Setting window of the temperature parameter.

boundaries of the model. In addition, Fig. 8.48 shows the minimum and/or maximum temperature points for both the monopole and chocked probes.

Fig. 8.48 shows that the maximum temperature for the monopole probe is 136.33°C, while the maximum temperature for the chocked probe is 73.28°C. Thereafter, the temperature values at different levels are determined using the isothermal contours as shown in its setting window in Fig. 8.49.

Fig. 8.49 illustrates that the temperature profile through the model can be divided into levels from the initial temperature to the maximum temperature, as shown in Fig. 8.50.

Figs. 8.50 and 8.51 validate that the temperature has a maximum value at the spot point, and gradually decreases with increasing distance from the spot point and to determine the

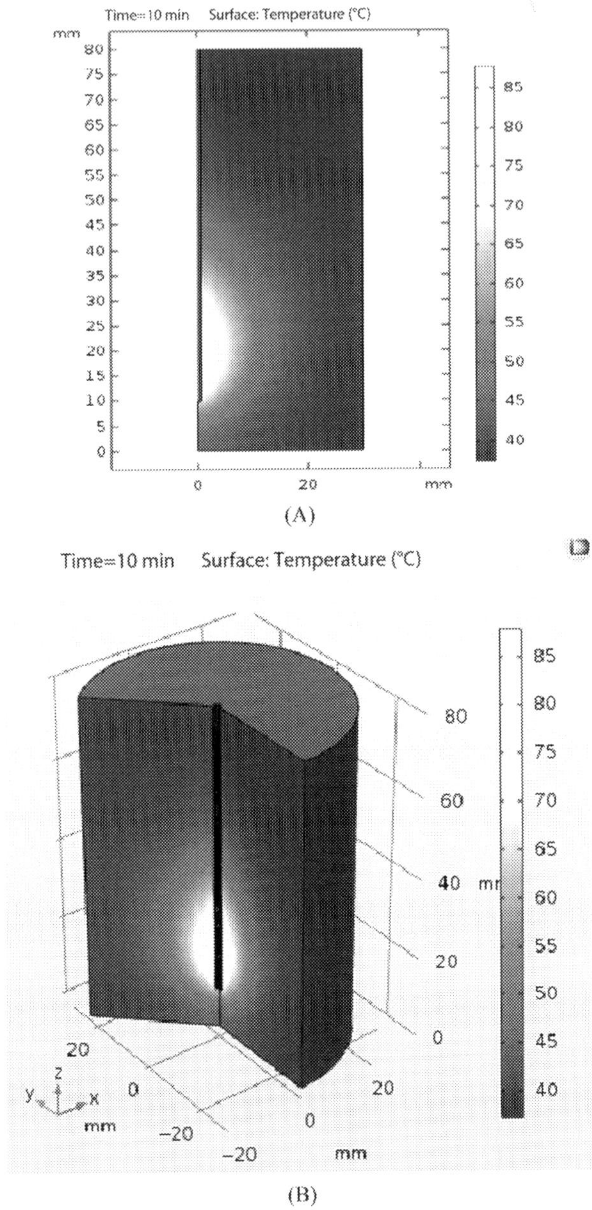

Figure 8.44 The graphic window for the temperature profile of the single-slot probe at a time of 10 min: (A) 2D plot and (B) 3D plot.

ablated region, which has a temperature greater than 45°C. The level of damage at several points inside the geometric model is explained using the 1D plot showing the setting window in Fig. 8.52.

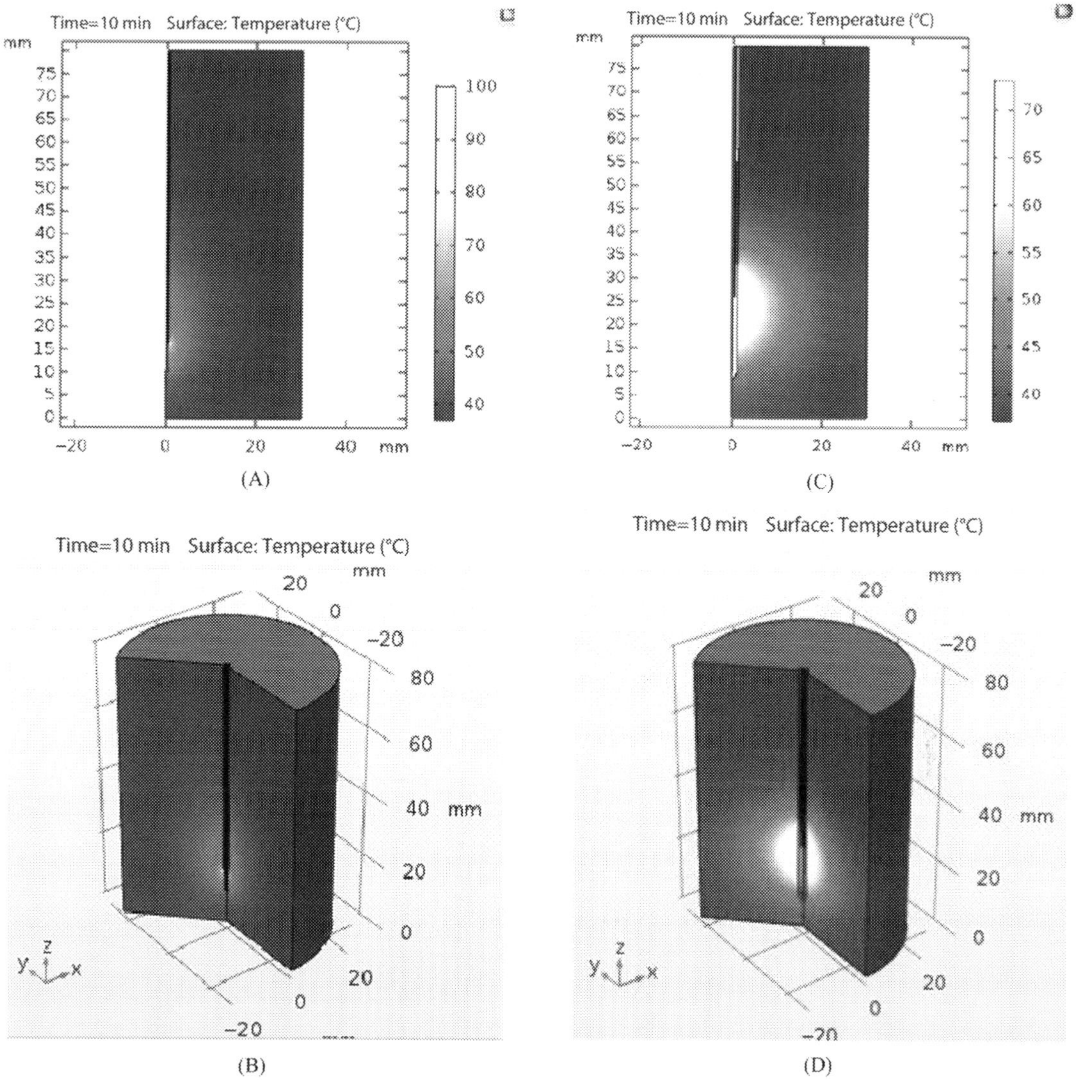

Figure 8.45 Graphic window for the temperature profile after 10 minutes, where (A,C) are the 2D and 3D plots of the monopole probe, and (B,D) are the 2D and 3D plots of the chocked probe.

Fig. 8.52 demonstrates that the level of damage with time is explained at four specified selected points inside the model, as shown in Figs. 8.53–8.55.

The difference in the MWA performance using the three probes is different in the speed of tissue destruction around the

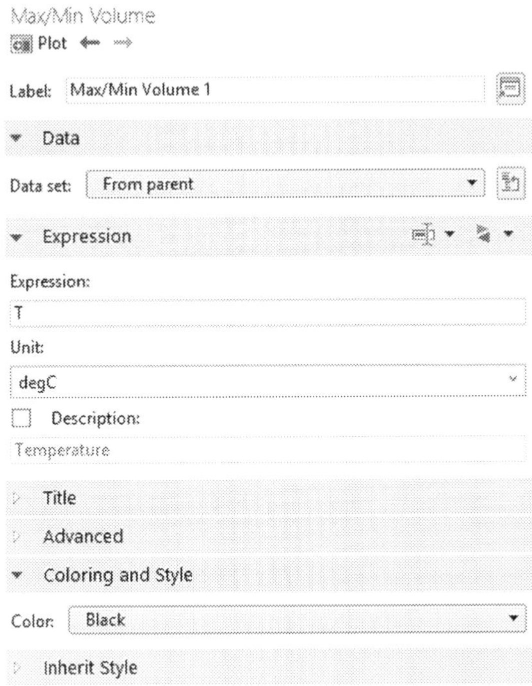

Figure 8.46 Setting window of the maximum and/or minimum temperature points.

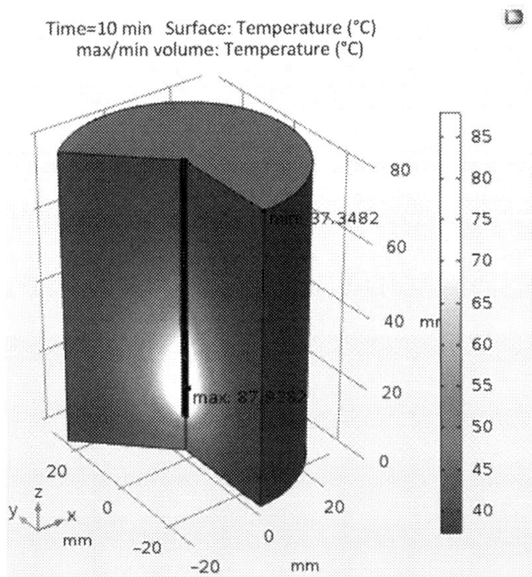

Figure 8.47 Geometric temperature window, including the maximum and minimum values, and their locations for the single-slot probe.

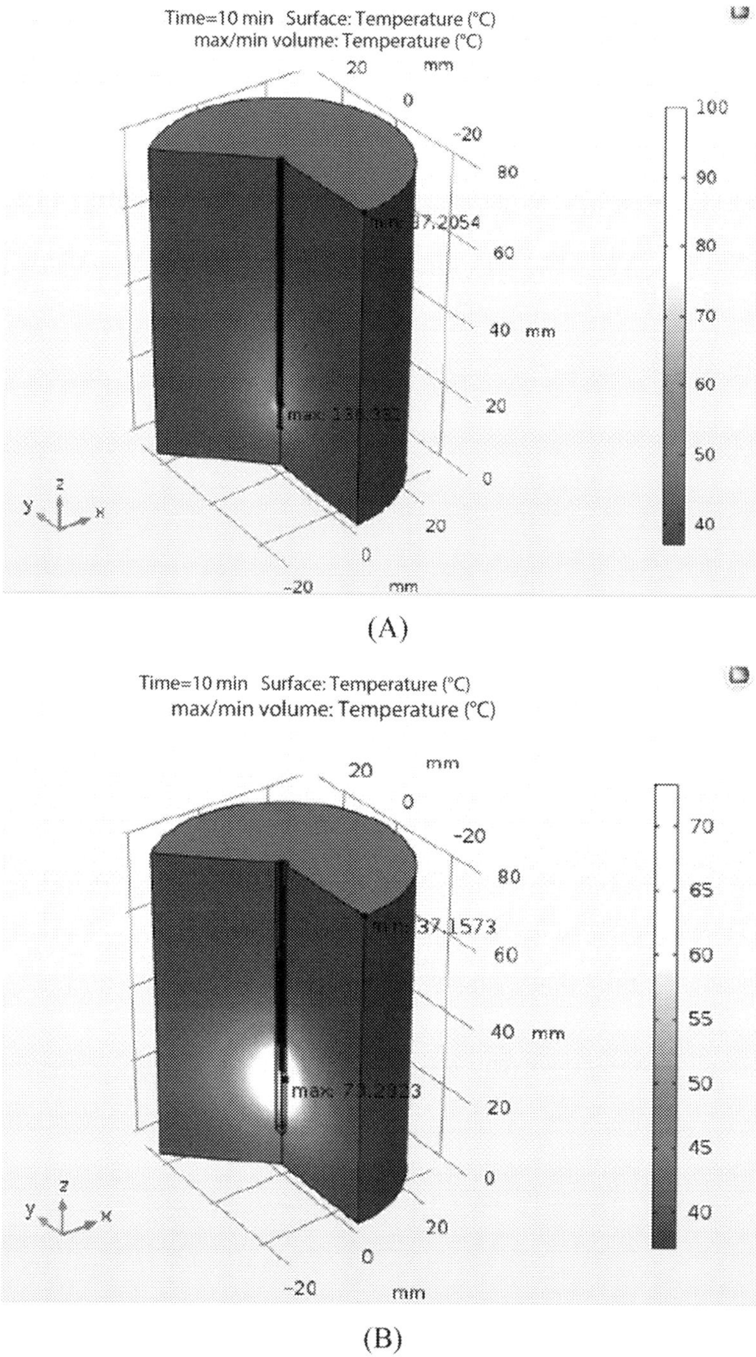

Figure 8.48 Geometric temperature window, including the maximum and minimum values, and their locations: (A) the monopole probe and (B) the chocked probe.

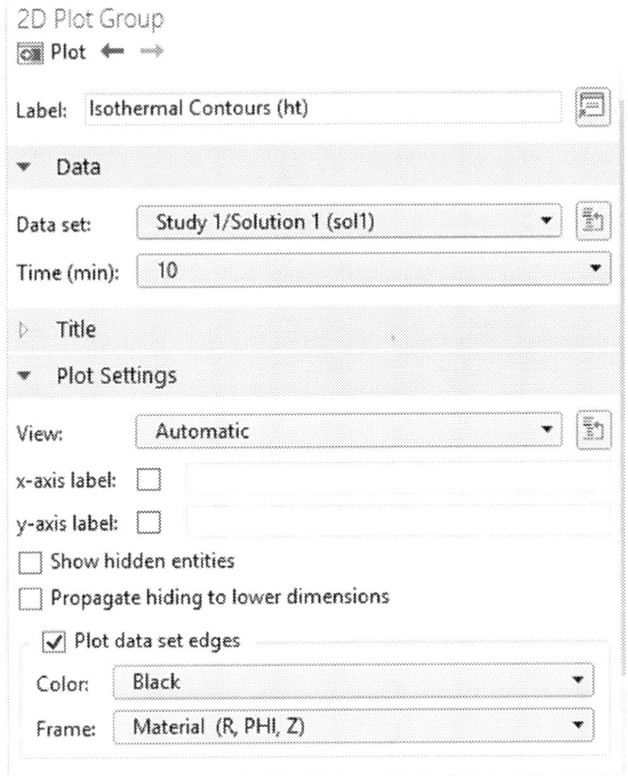

Figure 8.49 Setting window of the isothermal contour.

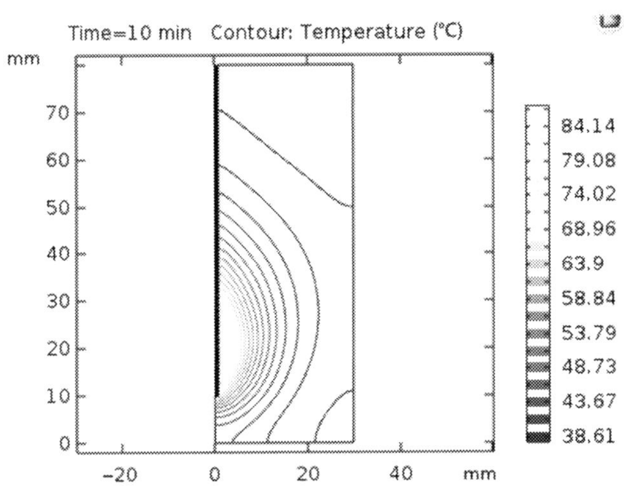

Figure 8.50 Geometric window of the isothermal contours for the single-slot probe.

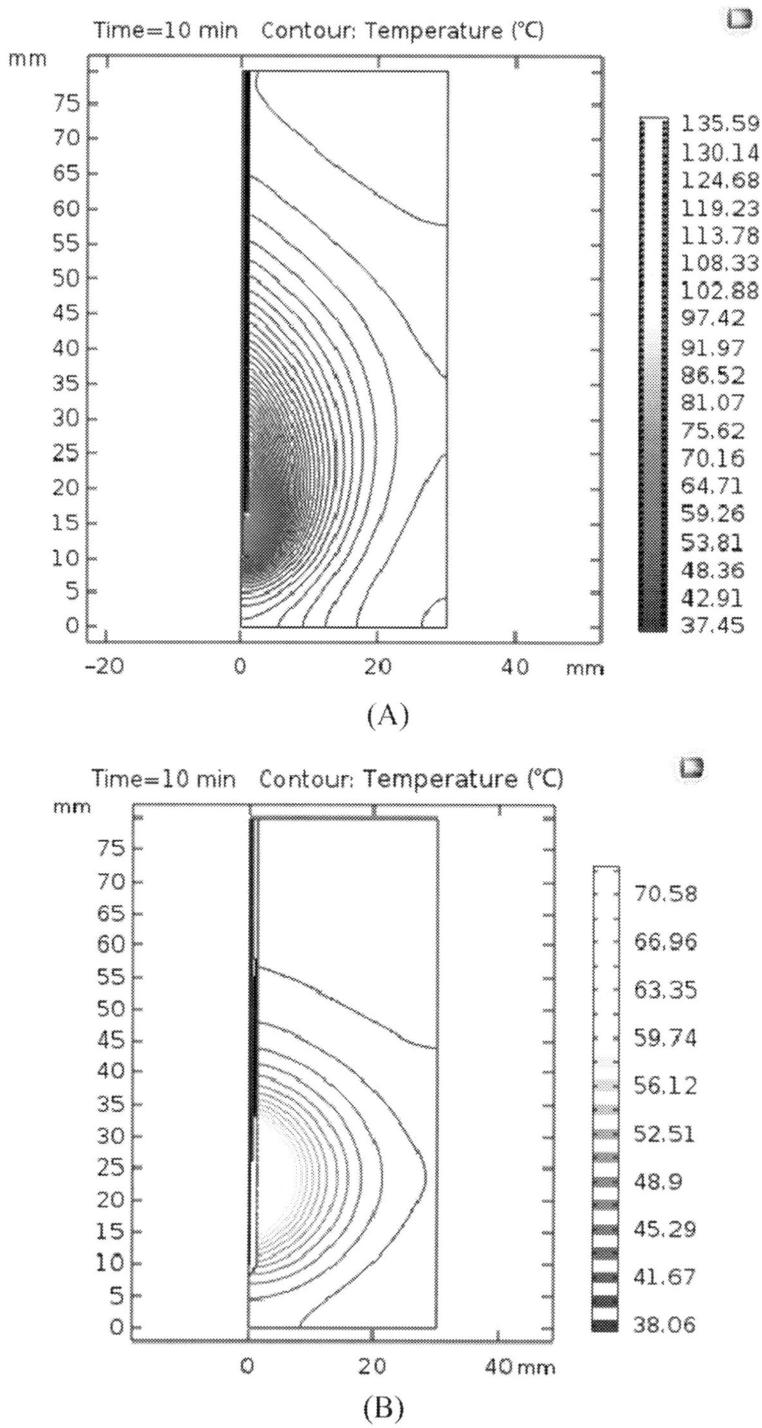

Figure 8.51 Geometric window of the isothermal contours for (A) the monopole probe and (B) the chocked probe.

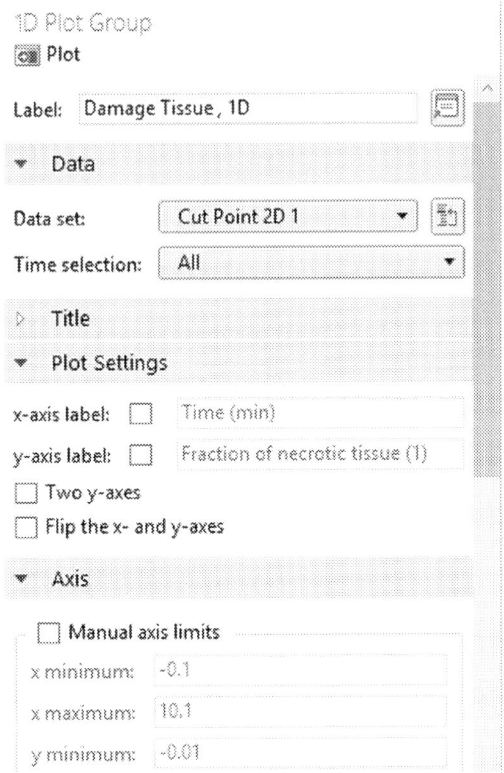

Figure 8.52 Setting windows for a 1D plot group for the relation between level of damage and time at some determined points around the probe.

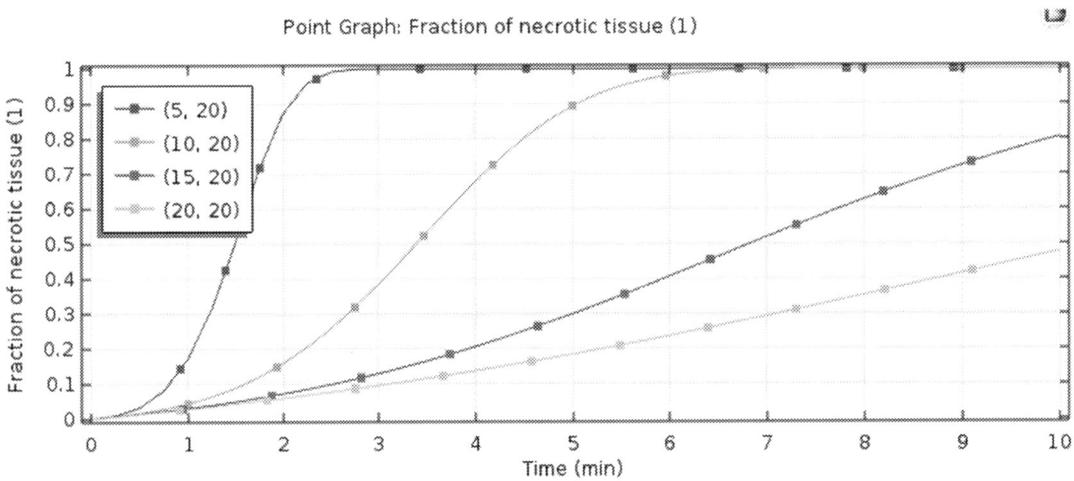

Figure 8.53 The relation between the level of damage and time at four points around the probe for the single-slot probe.

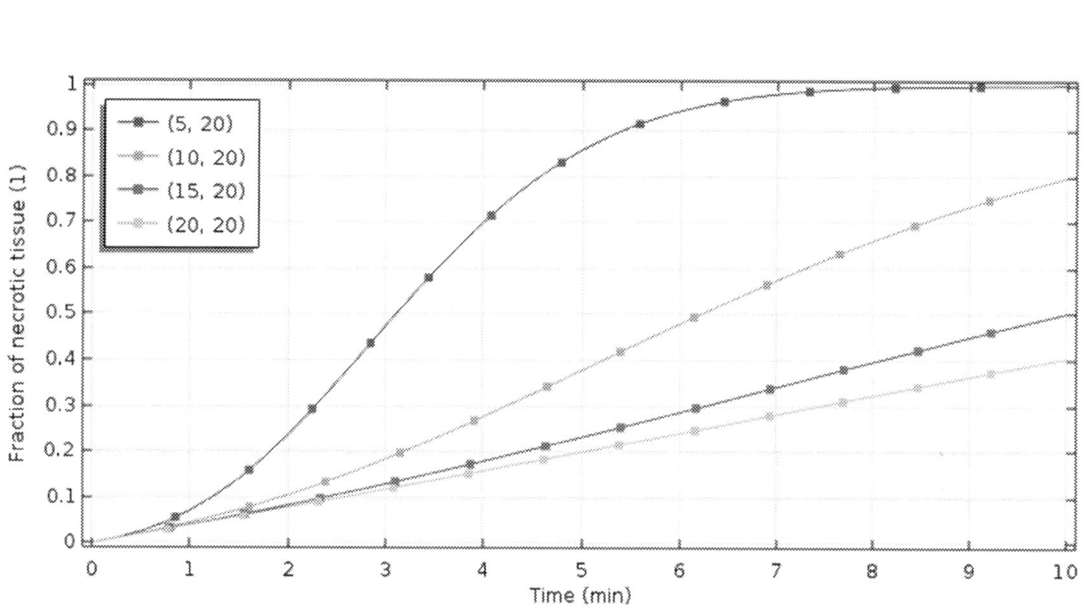

Figure 8.54 The relation between the level of damage and time at four points around the probe for the monopole probe.

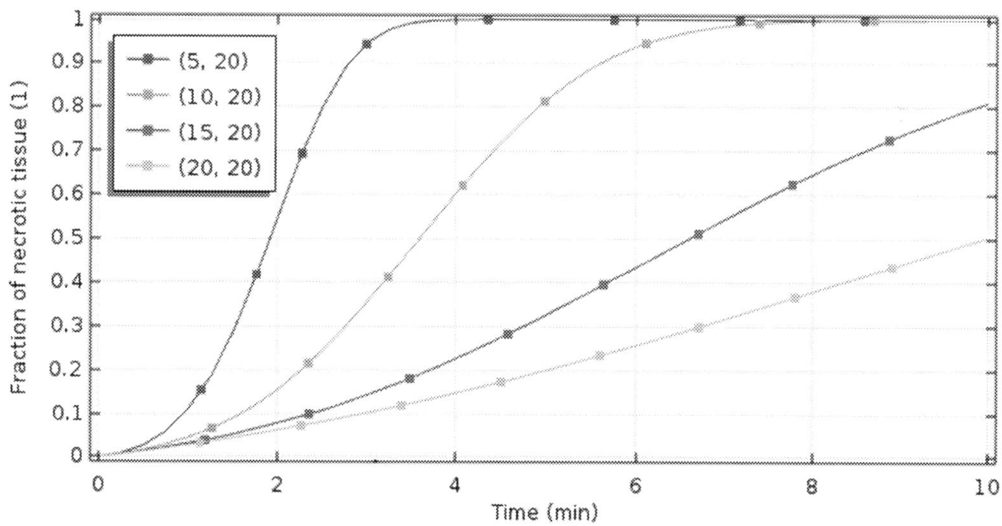

Figure 8.55 Relation between the level of damage and time at four points around the probe for the chocked probe.

probes, which takes around 2.5, 8.5, and 4 minutes to ablate the closest point of the probe at (5,20) using the single-slot probe, monopole probe, and chocked probe, respectively. The damaged tissue shape can be displayed in a 3D plot, as shown in the setting window in Fig. 8.56.

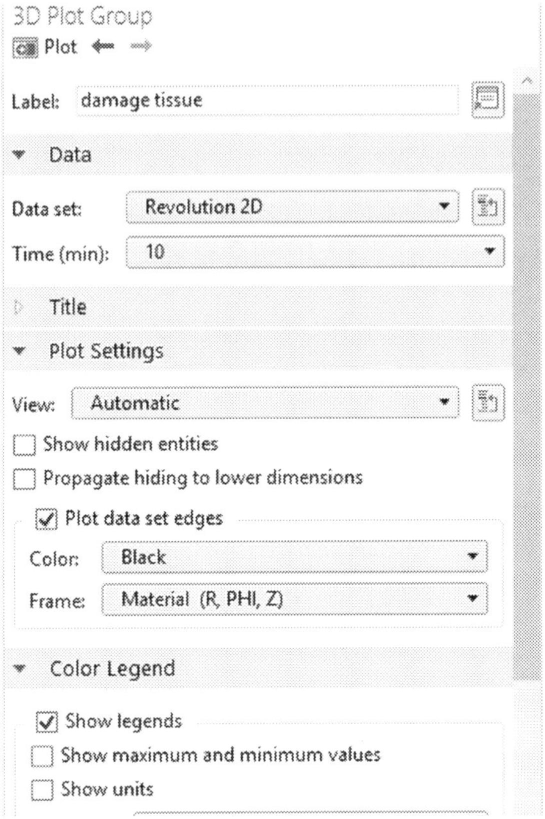

Figure 8.56 Setting window of the damaged tissue 3D plot group at time 10 min.

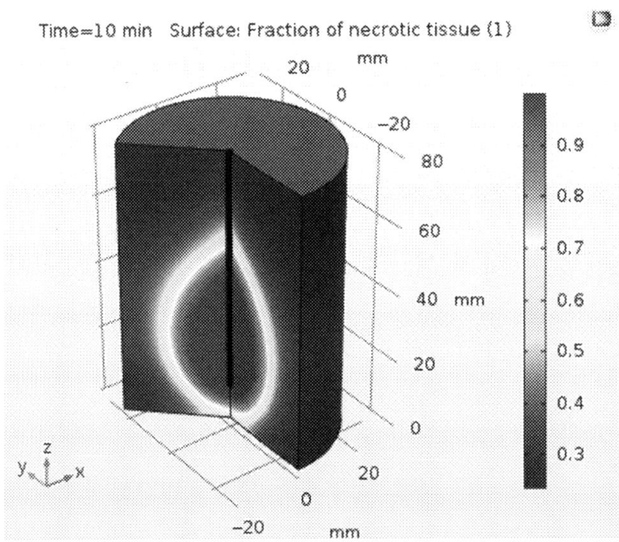

Figure 8.57 Graphic window for the damaged tissues at time 10 min for the single-slot probe.

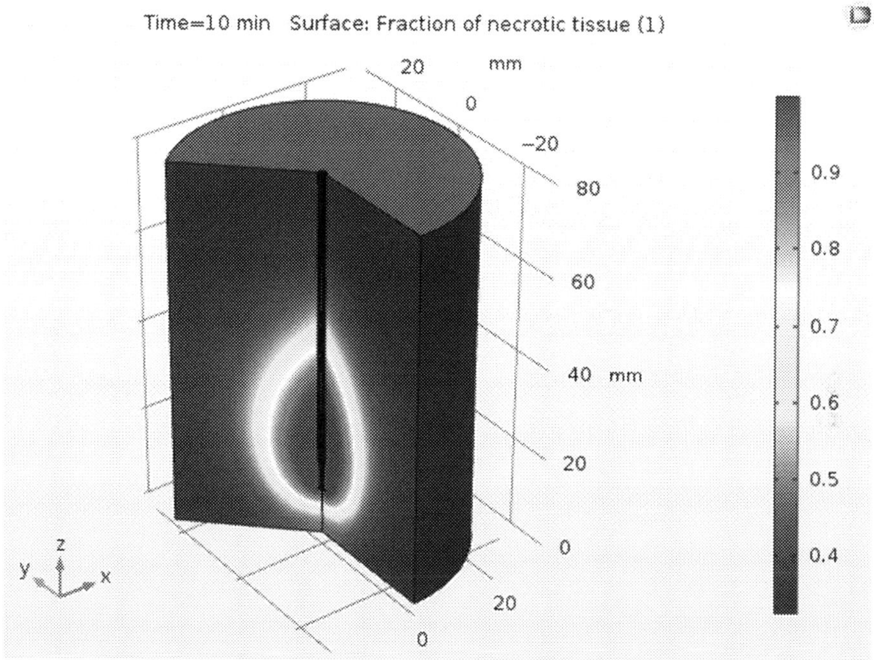

Figure 8.58 Graphic window for the damaged tissues at time 10 min for monopole probe.

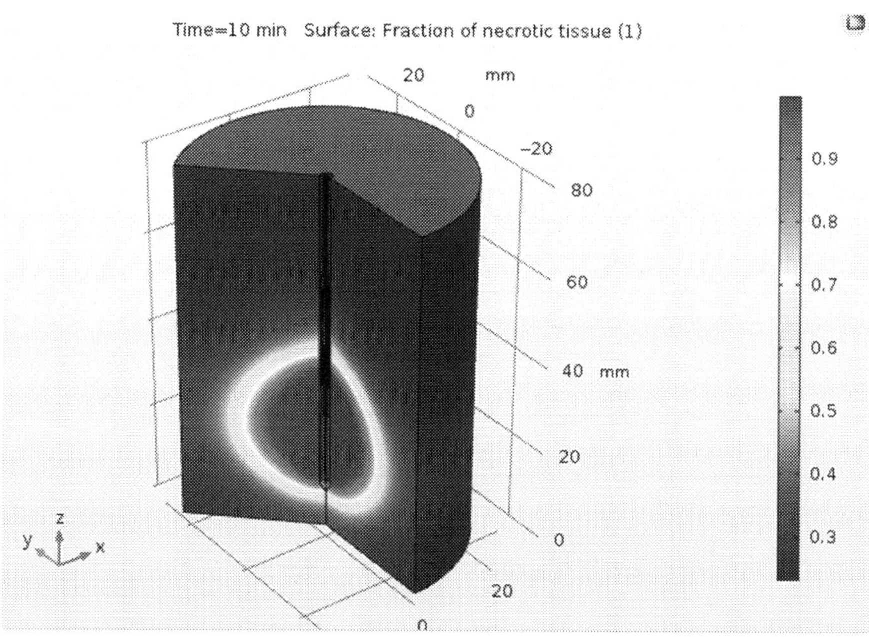

Figure 8.59 Graphic window for the damaged tissues at time 10 min for the chocked probe.

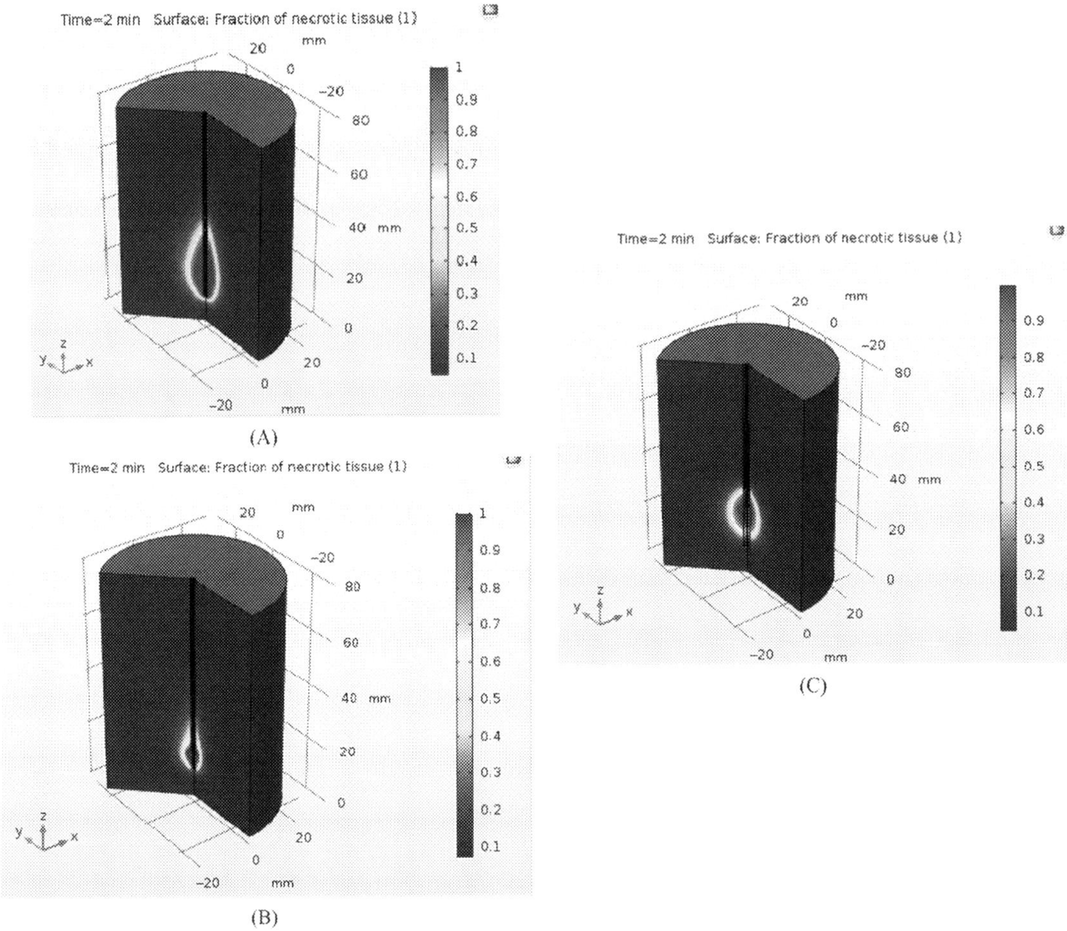

Figure 8.60 Graphic windows for the damaged tissues after 2 min for the three probes: (A) the single-slot probe, (B) the monopole probe, and (C) the chocked probe.

Figs. 8.57–8.59 demonstrate that the level of damage is maximum close to the probe's active part and decreases with increasing distance from the probe, and is greatly affected by time. Likewise, the shape of the ablated part varies according to the used probe. Under the same conditions, the three probes use 2.45 GHz frequency and a power of 10 watts. The results indicate that the chocked probe ablated larger tumor zone compared to the other two probes with

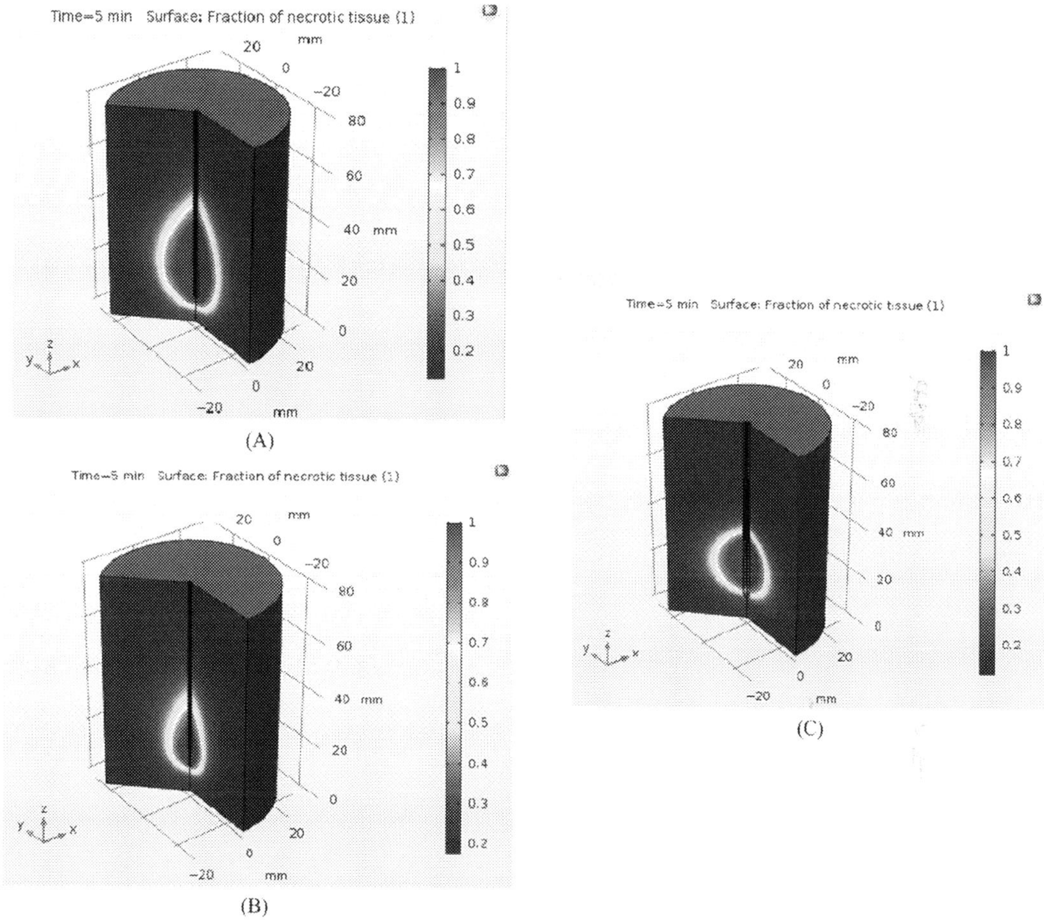

Figure 8.61 Graphic windows for the damaged tissues after 5 min for the three probes: (A) the single-slot probe, (B) the monopole probe, and (C) the chocked probe.

the more spherical shape around the probe. These results are due to less backward heating using the chocked probe, while the backward heating increased using the other two probes. For example, when the return currents are present on the axis of the probe, backward heating is generated that makes the ablative shape of the tumor tissue irregular and has a tail. Hence, the damaged tissue at the different ablating times is observed for the single-slot probe as shown in Figs. 8.60–8.62.

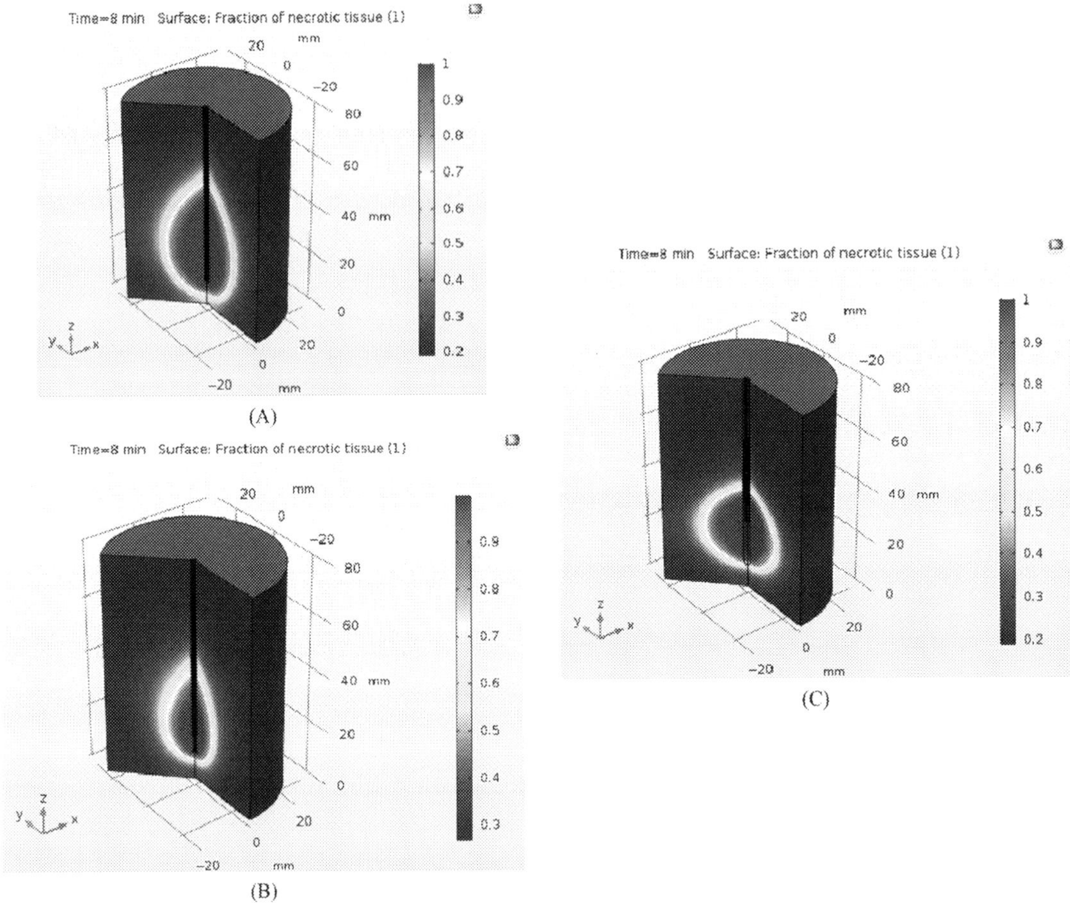

Figure 8.62 Graphic windows for the damaged tissues after 8 min for the three probes: (A) the single-slot probe, (B) the monopole probe, and (C) the chocked probe.

The results illustrated that the ablation time increases lead to increased ablation size using each probe. However, the shape and size of the ablative portion varied after the same time in the ablation process using the three probes under study. Figure 8.63 shows the required setting to display the damaged tissue in a 2D plot representation after 10 minutes using a single slot probe, a monopole probe, and a chocked probe as illustrated in Figs. 8.64–8.66, respectively.

The damaged area can be calculated using the surface integration module, as shown in Fig. 8.67.

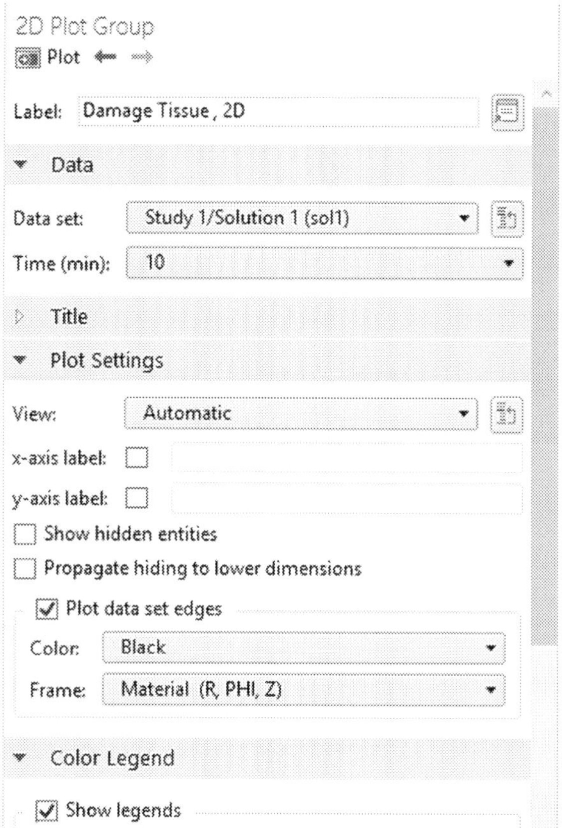

Figure 8.63 Setting window of the damaged tissue 2D plot group at 10 min.

Fig. 8.67 shows that the tumor is ablated when the level of damage at all points is greater than or equal to 1. Therefore the damaged area is computed at several periods for the three probes, as shown in Figs. 8.68–8.70.

Figs. 8.68–8.70 report the necrotic area inside the tumor tissue region. Although the chocked probe gives a more spherical shape of the damaged tissues around the probe. It provides less eradicated tissue area compared to the other two probes. The single-slot probe offers the largest area of tissue eradicated although the eradicated portion around the probe is irregular in shape due to the presence of backward heating.

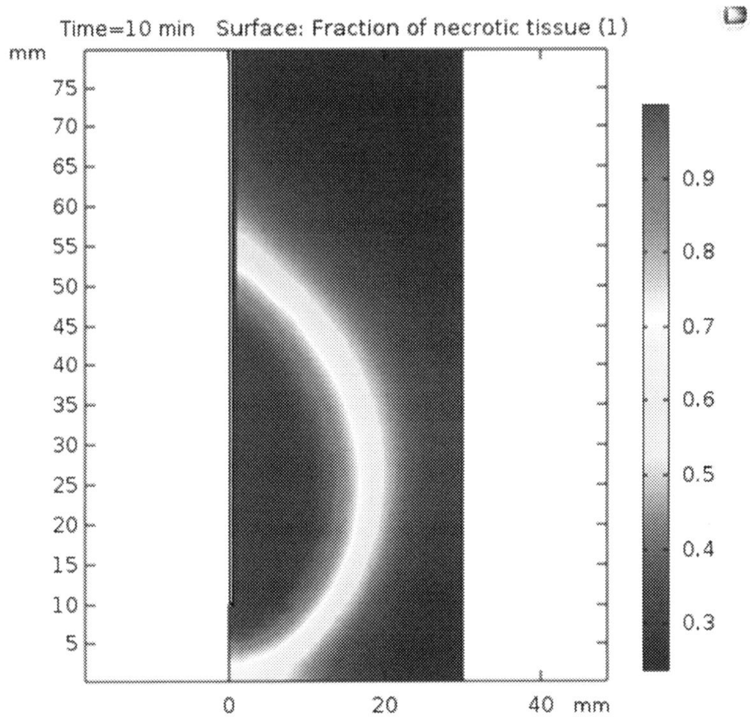

Figure 8.64 Graphic window for the damaged tissues at 10 min for the single-slot probe.

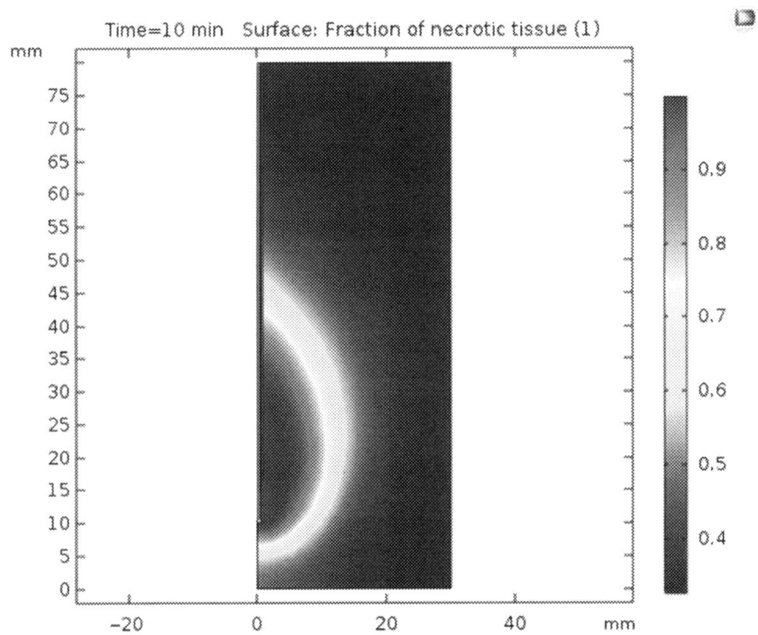

Figure 8.65 Graphic window for the damaged tissues at 10 min for the monopole probe.

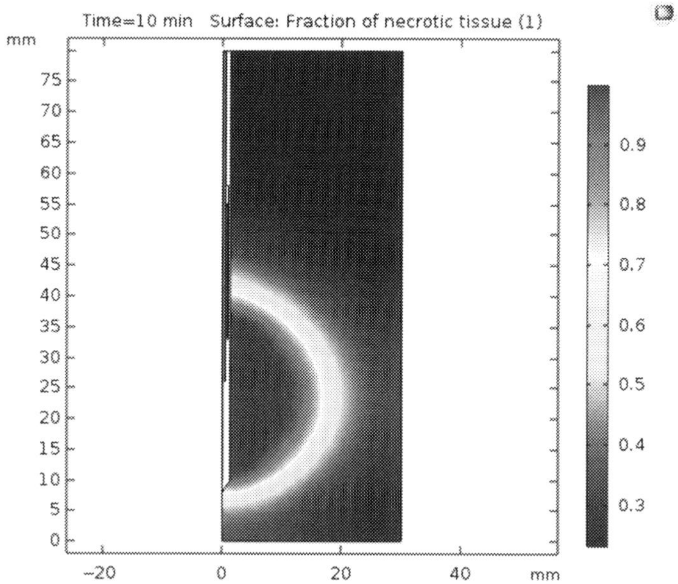

Figure 8.66 Graphic window for the damaged tissues at time 10 min for the chocked probe.

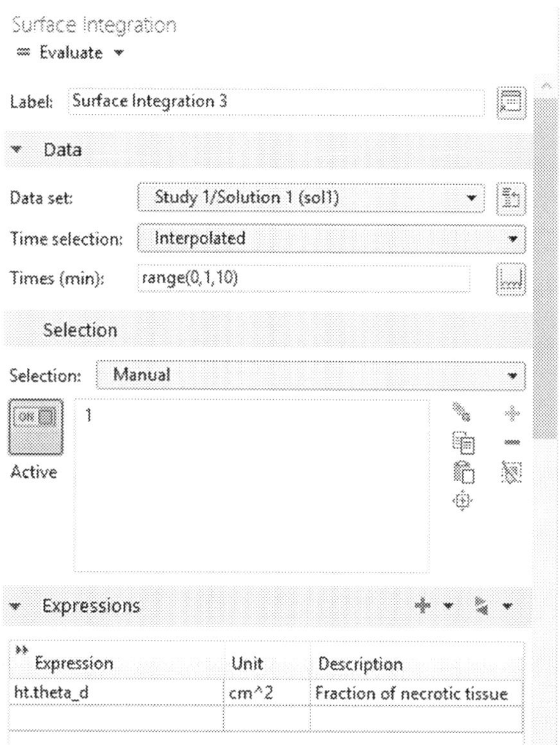

Figure 8.67 Setting window of the surface integration using the necrotic tissue equation.

Time (min)	Fraction of necrotic tissue (cm^2)
0.0000	1.1969E-5
1.0000	1.0821
2.0000	2.5842
3.0000	4.0002
4.0000	5.3143
5.0000	6.5285
6.0000	7.6551
7.0000	8.6993
8.0000	9.6693
9.0000	10.571
10.000	11.410

Figure 8.68 Ablated volumes at different ablation times for the single-slot probe.

Time (min)	Fraction of necrotic tissue (cm^2)
0.0000	1.7835E-5
1.0000	1.0573
2.0000	2.2722
3.0000	3.5116
4.0000	4.7167
5.0000	5.8664
6.0000	6.9540
7.0000	7.9786
8.0000	8.9413
9.0000	9.8447
10.000	10.692

Figure 8.69 Ablated volumes at different ablation times for the monopole probe.

Time (min)	Fraction of necrotic tissue (cm^2)
0.0000	1.1800E-5
1.0000	0.78819
2.0000	1.9926
3.0000	3.1604
4.0000	4.2503
5.0000	5.2729
6.0000	6.2350
7.0000	7.1431
8.0000	8.0005
9.0000	8.8104
10.000	9.5754

Figure 8.70 Ablated volumes at different ablation times for the chocked probe.

8.3 Conclusions

This chapter models three MWA probes using COMSOL Multiphysics software to provide data analysis and results of the predicted ablated zones and temperature distribution. The chapter explained in detail the design steps for simulating an MWA system using three MW probes, namely the single-slot microwave probe, the chocked probe, and the monopole probe, as a case study. Both Maxwell's equations and the bioheat transfer equations were applied as governing equations of the simulation process to show the performance of the ablation process. Several factors that affect the MWA performance include the tumor location, probe type, the organ to be ablated, and the temperature. The main target of the simulation modeling is to find the best probe design and required settings to provide complete ablation of the tumor.

References

Brace, C.L., 2009. Radiofrequency and microwave ablation of the liver, lung, kidney, and bone: what are the differences? Curr. Probl. Diagn. Radiol. 38, 135–143.

Carrafiello, G., Lagana, D., Mangini, M., Fontana, F., Dionigi, G., Boni, L., et al., 2008. Microwave tumors ablation: principles, clinical applications and review of preliminary experiences. Int. J. Surg. 6, S65–S69.

Clark, P.E., Woodruff, R.D., Zagoria, R.J., Hall, M.C., 2007. Microwave ablation of renal parenchymal tumors before nephrectomy: phase I study. Am. J. Roentgenol. 188, 1212–1214.

Curto, S., Taj-Eldin, M., Fairchild, D., Prakash, P., 2015. Microwave ablation at 915 MHz vs 2.45 GHz: a theoretical and experimental investigation. Med. Phys. 42, 6152–6161.

Ibitoye, A., Orotoye, T., Nwoye, E., Aweda, M., 2018. Analysis of efficiency of different antennas for microwave ablation using simulation and experimental methods. Egypt. J. Basic. Appl. Sci. 5, 24–30.

Izzo, F., Granata, V., Grassi, R., Fusco, R., Palaia, R., Delrio, P., et al., 2019. Radiofrequency ablation and microwave ablation in liver tumors: an update. Oncologist 24, e990.

Jiang, Y., Zhao, J., Li, W., Yang, Y., Liu, J., Qian, Z., 2017. A coaxial slot antenna with frequency of 433 MHz for microwave ablation therapies: design, simulation, and experimental research. Med. Biol. Eng. Comput. 55, 2027–2036.

Jones, C., Badger, S., Ellis, G., 2011. The role of microwave ablation in the management of hepatic colorectal metastases. Surgeon 9, 33–37.

Lorentzen, T., Skjoldbye, B., Nolsoe, C., 2011. Microwave ablation of liver metastases guided by contrast-enhanced ultrasound: experience with 125 metastases in 39 patients. Ultraschall in der Medizin-European J. Ultrasound 32, 492–496.

Meloni, M.F., Chiang, J., Laeseke, P.F., Dietrich, C.F., Sannino, A., Solbiati, M., et al., 2017. Microwave ablation in primary and secondary liver tumours: technical and clinical approaches. Int. J. Hyperth. 33, 15–24.

Poulou, L.S., Botsa, E., Thanou, I., Ziakas, P.D., Thanos, L., 2015. Percutaneous microwave ablation vs radiofrequency ablation in the treatment of hepatocellular carcinoma. World J. Hepatol. 7, 1054.

Rattanadecho, P., Keangin, P., 2013. Numerical study of heat transfer and blood flow in two-layered porous liver tissue during microwave ablation process using single and double slot antenna. Int. J. Heat. Mass. Transf. 58, 457–470.

Sawicki, J.F., Shea, J.D., Behdad, N., Hagness, S.C., 2017. The impact of frequency on the performance of microwave ablation. Int. J. Hyperth. 33, 61–68.

Simon, C.J., Dupuy, D.E., Mayo-Smith, W.W., 2005. Microwave ablation: principles and applications. Radiographics 25, S69–S83.

Thamtorawat, S., Hicks, R.M., Yu, J., Siripongsakun, S., Lin, W.-C., Raman, S.S., et al., 2016. Preliminary outcome of microwave ablation of hepatocellular carcinoma: breaking the 3-cm barrier? J. Vasc. Interv. Radiol. 27, 623–630.

Veltri, A., Gazzera, C., Calandri, M., Marenco, F., Breatta, A.D., Fonio, P., et al., 2015. Percutaneous treatment of Hepatocellular carcinoma exceeding 3 cm: combined therapy or microwave ablation? Preliminary results. Radiol. Med. 120, 1177–1183.

Wu, X., Liu, B., Xu, B., 2016. Theoretical evaluation of high frequency microwave ablation applied in cancer therapy. Appl. Therm. Eng. 107, 501–507.

9

Image-guided thermal ablation therapy

Chapter outline

9.1 Introduction

Despite the availability of surgical resection and advances in invasive surgical techniques, many patients are not suitable for these interventions because of their poor cardiopulmonary function, age, or other medical conditions. For such patients, thermal ablation techniques have been increasingly used. Recently, as one of the new cancer treatment approaches, tumor ablation has received growing attention from the medical community. Thermal ablation therapy relies on the principle that the tumor is destroyed by elevating the temperature within the tumorous tissues above a threshold and has appeared as a

Thermal Ablation Therapy. DOI: https://doi.org/10.1016/B978-0-12-819544-4.00010-1
© 2021 Elsevier Inc. All rights reserved.

very promising technique for cancer treatment. Minimally invasive thermal ablation procedures are particularly well tolerated by patients and require reduced hospital stays. Thermal ablation destruction of tissue by application of heat has been applied to a variety of lesions, including those that arise in the liver, kidney, breast, bone, and adrenal glands.

Heat can destroy tissues by raising the temperature in a target tissue above a lethal threshold ($>60^{\circ}C$) to ensure cellular death within seconds. The physical means to raise the temperature within a tissue are numerous and several thermal ablation techniques have been developed and evaluated. The most widely used thermal ablation technique is radiofrequency ablation (RFA). It inserts a needle-like applicator into the tumor. The applicator is connected to a radiofrequency (RF) current generator and the return circuit is usually made of a ground pad attached to the leg of the patient. It is commonly used in the case of liver, lung, and bone tumors. The heating is produced by the displacement of the ions around the tip of the applicator needle. In addition, microwaves (MW) are also used to heat a target tissue, but are limited partly due to the limited availability of microwave ablation devices.

The laser is used as an energy source to heat tissues. An optic fiber with a diffuser tip is inserted in the tumor and the infrared laser raises the temperature around the diffuser. Laser ablation suffers from a limited range of heating away from the optic fiber, which limits its use to small tumors. In addition, ultrasound ablation is one of the most promising ablation techniques due to its noninvasive nature, and it is also called focused ultrasound surgery (FUS) or high-intensity focused ultrasound. A transducer located outside of the patient generates ultrasound and focuses ultrasound at the tumor location, rapidly raising the temperature at the focal point above the lethal threshold (Marien et al., 2014).

To observe the tumor for ablation, assess the treatment response, and determine the location of inserting the ablation applicator, an imaging device is used (Gianfelice et al., 2003). Ideally, the imaging technique for guiding the thermal ablation procedure should provide a target definition, anatomical information, and temperature feedback (de Senneville et al., 2007; Crocetti et al., 2008). Imaging plays a crucial role in the application of image-guided ablation techniques, where different imaging techniques [e.g., fluoroscopy, ultrasound, computed tomography (CT), and magnetic resonance imaging (MRI)] are used during the ablation procedure (Callstrom et al., 2002). Imaging has an important role before, during, and after the thermal therapy process. Before a procedure, imaging techniques are vital in surgery planning. During the actual procedure, imaging has a triple role in targeting, monitoring, and controlling the treatment. Finally, the assessment of treatment relies on imaging results (Dupuy, 2011).

Image-guided therapy has been one of the most rapidly expanding fields of medicine over the past two decades, providing a minimally invasive approach in the treatment of various tumors. Image-guided ablative therapy, encompassing the expertise in imaging technology and various tumor ablation techniques, is one of the most rapidly expanding fields in radiology and medicine (Shen et al., 2009). It uses any form of medical imaging to plan, perform, and evaluate surgical procedures and therapeutic interventions. The image-guided therapy technique helps to make surgeries less invasive and more precise, and it can lead to shorter hospital stays (Nakajima et al., 1997). For many surgical procedures, it has been accepted as a less invasive alternative. In this therapy, tumor ablation typically occurs when a clinician introduces it to a rapid thermal change.

Image-guided thermal ablation offers clinicians and patients a repeatable, effective, low-cost, and safe treatment. It has shown rapid growth in utilization in tumor treatment and has been adopted by the radiology community. The most commonly used imaging modalities in image-guided therapy include CT, MRI, and ultrasound. These procedures are also supported by surgical navigation equipment, tracking tools, and integration software.

9.2 Imaging modalities

9.2.1 Computed tomography

CT is an imaging modality that utilizes X-ray photons and reconstruction algorithms for image acquisition. The CT scanner essentially consists of an X-ray tube and detectors. The X-ray tube produces an X-ray beam that passes through the patient, and then the beam is captured by the detectors and reconstructed to create a two-dimensional image or three-dimensional volume image. The data captured by the scanner are reconstructed images by various algorithms, which represent a cross-sectional slice through the patient at that location (Yang et al., 2007; Seeram, 2015). Each image is acquired at a slightly different angle and results from a different reconstruction algorithm. The individual volume elements that make up the image are each displayed as a two-dimensional pixel with a designation of density or attenuation, represented by a Hounsfield unit. CT takes a distinctive place in radiology due to its capability of imaging tissues including soft tissue, bone, and vasculature (Lusic and Grinstaff, 2013).

During a CT scan, patients usually lie down within the scanning device, and the machine passes X-ray beams through the body to produce an image. CT scans can produce a detailed picture of bone, soft tissue, muscles, internal organs, and, crucially, tumors, or other irregularities. While CT scans are a very widely used type of diagnostic imaging, they also have some drawbacks for patients.

Of all the internal imaging procedures available to physicians, the CT scan is the most detailed and can give a doctor the most complete picture of what is happening inside a patient's body. They are particularly useful and widely used in diagnosing cancer. The CT scan procedure is noninvasive and painless and generally quick and convenient for most patients. It is widely available at a range of different treatment centers. Because a CT scan gives a doctor a very clear picture of where a tumor or other problem is located and whether it has spread, it is used in planning a biopsy, surgery, radiation, or other treatment with more precision.

Compared to other diagnostic tests, CT scans deliver a relatively high dose of radiation to the patient. While this is not usually a problem for a single scan, patients who need to undergo repeated exams can be subjected to a significant level of radiation, increasing their cancer risk. Because a CT scan can provide so detailed information, it can sometimes alert doctors to minor abnormalities in the body that do not have symptoms related to them and that in the normal course of life would not have caused the patient any problems. However, doctors may feel the obligation to divulge this information to patients, which can cause anxiety and possibly unnecessary follow-up tests or treatments.

9.2.2 Magnetic resonance imaging

MRI, a medical imaging technique, uses magnetic radiation to visualize detailed internal structures in radiology (Yu et al., 2011). It provides three-dimensional views of body organs with good soft-tissue contrast and visualizes the structure and function of the brain, heart, liver, soft tissues, and inside of the bones. It can be used to diagnose cancers, brain diseases, and cardiovascular conditions. MRI technologists are unique in the radiology field in that the machines they operate use magnets and radio waves rather than radiation. The MRI scan was previously called nuclear magnetic imaging and, in the mid-1980s, researchers changed the name from nuclear to magnetic due to the negative reaction from the public (Dowling et al., 2012). The MRI scanner is a cylindrical tube surrounded by a circular magnet that contains the magnet and detectors. The patient lies on a moveable examination table that slides in and out of the center of the magnet. A specialized computer is used to process all the imaging information and construct three-dimensional images. Patients must remove all metals before being placed into the scanner. Otherwise, the images of patients with any type of metal will not be clear.

MRI scans do not use X-rays unlike CT scans, and instead use a powerful magnetic field and pulses of radio wave energy to provide detailed images of organs and other structures in the body. The magnets in the magnetic field interact with the hydrogen atoms in human bodies, which are found in the water stored in the body. This allows the clinicians to see detailed images of the structure of the body part by using a specialized computer along with the help of the magnetic field and radio frequencies. MRI findings are based on a compilation of sequences that are an ordered combination of radiofrequency and gradient pulses designed to acquire the data to form the image. MRI is cost-effective and a better way to view fetal imaging since it can provide much more information than ultrasound and does not use ionizing radiation like X-rays. MRI can also be used for needle guidance during interventional procedures such as tissue biopsy and for determining how far along a tumor is (Xu et al., 2008; Pondman et al., 2008; Kokes et al., 2009).

The main advantages of the MRI scan include:

- No radiation and safely used in pregnant women and babies;
- Useful for showing soft-tissue structures, such as ligaments and cartilage, and organs, such as the brain, heart, and eyes; and
- Provides information on the blood circulation about how the blood moves through certain organs and blood vessels.

The main disadvantages of the MRI system are summarized as:

- It is a small and enclosed space and the loud noises made by the magnets can make some people feel claustrophobic during an MRI scan;
- Its scanner is affected by movement and is unsuitable to examine problems as mouth tumors due to coughing or swallowing, which makes the images blurred; and
- It cannot show the bones and calcium clearly.

9.2.3 Sonography

Sonography, more commonly known as ultrasound, is a diagnostic medical procedure that uses sound waves to produce images of different parts of the anatomy and visualize anatomical structures inside a body (Smith and Finnoff, 2009). This technique involves using high-frequency sound waves to produce dynamic visual images of blood flow, tissues, and organs (Berg et al., 2008). Sonography is often used in the general abdomen, vascular, cardiac, or obstetrics. One common use for sonography is in abdominal ultrasound, which allows medical professionals and mothers-to-be to check up on the growth and health of a developing fetus. In the ultrasound imaging system, a transducer emits sound waves (echoes) at a certain frequency and captures the returning echoes at frequencies dependent on the tissues through which the waves traverse. The sound wave that is returned to the transducer is digitized to appear as echoes or dots on the screen. Images are acquired in real time and can be acquired in any imaging plane.

Ultrasound may be conventional in real-time or Doppler. Doppler ultrasound is used for vascular studies using the Doppler shift phenomenon. Ultrasound is cheap, easy to perform, can be moved to remote places, and does not use ionizing radiation. The outcome of the study is operator dependent. Ultrasound has advantages including being noninvasive, a quick procedure, and good for examining soft tissues and fetuses. The main disadvantage of sonography is its inability to penetrate bone, which makes visualization of some areas of the body very difficult (Culjat et al., 2010). The presence of gas also affects the visual quality of ultrasound images, as gas induces poor-quality image output. Sonography visualization is poor when the target area is deep within the body. Obese patients are not suitable for sonography because of this problem. In addition, sonography is completely dependent on the skill of the operator. A diagnostic procedure performed by different levels of ultrasound technician can lead to inconsistent and incorrect results.

9.3 Image navigation system

The image navigation system has greatly optimized the process of interventional tumor treatment and provided a reliable guarantee for improving the precision and safety of surgery by dynamically guiding surgical instruments to operate on lesions through intraoperative images in real time.

9.3.1 History of the image navigation system development

The image navigation system has undergone a relatively rapid development, from the initial mechanical positioning navigation system to an optical positioning navigation system, and then developed to the latest electromagnetic positioning navigation system over a period of about 20 years (Holly and Foley, 2003). The earliest, in 1986 (Roberts et al., 1986), took the lead in combining the operating microscope and CT images, and used ultrasound to achieve success in intraoperative positioning, which is the prototype of the image navigation system (Black et al., 1997). The same year, Roberts et al. invented the robotic positioning navigation system, and reported for the first time its use in craniocerebral surgery, which was the earliest mechanical positioning navigation system used in the clinic. In 1992, the optical navigation system appeared. Heilbrun et al. (1992), using the binocular stereovision principle, developed a navigation system with infrared tracking technology, to achieve positioning of the spatial target. In 1991, Kato et al. (1991) launched the electromagnetic positioning navigation system, which generates a magnetic field through electromagnetic field generator, and then used it to transmit and receive electromagnetic signals to determine the spatial position of the target.

A Canadian company, NDI, developed the Auroca electromagnetic navigation system, which has now been successfully applied to interventional therapy, endoscopic diagnosis, cardiac surgery, neurosurgery, radiation oncology, and many other fields. In 2006, Veran developed the first image navigation system for minimally invasive thoracic and abdominal interventional surgery (Edwards, 2008). It is based on the principle of global positioning satellite (GPS) technology and navigates through electromagnetic induction, and the system's unique auto-matching and respiratory gating technology further improves the accuracy of the electromagnetic navigation system. The electromagnetic positioning navigation system is a new navigation system currently under development, and its introduction has caused the medical community's general interest in image navigation systems. So far, the concept of navigation system-assisted minimally invasive interventional surgery has basically taken shape, and the image navigation system has been widely promoted and applied in the clinic (Baert et al., 2012).

9.3.2 Product introduction of the video navigation system

An image navigation system can be divided into the mechanical positioning navigation system, optical positioning navigation system, and electromagnetic positioning navigation system according to different spatial positioning principles (Grunert et al., 2003). These three types of navigation systems have advantages and disadvantages.

9.3.2.1 *Mechanical positioning and navigation systems*

Mechanical positioning and navigation systems were the first of the three types to be used in clinical practice and consisted of an integrated planning workstation, a mechanical guiding arm, and a display panel (Cleary and Nguyen, 2001). In the early stage, the frame-positioning device was used to guide the needle by fixing the position and direction of the puncture needle. As the frame positioning system is bulky and inconvenient to move and operate, it was gradually changed to the robotic arm positioning method, in which the robotic arm positioning system guides the puncture positioning through the surgical instruments fixed on the end of the robotic arm.

The advantages of the robotic arm are flexibility, high space utilization, greatly improved positioning system of the frame type, and not affected by the environment. The main disadvantage is that the robotic arm guidance is more accurate only for rigid body puncture, and for soft tissue puncture, the pushing and pushing of the tissue may produce large errors. Second, the mobility of the robotic arm is still limited, and is mainly used to guide the puncture in a vertical direction or with a small angle, and the direction of the puncture cannot be chosen arbitrarily, which has certain limitations.

9.3.2.2 *Optical positioning and navigation systems*

Optical navigation equipment consists of navigation markers, cameras, and computer workstations. Through the navigation markers fixed in the tail end of the surgical instruments, the camera captures the optical signal reflected from the markers, and then through the program, the spatial coordinates of the surgical instrument's information is calculated (Cai et al., 2015). The advantages of an optical navigation system are that it uses infrared as the signal source, has high positioning accuracy, is completely free from electromagnetic interference, and is widely used. The disadvantage is that infrared is often easily obstructed by obstacles, which leads to inflexible operation and poor stability. Widmann et al. performed 145 needle CT-guided microwave ablation surgeries with optical navigation on 20 liver tumor patients, and the results showed that after the application of optical navigation, high accuracy of multineedle stereo positioning was obtained (Widmann et al., 2007). At present, the optical navigation system is more often used in neurosurgery, orthopedic surgery, and other surgeries, but there are fewer reports on guided puncture ablation and radioactive particle implantation.

9.3.2.3 Electromagnetic positioning and navigation system

The electromagnetic positioning and navigation system mainly consists of a computer monitor, an electromagnetic field transmitter, an electromagnetic field receiver, and an electromagnetic tracking adapter (Barai and Momin, 2020). Through the electromagnetic field, the generator produces a magnetic field, and then use its emission and reception of electromagnetic signals to determine the spatial position of the target (Mercier et al., 2005). The advantage of the electromagnetic navigation system is that electromagnetic waves, as the signal medium, mean the operator does not have to worry about the blocking of light, there is flexible and convenient operation that can be punctured at any angle, and puncture accuracy is high. The disadvantages of electromagnetic navigation are the poor adaptability of the puncture needle, accessories need to be reused, longer preparation time for surgery, it is easy to interfere with the magnetic field, and there is poor stability.

9.4 Image-guided thermal ablation therapy using different imaging modalities

During thermal therapies, the magnitude and spatial distribution of the temperature changes depend heavily on both the delivery parameters (intensity, duration, etc.) of the exposure as well as the tissue properties (absorption, blood perfusion, and flow, etc.) (McDannold and Jolesz, 2000, Mast et al., 2005). These properties are major unknowns in any treatment. Therefore the outcome of such treatments is difficult to control, even after the treatment device has been located successfully and targeted in the tissue. Guidance (monitoring and control) of thermal therapies is a necessity (McDannold and Jolesz, 2000). An ideal image guidance method would provide the following capabilities (McDannold and Jolesz, 2000): target localization, targeting of the treatment device and the heat/cold zone, monitoring of temperature quantification, control (feedback either to maintain the desired temperature or capability to image the resulting tissue damage), and rapid image acquisition (temporal resolution matched to the time constant of thermal effects).

Image-guided thermal therapy, encompassing the expertise in imaging technology and various tumor ablation techniques, is one of the most rapidly expanding fields in radiology and medicine. For many surgical procedures, it has been accepted as a less invasive alternative. In thermal ablation, tumor cells are destroyed with either extreme heat (radiofrequency and microwave ablation) or extreme cold (cryoablation). To locate the lesion for ablation and assess the treatment response, the clinician uses imaging. Ideally, this imaging technique for guiding thermal ablation should provide a target definition, anatomical information, and temperature feedback.

During the therapy, different imaging modalities, such as sonography, CT, or MRI are used to help guide the needle into the tumor. Image-guided tumor ablation is a minimally invasive procedure and is used to treat tumors from liver cancer, kidney cancer, lung cancer, and bone cancer. Image-guided ablation may also be used to reduce the size of a tumor so it can be more easily eliminated by conventional treatment such as chemotherapy or radiation. Image-guided ablation may also provide relief when a tumor invades the chest wall and causes pain. Image-guided ablation can be performed under ultrasound, CT, or MRI guidance. The tumor should be well visualized on the imaging modality planned for the procedure (Maybody, 2010).

9.4.1 Sonography-guided thermal ablation techniques

Sonography plays an important role in the guidance of both percutaneous and intraoperative thermal ablation procedures. It is frequently used for guidance in electrode placement because of its flexibility, relative availability, low-cost, and real-time imaging capabilities. The role of sonography may be limited in some circumstances, such as for targeting the nidus of osteoid osteoma or for targeting intrapulmonary lesions (Chiou et al., 2007). Sonography provides real-time images for positioning the electrode and monitoring the thermal lesion.

The electrode needle, as well as the deployed prongs or tines, can be shown by sonography. Sometimes it may be difficult to know the precise positions of the prongs or tines in an organ being ablated with standard sonography, and such a condition can lead to problems, especially when a prong or a tine is positioned near a vulnerable structure. In these circumstances, three-dimensional sonography may provide more information about the location of the expanded prongs and the adjacent structures (Chiou et al., 2007).

Doppler flow imaging is helpful to show the vascular structure adjacent to the tumor, which should always be carefully evaluated before the procedure for needle route planning and consideration of a "heat sink" effect related to larger vessels nearby. On a sonogram, the ablated area shows increased echogenicity as the temperatures rise and then tissue coagulation necrosis occurs. When tissue is heated, dissolved gases are released to form microbubbles in the tissue, which appear as hyperechoic foci and sometimes escape into adjacent vessels. It is hard to predict the size of the necrosis based on this information. The limitation of monitoring ablation outcomes during the procedure could be improved by using a contrast-enhanced technique.

The advantages of ultrasound are its "real-time" capability and lack of ionizing radiation. Ultrasound can be used for accurate placement of applicators before monitoring using CT or MRI. Currently, sonography is most often used for guidance in needle placement during the RFA procedure, especially for hepatic neoplasms and soft-tissue tumors. This is because of the widespread availability and real-time imaging capabilities of sonography. However, for intrapulmonary lesions, intramedullary bone lesions, and some intraabdominal lesions, which are difficult to approach under sonographic guidance, CT or MRI can provide ideal imaging guidance (Chiou et al., 2007). In addition, imaging with ultrasound is highly operator dependent and may be compromised in certain settings, such as patients with large body habitus, the presence of abundant bowel gas, and when the tumor is near a focal loop of bowel that needs to be avoided. Another drawback to ultrasound is the degradation of landmarks during ablation. In cryoablation, image degradation is caused by acoustic shadowing on the far side of the ice ball, and in RFA it is produced by microbubbles (Ortiz-Alvarado and Anderson, 2010).

Recent advancements in ultrasound technology are also aiding guidance during percutaneous ablation. Contrast-enhanced flow imaging and four-dimensional ultrasonography (US) allow the operator to display the accurate positioning of the ablation needle in the target of hypervascular tumors, and to better evaluate the completeness of ablation posttherapy, with good correlation to contrast-enhanced CT. Real-time temperature estimation using ultrasound appears feasible in tissue phantoms but has been limited clinically by motion artifacts (Hotta et al., 2006).

After thermal ablation therapy, the follow-up imaging is used in determining whether the ablation is complete and for early recurrences. For most intraabdominal tumors, this is most frequently performed by contrast-enhanced CT or dynamic MR imaging. The size of the necrotic defect immediately after treatment should be larger than that of the pretreatment tumor, and the margin of the thermal lesion should be sharp. Peripheral enhancement may be present, appearing as a regular rim along the boundary of the necrotic area. A residual viable tumor is reliably depicted as it stands out on arterial phase images against the unenhanced areas of coagulation necrosis. Signs of recurrence on MR imaging include new enhancement areas, an increase in lesion size, and the development of T1-weighted hypointense and T2-weighted hyperintense areas. As with enhanced CT and MR imaging, contrast-enhanced sonography can be used for the evaluation of thermal lesions after RFA (Chiou et al., 2007).

9.4.2 CT-guided thermal ablation techniques

CT is not as operator dependent as ultrasound. It covers the critical organs and structures that need to be avoided. CT scanning is much less sensitive to body habitus than ultrasound and images are not affected by bowel gas. The target lesion should ideally be visible on a noncontrast CT examination. Ablation can be performed using a conventional CT scanner or a CT scanner with real-time fluoroscopic capability (Ortiz-Alvarado and Anderson, 2010).

Ablation is most often performed under ultrasound or noncontrast CT guidance. These modalities are sufficient for lesion detection and probe placement in most cases, but neither can depict the exact ablation margins during treatment. The real-time CT–ultrasound fusion imaging matches a preprocedural volumetric CT to real-time ultrasound images (Mcwilliams et al., 2010). An electromagnetic tracking system mounted on the ultrasound probe provides the position and orientation of the probe and permits representation of the corresponding multiplanar reformatted CT image in the same plane and position. A feasibility study showed high and consistent levels of matching accuracy between preprocedural CT and real-time ultrasound imaging in an in vitro liver model, with a mean registration error of only 3 mm (Crocetti et al., 2008).

The fusion imaging technology uses computer programs to analyze preprocedural CT, segment the tumor volume to be treated, design the desired needle approaches for the necessary number of overlapping ablation spheres, and provide real-time guidance during needle placement. CT-integrated robots that partially or fully automate the needle placement process based on a user-defined treatment plan have been designed in tumor ablation (Wood et al., 2007).

9.4.3 MRI-guided thermal ablation techniques

Thermal ablation is most commonly performed under ultrasound or noncontrast CT guidance. These modalities are sufficient for lesion detection and probe placement in most cases, but neither can depict the ablation margins during treatment (Mcwilliams et al., 2010). MRI-based monitoring has been shown in recent years to enhance the effectiveness of minimally or noninvasive thermal therapy techniques. MRI's unique soft-tissue contrast and ability to image in three dimensions and any orientation make it extremely useful for treatment planning and probe localization. The temperature sensitivity of several intrinsic parameters enables MRI to visualize and quantify the progress of ongoing thermal treatment. MRI is sensitive to thermally induced tissue changes resulting from the therapies, giving the physician a method to determine the success or failure of the treatment (McDannold and Jolesz, 2000).

MRI provides excellent soft-tissue resolution with multiplanar imaging capability and lacks ionizing radiation. MR fluoroscopic sequences can be used for real-time guidance. MR thermography can assess cytotoxic tissue temperatures noninvasively. Image-guided thermal ablation is performed using a dedicated interventional magnet, a conventional solenoid magnet, or an open magnet. The ice ball is visualized as a zone of decreased signal intensity on T1- and T2-weighted images (Maybody, 2010).

The chemical environment and relaxation properties of nuclei are the source of the signal in MR. MR is also sensitive to Brownian motion and associated molecular tumbling rates; thus, MRI techniques are intrinsically sensitive to temperature. As the technical and engineering research has advanced, a new field of MR thermometry has been developed. The testing and validating of MRI-based temperature methods have demonstrated the accuracy of the current methods. Using MR thermometry, it is possible to localize the focal spot and to monitor the ablated zone in real time (de Senneville et al., 2007). The development of image-based real-time "MR thermometry" methods imaging allows for a complete system for guided and monitored thermal ablation therapy. By choosing the correct parameters, MR thermometry can provide all the requirements for thermal imaging during ablation, including temperature sensitivity, the linearity of temperature effect, and adequate spatial and temporal resolution, with minimal artifact (de Senneville et al., 2007).

With the continuous development of image-guided navigation systems, tumor interventional diagnosis and treatment are completed from preoperative images to virtual real-time image guidance, single-needle ablation to multineedle combined ablation, and physician manual operation to navigation robot-assisted guidance. At present, we tend to rely on a single imaging modality for guidance, however, whether it is ultrasound, CT, or MRI, their different imaging characteristics have their own advantages in guidance. It is foreseeable that giving full play to the advantages of various imaging technologies to form a multimodal image fusion navigation technology is one of the directions of future development.

9.5 Image processing during the imaging-guidance process

With the rapid development of computer technology, medical imaging technology is also developing fast. It provides a wide range of morphological and functional imaging signals for clinical medicine. However, the various imaging techniques and screening methods have their advantages and disadvantages. N single imaging technique can be applied to all human organs and no one imaging technology can replace another; they are complementary and mutually reinforcing.

The information provided by medical images is divided into two main categories: images of anatomical structures (CT, MRI, B-ultrasound, etc.), and functional images (SPECT, PET, etc.). Anatomical images provide anatomical and morphological information about the organ at a high resolution, but they do not reflect the functional status of the organ or lesion. Although functional images do not provide anatomical details of the organ or lesion and have poor resolution, they can provide functional metabolic information about organs.

In clinical situations, usually multiple sessions are needed for the same patient using one model or multiple model images, that is, getting information from several images at the same time. This information is used for comprehensive analysis. In addition, the use of the same imaging equipment in different time-based imaging allows lesion growth to be observed, and the results of treatment before and after surgery to be compared, etc. When the information provided by an imaging device is not sufficient, multiple modules can be used. For example, looking at bone tissue on CT, while obtaining soft-tissue information from MRI; or combining the functional information derived from PET and SPECT, with the anatomical information from CT, MRI, and other sources.

In some complicated conditions, only one imaging modality is not sufficient to observe the lesions. Multiple imaging modalities are combined to provide more information on lesions. In clinical practice during image-guided thermal therapy, a major problem using multiple modalities is the task of image registration for different tissues, where the organ shape changes occurring between preprocedural and intraprocedural imaging pose significant challenges (Collins et al., 2017; Chu and Dupuy, 2014). To deliver thermal ablation with satisfactory clinical outcomes, accurately targeting the tumor area is critical. Typically, intraprocedural images are captured to track and guide the probe puncture, which often need to be aligned with a preprocedural image. The latter usually has high-resolution and high-quality delineation of regions of interest (ROIs), such that the probe can be more easily positioned (Maybody, 2010).

The most effective solution is the use of information based on medical image registration technology. The fusion technology combines these two types of images, taking advantage of their respective information advantages, to express multiple information from the human body in one image at the same time. It provides more intuitive information on human anatomy, physiology, and pathology. In particular, the registration technology is a prerequisite for image fusion and must be aligned before accurate integration.

The captured images using any of the image-guidance modality during the ablation process are passed through the well-known image processing stages, including image enhancement, segmentation, clustering, feature extraction, features selection, classification, and fusion (McCreedy et al., 2006, Giesel et al., 2009; Penney et al., 2001; Nicolau et al., 2009;

Ahmad et al., 2017; Ashour et al., 2015). Accordingly, an overview of such stages is described next.

9.5.1 Medical image segmentation

Medical image segmentation methods are often derived from the state-of-the-art techniques of image segmentation (Elnakib et al., 2011; Dey and Ashour, 2018; Hore et al., 2016; Dey et al., 2018; Rajinikanth et al., 2018). Different segmentation techniques include (1) traditional methods: threshold segmentation, region growth segmentation, (2) classification and clustering methods: K-nearest neighbors (KNN), fuzzy C-means (FCM), Monte Carlo random field model (MRF), dataset-based guided segmentation, expectation maximization (EM), Bayesian methods, support vector machines (SVM), artificial neural networks (ANNs), random forest methods, and convolutional neural networks (DNNs), and (3) deformation model methods: parametric deformation models, geometric deformation models.

9.5.2 Medical image registration

Medical image registration is the process of seeking a single or a series of spatial transformations to reach a correspondence point with another medical image. This spatial congruence is defined as the same anatomical point on the human body in two matching images that have the same spatial position (same position, same angle, same size). The result of the registration should be such that all the anatomical points on both images, or at least all points of diagnostic significance and surgical interest, are matched. For a quantitative analysis of several different images, the first step is to combine the images. The image is transformed into a common coordinate frame for future study. This is the strict alignment of the image. Medical image registration is the search for some kind of spatial transformation to achieve complete agreement in the spatial position and anatomical structure of the features of two images, requiring that the result of the alignment match all anatomical points on the two images or at least all points of diagnostic significance and within the surgical area. It seeks spatial transformations for a medical image, through the spatial agreement of corresponding points with another medical image. This concordance is the same on the human body in two matching pictures. The result of the registration should be that all anatomical points on the two images, or those up to and including the anatomical point, have the same spatial position on the image.

Registration addresses the correspondence of positions of the same anatomical structure in images acquired at different angles, times, or imaging devices. It includes several steps such as data input, image feature extraction, image transformation, region of interest calculation, and optimization processes. It also has steps for dimensional consistency of the images, either by projecting 3D images into 2D space or by converting 2D images into 3D space.

Image mapping transformation for image feature deformation solves image distortion problems such as translation, telescoping, and rotation in rigid body transformation (Chakraborty et al., 2016, 2017, 2020). In addition, there are a significant number of algorithms that combine the manual description of features with automated feature extraction to help the computer unfold image registration more accurately by manually or automatically extracting obvious structural features of the image, such as the edges of bones, bifurcations of blood vessels, and other features. A typical flow chart of image registration is shown in Fig. 9.1.

Multimodality medical image alignment means that the two images to be aligned are derived from different imaging devices. For example, both CT and MR images have high spatial resolution, while the former works better with denser tissues and the latter can identify soft tissues. SPECT and PET can provide information on body functions and metabolism, but spatial resolution is poor. Therefore it is important to use PET in clinical practice to be registered with CT (or MR). The combination provides both functional and anatomical information that is of clinical value. Because of the scanning setup, the scanning parameters of the two tomographic images do not exist because of the different principles and conditions of the scanning parameters, and there is not a simple one-to-one correspondence. Multimodal medical image registration is the focus of medical image registration research. Medical image registration has important clinical applications. The registration of medical images obtained from different or identical imaging modalities can not only be used for medical diagnostics, surgical planning, radiotherapy, but also takes a variety of measures, including planning, follow-up of pathological changes, and evaluation of treatment effects.

According to the features used in the registration methods, they are divided into eight groups as follows:

- Point method

  The point method is divided into the intrinsic points and extrinsic points methods. Internal points are derived from

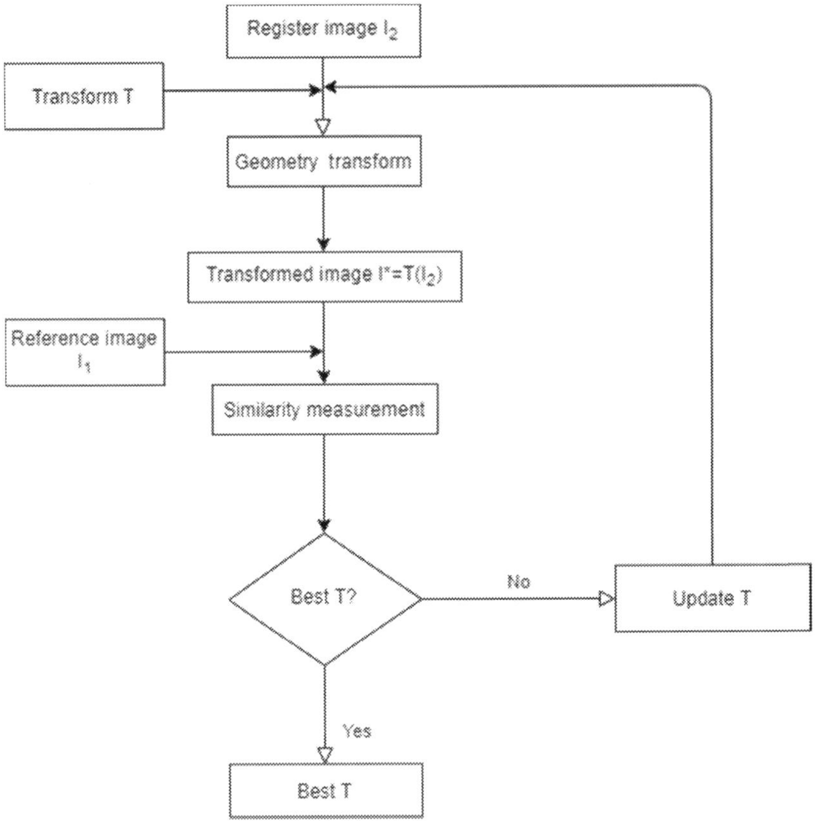

Figure 9.1 Flow chart of image registration.

patient-related image properties such as anatomical landmark points. External points are screws embedded in the subject's skull, markings on the skin, or other features that can be detected in both images. Once both internal and external points have been identified, the problem of aligning the two images boils down to a simple matter of finding the best match. The rigid body transform of the corresponding set of points is solved. With these markers aligned, the two images are now aligned (Collins et al., 2017).

- Curve method

  For 2D projected radiographs, we first manually find the corresponding open curves in the two images, and then find the open curves in the two images. The best fit for the local curvature of a curve is to find a set of corresponding points at the same sampling rate and use the point method to register two images (Xuan et al., 2006).

- Surface method

  A typical example of a surface-based registration technique is the "hood method." The set of points extracted from an image contour is called a "hat." A surface model extracted from another image rotunda is called the "head." The larger patient images, or higher resolution images when the image volume is of similar size, are used to generate a head surface model. The Powell search algorithm is used to find the desired geometric transformation to minimize the mean squared value of the distance between the cap and the head surface (Makela et al., 2002).
- Moment and principal axes method

  Using the concept of mass distribution in classical mechanics, we calculate the center of mass and principal axes of the two image pixels, and then pass the pan and rotate to align the centers of mass and the principal axes of the two images to achieve the alignment (Alpert et al., 1990).
- Correlation method

  For images of the same object due to differences in image acquisition conditions or small changes in the object sequence itself, the principle of maximizing the similarity between images is used to achieve interimage alignment. The similarity measurement can be of various types, such as a correlation function, a correlation coefficient, the squared sum of differences, or the absolute sum of differences (Hill and Batchelor, 2001).
- Maximization of mutual information

  Reciprocal information is a basic concept of information theory, a measure of the statistical correlation between two random variables. The mutual information is used as a measure of the alignment of multimode medical images. If the two images are geometrically aligned, the correspond voxel pair's intensity values have the greatest mutual information (Tsao, 2003).
- Atlas method

  Alignment of different brain images is far more difficult than alignment of different modal images of the same person because the human brain varies greatly in shape and size. If we transform the brain image to a certain scale, and measure a deep shape of the anatomical structures of different human brains, we find that there are some commonalities when the internal structures of the brain are properly oriented, which makes it possible to construct an anatomical map that presupposes a topology of the intersubject brain. Due to the direct involvement of neuroanatomical experts in the construction of brain maps, the use of brain maps can be used as a tool for the identification of the structure. If the spectra are aligned, the prior knowledge contained in the spectra can be used to automatically identify images of the patient or others (Vemuri et al., 2003).

- Nonlinear transformation techniques

In order to study the distortion of organs and tissues or the alignment of images between different human brains, the use of nonlinear transform alignments is required. The main techniques are physical model-based and spatial transformation-based registration methods (Lester and Arridge, 1999).

9.5.3 Medical image fusion

The integration of multimodal image information is called fusion. Image fusion methods can be divided into three groups: pixel-level image fusion, feature-level image fusion, and decision-level image fusion. The main methods are the neural network method, Bayesian estimation method, fuzzy clustering method, and expert system method (Li et al., 2017a, Yang and Li, 2012; Shahdoosti and Tabatabaei, 2019). Medical image fusion refers to the fusion of two (or more) images from different imaging modalities, or a calibrated image acquired at different times and combining the strengths or complementarities of the individual maps organically to gain confidence and a more informative map.

According to the fusion image imaging method, there is monomodality and multimodality fusion. According to the different fusion objects, there are single-sample time fusion, single-sample fusion, single-sample spatial fusion, and template fusion. Single-sample time fusion refers to following a patient for a period and fusing images from the same examination of the same organ. Single-sample spatial fusion refers to the fusion of images from the same examination of the same patient over the same period from several examinations performed on the organ. Template fusion creates a series of templates from many studies of healthy people and fuses the patient's image with the template image. The image fusion methods can be categorized into two groups: image fusion based on spatial domain and image fusion based on transfer domain.

9.6 Image-guidance-based thermal ablation

Image guidance using US, CT, or MRI has a critical role during each stage in the ablation procedure, starting from planning and ending with the validation of the success of the treatment plan using ablation. Typically, the thermal ablation techniques proceed in four sequential phases: treatment planning, targeting the tumor, heating for ablation, and validation of the ablation completeness, where image guidance is used during each phase, as shown in Fig. 9.2.

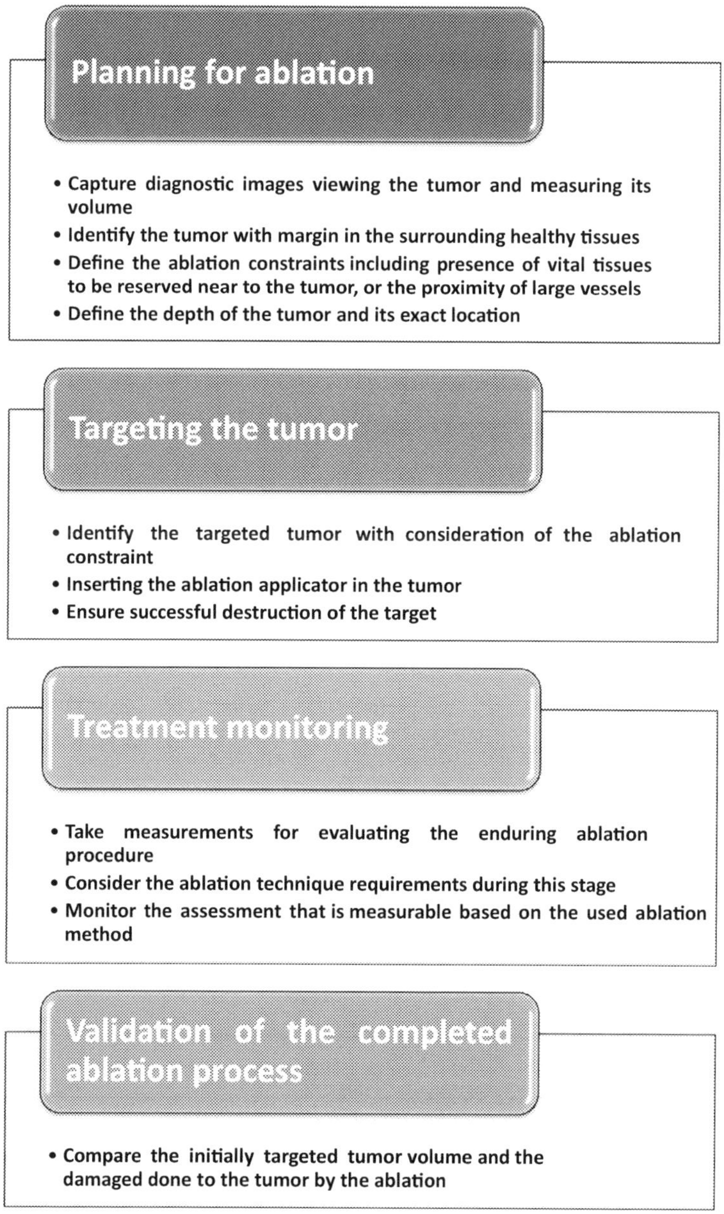

Figure 9.2 Image guidance-based thermal ablation techniques in its four sequential phases.

In monitoring of the treatment, the quantities to be measured for evaluating the ablation process depend on the applied ablation process. For example, some methods rely on measurement of the electrical impedance for the assessment of the treatment plan, while others (such as RF) measure the temperature at the inserted needle to evaluate the ablation progress. Since the thermal dose is the best gauge of final tumor size during the ablation process, to measure the required quantities to monitor the treatment procedure, not all imaging modalities can be used during this phase. For example:

- The CT scan cannot be used in monitoring of the treatment phase;
- The ultrasound provides only qualitative information, without providing any temperature maps.

On the other hand, the MR can provide monitoring information about the progress of the treatment, such as the thermal dose maps and the temperature maps.

At the end phase of the ablation process, the validation and assessment of the measurements performed during the procedure or the pre-established ablation protocols are carried out. Tactlessly assessing the ablated volumes using imaging modalities immediately after the end of the ablation process is untrustworthy due to the occurrence of some physiological changes, such as edema and/or hemorrhage, around the treated volume. Accordingly, from a clinical point of view, it is recommended to validate the ablation procedure about 3 months after the ablation treatment.

Generally, using image guidance during the previously mentioned four phases is important. However, the main task of image guidance is during the targeting phase, which includes the following stages, namely an energy source based on the used modality method (i.e., RF, MW, laser, or US), the ablation antenna which delivers the power from the energy source generator, feedback system using an embedded system to control the energy delivery using algorithms, and the imaging-guidance modality to guide the ablation process during the needle insertion in the tumor. Sometimes, cooling systems may be used to cool the tissues close to the targeted tumorous tissues. A general block diagram of the thermal ablation system using imaging guidance is demonstrated in Fig. 9.3.

In the general ablation system, the generated power from the energy source based on the used ablation modality is passed to the targeted tumorous tissues through the ablation probe/applicator. The imaging-guidance system is used to support the clinician during the insertion process of the applicator

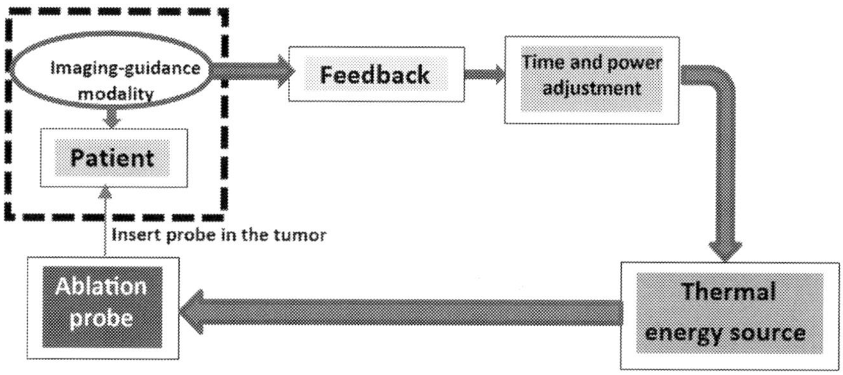

Figure 9.3 Block diagram of thermal ablation system using imaging guidance.

in the tumor tissues. It also can be used to assist the physician while selecting the parameters of the ablation probe based on the tumor size and allowable temperature calculated from the tumor location in a specific organ. For example, if the tumor is in the brain, the allowed power is low, thus, based on the control algorithm, the control of the output power from the generator will consider this limitation based on the CT imaging. The imaging modality can be integrated with an additional sensor within the antenna to assess temperature within the tumor and its surrounding healthy tissues, as well as to assess the treatment improvement. To improve the ablation process, the imaging guidance can consider the electrical and thermal properties of different tissues. The imaging modality, such as the CT, MRI, or US, can be used in the assessment of several factors, including:

- Damage ratio in the tumor and its nearby tissues;
- The zone where the ablation energy is deposited;
- Changes in the reflected power, which indicate the probe impedance mismatch;
- Ablation pattern of the tumor;
- Temperature; and
- Thermal dose.

Based on the exhibited images from the imaging modality, feedback is used to control other adjustment parameters, such as:

- Ablation duration;
- Positioning of the applicator; and
- Applied power level.

To date, several stages in the ablation system remain under investigation for superior, successful treatment using the ablation process during real-time monitoring. Typically, the thermal dose indicates the temperature history during the ablation process, which is the best indicator of the heat-induced tissue destruction, and when the ablated endpoint is reached. Thus, temperature monitoring is significant in guaranteeing that the healthy tissues surrounding the tumor are secure.

Magnetic resonance is a preferable imaging modality for guiding the thermal ablation process as it has an inimitable ability to produce temperature images of internal organs. This ability makes it recommended for real-time monitoring of the ablation procedure. Generally, MRI has several computable temperature-dependent measures, such as the proton resonance frequency (PRF), the T1 relaxation time, and the water molecule apparent diffusion coefficient (ADC). To measure the temperature from the MRI images, the proton resonance frequency is considered as it is also tissue independent for assisting the control of the treatment process (Denis de Senneville et al., 2005, Quesson et al., 2000; Fuentes et al., 2011). In in vivo studies, these quantities are used to generate temperature images for monitoring the ablation process, as originated by Parker et al. (1983). Since MRI provides volumetric views of the organ under monitoring with superior contrast compared to other modalities, it is recommended for temperature imaging during the focused ultrasound ablation. In addition, while using the focused ultrasounds ablation technique, the applicator is inserted at a location which allows acceptable ultrasound propagation to the targeted tumor tissues. This position is usually outside the patient's body. To validate the correct positioning of the applicator in the tumor, a low-power measurable test shot is applied, and a focal spot place is confirmed before starting the treatment phase. Accordingly, to date, the MRI of the temperature is considered the only usable imaging guidance for measuring the position of such a test shot, which is vital for the patient's treatment and safety. However, generally, MRI is enormously sensitive to magnetic susceptibility variations and RF interferences.

Several studies are striving to provide developed ablation systems by considering the main guidelines of the therapeutic device design, namely clinical efficiency, patient safety, ablation treatment consistency, cost effectiveness, an easy learning curve to guide the physician, and fast ablation process by real-time optimization of the heat pattern. Such designs should include volumetric ablation for complete tumor damage, temperature imaging for real-time monitoring during the treatment procedure, and adaptive control based on the measured feedback of the treatment parameters, such as temperature.

9.7 Conclusions

Image-guided therapy supports medical devices during cancer treatment. Recently, from the medical community point of view, thermal ablation, which is a local tumor treatment method, has received increased attention. Compared to surgical resection, thermal ablation is considered a promising treatment method for localized cancer treatment as it does not remove the tumor, but rather damages it in situ by raising the temperature above a lethal threshold. These minimally invasive ablation methods generally require a reduced stay in hospital. In this context, image-guided tools are used to support the ablation devices. Accordingly, image-guided thermal ablation supplants surgery for small local tumors. In thermal ablation therapy, the needle-like applicators are placed directly into tumors using imaging guidance and destroy tumors by the application of either intense heat or cold. In this chapter, the latest advancements in image-guided thermal ablation technologies are summarized and discussed, and different imaging techniques such as imaging modalities, medical image segmentation, registration, and fusion are discussed. Usually, tumors are visible with a good conspicuity with the use of contrast media. The main three imaging modalities which are used with the thermal ablation procedures are MRI, US, and CT.

- MRI is considered the superior diagnosis imaging modality with different contrast sources. Compared to the other imaging modalities, MRI can monitor the therapeutic ablation procedure by providing temperature images in real time during the treatment procedure. Nevertheless, under MRI, positioning the applicator is problematic due to limited accessibility to the patient. Thus, for example, RF ablation under MRI guidance is difficult. Conversely, FUS ablation does not need bulky applicator positioning, but instead a simple mechanical translation of the antenna to face the tumorous zone.
- US is a cheap, widely available, and easy imaging modality with inherent interactivity. It is a radiation-free real-time imaging modality, which provides high-resolution nonaxial imaging that can be used for continuous monitoring of the ablation. It is suitable during the insertion of the RF applicator during the RFA process. Since it has limited availability to image the temperature maps during the ablation process as it provides only qualitative information about the treatment, it is not recommended for monitoring the thermal ablation procedure. However, recent US generations can be equipped

with imaging fusion facilities to guarantee correct tumor targeting (Andersson et al., 2015; Li et al., 2017b).

- CT is used as imaging guidance when it is impossible to use US, such as when the tumor is not detectable with US in the case of obese patients, or organs without good acoustic windows. It offers a larger field of view and superior visualization of structures holding air and tumors, compared to US. However, CT uses ionizing radiation, leading to axial view visualization with the use of contrast media. Such cases make the applicator insertion cumbersome, and ablation monitoring impossible. Thus, CT imaging can be used post-treatment within a 3-month follow-up due to its wide availability compared to MRI. Also, ablation systems which are supported by both CT images and real-time US fusion may be extremely efficient for image-guided tumor ablation.

The clinical studies consider MRI as the superior imaging guidance to support the ablation procedure.

References

Ahmad, I., Gribble, A., Murtza, I., Ikram, M., Pop, M., Vitkin, A., 2017. Polarization image segmentation of radiofrequency ablated porcine myocardial tissue. PLoS One 12, e0175173.

Alpert, N., Bradshaw, J., Kennedy, D., Correia, J., 1990. The principal axes transformation—a method for image registration. J. Nucl. Med. 31, 1717–1722.

Andersson, M., Hashimi, F., Lyrdal, D., Lundstam, S., Hellström, M., 2015. Improved outcome with combined US/CT guidance as compared to US guidance in percutaneous radiofrequency ablation of small renal masses. Acta Radiol. 56, 1519–1526.

Ashour, A.S., Samanta, S., Dey, N., Kausar, N., Abdessalemkaraa, W.B., Hassanien, A.E., 2015. Computed tomography image enhancement using cuckoo search: a log transform based approach. J. Signal. Inf. Proc. 6, 244.

Baert, A., Gunther, R., Von Schulthess, G., 2012. Interventional Magnetic Resonance Imaging. Springer Science & Business Media.

Barai, S., Momin, M., 2020. Outside-in electromagnetic tracking method for augmented and virtual reality 6-degree of freedom head-mounted displays. 2020 4th International Conference on Intelligent Computing and Control Systems (ICICCS). IEEE, pp. 467–476.

Berg, D., Godau, J., Walter, U., 2008. Transcranial sonography in movement disorders. Lancet Neurol. 7, 1044–1055.

Black, P.M., Moriarty, T., Alexander, E., Stieg, P., Woodard, E.J., Gleason, P.L., et al., 1997. Development and implementation of intraoperative magnetic resonance imaging and its neurosurgical applications. Neurosurgery 41, 831–845.

Cai, K., Yang, R., Ning, H., Ou, S., Zeng, Z., 2015. An automatic algorithm for distinguishing optical navigation markers used during surgery. Dyna 90, 203–209.

Callstrom, M.R., Charboneau, J.W., Goetz, M.P., Rubin, J., Wong, G.Y., Sloan, J.A., et al., 2002. Painful metastases involving bone: feasibility of percutaneous CT- and US-guided radio-frequency ablation. Radiology 224, 87–97.

Chakraborty, S., Dey, N., Samanta, S., Ashour, A.S., Balas, V.E., 2016. Firefly algorithm for optimized nonrigid demons registration. Bio-Inspired Computation and Applications in Image Processing. Elsevier.

Chakraborty, S., Dey, N., Samanta, S., Ashour, A.S., Barna, C., Balas, M., 2017. Optimization of non-rigid demons registration using cuckoo search algorithm. Cognit. Comput. 9, 817–826.

Chakraborty, S., Pradhan, R., Deyb, N., Ashour, A.S., 2020. Optimized Tang's algorithm for retinal image registration, Information Technology and Intelligent Transportation Systems, 323. p. 142.

Chiou, S.-Y., Liu, J.-B., Needleman, L., 2007. Current status of sonographically guided radiofrequency ablation techniques. J. Ultrasound Med. 26, 487–499.

Chu, K.F., Dupuy, D.E., 2014. Thermal ablation of tumours: biological mechanisms and advances in therapy. Nat. Rev. Cancer 14, 199–208.

Cleary, K., Nguyen, C., 2001. State of the art in surgical robotics: clinical applications and technology challenges. Comput. Aided Surg. 6, 312–328.

Collins, J.A., Weis, J.A., Heiselman, J.S., Clements, L.W., Simpson, A.L., Jarnagin, W.R., et al., 2017. Improving registration robustness for image-guided liver surgery in a novel human-to-phantom data framework. IEEE Trans. Med. Imaging 36, 1502–1510.

Crocetti, L., Lencioni, R., Debeni, S., See, T.C., Della Pina, C., Bartolozzi, C., 2008. Targeting liver lesions for radiofrequency ablation: an experimental feasibility study using a CT–US fusion imaging system. Investig. Radiol. 43, 33–39.

Culjat, M.O., Goldenberg, D., Tewari, P., Singh, R.S., 2010. A review of tissue substitutes for ultrasound imaging. Ultrasound Med. Biol. 36, 861–873.

de Senneville, B.D., Mougenot, C., Quesson, B., Dragonu, I., Grenier, N., Moonen, C.T., 2007. MR thermometry for monitoring tumor ablation. Eur. Radiol. 17, 2401–2410.

Denis de Senneville, B., Quesson, B., Moonen, C.T., 2005. Magnetic resonance temperature imaging. Int. J. Hyperth. 21, 515–531.

Dey, N., Ashour, A.S., 2018. Meta-heuristic algorithms in medical image segmentation: a review. Advancements in Applied Metaheuristic Computing. IGI Global.

Dey, N., Ashour, A.S., Shi, F., Balas, V.E., 2018. Soft Computing Based Medical Image Analysis. Academic Press.

Dowling, J.A., Lambert, J., Parker, J., Salvado, O., Fripp, J., Capp, A., et al., 2012. An atlas-based electron density mapping method for magnetic resonance imaging (MRI)-alone treatment planning and adaptive MRI-based prostate radiation therapy. Int. J. Radiat. Oncol. Biol. Phys. 83, e5–e11.

Dupuy, D.E., 2011. Image-guided thermal ablation of lung malignancies. Radiology 260, 633–655.

Edwards, J.R., 2008. Methods, apparatuses, and systems useful in conducting image guided interventions. Google Patents.

Elnakib, A., Gimel'farb, G., Suri, J.S., El-Baz, A., 2011. Medical image segmentation: a brief survey. Multi Modality State-of-the-Art Medical Image Segmentation and Registration Methodologies. Springer.

Fuentes, D., Walker, C., Elliott, A., Shetty, A., Hazle, J.D., Stafford, R., 2011. Magnetic resonance temperature imaging validation of a bioheat transfer model for laser-induced thermal therapy. Int. J. Hyperth. 27, 453–464.

Gianfelice, D., Khiat, A., Amara, M., Belblidia, A., Boulanger, Y., 2003. MR imaging–guided focused US ablation of breast cancer: histopathologic assessment of effectiveness—initial experience. Radiology 227, 849–855.

Giesel, F., Mehndiratta, A., Locklin, J., Mcauliffe, M., White, S., Choyke, P., et al., 2009. Image fusion using CT, MRI and PET for treatment planning, navigation and follow up in percutaneous RFA. Exp. Oncol. 31, 106.

Grunert, P., Darabi, K., Espinosa, J., Filippi, R., 2003. Computer-aided navigation in neurosurgery. Neurosurg. Rev. 26, 73–99.

Heilbrun, M.P., Mcdonald, P., Wiker, C., Koehler, S., Peters, W., 1992. Stereotactic localization and guidance using a machine vision technique. Stereotact. Funct. Neurosurg. 58, 94–98.

Hill, D.L., Batchelor, P., 2001. Registration methodology: concepts and algorithms. Medical Image Registration. pp. 39–70.

Holly, L.T., Foley, K.T., 2003. Intraoperative spinal navigation. Spine 28, S54–S61.

Hore, S., Chakraborty, S., Chatterjee, S., Dey, N., Ashour, A.S., Van Chung, L., et al., 2016. An integrated interactive technique for image segmentation using stack based seeded region growing and thresholding. Int. J. Electr. Comput. Eng. 2088–8708, 6.

Hotta, N., Maeno, T., Ayada, M., Sato, K., Ishikawa, T., Okumura, A., et al., 2006. Four-dimensional ultrasonography for therapeutic radiofrequency ablation for hepatocellular carcinoma. Hepatogastroenterology 53, 521–525.

Kato, A., Yoshimine, T., Hayakawa, T., Tomita, Y., Ikeda, T., Mitomo, M., et al., 1991. Computer assisted neurosurgery: development of a frameless and armless navigation system (CNS navigator). No shinkei geka. Neurol. Surg. 19, 137–142.

Kokes, R., Lister, K., Gullapalli, R., Zhang, B., Macmillan, A., Richard, H., et al., 2009. Towards a teleoperated needle driver robot with haptic feedback for RFA of breast tumors under continuous MRI. Med. Image Anal. 13, 445–455.

Lester, H., Arridge, S.R., 1999. A survey of hierarchical non-linear medical image registration. Pattern Recognit. 32, 129–149.

Li, S., Kang, X., Fang, L., Hu, J., Yin, H., 2017a. Pixel-level image fusion: a survey of the state of the art. Inf. Fusion 33, 100–112.

Li, W., Bai, Y., Wu, M., Shen, L., Shi, F., Sun, X., et al., 2017b. Combined CT-guided radiofrequency ablation with systemic chemotherapy improves the survival for nasopharyngeal carcinoma with oligometastasis in liver: propensity score matching analysis. Oncotarget 8, 52132.

Lusic, H., Grinstaff, M.W., 2013. X-ray-computed tomography contrast agents. Chem. Rev. 113, 1641–1666.

Makela, T., Clarysse, P., Sipila, O., Pauna, N., Pham, Q.C., Katila, T., et al., 2002. A review of cardiac image registration methods. IEEE Trans. Med. Imaging 21, 1011–1021.

Marien, A., Gill, I., Ukimura, O., Nacim, B., Villers, A., 2014. Target ablation—image-guided therapy in prostate cancer. Urologic Oncology: Seminars and Original Investigations. Elsevier, pp. 912–923.

Mast, T.D., Makin, I.R.S., Faidi, W., Runk, M.M., Barthe, P.G., Slayton, M.H., 2005. Bulk ablation of soft tissue with intense ultrasound: modeling and experiments. J. Acoust. Soc. Am. 118, 2715–2724.

Maybody, M., 2010. An overview of image-guided percutaneous ablation of renal tumors. Seminars in Interventional Radiology. Thieme Medical Publishers, p. 261.

McCreedy, E.S., Cheng, R., Hemler, P.F., Viswanathan, A., Wood, B.J., Mcauliffe, M.J., 2006. Radio frequency ablation registration, segmentation, and fusion tool. IEEE Trans. Inf. Technol. Biomed. 10, 490–496.

McDannold, N.J., Jolesz, F.A., 2000. Magnetic resonance image-guided thermal ablations. Top. Magn. Reson. Imaging 11, 191–202.

Mcwilliams, J.P., Lee, E.W., Yamamoto, S., Loh, C.T., Kee, S.T., 2010. Image-guided tumor ablation: emerging technologies and future directions. Seminars in Interventional Radiology. Thieme Medical Publishers, p. 302.

Mercier, L., Langø, T., Lindseth, F., Collins, D.L., 2005. A review of calibration techniques for freehand 3-D ultrasound systems. Ultrasound Med. Biol. 31, 449–471.

Nakajima, S., Kikinis, R., Black, P.M., Atsumi, H., Leventon, M.E., Hata, N., et al., 1997. Image-guided neurosurgery at Brigham and Women's Hospital. Computer-Assisted Neurosurgery. Springer.

Nicolau, S., Pennec, X., Soler, L., Buy, X., Gangi, A., Ayache, N., et al., 2009. An augmented reality system for liver thermal ablation: design and evaluation on clinical cases. Med. Image Anal. 13, 494–506.

Ortiz-Alvarado, O., Anderson, J.K., 2010. The role of radiologic imaging and biopsy in renal tumor ablation. World J. Urol. 28, 551–557.

Parker, D.L., Smith, V., Sheldon, P., Crooks, L.E., Fussell, L., 1983. Temperature distribution measurements in two-dimensional NMR imaging. Med. Phys. 10, 321–325.

Penney, G.P., Blackall, J.M., Hayashi, D., Sabharwal, T., Adam, A., Hawkes, D.J., 2001. Overview of an ultrasound to CT or MR registration system for use in thermal ablation of liver metastases. Proc. Med. Image Underst. Anal. 6568.

Pondman, K.M., Futterer, J.J., Ten Haken, B., Kool, L.J.S., Witjes, J.A., Hambrock, T., et al., 2008. MR-guided biopsy of the prostate: an overview of techniques and a systematic review. Eur. Urol. 54, 517–527.

Quesson, B., De Zwart, J.A., Moonen, C.T., 2000. Magnetic resonance temperature imaging for guidance of thermotherapy. J. Magn. Reson. Imaging: Off. J. Int. Soc. Magn. Reson. Med. 12, 525–533.

Rajinikanth, V., Dey, N., Satapathy, S.C., Ashour, A.S., 2018. An approach to examine magnetic resonance angiography based on Tsallis entropy and deformable snake model. Future Gener. Comput. Syst. 85, 160–172.

Roberts, D.W., Strohbehn, J.W., Hatch, J.F., Murray, W., Kettenberger, H., 1986. A frameless stereotaxic integration of computerized tomographic imaging and the operating microscope. J. Neurosurg. 65, 545–549.

Seeram, E., 2015. *Computed Tomography: Physical Principles, Clinical Applications, and Quality Control*. Elsevier Health Sciences.

Shahdoosti, H.R., Tabatabaei, Z., 2019. MRI and PET/SPECT image fusion at feature level using ant colony based segmentation. Biomed. Signal. Process. Control. 47, 63–74.

Shen, S.-H., Fennessy, F., McDannold, N., Jolesz, F., Tempany, C., 2009. Image-guided thermal therapy of uterine fibroids. Seminars in Ultrasound, CT and MRI. Elsevier, pp. 91–104.

Smith, J., Finnoff, J.T., 2009. Diagnostic and interventional musculoskeletal ultrasound: Part 1. Fundamentals. PM&R 1, 64–75.

Tsao, J., 2003. Interpolation artifacts in multimodality image registration based on maximization of mutual information. IEEE Trans. Med. Imaging 22, 854–864.

Vemuri, B.C., Ye, J., Chen, Y., Leonard, C.M., 2003. Image registration via level-set motion: applications to atlas-based segmentation. Med. Image Anal. 7, 1–20.

Widmann, G., Widmann, R., Widmann, E., Jaschke, W., Bale, R., 2007. Use of a surgical navigation system for CT-guided template production. Int. J. Oral. Maxillofac. Implant. 22.

Wood, B.J., Locklin, J.K., Viswanathan, A., Kruecker, J., Haemmerich, D., Cebral, J., et al., 2007. Technologies for guidance of radiofrequency ablation in the multimodality interventional suite of the future. J. Vasc. Interv. Radiol. 18, 9–24.

Xu, S., Kruecker, J., Turkbey, B., Glossop, N., Singh, A.K., Choyke, P., et al., 2008. Real-time MRI-TRUS fusion for guidance of targeted prostate biopsies. Comput. Aided Surg. 13, 255–264.

Xuan, J., Wang, Y., Freedman, M.T., Adali, T., Shields, P., 2006. Nonrigid medical image registration by finite-element deformable sheet-curve models. Int. J. Biomed. Imaging 2006.

Yang, B., Li, S., 2012. Pixel-level image fusion with simultaneous orthogonal matching pursuit. Inf. Fusion 13, 10–19.

Yang, G., Li, L., Liang, Q., Zhu, Z.-J., 2007. Working principle and clinic application of dual source 64 slices computed tomography. Inf. Med. Equip. 10.

Yu, N., Gassert, R., Riener, R., 2011. Mutual interferences and design principles for mechatronic devices in magnetic resonance imaging. Int. J. Comput. Assist. Radiol. Surg. 6, 473–488.

Index

Note: Page numbers followed by "*f*" and "*t*" refer to figures and tables, respectively.

D

W

X

Printed in the United States
by Baker & Taylor Publisher Services